Climatic Change
in the
Intra-Americas Sea

Climatic Change in the Intra-Americas Sea

Implications of future climate on the ecosystems and socio-economic structure in the marine and coastal regions of the Caribbean Sea, Gulf of Mexico, Bahamas, and the northeast coast of South America

Edited by

George A. Maul

for the

United Nations Environment Programme
(Wider Caribbean Region)

and

Intergovernmental Oceanographic Commission
(Caribbean and Adjacent Regions)

Edward Arnold
A division of Hodder & Stoughton
LONDON NEW YORK MELBOURNE AUCKLAND

First published in Great Britain 1993

Distributed in the USA by Routledge, Chapman and Hall, Inc.
29 West 35th Street, New York, NY 10001

British Library Cataloguing in Publication Data

Maul, G. A.
Climatic Change in the Intra-Americas Sea: Implications of Future Climate on the Ecosystems and Socio-economic Structure in the Marine and Coastal Regions of the Caribbean Sea, Gulf of Mexico, Bahamas and N.E. Coast of S. America
I. Title
551.657

ISBN 0-340-58981-7

Library of Congress Cataloging-in-Publication Data

Climatic change in the Intra-Americas Sea / edited by George A. Maul for the United Nations Environment Programme (wider Caribbean region) and Intergovernmental Oceanographic Commission (Caribbean and adjacent regions)
p. cm.
Includes bibliographical references and index.
ISBN 0–340–58981–7
1. Climatic changes—Environmental aspects—North Atlantic Ocean.
2. Climatic changes—Environmental aspects—Caribbean Area.
3. Global warming—Environmental aspects—North Atlantic Ocean.
4. Global warming—Environmental aspects—Caribbean Area. I. Maul. George A. II. United Nations Environment Programme.
III. Intergovernmental Oceanographic Commission.
QC981.8.C5C5156 1993
551.69'163'1—dc20 93–6852
CIP

Typeset in 10/11 point Palatino by Anneset, Weston-super-Mare.
Printed and bound in Great Britain for Edward Arnold, a division of Hodder and Stoughton Limited, Mill Road, Dunton Green, Sevenoaks, Kent TN13 2YA by Butler & Tanner Ltd, Frome, Somerset.

Preface

In response to the concern expressed in certain regions about the possible impacts of sea-level and sea-water temperature changes, induced by expected climatic changes and, in particular, those related to human activities, the United Nations Environment Programme (UNEP) initiated, in cooperation with other governmental and non-governmental bodies, a series of regional studies on this matter for selected regions participating in the Regional Seas Programme. This report constitutes the result of the regional study for the Caribbean Sea, the Gulf of Mexico, the Florida-Bahamas area of the Atlantic Ocean, Bermuda, and the north-east coast of South America, that was prepared in cooperation with the Intergovernmental Oceanographic Commission (IOC) of the United Nations Educational Scientific and Cultural Organization (UNESCO). The 'region' is called the 'Wider Caribbean Region' by UNEP, and the 'Caribbean and Adjacent Regions' by the IOC. Early European oceanographers mistakenly called it the 'American Mediterranean,' but members of the Task Team prefer the term 'Intra-Americas Sea' to emphasize its unique and interrelated geography, climate, and culture.

The study covers the marine environment and the adjacent coastal areas influenced by or influencing the marine environment. The *Terms of Reference* of the study are:

1) to examine the possible effects of sea-level changes on the coastal ecosystems (deltas, estuaries, wetlands, coastal plains, coral reefs, mangroves, etc.);
2) to examine the possible effects of temperature elevations on the terrestrial and aquatic ecosystems, including possible effects on economically important species;
3) to examine the possible effects of climatic, physiographic and ecological changes on the socio-economic structure and activities;
4) to determine areas or systems most vulnerable to the above changes and to prepare a comprehensive, well-documented report reflecting the points above.

This and other studies are based on:

1) the best available existing knowledge about and insight into the problems relevant to the subject of the study;
2) assumptions accepted at the International Conference in Villach, 9–15 October 1985, i.e. increased temperature of 1.5–4.5°C and sea-level rise of 20–140 cm before the end of the 21st century (for the time being temperature elevations of 1.5°C and sea-level rise of 20 cm by the year 2025 will be considered, with the understanding that these estimates will be revised on the basis of regional scenarios yet to be developed); and

3) several detailed case studies included herein, which constitute in part, the material used to prepare the Overview (Chapter 1) of this book.

In consultations between UNEP, the UNEP Regional Coordinating Unit for the Caribbean Area Region (RCU/CAR), IOC and the IOC Sub-Commission for the Caribbean and Adjacent Regions (IOCARIBE), a regional Task Team was established to prepare the study. The RCU for the Caribbean Environment Programme of UNEP invited the Task Team to prepare a report. A first meeting of the Task Team was held in Kingston, Jamaica, 30 July–1 August 1987 to define the necessary inputs to this report. A second meeting and public forum was held 2–4 March 1988 in Miami, Florida, USA to report our preliminary results; the third meeting was held 26–28 June 1991 in Miami, and input from the last meeting comprises the major portion of this book.

The book was prepared at the University of Miami – National Oceanic and Atmospheric Administration's Cooperative Institute for Marine and Atmospheric Studies, located on the campus of the Rosenstiel School of Marine and Atmospheric Science. Many of the figures were redrawn by Ms Jean W. Carpenter and by Mr David A. Senn, library support was given by Ms Linda Pikula, and all the document processing was done by Ms Jill A. Reed; we are grateful to them for their cooperation and persistence in seeing this work to completion. We also wish to acknowledge Mr Salvano Briceño, Ms Beverly A. Miller and all the UNEP staff at the RCU in Kingston, Jamaica, and Dr Fernando L.E. Robles and Mr Anders Alm at the IOCARIBE Secretariat in Cartagena, Colombia for their support and encouragement. Finally, we express our thanks to the UNEP series editor, Dr John D. Milliman and to Ms Diane Leadbetter – Conway and the publishers for their many contributions to this book.

GEORGE A. MAUL
February, 1993
Virginia Key, Florida

List of Contributors

Anders Alm IOC Secretariat for IOCARIBE, Cartagena, Colombia

Ruben Aparicio Universidad de Oriente, Cumana, Venezuela

Stephen R Baig National Hurricane Center, Coral Gables, Florida, USA

Erik Blommestein ECLAC, Port of Spain, Trinidad and Tobago

A V Botello Universidad Nacional Autonoma de Mexico, Mexico DF, Mexico

James M Broadus Woods Hole Oceanographic Institution, Woods Hole, MA, USA

Steven Citrorn Universidad Nacional Autonoma Mexico, Mexico DF, Mexico

Donald P De Sylva University of Miami/RSMAS, Miami, Florida, USA

Guy Engelen Research Institute for Knowledge Systems, Maastricht, The Netherlands

James C Evans University of South Carolina, Colombia, SC, USA

Augustin Fernandez Universidad Nacional Autonoma de Mexico, Mexico DF, Mexico

Frank J Gable Woods Hole Oceanographic Institution, Woods Hole, MA, USA

Artemio Gallegos Universidad Nacional Autonoma de Mexico, Mexico DF, Mexico

Calvin Gray National Meteorological Service, Kingston, Jamaica

Kirby Hanson 11412 SW 82 Avenue, Ocala, Florida, USA

Malcolm Hendry University of the West Indies, Cove Hill Campus, Barbados

George A Maul National Oceanic and Atmospheric Administration and University of Miami/Cooperative Institute for Marine and Atmospheric Studies, Miami, Florida, USA

Aurelio Mercado University of Puerto Rico, Mayaguez, Puerto Rico

Frank E Müller-Karger Ocean Biochemistry, NASA Headquarters, Washington, DC, USA

John D Milliman Woods Hole Oceanographic Institution, Woods Hole, MA, USA

B D Santer Max Planck Institut für Meteorologie, Hamburg, Germany

N C Singh Caribbean Environmental Health Institute, Castnes, St Lucia

Samuel C Snedaker University of Miami/RSMAS, Miami, Florida, USA

J Dana Thompson Formerly, Naval Oceanographic and Atmospheric Research Laboratory, Stennis Space Centre, MS, USA

Inge Ulljee Research Institute for Knowledge Systems, Maastricht, The Netherlands

V P Vicente NOAA, National Marine Fisheries Services, Hatorey Puerto Rico

Roger White Memorial University, St John's, Newfoundland, Canada

T M L Wigley Climatic Research Unit, University of East Anglia, Norwich, UK

Jorge Zavala Universidad Nacional Autonoma de Mexico, Mexico DF, Mexico

Contents

Part 1
Overview

1

Implications of Future Climate on The Ecosystems and Socio-Economic Structure in the Marine and Coastal Regions of the Intra-Americas Sea

George A. Maul[1]
Task Team Chairman

Abstract

Global climate change, and particularly the impact of human activities on Earth's biogeographical environment, is of enormous socio-economic and ecological importance. It is the regional effect of global change, however, that weighs most heavily on individual lives because of the complexity of local response to a world-wide phenomenon. This chapter summarizes the opinion of a Task Team of 29 experts concerning the implications of climate change on the Gulf of Mexico–Caribbean Sea–Bahamas–Bermuda–Guianas region, of a global 1.5°C temperature and 20 cm sea-level rise by the year 2025. For some ecosystems in the Intra-Americas Sea the effect of temperature rise is much more important than sea-level rise, and *vice versa* for others; for some neither is important; for others both are important. Of the 14 ecosystems considered, the most heavily impacted are expected to be deltas and beaches, both because of sea-level rise; neither are particularly vulnerable to a modest temperature rise. Estuaries, wetlands, lagoons and seagrass beds will all be moderately affected by both the 1.5°C and 20 cm scenarios. The other two very important ecosystems, mangroves and coral reefs, are expected to have a low-to-moderate vulnerability to climate change *per se*, but both are expected to experience extreme stress due to local anthropogenic activities such as deforestation, coastal development, runoff, overfishing and tourism. Seven socio-economic issues were also studied in the context of local response to global change; tourism and the influence of tropical storms are considered most important *vis a vis* levels of vulnerability. As with the ecosystems, some other socio-economic issues are more affected by sea-level rise (e.g., settlements and structures and cultural heritage) than temperature rise (which mostly affects coastal zones, public health and human migration). In addition to evaluating the effects of 1.5°C and 20 cm global rises, the Task Team discussed the potential local rates of temperature and sea-level rise and found that for the Intra-Americas Sea, less of a climate change is expected than for other areas of Earth, but that human population pressure will significantly stress the region's environment. Finally, we report on new computer-based decision-making tools for evaluating the effects

---

[1]National Oceanic and Atmospheric Administration, and University of Miami/Cooperative Institute for Marine and Atmospheric Studies, Miami, FL 33149, USA

of climatic change, tools that will give decision-makers quantitative information upon which to base new policies for management.

1 Introduction

The history of modern civilization is inexorably related to Earth's climate. Climatic changes have influenced our literature, raised and toppled empires, altered our religious views, modified economies, forced mass migrations of both humans and animals, caused hunger and starvation; the list is nearly endless (e.g., Bryson and Murray, 1977). Yet in the quincentennial year of the European discovery of the Caribbean Sea (but see Van Sertima, 1976; 1992), little more is known about climate in the region than what the early explorers told their sponsors. Assessing the impact of climate *change* then becomes a particularly challenging problem.

This study emphasizes the marine and coastal environment and addresses the implications of climatic change in the Intra-Americas Sea. As was the perspective of the early explorers, it is limited to a sailor's view of the coastline, reefs, passages, harbours, deltas, estuaries, and deep waters of the Caribbean Sea, the northeast coast of South America (excluding Brazil), the Gulf of Mexico and the Florida-Bahamas region of the western North Atlantic Ocean (Fig. 1.1). Nevertheless, such a perspective presents a formidable challenge involving meteorology, oceanography, geology, economics, sociology, medicine, law and ecology.

To address such challenges, the United Nations Environment Programme (UNEP) was founded in 1972, and within two years UNEP established its Regional Seas Programme. An action plan for the Caribbean Environment Programme was adopted in 1981, and five years later the Regional Coordinating Unit (RCU), in consultation with the Intergovernmental Oceanographic Commission (IOC) and its Subcommission for the Caribbean and Adjacent Regions (IOCARIBE), began addressing marine environmental issues from the RCU's new offices in Kingston, Jamaica. In concert with the recommendations of the World Meteorological Organization (WMO)/International Council of Scientific Unions (ICSU)/UNEP-sponsored meeting in Villach, Austria (WMO/ICSU/UNEP, 1985), the RCU extended its marine environmental interests to include questioning the impact of climate change in the region (UNEP, 1987). Similar programmes are active in five other marginal seas under the Regional Seas Programme.

There is little doubt that climate is changing, but there is an important difference in the climate change that is now understood to be taking place: human activity may be involved. The effects of such anthropogenic activity on the region are difficult to isolate from other natural oscillations in Earth's climate. Nevertheless, the arguments are accepted that regional climatic scenarios (Lamb, 1987) are valuable, with the understanding that they are *not* a prediction of future climate but an internally consistent view of a plausible climatic future. However, with the influence of human activity, there is even more uncertainty in developing climatic scenarios, and clearly a great deal of scientific research is still necessary. It is with these caveats, like the sailors 500 years ago, that we explore the uncharted seas of the implications of climate change.

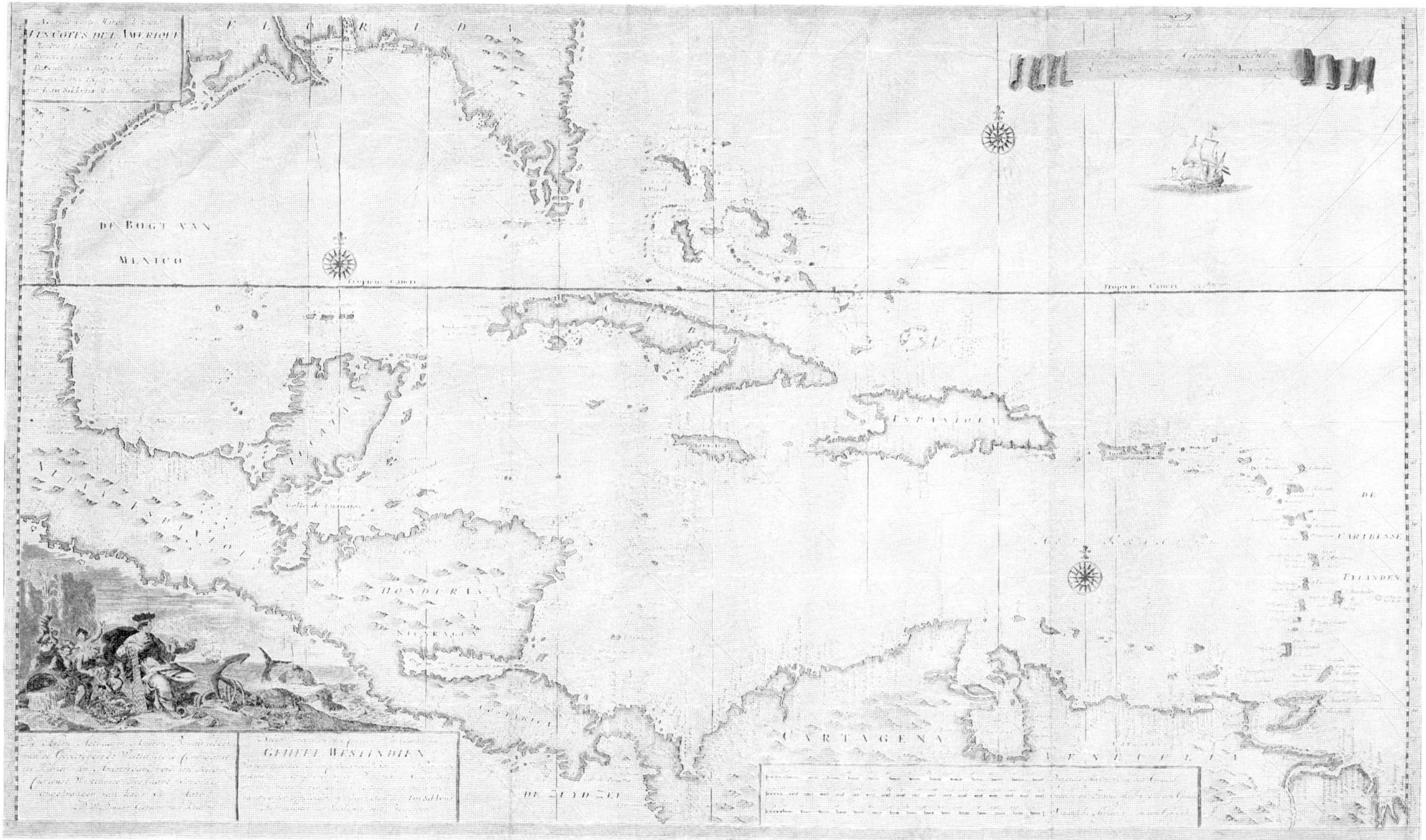

Fig. 1.1 Reproduction of early 18th century sea chart published by Gerard van Keulen. Translated from the French, the title reads: 'New Sea Chart of all THE COASTS OF AMERICA Showing all the Islands Bays and Rivers, as well as all the Rocks and Deeps, entirely composed from many Accounts of Very Experienced navigators by Jean Sikkena Master of Mathematics'. © 1991 Friends of the University of Miami Library; used by permission. See colour plates between pages 210 and 211.

2 Terms of Reference

Each Task Team involved in the UNEP Regional Seas Programme has used a common format in assessing the implications of climate change. The common format is called the *Terms of Reference*, developed at the WMO/ICSU/UNEP (1985) meeting in Villach (see Preface). At the meeting, an *equilibrium* global warming of 1.5–4.5°C and a global sea-level rise of 20–140 cm was adapted based on an expected doubling of the greenhouse gases between the beginning of the Industrial Revolution and the year 2030. The determinations to be made by each Task Team involve a common scenario, and although not specified in the *Terms of Reference*, this Task Team chose to question separately the validity of the scenario as it applies to the local marine and coastal environment.

Many climate-change scenarios have been made, as a reading of the references in Lamb (1987) and others will show. For this UNEP/IOC study, a rise in temperature of 1.5°C and a rise in sea level of 20 cm by the year 2025 (WMO/ICSU/UNEP, 1985) is the baseline scenario. The lower values are chosen because by 2025 it is not expected that climatic equilibrium due to CO_2-doubling will have been reached because of the thermal inertia of the oceans. Deliberations by the Task Team emphasized the point that 1.5°C and 20 cm are a *global*-change scenario, which is interpreted to mean a global average change, that may or may not be realistic for the Intra-Americas Sea. While the Task Team used these baseline values in addressing the implications, one report (Chapter 9) asks 'Does the historical record support such predictions for the region?'. The answer is ambiguous.

Since 1985, projections of future climate have been hotly debated in the scientific literature and (sometimes unfortunately) in the popular press as well. In an effort to bring about a scientific consensus, UNEP and the WMO co-sponsored the Intergovernmental Panel on Climate Change (IPCC). The 'best' IPCC (1990a) estimate is that by the year 2100, sea level will rise 50 cm, with a 'high' estimate of 100 cm. The conclusions of the Second World Climate Conference (Geneva, 1990) are that 'global warming is predicted to reach 2 to 5°C over the next century . . . accompanied by sea-level rise of 65 ± 35 cm by the end of the next century'. These are not contradictory statements, but clearly the issue continues to be debated (Maul, 1992), and requires close attention by scientists and administrators alike.

The last issue in interpreting the *Terms of Reference* is defining how the Task Team understands the terms 'climate' and 'climate change'. *Climate* is understood to be limited to the multi-decadal time-averaged meteorological and oceanographic conditions of the marine and coastal environment; although agriculture and forestry are briefly discussed, they are only considered in their relation to the ocean. *Climate change* is understood to mean the decadal-scale scenario of a 1.5°C temperature rise and a 20 cm sea-level rise; short time-scales, such as seasonal and high-frequency aspects of interannual change, are not considered, although certain aspects of El Niño-Southern Oscillation (ENSO) events are explored.

2.1 Temperature Rise of 1.5°C by 2025

Climatologists use a variety of means to describe Earth's past climate. In the US National Academy of Science report *Understanding Climatic Change* (NAS, 1975), a wealth of information is given on the subject. To illustrate current knowledge about temperature change, temperature records for the last 10,000 and 1000 years (Fig. 1.2), show that Earth's temperatures have varied significantly since the 15th century European discoveries of North America. In fact, when the Caribbean Sea was discovered, Europe was in a cold period known as the 'Little Ice Age'. At present, Earth's climate is warmer than it has been in the last 1000 years, but by no means is it as warm as it was in several past epochs, as the longer records (last 1,000,000 years) indicate. So while an interpretation of the last 100 years shows increasing global temperatures on average (Hansen and Lebedeff, 1988; lower panel in Fig. 1.2), the decades of 1940–1970 had declining temperatures. Translating the global records such as shown in Fig. 1.2 to the regional level was one important challenge for the Task Team.

Fortunately, geochemists are constructing climatic histories of the Intra-Americas Sea region, and are showing that much of the variability shown in the upper panel of Fig. 1.2 applies to the Caribbean Sea and Gulf of Mexico. Hodell *et al.* (1991) published a 10,000-year history of oxygen isotope ($\delta^{18}0$) measurements taken from Lake Miragoane, Haiti (not shown), which has many of the characteristics of this 'global' temperature curve; most especially they conclude that $\delta^{18}0$ roughly follows the Milankovitch orbitally induced insolation curve. But they note that superimposed on the orbitally forced climate trend (cooling for the next 5000 years) are abrupt events resulting from non-linear ocean-atmosphere interactions. So while the Task Team debated the 'historical record' (last ~150 years), they were cognizant of the progress and uncertainties in applying global-change arguments to a specific region.

Does the historical record support the WMO/ICSU/UNEP (1985) scenario of a 1.5°C temperature rise in the region by the year 2025? Data to assess a rise of sea-surface temperature were considered scarce, so Hanson and Maul (Chapter 9) decided to analyse air temperature at Key West, Florida. The 136-year record gives evidence that a warming has occurred between 1890–1950, but the last 30 years or so have been relatively steady at +0.3°C above the long-term mean; a similar analysis of air temperature from ship reports in the Straits of Florida shows no deviation from constancy of the mean. Gray (Chapter 5) found that the maximum air temperatures in Jamaica and in Trinidad and Tobago increased during the last 10 years, but that evaporation had decreased (which is inconsistent with a temperature increase); it is unclear if these changes are due to climatic change or to other factors. Aparicio (Chapter 6) reports an air-temperature trend of +0.1°C per decade in Venezuela, but other data from volunteer observing ships in the central Caribbean Sea suggest sea-surface temperatures have been decreasing since 1950. Linear extrapolation of these case studies leads one to see some suggestion of an air-temperature rise in the region, but that 1.5°C seems to be too high; less than 1.0°C rise by 2025 appears to be a more plausible picture of our future temperature.

2.2 Sea-Level Rise of 20 cm by 2025

As with questioning the validity of the global temperature change for the region, the Task Team looked into the historical sea-level record to put the WMO/ICSU/UNEP (1985) scenario into perspective. The remarkably slow rate of sea-level rise in South Florida and in Jamaica during the last 3200 years (only about 0.04 cm/yr; Fig. 1.3) allowed many shorelines to stabilize or to expand, and many shallow marine environments to build. However, since the early 1930s, sea-level records from many sites around Florida show much faster rates of sea-level rise, very similar to the rate during the period 3200–5500 years ago, when there was a rapid retreat of the shoreline. The data in Fig. 1.3 gave the Task Team a benchmark against which to judge the WMO/ICSU/UNEP global scenario of 50 cm rise per 100 years.

Does the modern historical record of the Intra-Americas Sea support the WMO/ICSU/UNEP (1985) scenario of a 20 cm sea-level rise by the year 2025? To answer this question, the highest quality (revised local reference)

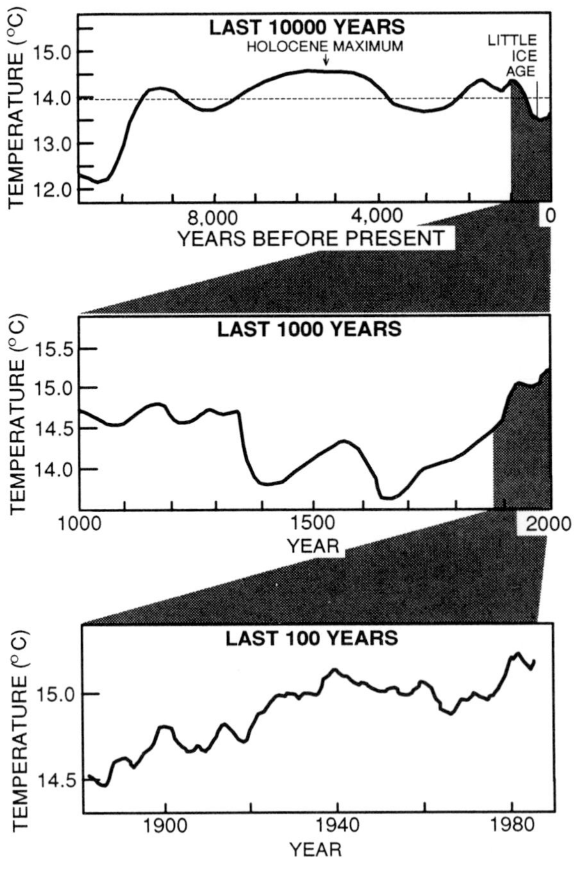

Fig. 1.2 Temperature change for the last 10,000 years and 1000 years based on records in Europe (upper two panels; redrawn from IPCC, 1990a), and a global estimate (lower panel; from Hansen and Lebedeff, 1988). Temperature scale for upper two panels is only approximate and is chosen to be in agreement with the record for the last 100 years which is based on direct observations. Maul (1992) discusses urbanization effects in Hanson and Lebedeff (1988) and gives references related to the issue (*cf.* Wigley and Santer, Chapter 2; Gallegos *et al.*, Fig. 3.5).

data on file with the Permanent Service for Mean Sea Level (PSMSL) were studied by Hanson and Maul (Chapter 9) for the Intra-Americas Sea, and from Venezuelan records by Aparicio (Chapter 6) for the southern Caribbean Sea. For the longest records, Hanson and Maul found that sea level is rising on average at about 0.36 cm/yr (±0.25 cm/yr) over the last 30 years, but due to complicated tectonic activity, subsidence, and petroleum/groundwater extraction, the values ranged from +1.0 cm/yr in Texas (rising sea level) to −0.3 cm/yr in Mexico (falling sea level). At Key West, Florida, a site of tectonic stability, the rise is 0.22 cm/yr (±0.01 cm/yr), based on the years 1913–1986. More important perhaps, is that sea-level rise due to temperature/salinity changes in the upper 1000 m of the water column east of Abaco Island, the Bahamas, for 1950–1987 was +0.14 cm/yr (Chapter 9), and there is no evidence of acceleration in the rate of rise. So, as with the temperature scenario, a lower value, perhaps 10 cm by 2025, may be a more plausible regional value, but the high spatial variability makes a regional average nearly meaningless; site-specific values are required for realistic assessments.

Gallegos *et al.* (Chapter 3) and Mercado *et al.* (Chapter 4) cautiously advise that many more long records are required in order to sort out the decadal and longer wave motions in the relative sea-level record.

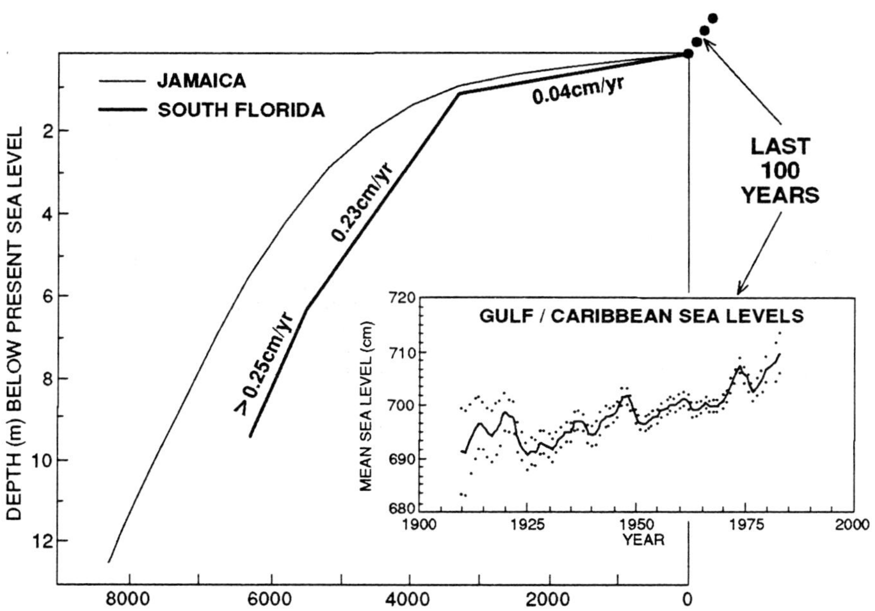

Fig. 1.3 Sea-level change based on geological records taken in south Florida by Wanless *et al.*, 1988, and from Jamaica by Digerfeldt and Hendry, 1987. The relative rise of 36 cm per 100 years shown in the inset is based on direct observations at all PSMSL (Pugh *et al.*, 1987) RLR sea-level stations in the region; dots show range of standard error of the mean. The shoreline retreat associated with sea-level rise occurs in storm-driven steps and is often not immediately seen as beach erosion or lowland inundation.

The physics of the very lowest frequencies in oceanic circulation are not well understood. Circulation, the three-dimensional movement of water with time, is affected by geological activity, the wind field, Rossby waves, periodic behaviour of the Sverdrup balance, interbasin modes of oscillation, and so forth. Progress in numerical modelling will give the resolution to determine submesoscale features and subdecadal oscillations (Mercado *et al.*, Chapter 4), but it may be 10 years before such calculations are possible. To further complicate the issue, Chao (1991) argues that sea level has fallen by about 7 cm in the last 100 years due to the building of reservoirs for irrigation and water control. In the interim, thoughtful extrapolation of the PSMSL observations, in concert with careful monitoring and a vigorous modelling activity, will give the most plausible estimates of future sea level.

3 Effects of Sea-Level Changes on Coastal Ecosystems

The first task in the *Terms of Reference* (*q.v.* Preface) is to examine the possible effects of the sea-level changes on the coastal ecosystems (*cf.* IPCC, 1990b). Gable (Chapter 10) gives an overview of the ecosystems in the region, and *Oceanus* (1987/88) is an issue devoted to Caribbean Sea marine science. The variable of interest in this section is relative sea-level (RSL) rise, that is the net effect of tectonic uplift or subsidence plus expansion or contraction of the water column. During the Holocene (last 10,000 years) in Jamaica for example (Hendry, Chapter 7), sea-level rise is less than the 0.5 cm/yr implied by the assumptions at the International Conference in Villach, and in the last 3000 years before the present century, RSL rise has been almost nil (Fig. 1.3). All other things being equal, 0.5 cm/yr (20 cm between 1985 and 2025) is expected to place stress on coastal ecosystems.

3.1 Deltas

In the region there are four major river deltas: the Mississippi (USA), the Rio Grande (Mexico/USA), the Magdalena (Colombia) and the Orinoco (Venezuela); in addition, waters of the Amazon River are known to flow into the Caribbean Sea (Muller-Karger, Chapter 8). Deltas are particularly vulnerable to erosion enhanced by sea-level rise because the sediments are unconsolidated muds subject to subsidence and compaction. One might expect, according to the Bruun Rule, a shoreline retreat up to several metres horizontally for each centimetre RSL rise; this translates into thousands of hectares of lost land. The problem is exacerbated by potential increased storm activity (Gray, Chapter 5), since most shoreline erosion occurs during storms, and by subsidence such as in the case of Louisiana. However, Mercado *et al.* (Chapter 4) argue that the 20 cm RSL rise scenario will be of no practical consequence on storm-surge *modelling*, as far as it might calculate increased surge heights. Delta benthic systems, particularly seagrass beds, would be most affected or destroyed by the expected RSL rise. In contradistinction, RSL near the Orinoco Delta may be falling, but more measurements are needed to document this preliminary result.

3.2 Estuaries

The RSL effect on estuaries, as with deltas and many other geomorphological features, must be considered on a case-by-case basis in order to make meaningful impact assessments in the region (Vicente *et al.*, Chapter 11). Because of local uplift, the following areas are expected to have reduced increase in RSL due to climate change: east coasts of the Cayman Islands; north coast of Jamaica; southeast coast of Cuba; north coast of Bahia, the southwest coast of Haiti; Barbados; north coast of the Dominican Republic; and the southwestern Gulf of Mexico (see Fig. 10.2 for locations). In addition to the subsidence experienced in deltas, other areas experiencing downwarping include: the Maracaibo region of Venezuela; the entire northern Gulf of Mexico from Texas to Alabama; the estuary of Port au Prince; and the western Gulf of Honduras. Coastal lagoons, salinas and estuaries (transient environments that owe their existence to sea-level rise), depending on their location, could suffer from saline intrusion. Lagoons, however, should be able to support their usual nurseries; salinas on the other hand (Vicente *et al.*, Chapter 11), could be flooded over continuously and change their economic value.

3.3 Wetlands

The ability of a wetland to sustain vertical growth is a balance between sedimentation and RSL rise. In the tectonically complicated Intra-Americas Sea, no single definitive statement is possible, but in the last 5000 years, many wetlands have been able to keep pace with rising sea level. In areas with marked subsidence, particularly if there is canalization of organic silts and clays away from the wetland into the marine environment (Hendry, Chapter 7), wetlands will be submerged and lost to productivity. Where wetlands are bounded by steep-sided basins, as is the case in many of the islands, it is unlikely that they will be replaced as sea-level rises; on gentler island and continental floodplains, such as the northern Gulf of Mexico, the problem may be less severe. Loss of some wetland economies such as shellfish industries is expected to occur with the 20 cm RSL rise scenario (Snedaker, Chapter 12).

3.4 Coastal Plains

The primary effect on coastal plains will be increased flooding during storms (from raised sea level and/or from heavy rainfall). Unfortunately many storm-surge models differ markedly in their predictions (the variability in their predictions being several orders of magnitude larger than the 20 cm scenario; Mercado *et al.*, Chapter 4). Shore migration (both erosion and accretion) will vary depending on the substrate, and sandy beaches will be more affected than rocky coasts. No single rule can be applied for the region as a whole, but modelling on a local scale is required to account for differences such as tectonic displacement, beach structure, offshore bottom topography, and storm frequency and magnitude. The concentration of human population in the poorly drained low-lying coastal plains is a source of concern in many countries. Special attention should be paid to areas where subsidence is evident, as it will exacerbate the flooding problem. Port au Prince, Haiti, Puerto Cortes, Honduras and the Galveston-centred area of the US Texas–Louisiana coast, are coastal plain areas most vulnerable to flooding from sea-level rise and storms.

3.5 Coral Reefs

The second largest coral reef system in the world dominates the offshore area of the western Caribbean Sea (Milliman, Chapter 13), and all areas except the northern Gulf of Mexico coast have extensive reef systems. Growth of individual coral organisms is estimated between 1–20 cm/yr (Vicente *et al.*, Chapter 11), and reef growth rates as a whole are known to be up to 1.5 cm/yr (Hendry, Chapter 7). Not all reefs accumulate at these rates, but if they did, they could keep pace with the rise in RSL of 20 cm by 2025 if other factors do not alter growth conditions. Environmental stress on the reefs from other variables (storms, sedimentation, disease, rainfall, radiation, turbidity, overfishing, mass mortality in algal grazers, etc.) may prevent some reefs from keeping pace with rising RSL, resulting in alteration of the nearshore hydrodynamics. The issue is further complicated by consideration of the type of reef, coastal geomorphology, reef depth and ecological state of the reef in question. Accurate predictions on the effect of RSL rise may be possible in reefs that have already been physically and biologically monitored, such as in Panama, Jamaica, and Puerto Rico.

3.6 Mangroves

Mangrove forests are a unique feature of protected coastal shorelines of the tropics and subtropics; their root systems (prop roots and pneumatophores) stabilize the sediment, dampen wave energy, provide habitat shelter for numerous organisms, and form the base of the nearshore marine foodweb (Vicente *et al.*, Chapter 11). The five species comprising the mangrove flora of the Intra-Americas Sea occupy an area of approximately 3.2 million hectares, or some 15% of the estimated world area of mangrove of 22 million hectares. Within the region, the best developed mangrove forests are associated with areas of high precipitation and upstream land runoff. Because mangroves grow best in moderately saline environments where the rate of peat production exceeds the anaerobic decomposition of peat by seawater sulphate-reducing microorganisms, it is postulated that mangroves can keep pace with RSL in rain-fed humid areas, but may be overstepped and abandoned in more arid areas particularly if inland retreat is not possible. Thus, in terms of global climate change, future changes in patterns of precipitation and catchment runoff may be more important than RSL (Snedaker, Chapter 12). Notwithstanding the current high rate of regional mangrove loss by overcutting, land clearing and habitat conversion, global climate change is a minor factor in consideration of the fate of this regionally important coastal habitat.

3.7 Seagrass Beds

Seagrasses are a benthic environment throughout the region that are important in stabilizing bottom sediments (Hendry, Chapter 7), serve as nurseries for juveniles, and for providing surfaces upon which many organisms attach. A 20 cm RSL rise *per se* is not expected to seriously affect the six common species (Vicente *et al.*, Chapter 11), but if there are other changes, such as in the quality of light, influence of herbivores, substrate, wave energy, or bottom slope, the beds may be impacted.

3.8 Fisheries

The impact of sea-level rise on fisheries is not expected to be great unless turbidity increases due to erosion from higher water or river runoff (Muller-Karger, Chapter 8). Turbidity increase could have a negative impact on fisheries particularly during the early life history stage (W. Richards, NOAA/NMFS, pers. comm.). Estuarine-dependent species in areas such as Mississippi, the Florida Everglades, Guyana and the Orinoco Delta, may be particularly vulnerable to sea-level rise, especially if salinity changes are involved. These ecosystems are also particularly vulnerable to increases in the discharge by rivers of pollutants, which may accumulate, and eventually become harmful to humans and other animals in the foodweb.

4 Effects of Temperature Elevations on Ecosystems

As discussed above, there is considerable question if trace gas-induced temperature elevation can be seen in the records at Key West, Venezuela, Jamaica or in Trinidad and Tobago. Temperature change, however, is only one aspect of the meteorology that will effect terrestrial and aquatic ecosystems. Hanson and Maul (Chapter 9) find no evidence for changes in precipitation at Key West during the last 101 years; Aparicio (Chapter 6) finds none along the southern Caribbean Sea; Muller-Karger (Chapter 8) does not detect trends in precipitation over the Mississippi, Orinoco, or Amazon drainage basins during the past 50-100 years. In the northern Caribbean Sea, Gray (Chapter 5) finds hints of decreased rainfall in the last 20 years, which is possibly associated with decreased hurricane activity. An increase of 1.5°C in sea-surface temperature could increase the number of hurricanes by as much as 40% (Shapiro, 1982), and the maximum wind speed by 8%; there is, however, a considerable uncertainty in these figures (40% increase means on average $+1.6 \pm 1.2$ hurricanes per year). Many other factors are important in hurricane analysis, and it may be that the storm-formation location and track are more important than changes in strength or frequency.

In the sense that Lamb (1987) develops climate-change scenarios as plausible future events, Gray (Chapter 5) assumes the following (*cf.* Gallegos *et al.*, Chapter 3; Aparicio, Chapter 6): rainfall will continue to decrease, air temperatures will continue to rise, surface wind speed and evaporation will continue to increase. Caution must be exercised in applying these changes as anything other than persistence forecasting. It is unknown, for example, if the decreased frequency of large hurricanes over the last two decades is really a long-term trend, if it is random, or part of some cycle as yet not understood. Hurricanes are an important contributor of rainfall; is the decrease in precipitation merely a reflection of fewer large storms? Increased temperature may affect the drag of wind on water, but Mercado *et al.* (Chapter 4) and Hendry (Chapter 7) see no clear indication of a significant change in storm surges or waves associated with elevated temperature. Clearly, however, any change in wind and precipitation patterns will impact the discharge of major rivers and the dispersal patterns of their water and its suspended/dissolved constituents (Muller-Karger, Chapter 8). With these thoughts in mind, the second item in the *Terms of Reference* (*q.v.* Preface), 'effects of temperature elevation on the ecosystem, including on economically important species' is considered (*cf.* IPCC, 1990b).

4.1 Agricultural Resources

Saline intrusion is expected to have more impact on agriculture in the coastal plains than elevated temperature, particularly on rice production along the Guyana coast. Soil erosion probably will increase, but poor management practices are probably more contributive than temperature elevation and saline intrusion. Vicente *et al.* (Chapter 11) argue that it is unlikely that inland and hilly forests will be affected much by increased temperature, although Gray (Chapter 5) warns of increased erosion due to increased winds and decreased precipitation. However, warmer temperatures could be a significant factor in forest fires, particularly if precipitation decreases. Human settlements are unlikely to be affected significantly by 1.5°C weather changes, except where RSL is important (Alm *et al.*, Chapter 15).

4.2 Coastal Systems

In the tropics, marine organisms live closer to their maximum thermal tolerance than those in more temperate climates. Although the 1.5°C temperature-rise scenario would raise the summertime mean temperature to 30.5°C over much of the region, most migratory organisms are expected to be able to tolerate such a change. Some corals will be affected (for example, the 1983 and 1987 bleaching events), but it is expected that other environmental stresses will be more important (Milliman, Chapter 13). Littoral-supralittoral organisms, such as mangroves, are adapted to withstand high temperature, and unless the 1.5°C increase affects the reproductive cycle, the temperature elevation will probably cause unmeasurable results (Snedaker, Chapter 12). Similarly, only seagrass beds already located in thermal stress situations (i.e., in shallow lagoons or near power plant effluents) are expected to become negatively affected by the projected WMO/ICSU/UNEP (1985) temperature rise.

4.3 Fisheries

The blue, clear waters of the region are relatively nutrient-poor, and most of the fisheries are concentrated on Campeche Bank and along the northern coast of the Gulf of Mexico, at the Mosquito Bank off Honduras and Nicaragua in the Caribbean Sea, and in the Gulf of Paria and the coastal waters of the Atlantic Ocean off Surinam, Guyana and French Guiana (Gable, Chapter 10). It is not expected that a modest increase in temperature will significantly affect the fisheries except in some shallow lagoons where hypersalinity may affect productivity, particularly if juveniles have a critical dependence on salinity or temperature. Increased alongshore winds, however, could lead to increased coastal upwelling along some continental coasts (Aparicio, Chapter 6) or other circulation changes (Gallegos *et al.*, Chapter 3), and thus to increased productivity (*cf.* Vicente *et al.*, Chapter 11).

Tropical fish eggs hatch very quickly (12-48 hours), and development is associated with temperature. Just as 'cold snaps' can be devastating, so can 'hot snaps', particularly during early juvenile stages. Extrema in temperature are usually averaged out in climate analysis, but with increased temperature, the likelihood of 'hot snaps' increases; the 1987 Caribbean Sea coral bleaching event was attributed to 'hot snaps' by some researchers (Milliman, Chapter 13). The complexities of the ecosystem could be greatly affected by slight temperature changes. It is unknown, for example, why fish stocks either decline or increase by orders of magnitude, except due to early life history events caused directly by the physical environment or indirectly through complex chains in ecosystem dynamics. Temperature effects on tropical fisheries remains an important and unanswered question, although there is some evidence of fish migration associated with increased coastal temperatures.

5 Possible Socio-Economic Changes

Climate change will have socio-economic impacts on both the micro-economic or localized level, and on the macro-economic or economy-wide level (*cf.* IPCC, 1990c). The complexity of these interactions is

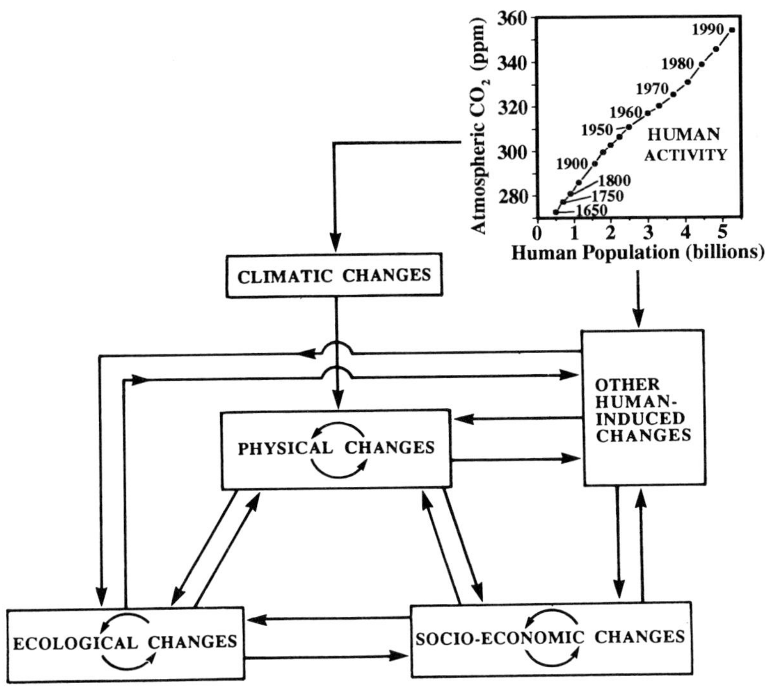

Fig. 1.4 Pathways in socio-economic aspects of the climate change problem (from Alm *et al.*, 1988; population vs. CO_2 (1650–1990 AD) from Idso, 1989). The full complexities are further discussed in Chapters 15 and 16, but the interested reader is also referred to IPCC (1990b) for impact assessments and IPCC (1990c) for response strategies.

summarized in Fig. 1.4, showing the numerous pathways possible in complex social systems (*cf.* Alm *et al.*, Chapter 15; Engelen *et al.*, Chapter 16) and showing that the generic effect of human activity is the strongly linear relation between human population and atmospheric CO_2 concentration (Idso, 1989). The smaller or the more coastal-oriented an economy is, the greater will be the impact of sea-level rise. The Intra-Americas Sea, with its many small island-based economies such as fishing and/or tourism, is particularly vulnerable to the physical changes associated with changing climate. Some climatic changes will benefit certain sectors of an economy (rising RSL may benefit the construction industry), while being detrimental to others (beach erosion may cause a loss in tourism). A climate change-induced benefit to the construction industry reflects a *transfer* of benefits and costs rather than the *creation* of new benefits and costs. The net sum of costs and benefits must be assessed on an individual basis because it is the true cost due to climate change only that is of interest (Engelen *et al.* Chapter 16).

5.1 Agriculture and Forestry

Islands usually have small, coastal aquifers, and sea-level rise will impact water quality in aquifers that have hydrological continuity with the sea. Loss of agricultural land in low-lying coastal plains will be a minor, but perceptible impact, particularly in those areas where saline intrusion affects the water supply, such as on the leeward side of small mountainous islands; continental areas are not expected to be seriously affected. Differing permeability in aquifers can cause great variability in the effect of rising RSL. Relocation of wells, construction of weirs, water storage schemes and barging of water are all possible socio-economic responses. In regards to forestry, as noted in an earlier section, the expected climate change impact is anticipated to be small compared to proper management policies on the industries and people involved.

5.2 Fisheries and Coastal Zones

Most fishing in the coastal zone in the region is artisanal except for a few larger industries such as shrimping and the menhaden fishery in the Gulf of Mexico. The WMO/ICSU/UNEP (1985) scenario of 1.5°C and 20 cm increases by 2025 are not expected to create any significant changes in the fisheries, although to the artisanal fisherman, a displacement in traditional fishing sites may be perceived as being important (Alm *et al.*, Chapter 15). There does remain an unanswered question of the effect on fisheries of extreme temperature events. Aquaculture in the region as a whole is considered undeveloped at the present. The critical issue of shoreline migration, which is the most important impact on the coastal zone, is discussed in the following sections.

5.3 Tourism

The single most important industry in the region is tourism, especially in Florida, the Bahamas, Cuba, Jamaica and the Lesser Antilles. Of all the possible climate change impacts that affect tourism, none is so clearly demonstrated as beach erosion (*q.v.* Hendry, Chapter 7). Shoreline migration will create new areas of economic benefit as new beaches are built, but the protection, replenishment and stabilization of existing beaches, at least until major existing tourist investments are amortized, represent a major socio-economic impact. It is difficult to estimate the impact of climate-induced sea-level rise, in addition to the erosion associated with the relentless interaction of the sea on the coast, that is not associated with climate change; in addition, certain sand-mining practices (such as in Trinidad and Tobago) are already considered important. A temperature rise of 1.5°C locally is probably of little consequence, but the scenario of a global 1.5°C rise probably means much higher temperatures at mid-latitudes (where most tourists come from), thus reducing the attractiveness of a warmer climate for vacationing. Indirect socio-economic effects on tourism due to increasing pollution, coral-reef mortality and storm damage are also involved.

5.4 Settlements and Structures

Up to a certain point, structures will be worth building to protect settlements and facilities. Navigation and port facilities normally have to be reconstructed and maintained, so the socio-economic impact of a 20 cm sea-level rise is not considered serious (Alm *et al.*, Chapter 15). Some nearshore roads, seawalls and bridges will have to be increasingly repaired, and if the RSL rise is augmented by increased storm activity, the impacts will be serious, particularly in countries with marginal economies. As with agriculture in low-lying lands that depend on well water, many municipal water supplies and drainage and sewage systems, will have to be modified; areas of particular concern in this regard are the coastal cities of Guyana and Belize (Vicente *et al.*, Chapter 11). The most damaging socio-economic aspect is climate change coupled with population growth (see Fig. 15.2) and migration to coastal cities. Often the population growth is into areas most likely to be impacted by water-level changes, and in periods of extreme weather events, serious public health impacts are probable in addition to physical danger (Hardin, 1971).

5.5 Public Health

Both temperature and sea-level rises are expected to have an effect on human health; temperature because many diseases and acute effects are associated with elevated temperatures, and with water levels because water is a principal agent for many diseases and organisms that carry disease (de Sylva, Chapter 14). If higher temperatures are coupled with higher humidity as Gray expects (Chapter 5), heat-related health stress and mortality will increase. Human health changes are related to a wide variety of considerations including: mortality and morbidity related to weather and climate; extreme weather events; airborne materials; seasonal diseases caused by microorganisms; parasitic diseases; nutrition; water quality and abundance; and changes in the marine environment including population shifts in dangerous fish, such as sharks, and toxic organisms (de Sylva, Chapter 14). Socio-economic effects relate not only to increased spread of tropical diseases and their associated shift in costs and benefits to the health industry, but also to potential losses in other industries due to health-related absenteeism. It is anticipated that transfer of costs and benefits will be associated with climatic change to public health in the region, but that the health-care delivery systems will keep pace with the climate-related aspects to the year 2025; whether or not the systems are capable of coping with other social changes is uncertain.

6 Most Vulnerable Areas or Systems

In order to determine areas or systems which appear to be most vulnerable to changes in sea-level and temperature (*q.v. Terms of Reference*; IPCC, 1990b), and their impact on ecological and socio-economic structures and activities, three broad topics are addressed: (1) physical processes; (2) ecological aspects; and (3) socio-economic issues. Much of the material in the following sections is drawn from notes and recommendations made during the deliberations of these three working groups of the Task Team.

It was clear that detailed information on the wide variety of areas or systems most vulnerable to climate change in the region could not be prepared without additional substantial effort and support. In order to ultimately provide such detail, the consensus was the following:

- Strengthen existing institutions rather than creating new ones.
- Improve communication and information exchange particularly through the use of electronic media and 'personal computer' technology.
- Reduce uncertainties in the regional impact of the global 1.5°C/20cm scenario by data generation, case studies, and modelling, obtaining probability estimates on sea-level rise and other climate change.
- Continue the interdisciplinary interaction of the Task Team in order to provide *quantitative* information to member states.

The latter point of quantifying results, based on the best physical or economic models, is considered the penultimate goal of this joint UNEP/IOC programme.

6.1 Physical Processes

Climate change involves much more than RSL rise and temperature increase; precipitation, evaporation, humidity, wind velocity, storms, cloudiness, insolation, ocean currents, waves, mixing, riverine input, etc., are all important variables. In order to strengthen quantitative information transfer to states, regional climate models nested in coupled ocean-atmosphere global circulation models are needed, along with a vigorous, stable, long-term *in situ* verification programme, coupled with an active multidisciplinary research effort which should include examination of the historical, geological and archeological records in order to supplement direct measurements. Understanding future shoreline migration is arguably the first priority based on current information, but if precipitation changes (for example) are markedly underestimated, the impact on agriculture and coastal ecosystems could be far more important. To this end, participation in efforts such as the World Climate Programme, with significant international visibility by scientists from the region, is absolutely necessary to improve the physical basis upon which quantitative information is provided to ecologists, sociologists, economists, politicians and managers.

6.2 Ecological Aspects

Identification of the most vulnerable ecosystems requires more of a microscale approach than the mesoscale thinking required of the physical processes discussed above. Preparation of a regional map with a classification scheme showing areas and ecosystems most vulnerable to climate change is a massive, but necessary undertaking. Seagrass beds, coral reefs, mangroves (particularly the black mangrove) and coastal lagoons are probably the most critical habitats to be mapped. Associated with the critical habitats are species that utilize them as feeding and/or nursery grounds. Of vital concern to these critical habitats are climate-related impacts from the sewage and toxic wastes of nearby human population centres and agricultural regions. Conversely, impacts of saline intrusion on local fresh-water supplies and inundation of seaside population centres, particularly during storms, are critical concerns to local residents; high population-density islands such as Barbados, and cities with rapidly rising RSL such as Galveston, Port au Prince, Puerto Cortes, New Orleans and Cartagena, are particularly vulnerable.

6.3 Socio-Economic Issues

Before effective socio-economic responses to climatic changes can be initiated, there is a need to reduce significantly the degree of uncertainty about the likelihood, extent and direction of such changes. The most vulnerable 'system' in the socio-economic and health sectors is the credibility of those making impact assessments. Governments and institutions will revert to procrastination as the most viable response to weak forecasting, rather than to improving information development and dissemination, risk spreading and diversification, or to reducing levels of fixed commitments. Some states, Costa Rica for example, have already established new building set-back laws for construction along the coasts; others, Florida for example, have locally opted for massive beach replenishment programmes. Most small island states, which numerically constitute a substantial fraction of governing units, do not have the financial resources nor the technical expertise to develop appropriate socio-economic responses to climate change; it is for this reason that initial efforts in socio-economic numerical modelling are directed to small islands (Engelen *et al.*, Chapter 16). Probably the greatest single socio-economic scenario that individual governments must prepare for is a significant international migration of populations from highly vulnerable locales to areas where safety and the quality of life is deemed to be better. To prepare for such future change, a catalogue of institutional responses needs to be developed along with specification of conditions under which those responses should be implemented.

6.4 Synthesis

H.L. Mencken once said 'For every complex problem there is a solution that is simple, neat, and wrong.' With this caveat in mind, an attempt to synthesize much of the implications of climatic changes in the region is given in Table 1.1. Three subjective levels of vulnerability to rises in sea level and temperature are chosen and assigned to the ecosystems and socio-economic topics outlined in the *Terms of Reference* (*q.v.* Preface).

While on a site-specific scale many of the estimated impact levels will be different, on a regional scale vulnerability to most climatic changes *per se* is judged low to moderate. However, due to other pressures on the marine environment, and to human efforts to deal with the effects of these pressures, the vulnerability of society to climatic changes increases. In many cases the future impacts on society of non-climatic factors may far exceed those due to climatic changes. It is important, therefore, for policy considerations, to view this synthesis in a proper context, which is, climatic changes will exacerbate environmental changes already ongoing and documented in other studies.

Table 1.1 Implications of climatic changes in the Intra-Americas Sea.*

Terms of Reference	*(a) RSL 20 cm*	*(b) SST 1.5°C*
Ecosystem	Level of vulnerability**	
Deltas	H	L
Estuaries	M	M
Wetlands	M	M
Coastal plains	M	L
Coral reefs	L	M
Lagoons	M	M
Mangroves	M	L
Seagrass beds	M	M
Fisheries	L	M
Agriculture	L	L
Forests	L	M
Rivers	L	M
Coastal lakes	L	L
Beaches	H	L
Socio-economic issues		
Coastal zones	L	M
Tourism	H	M
Settlements and structures	M	L
Public health	L	M
Tropical storms	L	H
Human migration	L	M
Cultural heritage	M	L

Notes

*These levels of vulnerability reflect only the WMO/ICSU/UNEP (1985) climate scenario detailed in the *Terms of Reference* (*q.v.* Preface), and must be considered as issues that exacerbate other problems such as population pressure, pollution, subsidence, coastal erosion, construction, warfare, etc.

**(L) Low impact, (M) Moderate impact, (H) High impact.

7 Models of Future Climate

In Chapter 2, Wigley and Santer give a very detailed discussion of the possible future climate of the Intra-Americas Sea (*cf.* IPCC, 1990a). They compare the results of four numerical models that predict future surface air temperature change and precipitation change for each of the four seasons. Each model calculates the effect on temperature and rainfall of doubling all the greenhouse gases, expressed as doubled CO_2 ($2\times CO_2$). All four numerical models are *global* models, but Wigley and Santer only report the *regional* results of interest herein. Fig. 1.5 from Wigley and Santer (pers. comm.) shows the model results for annual mean temperature and annual mean precipitation. Results from atmospheric General Circulation Models (GCMs) for future climate on a regional scale must be interpreted very cautiously because of the limitations in numerically simulating such a complex problem as climate. Cautiously then, the range of annual average *modelled* temperature change (Fig. 1.5a) and annual average *modelled* precipitation change (Fig. 1.5b), are discussed below.

The annual average temperature change caused by effective CO_2-doubling shows a fairly consistent result in each GCM: an increase of 2–4°C is calculated. Details of increased annual average temperature change are different from GCM to GCM, but in general the continental boundaries of the Intra-Americas Sea are modelled to have higher annual average temperatures than the islands. Annual average precipitation changes due to effective CO_2-doubling also shows significant variability between GCMs, but each model shows that the zero contour (in millimetres per day), which runs through the centre of the Caribbean Sea, is the dominate feature of the calculation. Precipitation in the region is strongly influenced by tropical storms, however, which are not in these GCMs.

Although the results shown in Fig. 1.5 vary considerably between GCMs, the general conclusion is in agreement with the WMO/ICSU/UNEP (1985) scenario of rising temperature. Climate, however, is the sum of many geophysical factors, the greenhouse gases only being one of them, and there may be competing factors (particularly on a regional scale) that can modify these modelled results. Human activities such as massive deforestation can alter the balance of factors that add up to Earth's climate, and the prudent observer will interpret the results shown in Fig. 1.5. cautiously

An alternative method of considering future climate is the scenario modelling (Lamb, 1987) discussed earlier. In Chapter 3, oceanographer Gallegos and his colleagues have applied the scenario model to the Intra-Americas Sea. As with Wigley and Santer (Chapter 2), Gallegos *et al.* have focused in on seasonal variability as potentially having a greater short-term implication than the mean (annual) changes. Based on analysis of actual data they foresee larger seasonal fluctuations than at present, a result not dissimilar from the numerical model results. Gallegos *et al.* carry the results of their scenario modelling further, and give indications of the effect of increased intra-season variability on the region's marine waters. They foresee that a few consecutive hot summers have the potential to: readjust coastal sea level, which may affect the fresh-water balance in coastal ecosystems; modify the location and magnitude of shoreline

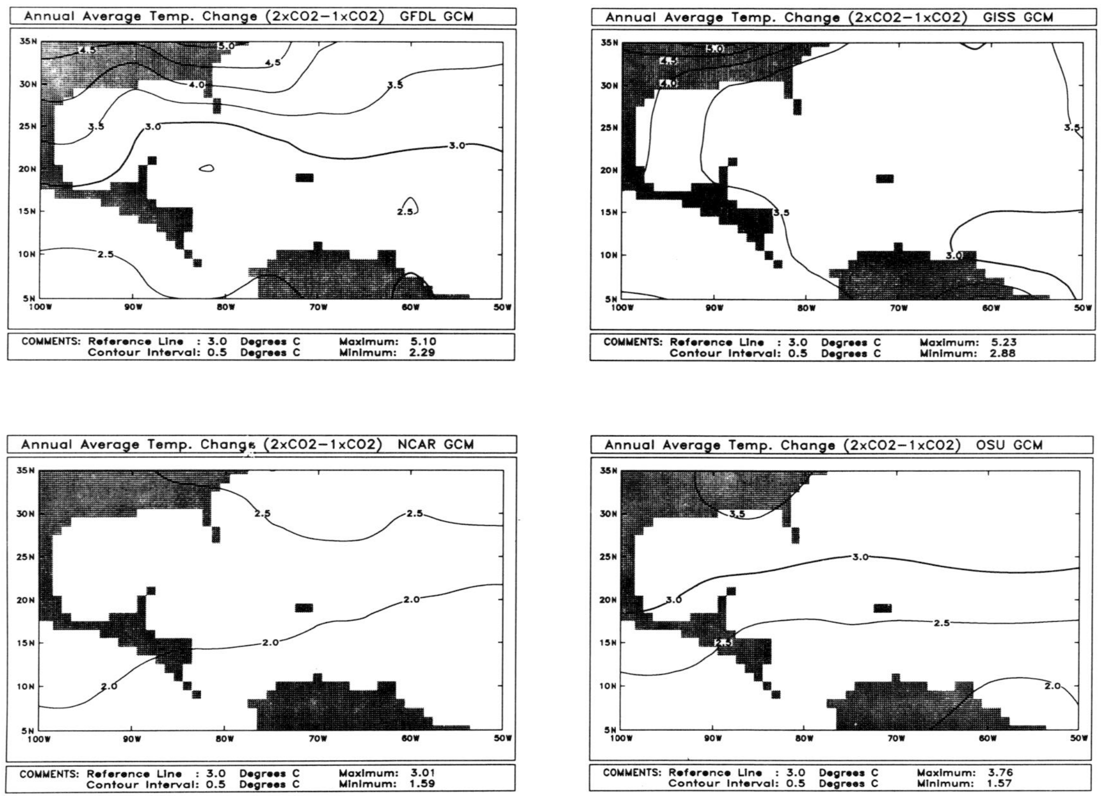

Fig. 1.5a Annual average temperature change due to effective CO_2-doubling from GCMs of the Geophysical Fluid Dynamics Laboratory (GFDL), Goddard Institute for Space Studies (GISS), Oregon State University (OSU), and the National Center for Atmospheric Research (NCAR). Maps courtesy of Wigley and Santer (*cf.* Chapter 2).

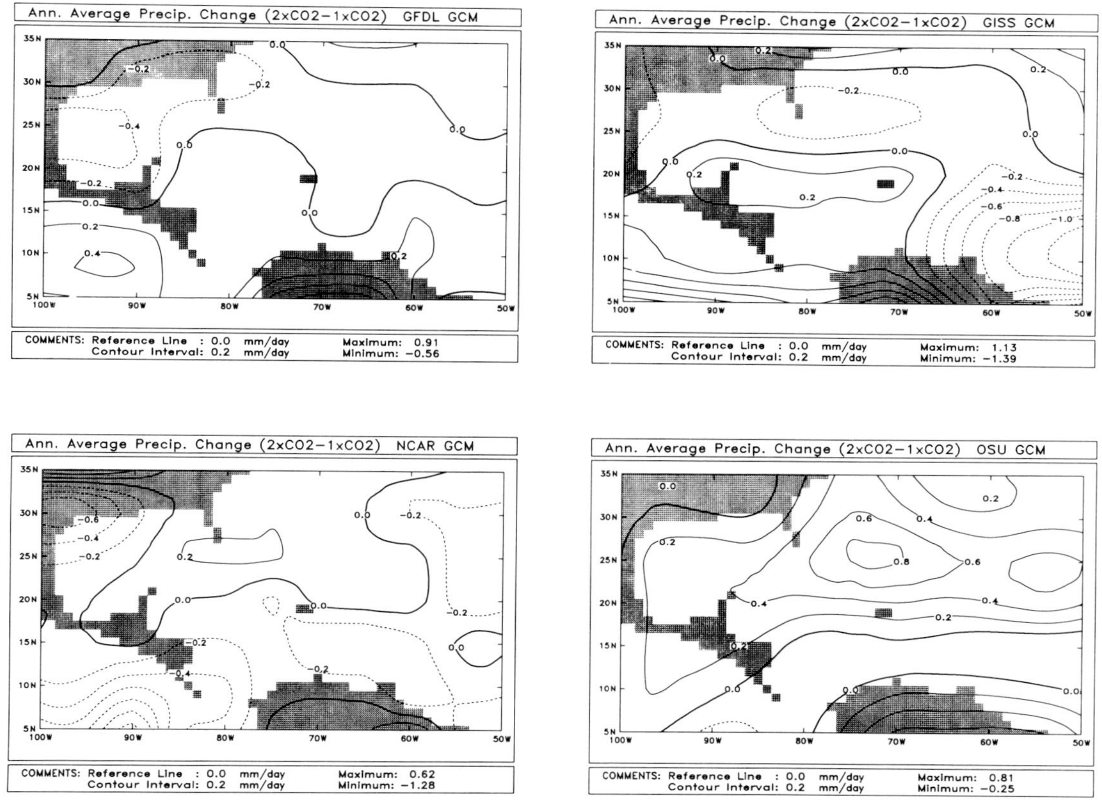

Fig. 1.5b Annual average precipitation change due to effective CO_2-doubling from GCMs of the Geophysical Fluid Dynamics Laboratory (GFDL), Goddard Institute for Space Studies (GISS), Oregon State University (OSU), and the National Center for Atmospheric Research (NCAR). Maps courtesy of Wigley and Santer (*cf.* Chapter 2).

migration; alter patterns of economically important marine species; cause sufficient changes in surface currents to effect marine transportation and contingency plans for spills of hazardous substances; and reorder air-sea interaction which may shift local weather patterns such as precipitation.

When comparing results from GCMs and from scenario models (*cf.* Chapters 2 and 3), there are caveats that must be considered. As eloquent as Lamb's (1987) arguments for scenario models are, there are questions as to whether or not past climate is a harbinger of the future. Similarly, the GCMs are well known to have limitations, and the parameterization of certain physics (notably clouds) and unmodelled effects of volcanism (AGU, 1992) are of concern in our confidence of the $2xCO_2-1xCO_2$ forecasts. Of particular importance is 'How is the global surface temperature change distributed?'.

MacCracken *et al.* (1990) have explored temperature patterns in the Northern Hemisphere using paleoclimate reconstructions from the time periods of relative global warmth. MacCracken (pers. comm.) has kindly provided two such reconstructions: the mid-Holocene (6000 ybp) minus the latter half of the 19th century (*q.v.* Fig. 1.2) and the Eemian interglacial optimum (125,000 ybp) minus the latter half of the 19th century; these are shown in Fig. 1.6 for winter (lower panel) and summer (upper panel) along with a similar meridional profile of predicted temperature change from the four GCMs (*q.v.* Fig. 1.5a). For the tropical/subtropical region with which this report is concerned, there are some remarkable differences.

All three profiles show that the surface temperature change will be larger at high latitudes, a pattern consistent with all IPCC forecasts (Houghton *et al.*, 1992). However, both the Holocene and the Eemian reconstructions show that south of about 35°N (i.e., in the tropics and subtropics) the temperature was cooler during periods of global warmth. The GCMs show quite the opposite: warming at all latitudes with $2\times CO_2-1\times CO_2$ predictions. Readers are reminded that the WMO/ICSU/UNEP (1985) scenario with which we are dealing is for a 1.5°C increase in the regions surrounding the Intra-Americas Sea that is part of a global 1.5°C increase. Fig. 1.6 leaves us with important questions that are unresolved.

Lest there remain ambivalence in the reader's mind concerning future climate, a global forecast of temperature and sea level to the year 2100 is given in Fig. 1.7. The stippled area for each projection represents the range of uncertainty in the 'best guess' IPCC scenario for 1992 (Houghton *et al.*, 1992) as calculated by Wigley and Raper (1992) for the global equilibrium temperature change ($\triangle T_{2x}$=2.5°C) due to the equivalent CO_2-doubling. Based on the revised IPCC estimates, the global temperature and sea level will be 2.5°C and 48 cm higher in 2100 than today, slightly lower values than in the IPCC (1990a) estimates. With respect to the WMO/ICSU/UNEP (1985) scenario dealt with herein (1.5°C and 20 cm by 2025 respectively), Fig. 1.7 suggests that the 1.5°C temperature rise is most likely to occur *c.* 2060, and the 20 cm sea-level rise *c.* 2050. Although the range of uncertainly is much larger at 2100 than at 2050, there is little doubt in the Wigley and Raper (1992) calculation that by the middle of the next century, a warmer Earth is expected, but forecasting the $\triangle T_{2x}$ scenario on a regional basis is frought with additional uncertainly.

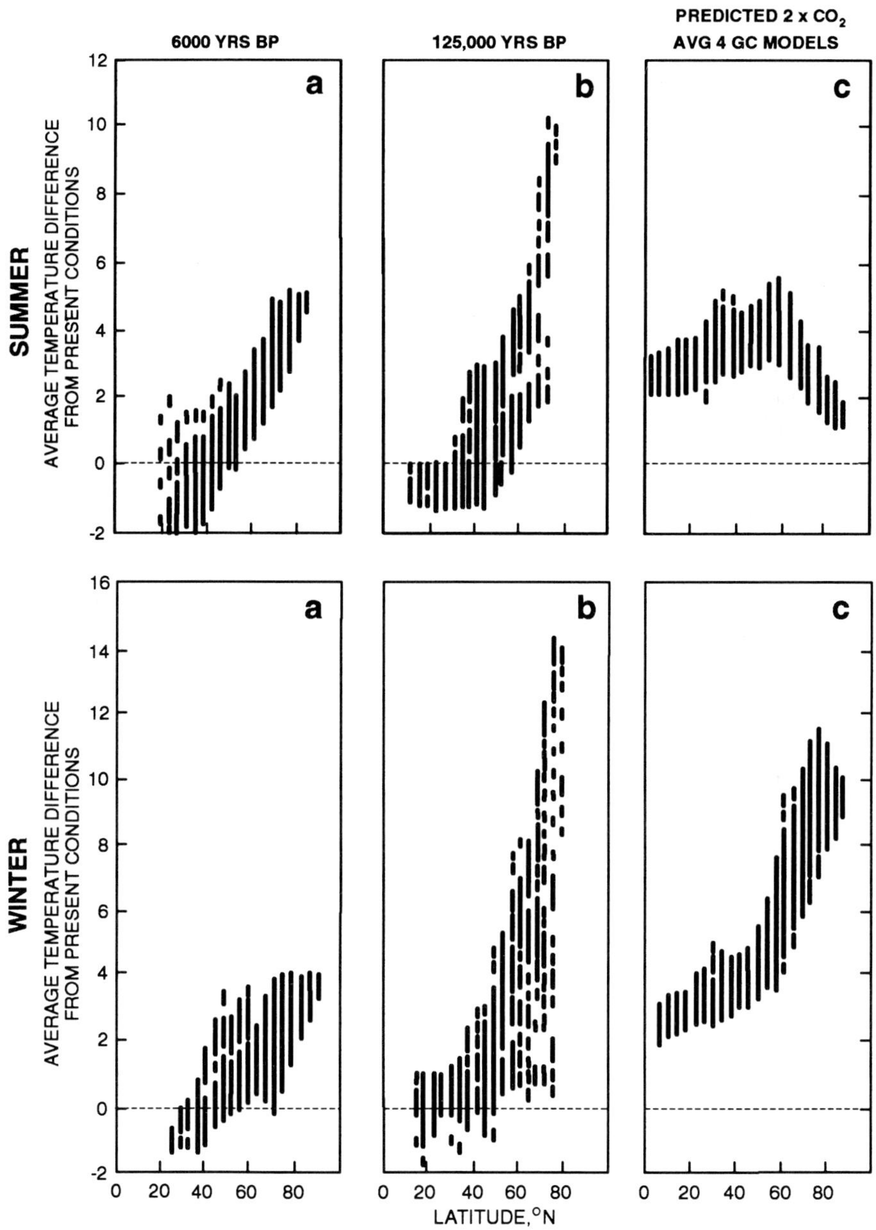

Fig. 1.6 Paleoclimate and greenhouse model comparisons of the meridional profile of seasonal average surface temperature in the Northern Hemisphere. (a) Holocene minus latter half of the 19th century; (b) Eemian minus latter half of the 19th century; (c) average predicted from four GCMs of $2\times CO_2 - 1\times CO_2$. From MacCracken (pers. comm.) following an analysis in MacCracken *et al.* (1990).

8 Conclusion

The atmospheric concentration of CO_2, the primary greenhouse gas, is undoubtedly increasing, and as the upper right hand panel of Fig. 1.4 shows, the increase is clearly associated with human population growth (Idso, 1989). Thus the box in Fig. 1.4 marked "human activity" not only contributes to global "climatic changes", it also causes "other human-induced changes" particularly on the local or regional level. In the near term, it is this local anthropogenic effect that dominates "physical", "ecological", and "socio-economic" change in the Intra-Americas Sea.

In order to understand these physical, ecological, and socio-economic interactions on the marine and coastal environment, six regional Task Teams on Implications of Climate Change have been organized by UNEP: the Mediterranean, Southeast Pacific, South Pacific, East Asian Seas, South Asian Seas, and the Wider Caribbean Region. Each area has unique problems, but each shares the common concern of changing air and water circulation, coastal geomorphology, coastal ecosystems, soil degradation, fresh-water resources, precipitation patterns, terrestrial ecosystems, coastal industries and settlements, and littoral zone population dynamics. The underlying thread often emphasizes negative aspects of climate change; this isn't necessarily universal. Whenever established patterns are disturbed, vested interests tend to exhibit a concern. Rising RSL is probably of more concern in the Intra-Americas Sea than rising temperature, but it is too early to be definitive.

Of primary concern, however, is the availability of adequate data. The sea-level network, which was briefly in good repair for earlier regional programmes such as BOMEX (the Barbados Oceanographic and Meteorological Experiment) in 1969, is now marginally adequate. From a climate perspective, a sea-level observing network must be modernized and include marine meteorological data, geodetic levelling data, sea water

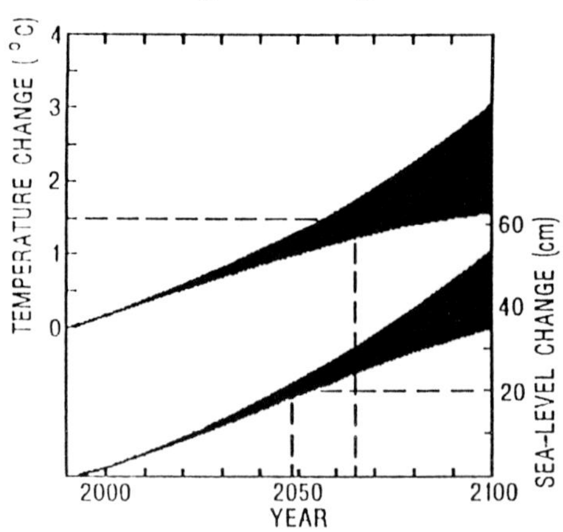

Fig. 1.7 Global temperature (upper curves) and sea-level (lower curves) projections for the various IPCC 1992 scenarios. A range of projections within each stippled area are calculated for the 'best-guess' sets of ice-melt and climate models. Redrawn from Wigley and Raper (1992), the projections only show the anthropogenic component of a future climate, with natural variability superimposed on them.

chemistry data, and ancillary site-specific information. Because of the many short records of sea level and weather, and the difficulty of making conclusions based on them, a concurrent programme of geological, archaeological, and historical data analysis is considered a cost-effective means of strengthening those conclusions. There must also be rapid and free exchange of the observations, a basin-wide commitment to common problems, a responsibility to calibrate and intercompare measurements, and adequate sustained funding. Establishing and *maintaining* a modern sea-level/weather observing network is absolutely necessary to document and ultimately forecast climate-change impacts. Of particular importance in such an observing system is the ability to record extrema in precipitation, sea level, and in temperature of both the water and air; it is in the extreme events that climate-change impact may be most noticeable.

Finally, we choose not to argue for wholly negative impacts. There is a realistic expectation that certain positive benefits may accrue; the local response to global change is simply not predictable at this time. What may be perceived as negative to one sector of society in the region may be beneficial to another. Two examples: (1) a change in precipitation associated with a temperature rise may allow the introduction of different crops but perhaps at the sacrifice of others; (2) an increase in the alongshore component of the wind could increase coastal upwelling and be a benefit to fisheries, yet it may be a cause for concern to agronomists dealing with aerial erosion. A truly challenging and interesting problem will be to identify and explore the legal and institutional implications under the diverse systems and governments which characterize a region that has been influenced by so many native, African, and European cultures.

9 Acknowledgements

The scope of the work summarized in this chapter could not have been possible without the unprecedented willingness of the members of the UNEP/IOC Task Team to share their work. In addition I wish to express my gratitude to the University of Miami's Cooperative Institute for Marine and Atmospheric Studies whose staff contributed so much and especially to Jill Reed; to Mike MacCracken, John Walton, and Stanley Grotch at the Lawrence Livermore National Laboratory whose work was used to construct Fig. 1.6; to the staff of the UNEP Regional Coordinating Unit and the IOC Subcommission IOCARIBE Secretariat who financially supported in part this work; and to H.F. Bezdek and D.V. Hansen of the NOAA Atlantic Oceanographic and Meteorological Laboratory who permitted this work through salary support and staff support.

10 References

AGU (American Geophysical Union) Special Report, 1992: *Volcanism and Climate Change*. AGU, Washington DC, 27 pp.

Alm, A., Blommestein, E. and Broadus, J. 1988: Climatic changes and socio-economic impacts in the Caribbean. In: *Implications of Climate Changes in the Wider Caribbean Region*, UNEP(OCA)/CAR WG.1/INF. 3, 107–127.

Bryson, R.A. and Murray, T.J., 1977: *Climates of Hunger*. Univ. Wisconsin Press, Madison, 171 pp.

Chao, B.F., 1991: Man, water, and global sea-level. EOS, **72(45)**, p. 491.
Digerfeldt G. and Hendry, M.D., 1987: An 8000 year Holocene sea-level record from Jamaica: Implications for the interpretation of the Caribbean reef and coastal history. *Coral Reefs*, **5**, 165–169.
Hansen, J. and Lebedeff, S. 1988: Global surface air temperatures: Update through 1987. *Geophys. Res. Lett.*, **15(4)**, 323–326.
Hardin, G., 1971: Nobody ever dies of overpopulation. *Science*, **171(3971)**, 527 pp.
Hodell, D.A., Curtis, J.H., Jones, G.A., Higuera-Gundy, A., Brenner, M., Binford, M.W. and Dorsey, K.T., 1991: Reconstruction of Caribbean climate change over the past 10,500 years. *Nature*, **352**, 790–793.
Houghton, J., Callander, B. and Varney, S., 1992: Climate Change 1992, the supplementary report to the IPCC Scientific Assessment. Cambridge Univ. Press, Cambridge, 224 pp.
Idso, S. B., 1989: *Carbon Dioxide and Global Change: Earth in Transition*, IBR Press, Tempe, Arizona, 292 pp.
IPCC (Intergovernmental Panel on Climate Change), 1990a: *Climate Change, the IPCC Scientific Assessment*. Cambridge Univ. Press, Cambridge, New York, Port Chester, Melbourne, Sydney, 365 pp.
IPCC, 1990b: *Climate Change, the IPCC Impacts Assessment*. Australian Gov't Publ. Serv., Canberra, 245 pp.
IPCC, 1990c: *Climate Change, the IPCC Response Strategies*. WMO/UNEP, Geneva, 270 pp.
Lamb, P.J., 1987: On the development of regional climatic scenarios for policy-oriented climatic-impact assessment. *Bull. Amer. Meteorol. Soc.*, **68(9)**, 1116–1123.
MacCracken, M.C., Budyko, M.I., Hecht, A.D. and Izrael, Y.A. (eds), 1990: *Prospects for Future Climate*. Lewis Pubs., Inc., Chelsea MI 48118, 270 pp.
Maul, G.A., 1992: Temperature and sea level change. In: *Global Warming: Physics and Facts*. American Institute of Physics, New York, pp. 78–112.
NAS, (National Academy of Science) 1975: *Understanding Climatic Change, A Program for Action*. National Academy Press, Washington, DC, 239 pp.
Oceanus, 1987/88: Vol. 30(4). Caribbean Marine Science. Woods Hole Oceanographic Institution, Woods Hole, MA 02543.
Pugh, D.T., Spencer, N.E. and Woodworth, P.L., 1987: Data Holdings of the Permanent Service for Mean Sea Level. Bidston Observatory, Birkenhead, Merseyside L43 7RA, UK, 156 pp.
Shapiro, L.J., 1982: Hurricane climatic fluctuations. Part II: Relation to large-scale circulation. *Mon. Wea. Rev.*, **110**, 1014–1023.
UNEP (United Nations Environment Programme), 1987: *Action Plan for the Caribbean Environment Programme*. UNEP Regional Coordinating Unit, 14–20 Port Royal Street, Kingston, Jamaica, 24 pp.
Van Sertima, I; 1976: *They Came before Columbus*. Random House, New York, 288 pp.
Van Sertima, I; 1976: *African presence in early America*. Transaction Publ., New Brunswick, 313 pp.
Wanless, H.R., Tedesco, L.P., Rossinsky, V. Jr. and Dravis, J.J. 1988: *Carbonate Environments and Sequences of Caicos Platform with an Introductory Evaluation of South Florida*. Amer. Geophys. Union, 28th International Geological Congress Trip Guidebook, T374, 75 pp.
Wigley, T.M.L. and Raper, S.C.B., 1992: Implications for climate and sea-level of revised IPCC emissions scenarios. *Nature*, **357**, 293–300.
WMO/ICSU/UNEP, 1985: Report of the International Conference on the Assessment of the role of Carbon Dioxide and of other Greenhouse Gases in Climate Variations and Associated Impacts, held in Villach, Austria, 9–15 October 1985. WMO Ref. No. 661, 1–4, 1986.

Part 2
Modelling climate change

2

Future Climate of the Gulf/Caribbean Basin from Atmospheric General Circulation Models

T.M.L. Wigley[1] *and B.D. Santer*[2]

Abstract

Changes in global climate over the past 100 years are reviewed as a backdrop to an assessment of future climatic change under greenhouse-gas forcing. The difference between equilibrium and transient temperature change is explained. The former is important in defining the potential warming commitment, while the latter defines the actual amount of warming likely to be experienced at any given future time. Transient global-mean warming between 1990 and 2025 is expected to range between 0.3–1.5°C. Regional-scale changes in climate can only be estimated using General Circulation Models (GCMs). For the Gulf/Caribbean basin, the only GCM result that one can place confidence in is a general warming. Allowing for uncertainties in the climate sensitivity of GCMs, the regional transient-response warming between now and 2040–2060 is likely to lie in the range of 1–3°C, with no evidence of any marked differences between seasons. At the low end of this range, it would be many decades before such regional-scale changes could be distinguished from the noise of natural variability. There is some evidence of decreased winter/spring precipitation in the northern part of the basin and increased winter/spring precipitation in the southern part. In summer and autumn the models show no consistent patterns of change. Based on the observed statistical link between hurricane frequency and sea-surface temperatures, it is possible that the mean number of hurricanes per year may increase.

1 Introduction

Changes in global climate between now and the middle of the 21st century are likely to be dominated by the influence of the greenhouse effect caused by increasing concentrations of carbon dioxide (CO_2), methane (CH_4), nitrous oxide (N_2O), ozone (O_3) and halocarbons. These greenhouse gases, individually and collectively, change the radiative balance of the atmosphere, trapping more heat near the Earth's surface and causing a rise in global-mean surface air-temperature. Substantial global warming is virtually certain; however, the attendant changes in climate at the regional

[1]Climatic Research Unit, University of East Anglia, Norwich NR4 7TJ, UK
[2]Max Planck Institut für Meteorologie, 2 Hamburg 13, FRG

level are highly uncertain. Except for a few regions, we can be sure only that large changes *will* occur. We cannot yet quantify these changes. Indeed, for precipitation we are generally unable to specify even the sign of regional changes with any reliability. Nevertheless, some useful statements can be made, and it is of some value to review our present state of knowledge.

We begin by considering the global scale, highlighting the important distinction between equilibrium and transient changes in global-mean temperature. We then review published projections of the future climate of the region based on general circulation climate models. After critically reviewing these projections, we give a consensus view of future climate (*cf.* Gallegos *et al.*, Chapter 3; Mercado *et al.*, Chapter 4).

2 Global-Mean Changes

2.1 Observed Climate Changes Over the Past 100 Years

Future changes in climate can be set in context by a brief description of changes that have occurred over the past 100 years. Both regionally and in terms of global-mean values, climatic conditions have fluctuated noticeably from year-to-year, decade-to-decade and on longer time scales. Consider the largest spatial scales first.

The near-surface air temperature averaged over the globe has increased by about 0.5°C since the late 19th century (Jones *et al.*, 1986a, b, c; Jones and Wigley, 1990; Fig. 2.1). Parallel changes in the temperature of the lower troposphere also have occurred; see Folland *et al.* (1990) for a recent review.

As is evident in Fig. 2.1, this warming has not been continuous. Nor has the warming been spatially homogeneous, and trends have varied substantially from region to region. Over the past 40 years, a large part of the North Pacific, along with much of the North Atlantic and western Europe, has undergone a noticeable cooling (Fig. 2.2, from Jones *et al.*, 1988). That such areas of cooling have occurred in spite of the fact that a *global-mean* warming (of 0.2–0.5°C over 1947–1986) would be expected on the basis of the greenhouse effect, demonstrates the degree to which external forcing influences can be masked by natural climatic variability, especially at the regional scale. Similar decadal and longer time-scale 'anomalies' (i.e., departures from greenhouse effect expectations) can be expected in the future. It will not be until the early decades of next century, or later, that the increasing strength of the expected greenhouse effect 'signal' (which will be quantified in a later section) will begin to dominate over this natural 'noise' at the regional level (Wigley and Barnett, 1990).

Noticeable changes in precipitation also have occurred this century on all spatial scales. Large-scale area-average precipitation changes are more difficult to quantify than temperature changes because of the higher spatial variability of precipitation and because of data homogeneity problems which are more difficult to overcome, and we have no information on trends over the oceans. For land-based data in the Northern Hemisphere, Bradley *et al.* (1987) show an upward trend from 1920 to the present in the mid-to-high latitudes (35–70°N) and a marked downward trend

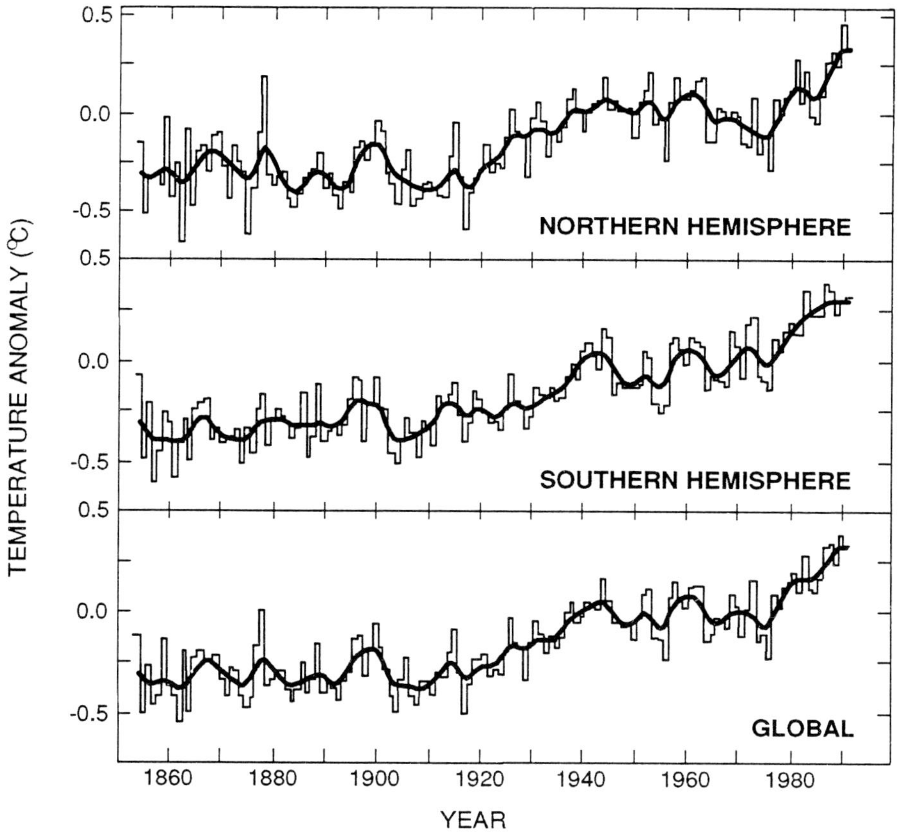

Fig. 2.1 Hemispheric and global annual-mean temperature changes based on land and marine data.

in tropical-to-subtropical latitudes (5-35°N, Fig. 2.3). On smaller spatial scales, some regions of the world have experienced marked changes on decadal time-scales. The Sahel region of Africa is a particularly striking example.

Almost all other climate variables show evidence of a continually changing climate. For example, there have been noticeable changes in the atmospheric circulation of the North Atlantic as measured by the pressure gradient between the Azores High and Iceland Low – a trend towards a decreasing gradient. The extent of mountain glaciers has decreased over the 20th century (Meier, 1984), albeit with marked regional differences and inter-decadal variations. This has contributed noticeably to a general rise in sea level (Warrick and Oerlemans, 1990). Of the 10–15 cm global-mean rise (the current best estimate), roughly similar contributions can be attributed to each of mountain glacier melting, oceanic thermal expansion, and melting from large ice sheets (Greenland and Antarctica), although this breakdown is subject to considerable uncertainty.

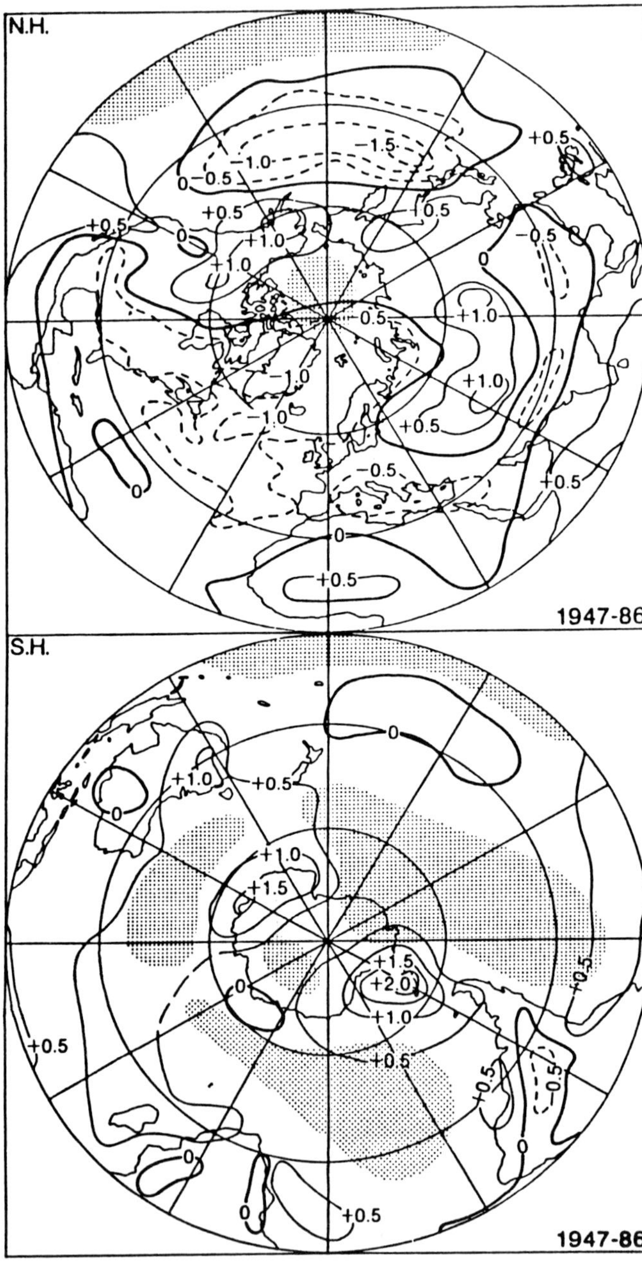

Fig. 2.2 Trends in annual mean temperature over the period 1947–1986. The isolines show the change in temperature (°C) over 1947–1986 accounted for by a linear trend fitted to each annual grid-point time-series. For the Antarctic region, the linear trend has been fitted to annual data for the 1947–1986 period. Shaded areas are regions with no data.

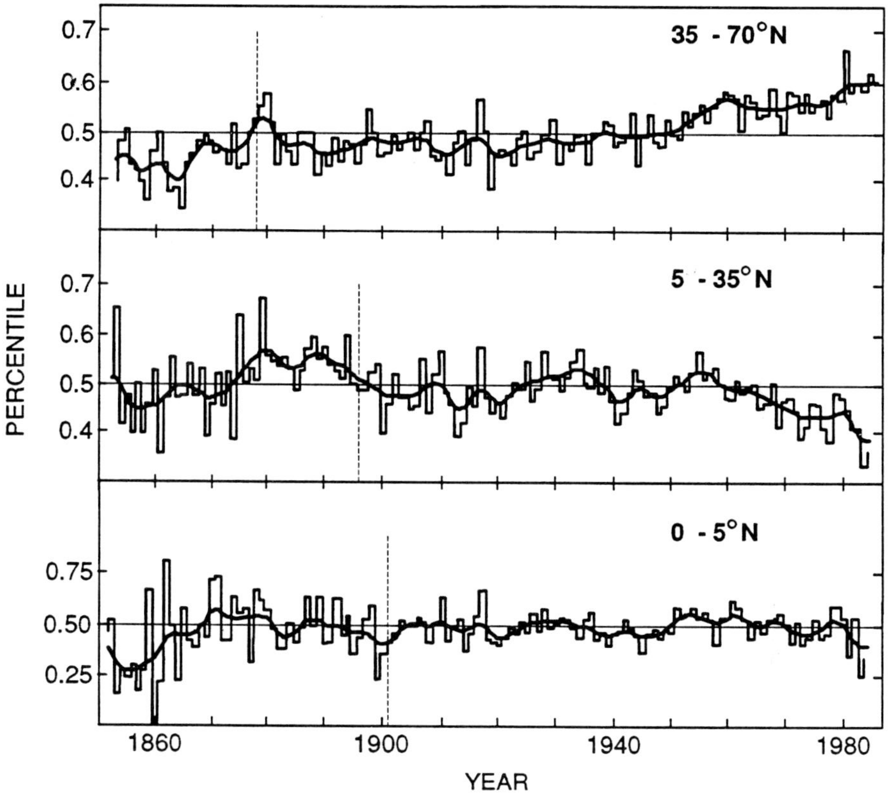

Fig. 2.3 Precipitation changes for different parts of the Northern Hemisphere (land data only).

In parallel with these changes in mean conditions, most measures of climate show changes in variability (as measured by variance and/or the frequencies of extreme events). Very few of these changes in variability (in contrast to changes in means) have been statistically significant. A notable exception is the area-average precipitation of England and Wales, where there has been a significant trend towards more wet extremes in spring and more dry extremes in summer (Wigley and Jones, 1987).

A leading question is: what is/are the cause/causes of these changes? In the absence of changes in external forcing factors, the climate system would be expected to show considerable natural variability on annual, decadal and longer time-scales simply because of the complexity of interactions between the oceans, cryosphere, atmosphere and land surface, and the range of time-scales associated with these components (Wigley and Raper, 1990). Indeed, at the regional level and perhaps even at the global-mean scale, much of the observed variability of climate over the past 100 years may be attributable to such natural variability. In addition, forcing undoubtedly results from: changes in the reflectivity (or albedo) of the Earth-atmosphere system associated with explosive volcanic eruptions (which produce dust and aerosol layers in the stratosphere), with changes in cloudiness and with changes in surface characteristics (vegetation, snow and ice cover, etc.); changes in solar irradiance; and changes in the concentrations of greenhouse gases (CO_2, CH_4, N_2O, O_3 and halocarbons) that perturb the radiation balance of the atmosphere (see Hansen and Lacis, 1990, and Shine *et al.*, 1990 for recent reviews). Man-made SO_2 emissions may also have influenced the climate, partially offsetting greenhouse-gas-induced radiative-forcing changes in the Northern Hemisphere (Wigley, 1989, 1991; Charlson *et al.*, 1991)

For the greenhouse gases, the global-mean radiative-forcing change over the past 100 years has been large (around 2 Wm^{-2}), and this would be expected to cause a global-mean warming. Although a warming has occurred, and its magnitude (*viz.* 0.4–0.6°C) is compatible with that expected to have resulted from the greenhouse effect (*viz.* 0.3–1.1°C, see below), we are not yet able to positively relate cause and effect. Nevertheless, the qualitative agreement and the realistic physical basis of current climate models demand that the possibility of substantial future global-mean warming, and the multitude of regional changes that would accompany such a warming, must be taken seriously.

2.2 Future Global-Mean Changes

The magnitude of future global-mean warming will depend on future atmospheric trace-gas concentrations, on the sensitivity of the climate system to external forcing and on delays due to the thermal inertia of the oceans. The enhanced greenhouse effect will cause changes in global-mean temperature on time-scales of decades to centuries, with associated trends in all climate variables at smaller spatial scales. These changes will also have, superimposed on them, inter-annual to inter-decadal variability that is a natural characteristic of the unperturbed climate system.

In considering future climatic change, it is important to distinguish between the equilibrium and transient response of the climate system. We consider the equilibrium response first. For any given future greenhouse-gas forcing (ΔQ) there will be a corresponding equilibrium global-mean temperature change (ΔT_e) which is the temperature change that is achieved when the climate system reaches a new steady state. The relationship between ΔQ and ΔT_e is determined by the climate sensitivity.

The climate sensitivity is usually expressed in terms of the equilibrium change in global-mean temperature (ΔT_e) that would occur for a forcing change of 1 Wm^{-2} at the top of the troposphere. Alternatively, it can be expressed as the ΔT_e value that would occur if the CO_2 level doubled (a doubling of the CO_2 concentration corresponds to a forcing change of about 4.4 Wm^{-2}), *viz* ΔT_{2x}. The higher the value of ΔT_{2x}, the higher the climate sensitivity. The value of ΔT_{2x} is uncertain, but it is thought to lie in the range 1.5–4.5°C (Mitchell *et al.*, 1990). The CO_2-doubling case is a standard case used in most modelling studies. The equilibrium change for an arbitrary concentration change from C_o to C can be related to this standard case using

$$\Delta T_e = \Delta T_{2x} (\ln [C/C_o]/\ln 2) \qquad (1)$$

Equation (1) arises because ΔQ for CO_2 is logarithmic in concentration,

$$\Delta Q = 6.3 \ln(C/C_o) \tag{2}$$

Since CO_2 is not the only greenhouse gas that must be considered in the forcing, it is often convenient to express the total radiative forcing (ΔQ_T) due to changes in all anthropogenic greenhouse gases in terms of an equivalent CO_2 concentration (C^*). This is defined by inverting equation (2) to give

$$C^* = C_o \exp(\Delta Q_T/6.3) \tag{3}$$

where, by convention, C_o is taken to be the pre-industrial level of CO_2 alone (i.e. the other greenhouse gases are not considered in defining this reference level).

Since pre-industrial times (~ 1750–1800), the concentrations of all greenhouse gases have increased substantially: CO_2, 279–354 ppmv; CH_4, 790–1720 ppbv; N_2O, 280–310 ppbv; CFC-11, 0–280 pptv; CFC-12, 0–480 pptv (Watson *et al.*, 1990). Pre-industrial levels of CO_2 and CH_4 are known to have varied naturally by small amounts, but at least over the past 10,000 years these natural variations have been minor compared with the more recent anthropogenic changes). Thus, relative to conditions in the mid- to late 18th century, the combined radiative forcing (ΔQ_T) is about 2.5 Wm^{-2} today, with 1.5 Wm^{-2} of this being due to CO_2 changes and the rest due to changes in the other trace gases, mainly methane and the halocarbons (Shine *et al.*, 1990). The 1990 value of the equivalent CO_2 concentration was around 410 ppmv (relative to a base level of 279 ppmv) compared with a concentration for CO_2 alone of 354 ppmv. A further forcing change of about 1.5 Wm^{-2} is expected between 1990 and 2025 (Houghton *et al.*, 1990, Annex, Scenario B). This corresponds to an equivalent CO_2 concentration of about 530 ppmv in 2025, approaching double the pre-industrial CO_2 level.

Using ΔT_{2x} = 1.5–4.5°C, the greenhouse forcing changes to 1990 imply an *equilibrium* warming of 0.8–2.5°C over the period since 1765. Over the period 1990–2025, based on the 1.5 Wm^{-2} projected forcing change, the implied equilibrium warming is 0.5–1.5°C. This projection incorporates the uncertainty that arises from uncertainties in the climate sensitivity, but it does not allow for uncertainties in future greenhouse-gas concentrations (i.e. future forcing uncertainties). Allowing for these widens the range of possible values for ΔT_e over 1990–2025 considerably. Even values at the lower end of the range of possibilities, however, correspond to very significant changes in climate.

The range 0.5–1.5°C represents the range of possible *equilibrium* warming over the interval 1990–2025 under a particular forcing scenario, namely the 'B' scenario given by the Intergovernmental Panel on Climate Change (IPCC; Houghton *et al.*, 1990). Although not labelled as such by the IPCC, this is often taken to be a 'best guess' scenario. The equilibrium warming, however, is not what would actually be observed over this interval (namely, the *transient-response* warming). This is because, due to the large thermal inertia of the oceans, observed changes lag behind the equilibrium changes. The amount of lag is determined by the rate of ocean mixing, by the climate sensitivity and by the rate of change of forcing. Because of this thermal inertia effect, only some 50–90% of the equilibrium warming may be observed at any given time.

Let us consider the consequences of this lag effect. Over 1880–1990, the radiative forcing change was about 2.2 Wm^{-2}, implying an expected equilibrium warming in the range of 0.7–2.3°C. However, the expected *transient* warming (i.e. the value which should be compared with observations) is much smaller, lying in the range 0.5–1.2°C (see Wigley and Barnett, 1990, Fig. 8.1). This latter range can be compared with the observed warming of about 0.5°C. Since 0.5°C lies at its lower end, one might conclude that the climate sensitivity must lie near the lower end of the range 1.5–4.5°C. There are, however, a number of other ways to explain this discrepancy, so we cannot yet dismiss the higher sensitivity values. Even 4.5°C could be consistent with observations, given uncertainties regarding the causes of past climatic change (Wigley and Raper, 1991), especially those related to sulfate aerosols (see Wigley, 1989, Fig. 1).

An important consequence of the equilibrium/transient difference is that, no matter what happens to future greenhouse-gas concentrations, we always will be committed to an additional warming as the climate system tends towards equilibrium. Thus, the difference between the observed 1880–1990 global-mean warming of 0.5°C and the ΔT_e range of 0.7–2.3°C represents a current warming *commitment* (a phrase introduced by Mintzer, 1987), in the sense that it would result even if the concentrations of all greenhouse gases could be held constant at their 1990 values.

3 Regional Changes

3.1 Introduction

In the above, we have concentrated on global-mean temperature as a key measure of the state of the climate system. Changes in all climate variables (precipitation, evaporation, wind patterns and strengths, cloudiness, etc.) will necessarily accompany changes in global-mean temperature. In the future, just as in the past, these changes will differ noticeably from region to region. Thus, for assessing impacts, global-mean quantities are mainly of academic interest, although they do show that future climatic changes will probably be large and probably will occur more rapidly than any previous changes. For direct assessments of impacts, regional-scale (or smaller) details of future changes are needed – not only for temperature, but for a variety of climate variables. At present, our capability for predicting these details is limited and we must resort to the use of scenarios.

3.2 Climate Scenarios

The most important method for obtaining information on possible future climates is to use an atmospheric General Circulation Model (GCM). However, because of deficiencies in current GCMs (described in more detail later), their outputs should only be considered as possible scenarios for future climatic change rather than predictions. A climate scenario, as defined in the literature, is intended to be an internally consistent picture of possible future climatic conditions. In addition to the use of

GCMs, scenarios can be constructed by a number of other means (see, for example, Pittock and Salinger, 1982; Palutikof *et al.*, 1984). However, except for Gallegos *et al.* (Chapter 3), these methods have not been applied to the basin, and we will concentrate here on GCM results. But first, it is important to keep in mind the limitations of GCMs. Note that, within the modelling community these are well recognized problems. They often, however, are not realized by analysts who make use of GCM output.

3.3 GCM Limitations

(See Mitchell *et al.*, 1990, for a recent review.)

1) Most GCM studies of the greenhouse effect consider only the equilibrium response to a doubled or quadrupled concentration of carbon dioxide. The regional patterns of climatic change for the transient response for the few coupled ocean–atmosphere GCM simulations that have been carried out to date, show differences from the equilibrium response patterns, but these are only large in high-latitude regions (Bretherton *et al.*, 1990).
2) Atmospheric GCMs still have a number of recognized deficiencies in the way they model clouds, sea-ice effects and land-surface processes.
3) Different GCM studies that attempt to describe the climate of a high-CO_2 world show regions of agreement and disagreement. Since regions of agreement can occur by chance, they cannot be necessarily accepted as regions where the predictions are more reliable.
4) The atmospheric models currently used in greenhouse-gas studies are unable to simulate the current ($1 \times CO_2$) atmospheric circulation at the regional level with any realism (see, for example, Santer and Wigley, 1990; Hulme, 1991), although they do perform reasonably well in describing some of the large-scale features of the general circulation. A reliable 'control-run' simulation is thought to be a necessary condition for reliability in any perturbation experiment.
5) While GCMs can produce output with very short time steps ($\sim$ 10 minutes), the realism of these short-time-step results is in doubt. For current models, the limit of temporal resolution on which results might have some meaning is of order one day. However, some models currently in use do not have a diurnal insolation cycle, which necessarily precludes realism even at the daily time-scale, and the strange results produced by some models on the *seasonal* time-scale, point to serious deficiencies on time-scales of months and longer.
6) The current spatial resolution of GCMs is too coarse for most impact studies. Coarse resolution means that the orography in the models is highly smoothed and small-scale weather systems are non-existent. This means that one would not expect a GCM to produce reliable results even at the grid-point scale. At best, one can only hope for reliability on the scale of the area covered by tens of grid points – but even this is an optimistic hope at present.

Despite the deficiencies, there are some broad-scale GCM results which can be accepted with some confidence as predictions of future climate.

These are summarized in Table 2.1.

In spite of the problems that plague current GCMs, they are the best tool we have for projecting future changes in climate at the regional level. A number of such projections have been used in impact studies, sometimes with implied faith in the reliability of the projections and other times more sceptically, with a primary aim being to examine climate sensitivities and to develop tools and methods for impact assessment (e.g., Parry *et al.*, 1988). Surprisingly, most such studies have relied on a single GCM, and very few detailed (i.e., regional scale) comparisons have been made of the projections of different GCMs. Just such a comparison is made later in this paper.

4 GCM Projections for the Basin

It is clear from the preceding section that the climate of the region is determined partly by larger-scale characteristics of the atmospheric circulation (the Azores subtropical high, the monsoon, easterly waves, the position of the subtropical jet stream, etc.), partly by interactions between the large-scale flow and orography and land-sea contrasts, and partly by more local effects. Greenhouse-gas changes will change the large-scale characteristics of climate and will affect temperatures and temperature

Table 2.1 A selection of model results from equilibrium GCM experiments for a doubling or quadrupling of atmospheric CO_2 concentration, together with an estimate of the confidence that can be placed in these results. 'Unknown' indicates that knowledge of possible future changes is zero. This applies to all items not mentioned.

Model results	*Confidence*
Global scale (i.e., global-mean values)	
Warming of lower troposphere	High
Increased precipitation	High
Cooling of stratosphere	High
Warming of upper troposphere (especially the tropics)	Moderate
Zonal-mean to regional scale	
Reduced sea ice	High
Enhanced polar warming in Northern Hemisphere (especially winter half year)	High
Increased P−E* in high latitudes	High
More absolute high temperature extremes	High
Increased continental summer dryness	Moderate
Stronger monsoon	Moderate
More tropical storms	Unknown
More/less blocking	Unknown
Greater/less interannual variability	Unknown
Spatial detail in general	Unknown
Rainfall extremes	Unknown

*Precipitation (P) minus evapotranspiration (E)

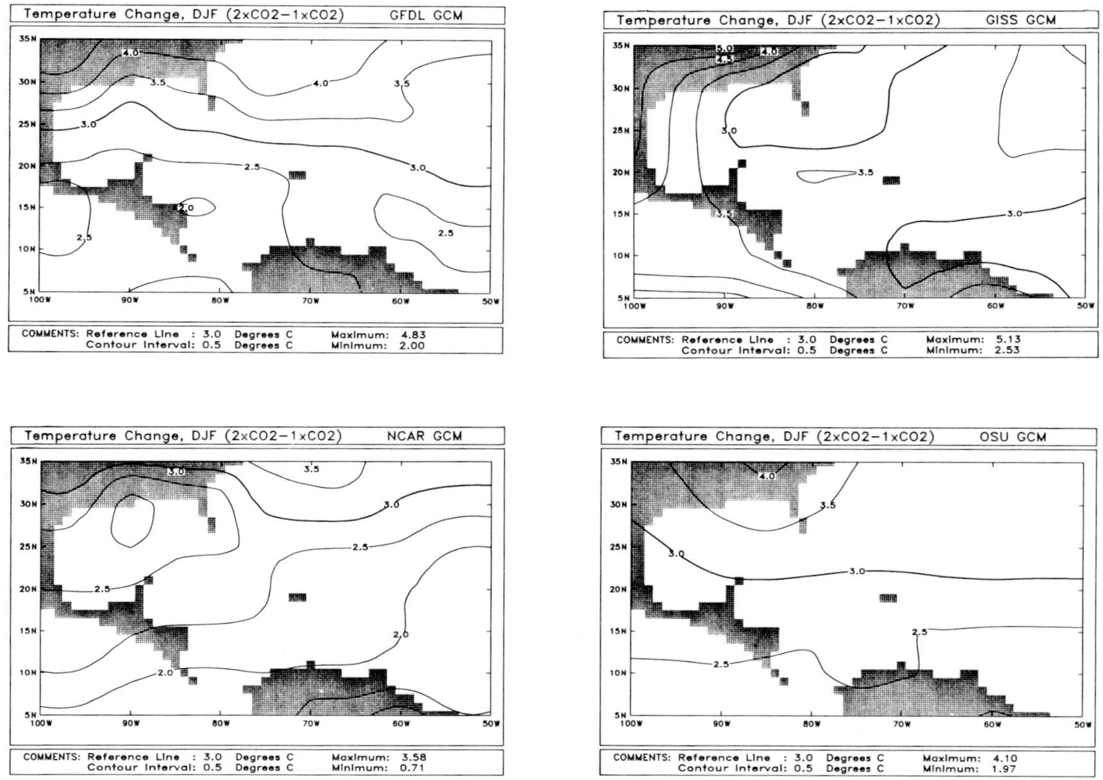

Fig. 2.4 Winter (December, January, February [DJF]) temperature changes (in °C) due to a doubling of the CO_2 concentration; results from four independent GCMs.

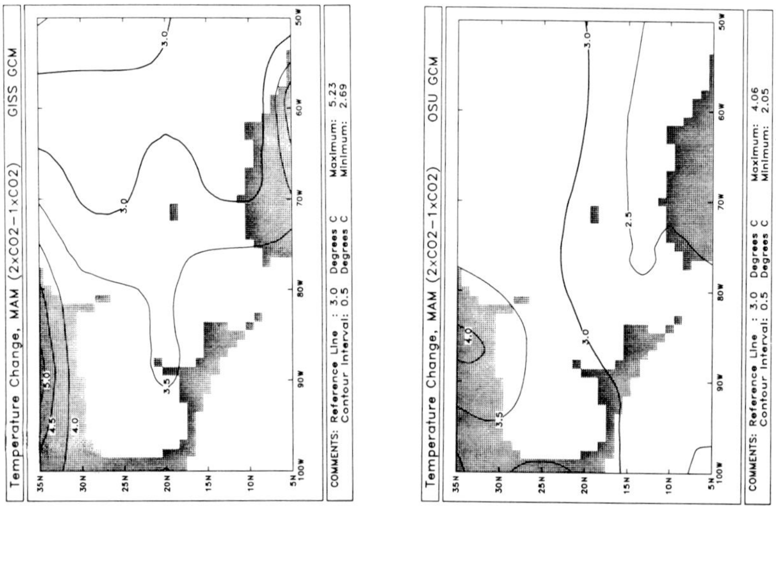

Fig. 2.5 Spring (March, April, May [MAM]) temperature changes (in °C) due to a doubling of the CO_2 concentration; results from four independent GCMs.

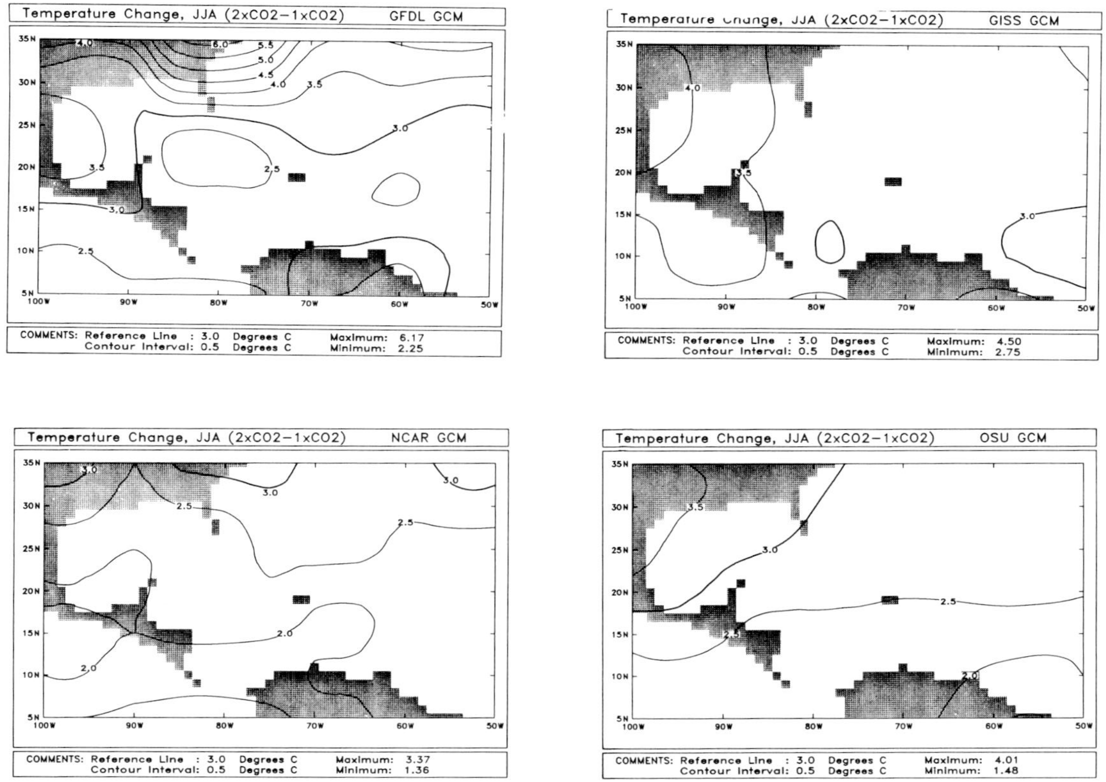

Fig. 2.6 Summer (June, July, August [JJA]) temperature changes (in °C) due to a doubling of the CO_2 concentration; results from four independent GCMs.

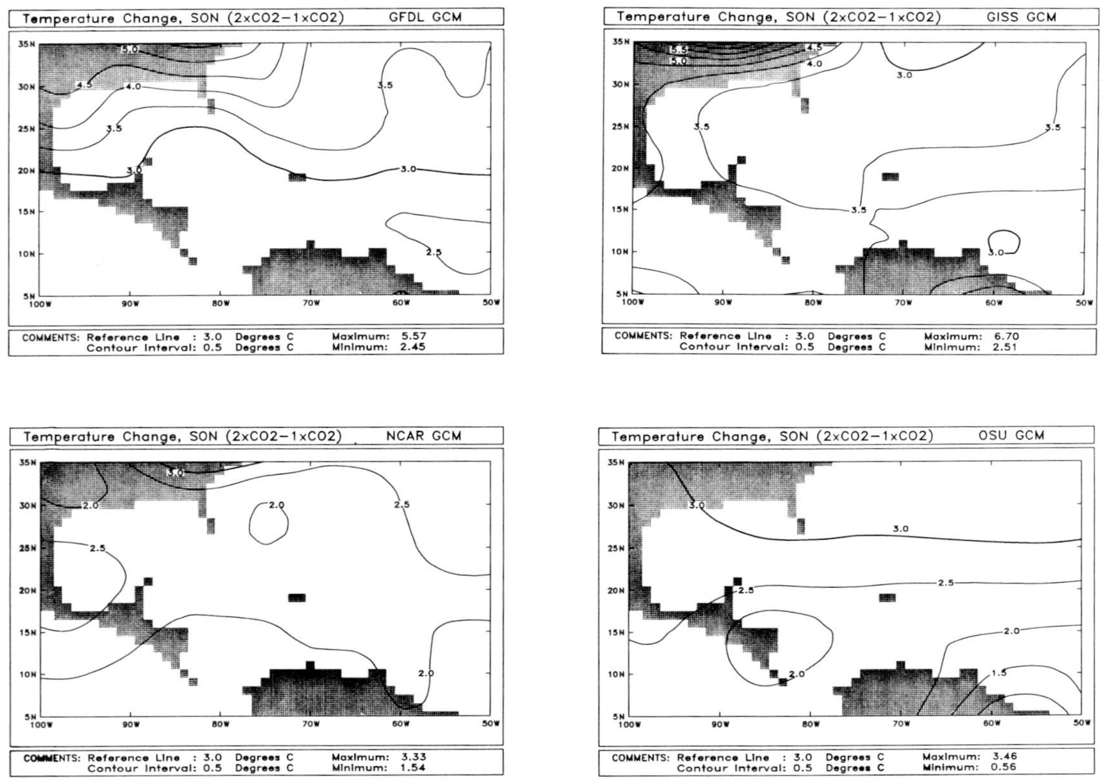

Fig. 2.7 Autumn (September, October, November [SON]) temperature changes (in °C) due to a doubling of the CO_2 concentration; results from four independent GCMs.

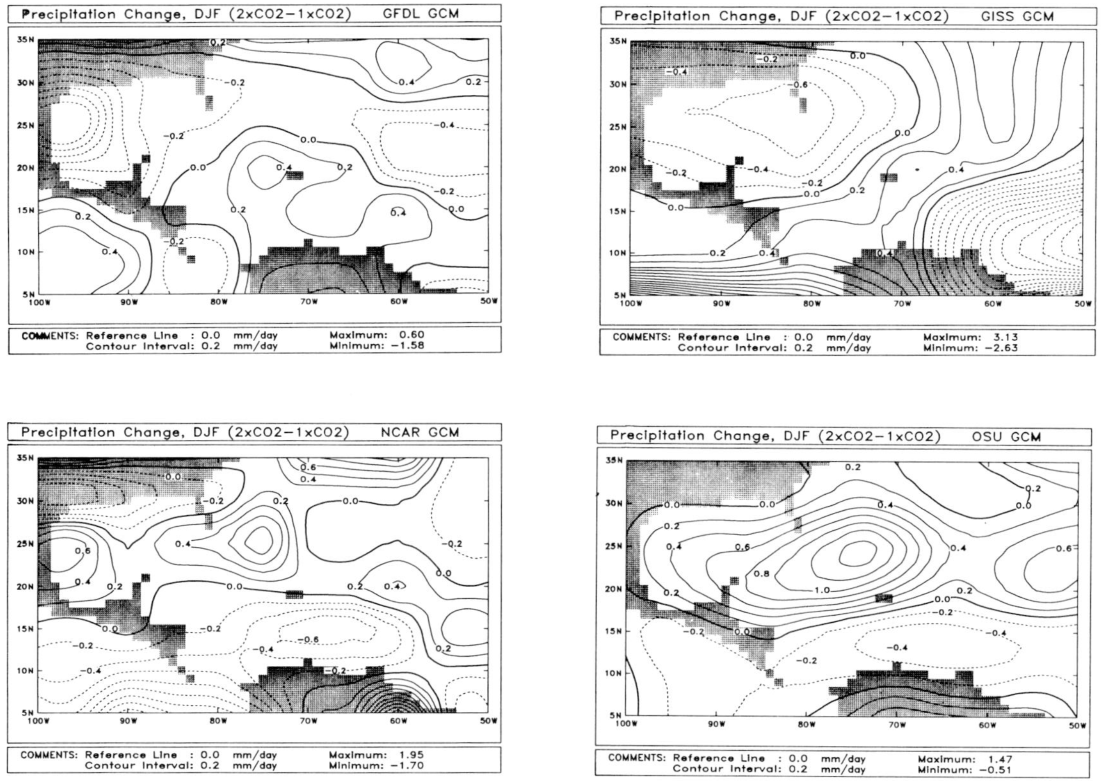

Fig. 2.8 Winter (December, January, February [DJF]) precipitation changes (in mm/day) due to a doubling of the CO_2 concentration; results from four independent GCMs.

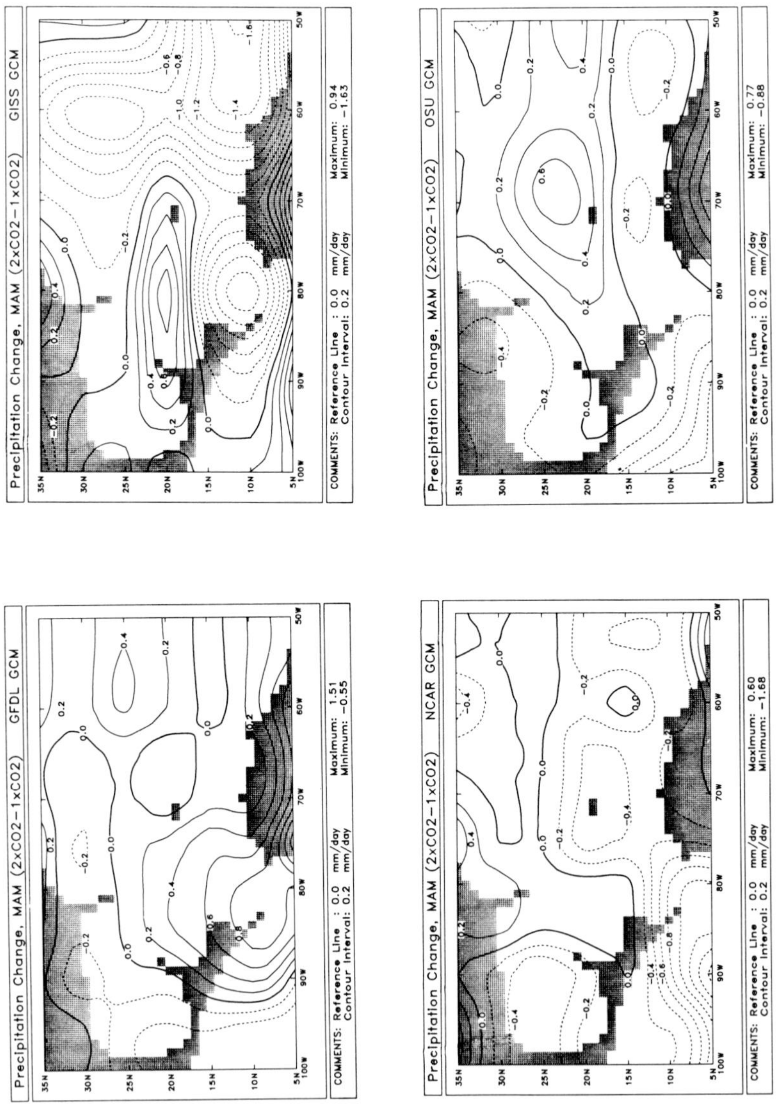

Fig. 2.9 Spring (March, April, May [MAM]) precipitation changes (in mm/day) due to a doubling of the CO_2 concentration; results from four independent GCMs.

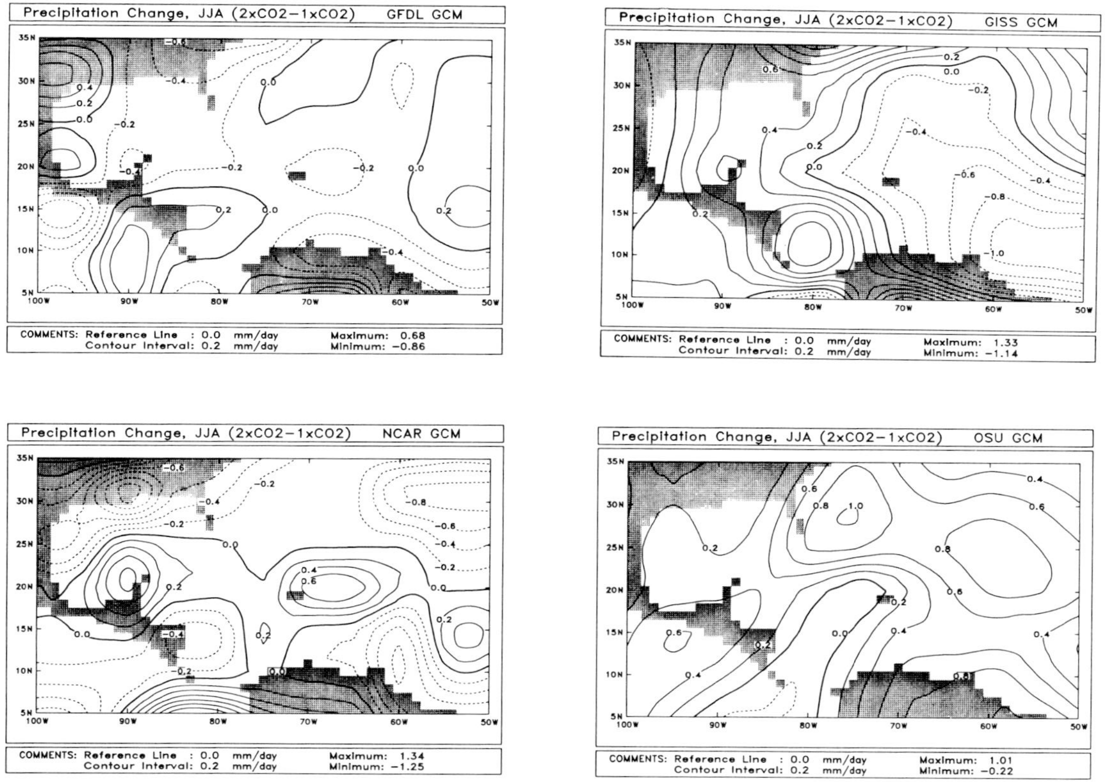

Fig. 2.10 Summer (June, July, August [JJA]) precipitation changes (in mm/day) due to a doubling of the CO_2 concentration; results from four independent GCMs.

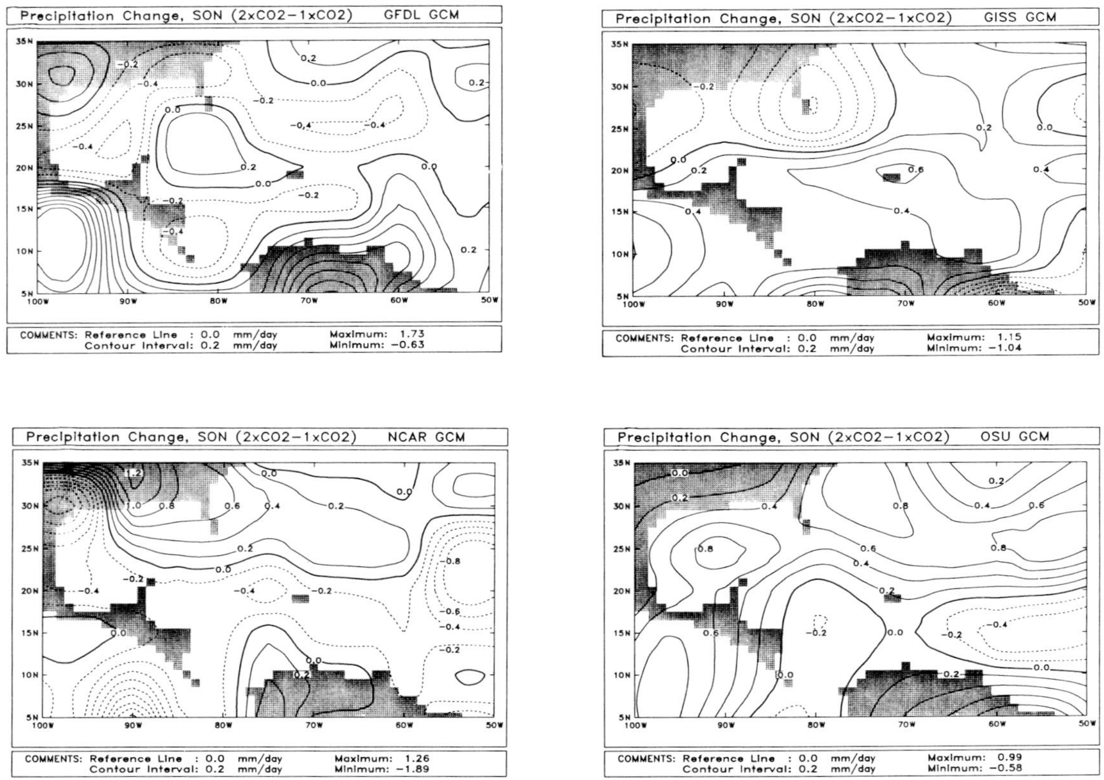

Fig. 2.11 Autumn (September, October, November [SON]) precipitation changes (in mm/day) due to a doubling of the CO_2 concentration; results from four independent GCMs.

patterns in the sea and surrounding land areas. The precise patterns of future climatic change will, however, be controlled to a considerable extent by the way these effects are modulated by geography.

In an earlier section, it was noted that many GCMs currently used in greenhouse-gas studies are unable to reproduce regional-scale features of present-day climate reliably. These GCMs, furthermore, have quite coarse resolution and they use highly smoothed orography (which is insufficiently detailed to be able to show orography with any semblance of realism). Since they cannot properly resolve many important regional-scale features of the general circulation (no less hurricanes; *cf.* Gray, Chapter 5), and they cannot possibly simulate the way orography interacts with the circulation in controlling local precipitation patterns, this means that they are unlikely to be able to produce realistic simulations of the present and/or the future climate of the basin. How, then, can we gain any insights into future changes in climates over the basin, given the inadequacy of our primary tool the General Circulation Model? The answer is that we cannot obtain any *detailed* predictive insights. We can examine GCM results (as will be done below), but we must be extremely cautious in interpreting and applying these results – i.e., they should be treated strictly as scenarios of a possible future climate and not as predictions.

Figs. 2.4–2.11 summarize the results from four independent GCM studies in terms of seasonal-mean values of the equilibrium changes in temperature and precipitation due to a doubling of the atmospheric CO_2 level. (Note that, for 'CO_2' we can read 'equivalent CO_2' in order to account for the other greenhouse gases.) The GCMs were 'state-of-the-art' models when this work was carried out (in 1989). Improved (e.g., with higher resolution) versions of some of the models have been developed subsequently, and other model results are now available. However, the general character of the results given here is not changed by these more recent developments. The models used are: the GISS model (Goddard Institute for Space Studies; Hansen *et al.*, 1984), the GFDL model (Geophysical Fluid Dynamics Laboratory; Manabe and Stouffer, 1980), the Community Climate Model of the National Center for Atmospheric Research (NCAR; Washington and Meehl, 1984) and the OSU model (Oregon State University, Schlesinger and Zhao, 1989). Further details of these models are given by Schlesinger and Mitchell (1987).

These four models are similar in their basic physics, but they have major differences in the way they handle sea ice, clouds and surface processes. They also differ in the way they solve the various partial differential equations (GISS and OSU solve these in grid-point form, while GFDL and NCAR use a spectral method) and in their resolution (OSU has four times the horizontal resolution of GISS, with GFDL and NCAR in between; but OSU has much coarser vertical resolution than any of the other models). The models are known to perform poorly in simulating present-day regional-scale climatic conditions (Santer and Wigley, 1990).

Before describing the results produced by these models, we need to assign some time in the future to which they might apply. The key point here is that they are *equilibrium* results for $2\times CO_2$. That is, they do not simulate the transient response to continually changing CO_2 levels. As noted earlier, in low and middle latitudes, the spatial details of the transient response are similar to those for the equilibrium response, so this is not a serious defect. The main equilibrium/transient difference is therefore one of timing. Thus, if an equivalent $2\times CO_2$ concentration level were reached in, say, 2025, the climate would not achieve the equilibrium configuration corresponding to this level for a number of years. The magnitude of the delay is uncertain, since it depends on properties of ocean heat transport which are presently rather poorly understood, but a delay of 15–35 years is a reasonable guess. This means that the climate changes shown in Figs. 2.4–2.11 should be representative of the change between pre-industrial times and some time around 2040–2060. Since the changes that have occurred between pre-industrial times and now are much less than many of those shown in Figs. 2.4–2.11, we can say that these Figures also represent the changes that might occur between now and 2040–2060.

But can we put any faith in these results? Certainly we cannot believe any of the spatial differences between different parts of the study area. Some of these differences are so large and they occur over such small distances, that they are virtually impossible – they serve only to expose defects in the physics of the models concerned. In all cases, the spatial patterns of change vary from model to model, particularly for precipitation. These model-to-model differences hardly instill any confidence in the performance of any individual model.

The only common feature in which we can have any faith, is a large-scale warming in all seasons. Warming tends to be larger in the north of the region. There are only small differences between the seasons, and a warming of about 3°C spread uniformly over the seasons would be the best guess. This is somewhat less than the global-mean changes produced by these particular models. In winter and spring, the models tend to show a decrease in precipitation in the north and an increase in the south (a striking exception is the OSU winter simulation, which shows an area of strong increase in the north). In summer and autumn, results from different models are often diametrically opposed. There is very little we can say, therefore, about future precipitation changes.

It is important to note that, on top of the uncertainties in the regional patterns of change, there is a further uncertainty in the magnitudes of change because all four models have similar climate sensitivities, namely about 4–5°C for a doubling of the CO_2 concentration. This range is at the top end of the accepted range of climate sensitivity, *viz.* $\Delta T_{2x} = 1.5–4.5°C$. Thus, the results shown in Figs. 2.4–2.11 probably represent upper limits to the changes likely to occur between now and the middle of the 21st century. The estimated warming of about 3°C by 2050 over the basin as a whole therefore should be translated to a warming *range* of about 1–3°C. At the low end of this range, the change is similar to the extent of natural variability observed over 1947–1986 (see Fig. 2.2). If this observed change is representative of the normal level of natural variability, and if the greenhouse-gas warming signal were a steady warming, then it would be many decades before the regional-scale signal could be distinguished convincingly from the noise of natural variability. On the other hand, if the signal were at the top end of the likely range, it would probably be clearly discernible by the early decades of the next century.

5 Summary and Conclusions

The greenhouse problem can be viewed as one involving large uncertainties but with risks that are potentially high. Even the most well-defined projections, those for global-mean temperature, are subject to considerable uncertainty. Global-mean warming projections for the interval 1990–2025 given by IPCC (Houghton *et al.*, 1990) lie in the range 0.3–1.5°C, a range that accounts for uncertainties in both the climate sensitivity and future greenhouse-gas concentrations. At the low end of the range, the change is comparable to the global-mean warming that has occurred already this century (although the future warming would be more rapid). However, the likely observed change in 2025 is not the crucial parameter; rather, it is the equally large additional warming *commitment* which needs closest consideration. Even at the low end of the range of projections, we will be committed to significant changes in regional climate. When coupled with changes in extreme-event frequencies which also must occur, the impacts could be considerable. At the high end of the commitment range (i.e., the low-probability/high-risk area), the projected global-mean temperature would far exceed anything previously experienced by humankind. There is clearly considerable cause for concern and a pressing need to reduce the uncertainties.

The impacts of any global-mean climatic change will depend on the regional details of changes in a wide variety of climate variables and in changes in the interannual variability of these variables. At present, we are unable to predict these changes. However, GCM results do give us data that can be used to develop scenarios of future changes. For the basin, GCM results point to a warming which would be somewhat less in magnitude than the global-mean value (see Fig. 1.5a), with no evidence for any marked seasonal differences in the warming (Figs. 2.4–2.7). Although the magnitude of this warming is uncertain, we can be fairly confident that, as a prediction, it is qualitatively correct (but see Fig. 1.6). Similarly, there are physical reasons to expect that the model result of slightly enhanced warming in the north of the basin is correct. It may be, however, many decades before the change can be statistically detected above the noise of natural, regional-scale climatic variability. Nevertheless, the existence of a background warming trend will still be of considerable importance. As time progresses, the probability of periods of extreme warmth will increase, with attendant effects on ecosystems and water resources. Furthermore, increased warmth will probably lead to greater evapotranspiration, adding to direct thermal stresses on vegetation.

Projected precipitation changes vary so much from model to model that one cannot say on the basis of model results alone whether precipitation will increase or decrease. The best guess would be no major precipitation changes. In winter and spring, precipitation decreases appear to be slightly more likely in the northern part of the basin, while the evidence for the southern part of the basin points marginally towards precipitation increases. Whether or not there will be changes in the finer-scale details of precipitation (precipitation intensity, extreme values, dry and wet day probabilities, etc.) is impossible to say on the basis of these model results.

One important possibility, which is outside present model capability to predict, requires special mention. This is that hurricane frequency may well increase. In the Atlantic hurricane region, hurricane frequency is positively correlated with sea-surface temperatures, so a global warming could, in this region, lead to more hurricanes (see, for example, Raper, 1992; Gray, Chapter 5).

All of the above suggestions must be taken as possibilities, rather than probabilities. They are, however, amenable to more detailed investigation using existing data and model results. Indeed, there is much that can be done to reduce uncertainties and to develop better methods for interpreting GCM results. At the same time, GCMs themselves are constantly being improved and much better results can be expected to appear within the next 5–10 years. The greenhouse effect is our most pressing environmental problem, one which presents major scientific challenges across a wide range of disciplines. Developing better regional-scale projections of future climate is at the forefront of these challenges.

6 Acknowledgement

The research that led to this paper was funded by the US Department of Energy, Carbon Dioxide Research Division.

7 References

Bradley, R.S., Diaz, H.F., Eischeid, J.K., Jones, P.D., Kelly, P.M. and Goodess, C.M., 1987: Precipitation fluctuations over Northern Hemisphere land areas since the mid-19th century. *Science*, **237**, 171–175.

Bretherton, F.P., Bryan, K. and Woods, J.D., 1990: Time-dependent greenhouse-gas-induced climate change. In: Houghton, J.T., Jenkins, G.J. and Ephraums, J.J. (eds), *Climate Change: The IPCC Scientific Assessment*, Cambridge Univ. Press, 173–193.

Charlson, R.J., Langner, J., Rodhe, H., Leovy, C.B. and Warren, S.G., 1991: Perturbation of the Northern Hemisphere radiative balance by backscattering of anthropogenic sulfate aerosols. *Tellus*, **43AB**, 152–163.

Folland, C.K., Karl, T.R. and Vinnikov, K.Ya., 1990: Observed climate variations and change. In: Houghton, J.T., Jenkins, G.H. and Ephraums, J.J. (eds), *Climate Change: The IPCC Scientific Assessment*, Cambridge Univ. Press, pp. 195–238.

Hansen, J.E. and Lacis, A.A., 1990: Sun and dust versus greenhouse gases; an assessment of their relative roles in global climate change. *Nature*, **346**, 713–719.

Hansen, J., Lacis, A., Rind, D., Russell, G., Stone, P., Fung, I., Ruedy, R. and Lerner, J., 1984: Climate sensitivity: analysis of feedback mechanisms. In: Hansen, J. and Takahashi, T. (eds), Maurice Ewing Series No. 5, American Geophysical Union, Washington, DC, pp. 130–163.

Houghton, J.T., Jenkins, G.J. and Ephraums, J.J. (eds), 1990: *Climate Change: The IPCC Scientific Assessment*. Cambridge Univ. Press, 365 pp.

Hulme, M., 1991: An intercomparison of model and observed global precipitation climatologies. *Geophys. Res. Lett.*, **18**, 1715–1718.

Jones, P.D., Raper, S.C.B., Bradley, R.S., Diaz, H.F., Kelly, P.M. and Wigley, T.M.L., 1986a: Northern Hemisphere surface air-temperature variations: 1851–1984. *J. Clim. App. Met.*, **25**, 161–179.

Jones, P.D., Raper, S.C.B. and Wigley, T.M.L., 1986b: Southern Hemisphere surface air-temperature variations: 1851–1984. *J. App. Met.*, **25**, 1213–1230.

Jones, P.D., Wigley, T.M.L. and Wright, P.B., 1986c: Global temperature variations between 1861 and 1984. *Nature*, **322**, 430–434.

Jones, P.D., Wigley, T.M.L., Folland, C.K. and Parker, D.E., 1988: Spatial patterns in recent worldwide temperature trends. *Climate Monitor.*, **16**, 175–185.

Jones, P.D. and Wigley, T.M.L., 1990: Global warming trends. *Scientific American*, **263**, 84–91.

Manabe, S. and Stouffer, R.J., 1980: Sensitivity of a global climate model to an increase of CO_2 concentration in the atmosphere. *J. Geophys. Res.*, **85**, 5529–5554.

Meier, M.F., 1984: Contribution of small glaciers to global sea level. *Science*, **225**, 1418–1421.

Mintzer, I.M., 1987: *A Matter of Degrees: The Potential for Controlling the Greenhouse Effect*, Research Report No. 5, World Resources Institute, Washington, DC, 60 pp.

Mitchell, J.F.B., Manabe, S., Meleshko, V. and Tokioka, T., 1990: Equilibrium climate change – and its implications for the future. In: Houghton, J.T., Jenkins, G.J. and Ephraums, J.J. (eds), *Climate Change: The IPCC Scientific Assessment*, Cambridge Univ. Press, pp. 131–164.

Palutikof, J.P., Wigley, T.M.L. and Lough, J.M., 1984: *Seasonal Scenarios for Europe and North America in a High-CO_2, Warmer World*, Technical Report TR012, US Dept. of Energy, Carbon Dioxide Research Div., Washington, DC, 70 pp.

Parry, M.L., Carter, T.R. and Konijn, N.T., (eds), 1988: *The Impact of Climatic Variations on Agriculture. Volume 1: Assessments in Cool Temperate and Cold Regions*, Kluwer Academic Publishers, Dordrecht, 876 pp.

Pittock, A.B. and Salinger, J.M., 1982: Towards regional scenarios for a CO_2-warmed Earth. *Climatic Change*, **4**, 23–40.

Raper, S.C.B., 1992: Observational data on relationships between climate change and the frequency and magnitude of severe tropical storms. In: Warrick, R.A., Barrow, E.M. and Wigley, T.M.L. (eds), *Climate and Sea Level Change: Observations, Projections and Implications*, Cambridge Univ. Press (in press).

Santer, B.D. and Wigley, T.M.L., 1990: Regional validation of means, variances and spatial patterns in GCM control runs. *J. Geophy. Res.*, **95**, 829–850.

Schlesinger, M.E. and Mitchell, J.F.B., 1987: Climate model simulations of the equilibrium climatic response to increased carbon dioxide. *Rev. Geophys.*, **25**, 760–798.

Schlesinger, M.E. and Zhao, Z.-C., 1989: Seasonal climate changes induced by doubled CO_2 as simulated by the OSU atmospheric GCM/mixed layer ocean model. *J. Climate*, **2**, 429–495.

Shine, K.P., Derwent, R.G., Wuebbles, D.J. and Morcrette, J.-J., 1990: Radiative Forcing of Climate. In: Houghton, J.T., Jenkins, G.J. and Ephraums, J.J. (eds), *Climate Change: The IPCC Scientific Assessment*, Cambridge Univ. Press, pp. 41–68.

Warrick, R.A. and Oerlemans, J., 1990: Sea level rise. In: Houghton, J.T., Jenkins, G.J. and Ephraums, J.J., (eds), *Climate Change: The IPCC Scientific Assessment*, Cambridge Univ. Press, pp. 257–282.

Washington, W.M. and Meehl, G.A., 1984: Seasonal cycle experiment on the climate sensitivity due to a double of CO_2 with an atmospheric general circulation model coupled to a simple mixed layer ocean. *J. Geophy. Res.*, **89**, , 9475–9503.

Watson, R.T., Rodhe, H., Oeschger, H. and Siegenthaler, U., 1990: Greenhouse gases and aerosols. In: Houghton, J.T., Jenkins, G.J. and Ephraums, J.J. (eds), *Climate Change: The IPCC Scientific Assessment*, Cambridge Univ. Press, pp. 1–40.

Wigley, T.M.L., 1989: Possible climate change due to SO_2-derived cloud condensation nuclei. *Nature*, **339**, 365–367.

Wigley, T.M.L., 1991: Could reducing fossil-fuel emissions cause global warming? *Nature*, **349**, 503–506.

Wigley, T.M.L. and Barnett, T.P., 1990: Detection of the greenhouse effect in the observations. In: Houghton, J.T., Jenkins, G.J. and Ephraums, J.J. (eds), *Climate Change: The IPCC Scientific Assessment*, Cambridge Univ. Press, pp. 239–256.

Wigley, T.M.L. and Jones, P.D., 1987: England and Wales precipitation: a discussion of recent changes in variability and an update to 1985. *J. Climatol.*, **7(3)**, 231–246.
Wigley, T.M.L. and Raper, S.C.B., 1990: Natural variability of the climate system and detection of the greenhouse effect. *Nature*, **344**, 324–327.
Wigley, T.M.L. and Raper, S.C.B., 1991: Detection of the enhanced greenhouse effect on climate. In: Jäger, J. and Ferguson, H.L. (eds), *Climate Change: Science Impacts and Policy*, Cambridge Univ. Press, pp. 231–242.

3

Scenario Modelling of Climate Change on the Ocean Circulation of the Intra-Americas Sea

Artemio Gallegos[1], *Steven Czitrom*[1], *Jorge Zavala*[1] *and Agustín Fernández*[1]

Abstract

Fisheries, offshore petroleum extraction, coastal settlements, tourism and commercial maritime navigation are the most important human ventures in the region with a strong dependence on the oceanographic conditions of the Intra-Americas Sea. Expectations are that by the year 2025 the equilibrium temperature of the biosphere will be 1.5°C warmer and the global-mean sea level will be 20 cm higher than today (WMO/ICSU/UNEP, 1985). Such increments will certainly bring about sensible changes in the marine environment of the Intra-Americas Sea along with socio-economic consequences. Using a climate scenario approach, we describe the physical effects of putative climate-induced wind and temperature changes on the ocean circulation of the Intra-Americas Sea. We foresee larger fluctuations in the surface oceanic circulation than at present with possible adverse effects on fisheries, ecologically important living resources, and other economic activities. A few consecutive hot summers have the potential to enhance the surface circulation to the extent that it might: (1) readjust coastal sea level which will affect the fresh-water/seawater balance in coastal ecosystems; (2) modify the location and magnitude of shoreline erosion with consequences on coastal settlements and development; (3) change migratory patterns of economically important species and distribution patterns of eggs and larvae; (4) alter maritime navigation routes and contingency plans for oil spills and similar environmental hazards; and (5) reorder the location and magnitude of the sea-to-air transport of latent heat in the Intra-Americas Sea, thus shifting the distribution of precipitation and evapotranspiration throughout the region.

1 Introduction

The Intra-Americas Sea, the name with which we will use to refer to the Caribbean Sea, the Gulf of Mexico and the adjacent waters of the tropical western North Atlantic Ocean between 5 and 30°N, is:

- the origin of at least half of the precipitable water over the bordering land and islands, and all significant rainfall occurring in major synoptic-scale storms;

[1]Universidad Nacional Autonoma de Mexico, Mexico, DF, Mexico

- the regulator of the marine and inland climates of the region, including Central America, northern South America and southeastern North America;
- a primary and in some cases the single source of animal protein for the populations of the region;
- an important medium for trade and cultural exchange;
- the receptacle of the urban and agro-industrial waste of the bordering states;
- the environment that harbours vast deposits of oil, gas and minerals; and
- an attractive tourism and recreational environment that produces appreciable earnings.

These facts explain and justify the growing concern national governments of the region, and intergovernmental agencies (UNEP, 1987), have about the socio-economic impacts that will occur in the Intra-Americas Sea induced by the imminent increase in the average equilibrium temperature of the global climate system and the consequent sea-level rise and atmospheric weather changes (WMO/ICSU/UNEP, 1985).

It has been nearly a century since the first warnings that an increase in atmospheric carbon dioxide due to fossil fuel burning could alter the optical characteristics of the atmosphere and induce a change in Earth's radiation balance (Arrhenius, 1896; Callendar, 1938). In the 1950s several scientists, notably Carl Gustav Rossby and Roger Revelle, revived the carbon dioxide question and stimulated the presentation of elaborate analyses and the first careful assessments of this problem (Revelle and Suess, 1957; Bolin and Eriksson, 1959). Their work encouraged a fundamental topic of scientific research – climate and global change – and gave evidence that human undertakings might disturb the natural evolution of the atmosphere, land and sea, as well as life on this planet. The years that followed witnessed the emergence of environmental science and vigorous empirical and theoretical research in scientific fields related to climate, climate variability and prediction of climate change (see Appendix for definitions of terms).

The most obvious manifestation of climate in a given locality on Earth is the time sequence of weather changes. Typically, these changes come in succession more or less in the same order, thus defining the seasons, and repeat year after year with some anomalous behaviour two or three times in a decade. One can find coherent descriptions of the global climate system (GCS) in the scientific literature (SCOPE, 1986; Schneider, 1989). Briefly, they all consider the sun as the primary and single energy source of the GCS. Solar radiation is differentially distributed and absorbed, creating strong and permanent horizontal and vertical generalized gradients which set the atmosphere and the ocean in motion, that redistributes mass, energy and momentum against such gradients. The GCS thus works towards a state of homogeneity and thermodynamic equilibrium that it will never attain, at least not before the external forcing disappears.

Some of these descriptions emphasize how the seasons arise, others stress the effect of the annual cycle, a few focus on modelling the anomalous years within a decade, and still others underline long-term, glacial–interglacial and millennial variations (Kraus, 1982). In all these descriptions the oceans play a principal role in the time evolution of climate, regulating it via three basic ocean–atmosphere energy-transfer processes: absorption of radiation, evaporation–precipitation, and transport of heat. In fact, the world ocean is envisaged in all cases as a kind of flywheel of the GCS. Climate is thus controlled to a large extent by the rates of energy transfer between the ocean and the atmosphere. These rates determine the lag times and the feedback loops and ultimately the character and magnitude of the climatic fluctuations in the ocean and in the other components of the GCS.

2 Statement of the Problem

In view of the potential rise of global-mean equilibrium temperature and global-mean sea level, it is imperative to conduct an assessment of the impact of wind and temperature changes on the ocean circulation of the Intra-Americas Sea. Such an evaluation depends heavily on a reliable prediction of the response of the GCS, which is a highly complex mechanical and thermodynamical system (Hasselmann, 1982; Wigley and Santer, Chapter 2).

In principle, any attempts to address this problem should look into (1) the knowledge of the present state of the oceanic circulation of the Intra-Americas Sea; (2) the information on the main forcing functions that determine the main features of the oceanic circulation of the Intra-Americas Sea (wind stress and wind-stress curl, seasonal evaporation minus precipitation, tidal forcing, river outflow, net radiation); and (3) the changes that these forcing functions will attain. Methodologically, it is necessary to be able to detect statistically significant changes in the oceanic circulation of the Intra-Americas Sea and to distinguish those changes specifically due to climate-induced global wind and temperature fluctuations. As will be shown below, however, this 'observational approach' cannot at present provide reliable information for predictions, and an alternate method based on climate scenarios is explored.

3 Oceanic Observations in the Intra-Americas Sea

Hydrographic surveys and ocean current measurements have been carried out in the Intra-Americas Sea for the past 100 years. Many of these studies focused on local circulation aspects, such as volume and mass transports across passages (e.g., Wennekens, 1959; Worthington, 1955, 1966 and 1971; Sturges, 1970 and 1975; Maul and Baig, 1983), while others concentrated on the circulation patterns within a single basin of the Intra-Americas Sea (e.g., Parr, 1935; Ichiye, 1962; Gordon, 1967; Armstrong and Grady, 1967; Febres-Ortega, 1972 and 1978; Hofmann and Worley, 1986; Aparicio, Chapter 6). As a result of this extensive hydrographic work, the origin and distribution of the bottom, deep and intermediate water masses that fill deep basins have been established (e.g., Wüst, 1964; Sturges, 1965; Nowlin, 1971). It has been ascertained that a large percentage of the waters above the average sill depth of the most important passages in the Intra-Americas Sea ($\sim$1500 m) enter via the passages of the Lesser Antilles, eventually flow through the Yucatan Channel and exit through the Straits of Florida (e.g., Armstrong, 1967; Brucks, 1971; Perloth, 1971; Maul, 1978; Kinder *et al.*, 1985). It is also well documented that the waters below sill depth are weakly stratified, neutrally stable, move very slowly and show relatively high dissolved oxygen concentrations due to intermittent renewal of deep waters from the North Atlantic Ocean (e.g., Wüst, 1964; Nowlin *et al.*, 1969; Metcalf, 1976; Morrison and Nowlin, 1982).

Assembling these oceanographic analyses results in a composite from which a reasonable approximation to the circulation of the water masses in the Intra-Americas Sea emerges. In addition, monthly estimates of the geographical distribution of important meteorological quantities (surface wind, wind stress and wind-stress curl, sea-surface temperature and near-surface air temperature, evaporation–precipitation rates, incident and back radiation, cloud cover and mixed-layer thickness), have been published (e.g., Keehn, 1968; Jenne, 1975; Stommel and Fieux, 1978; Ropelewski *et al.*, 1980; Tucker and Barry, 1984). All these quantities characterize air–sea interaction processes that act as forcing functions for the surface and subsurface circulation of the Intra-Americas Sea.

In spite of this large body of information there are still considerable uncertainties in the space-time scales of the fluctuations of the velocity field of the Intra-Americas Sea. Fig. 3.1 is a composite of some satellite-tracked buoy trajectories from a variety of sources. Many more have been tracked in the Gulf of Mexico than those which are shown, but the sense of ubiquitous basin-scale and mesoscale variability in the surface currents is readily seen for both basins. Satellite pictures are now providing data that permits tracing of surface currents and water associated with river or upwelling plumes (Muller-Karger, Chapter 8). Diverse current measurements (e.g., Stalcup and Metcalf, 1972; Smith, 1977; Molinari and Mayer, 1982; Maul and Baig, 1983; Kinder *et al.*, 1985) also indicate that a situation not very different from that shown in Fig. 3.1 can be expected for motions below the mixed layer and down to several hundred metres. This implies that the space-time-scales and magnitude of important physical quantities such as volume, mass and heat transports are still poorly known, probably underestimated, and their space-time correlations with local forcing functions yet to be determined. Studies of this nature have only been carried out in the Straits of Florida (e.g., Niiler and Richardson, 1973; Niiler, 1976; Molinari *et al.*, 1985a). At present the perception and understanding of the circulation in the Intra-Americas Sea still remains more conceptual than detailed, and the capability to predict changes is still limited.

The detection of statistically significant changes in the circulation of the Intra-Americas Sea is difficult because the standard deviations are large enough to hide any change of the average (climatological mean) circulation. In other words, the 'signal-to-noise ratio' in the actual observations is too small to discriminate between those due to specific causes and random variability, particularly those variations that are expected from climate-induced wind and temperature changes. Notwithstanding past and present efforts, and considering present needs, ocean monitoring is still insufficient. Indeed, for an observational approach present-day field programmes are still inadequate for the basic prediction problem. The resolution of the frequency and wave number components of the oceanic velocity field, whose domain is 0.1–0.001 cycles/day and 0.1–0.001 cycles/km (e.g., Gill, 1982), imposes extreme and burdensome conditions on any sampling scheme. It thus appears that the prediction of changes in the mean circulation of the Intra-Americas Sea produced by climate-induced wind and temperature changes is an intractable problem. Fortunately, there are ways to complement the strict observational approach and overcome its rigorous demands.

4 Models of the Circulation of the Ocean

Ocean modelling is fundamental in modern oceanographic research. It is an expanding methodology that keeps pace with advances in computer technology and numerical methods. There is no area of marine science that has not been approached through modelling techniques and the circulation of the ocean might be the most common topic (e.g., Holland, 1977). In fact, the family of ocean circulation models is large, from global to sub regional scales, and the growing set of ocean circulation models of the Intra-Americas Sea is but one more of the members of this family (IOCARIBE, 1989).

General results from ocean circulation models of the Intra-Americas Sea emphasize the need for higher spatial resolution and for improved bottom

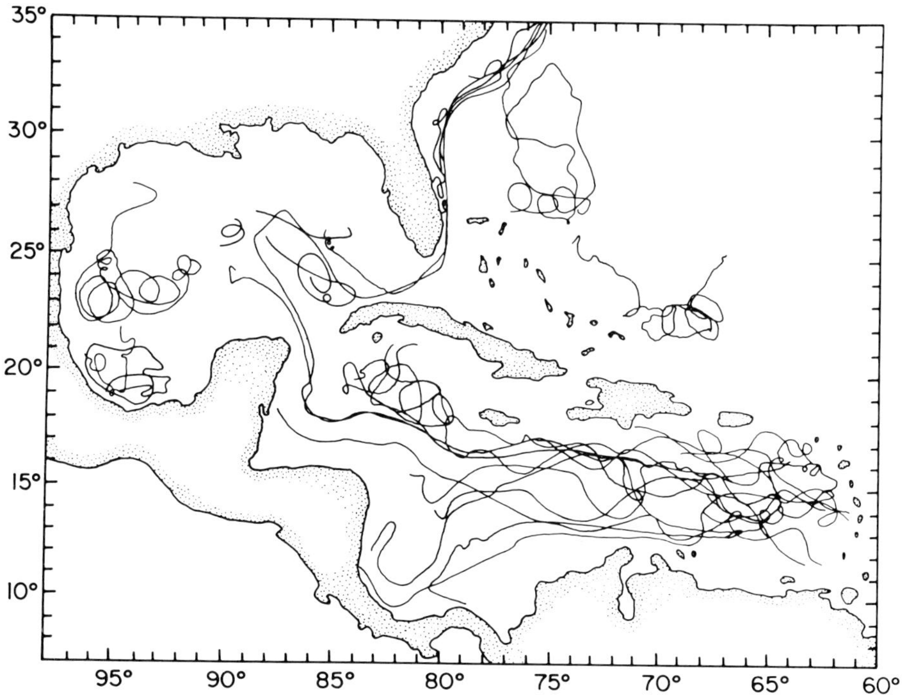

Fig. 3.1 Composite of satellite-tracked drifter trajectories in the Intra-Americas Sea. This chart depicts a mean surface circulation and a sense of its variability (G. Maul, pers. comm.); see also Gable (Fig. 10.5).

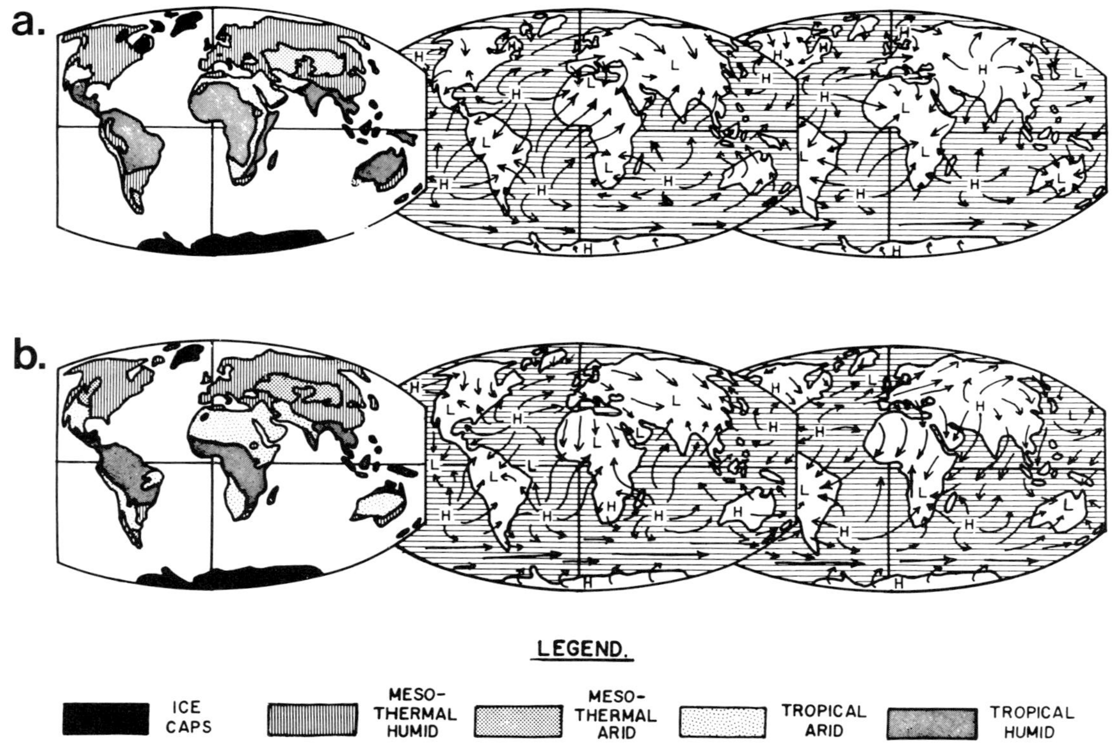

Fig. 3.2 Climatic land types, pressure systems and windfields for two climate scenarios known as (a) 'moist interglacial', and (b) 'dry interglacial'. The first map of each group shows the distribution of climatic types. The second and third maps show the July and January distribution of surface atmospheric pressure systems and surface wind fields, respectively (from Harris and Fairbridge, 1967).

topography (*cf.* Mercado *et al.*, Chapter 4). Some model experiments show realistic amplitude and phase for the signal of the annual mass transport, and are not very far from resolving mesoscale eddies like the ones shown in Fig. 3.1. Models are used to examine the physics of dynamical and thermal adjustments at several time and space scales. In particular, ocean circulation models simulate currents and transports under varying conditions of surface wind and mixed-layer temperature. In addition to improved resolution, ocean circulation models now incorporate data assimilation procedures and ingest *in situ* data (hydrography, satellite altimetry, sea level, precipitation, cloudiness, satellite scatterometry, net radiation, XBT, etc.). The benefit of these advances will be more realistic and stable simulations. Models, whose wide implementation should be pursued, are useful prediction tools with obvious applications to global climate-change impact studies in the Intra-Americas Sea.

5 Climate-Scenario Approach

Climate-scenario modelling is a concept that emerges from the study of the environmental conditions over the surface of Earth during past times, at regional and global scales (Lamb, 1987). It is based on the argument that past climate patterns can be used as future climate analogues. This method uses both paleoclimatic and environmental data to make internally consistent reconstructions of remote and recent past climate conditions that may reasonably be expected to occur again in the future,

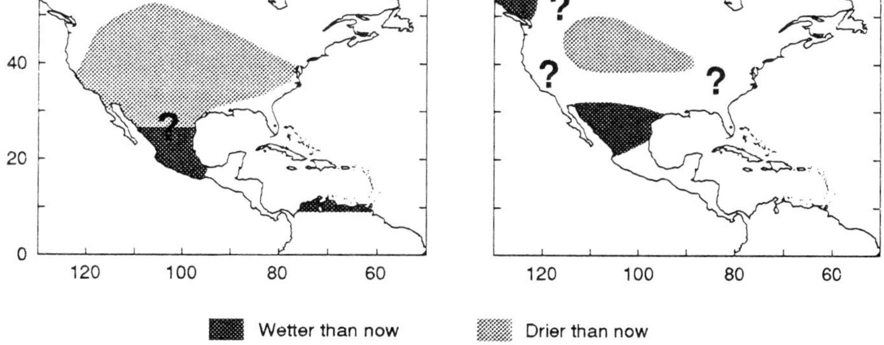

Fig. 3.3 Climate scenarios corresponding to the Altithermal Period (*c.* 8000–4500 ybp), showing areas where the conditions were (a) drier or (b) wetter than present. Blank areas denote scarce or null relevant climatic data. Note that both climate scenarios agree in most of the area within the Intra-Americas Sea (from Kellogg and Schware, 1981).

particularly in the case where the atmosphere attains high concentrations of greenhouse gases. Climate predictions based on the scenario method are reliable only to the extent that causes of recent and distant past climate changes are understood. This is not a disadvantage with numerical models of the global climate system, but which, at their present stage of development, may not produce reliable predictions of climatic change at regional and seasonal scales (Wigley and Santer, Chapter 2).

Within the past million years Earth has completed eight glacial cycles, oscillating irregularly between a 'main glacial' period, when Earth's global average environmental conditions were dry, cold, icy and with a relatively low sea level, and a 'moist interglacial' period, when conditions were moist, warm, almost ice-free, and had a relatively high sea level (Broecker and Denton, 1990). Present-day climate is somewhere between these climatic extremes and advances inexorably towards one of them. Superimposed on these millennial global-scale glacial cycles are other climatic fluctuations of considerably smaller time and space scales. An example is the 'Little Climatic Optimum' (800–1200 AD), also known as the 'Medieval Warm Epoch', with a sea-level rise of 0.5 m with respect to the present, and an average temperature 1°C warmer than today (probably the warmest period in the last 2000 years; *q.v.* Fig. 1.2). It was mostly felt in the North Atlantic region (Williams and Wigley, 1983). Regional climate changes such as this might now be triggered by increasing concentrations of greenhouse gases in the atmosphere and are the types of climate changes that demand our attention in this chapter.

The world ocean influences climate change because it is the foundation of Earth's hydrologic cycle and has the ability to store, move and deliver vast amounts of heat. Yet, the ocean is subordinate to changes in the atmosphere, specifically to fluctuating patterns of surface winds, evaporation–precipitation, and net radiation. In this coupled duality of the ocean–atmosphere system we concentrate first on the wind, evaporation–precipitation and net radiation fields that prevailed over the North Atlantic region (including the Intra-Americas Sea) in an Earth warmer than present, such as discussed in the following studies on remote and recent past climate:

1) The geographic distribution of climatic land types, pressure systems and wind fields (for July and January) for two different climate states, shown in Fig. 3.2(a) and (b), are due to Butzer (1964) and Harris and Fairbridge (1967) and correspond to climate states known as (a) 'moist interglacial' when ambient conditions are moist, warm and almost ice-free, and (b) 'dry interglacial' when Earth is drier, cooler and has moderate to small polar ice-caps, as it is at present.
2) Kellogg (1977) and Butzer (1980), working separately and based on surveys of the literature on paleoclimates, presented reconstructions of the 'Altithermal Period' (*c.* 8000–4500 ybp), an epoch when Earth was several degrees warmer than today. Figs. 3.3(a) and (b) show those land areas that were wetter or drier than now. Their reconstructions agree with each other over a large portion of the land within the Intra-Americas Sea.
3) Palutikof *et al.* (1984), noting that the 1901–1920 was a period 0.5°C

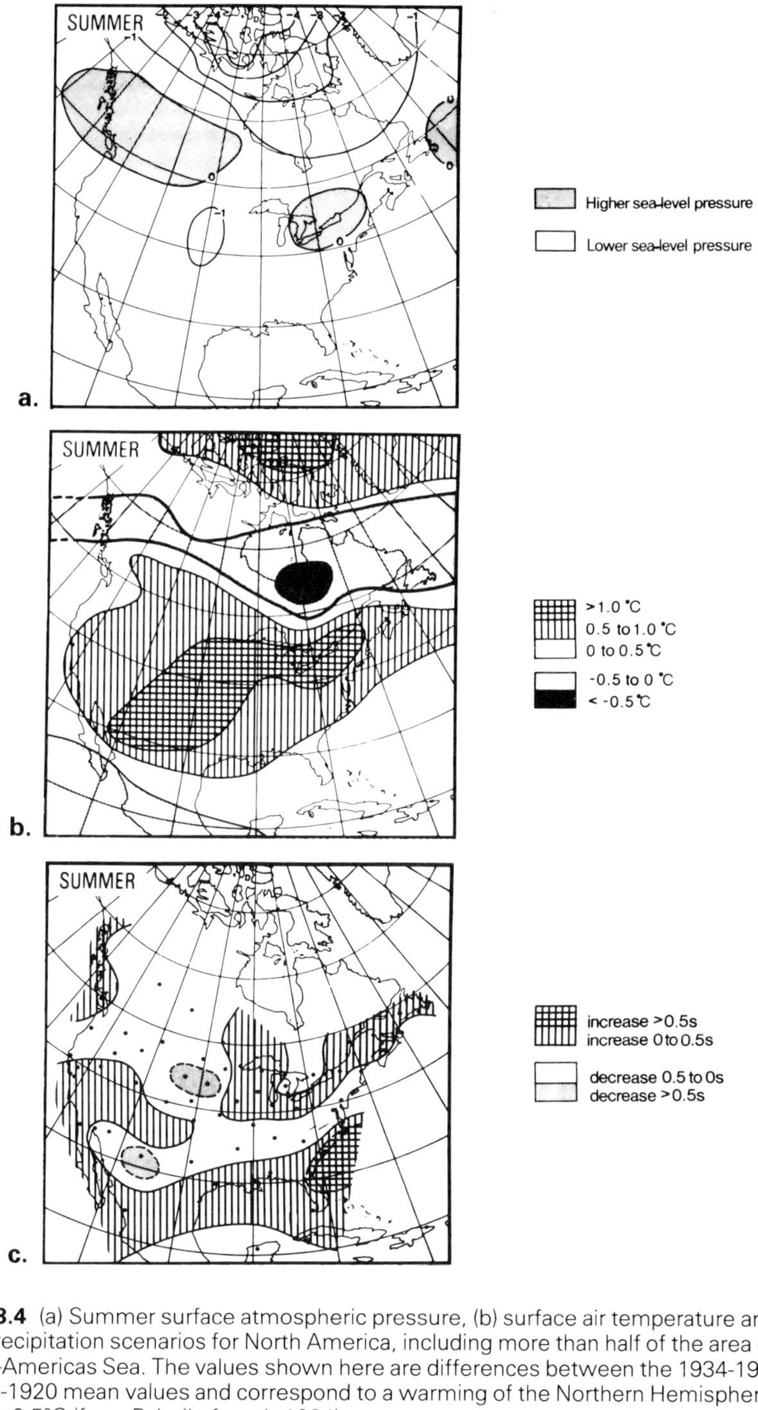

Fig. 3.4 (a) Summer surface atmospheric pressure, (b) surface air temperature and (c) precipitation scenarios for North America, including more than half of the area of the Intra-Americas Sea. The values shown here are differences between the 1934-1953 and 1901-1920 mean values and correspond to a warming of the Northern Hemisphere of about 0.5°C (from Palutikof *et al.*, 1984).

colder than the 1934–1953 period, compared time-averaged fields of surface atmospheric pressure, surface air temperature and precipitation. They mapped the corresponding differences between these two periods for the summer season over a geographic area that includes a large part of the region, shown in Fig. 3.4(a), (b) and (c). A recent analysis of a large set of historical surface air-temperature data made by Jones and Wigley (1990) confirms that during the first 20 years of this century the average equilibrium temperature over the land and oceans was nearly 0.5°C below that during the period 1934–1953 (Fig. 3.5). Both reports indicate that significant, even if small, changes in the equilibrium temperature of the GCS have occurred in the relatively brief time interval of 12 years (1921–1933).

4) Additional information on environmental conditions that persisted during relatively warm periods is also found in Williams (1979). He conducted an analysis on a 70-year-long meteorological record, comparing the geographical distribution of precipitation over the North American continent during the 10 warmest summers and 10 warmest winters against the distribution of precipitation averaged over the total length of the record over the same area. His results, expressed in terms of deviations from the mean, are shown in Fig. 3.6(a) and (b).
5) Fairbridge (1964), based on studies of wind-forced movements of sand dunes in North Africa, showed that the transition zone between the Trade Winds and the Westerlies over the North Atlantic Ocean is displaced latitudinally during the warm/cold climatic cycles; it moves polewards during warm periods and moves to lower latitudes during cold epochs.
6) Lamb (1964) and Butzer (1964) studied the latitudinal displacement of the atmospheric pressure belts over the North Atlantic Ocean. Lamb (1964) concluded that during interglacials (warm periods) westerly storms crossed the North Atlantic Ocean at approximately 70°N, while

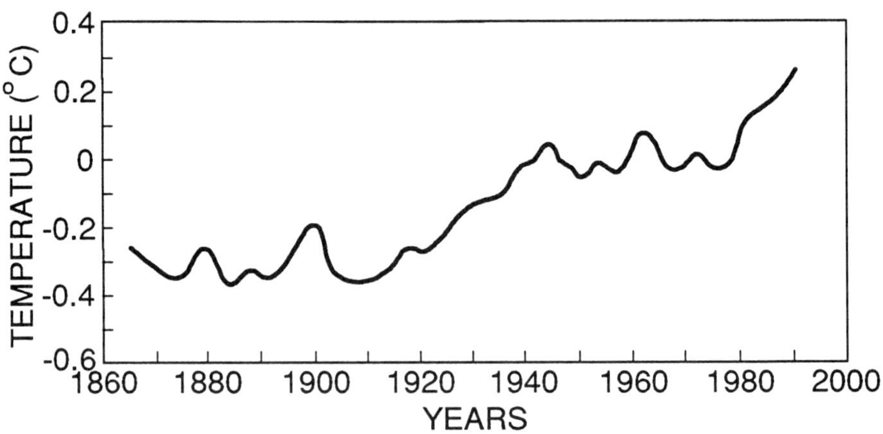

Fig. 3.5 Variability of the global equilibrium temperature of Earth for the last 120 years. The temperature curve was obtained from annual temperatures after applying a 10-year moving average. Note that the temperature difference between the 1901-1920 (colder) and 1934-1953 (warmer) mean values is about 0.5°C (from Jones and Wigley, 1990).

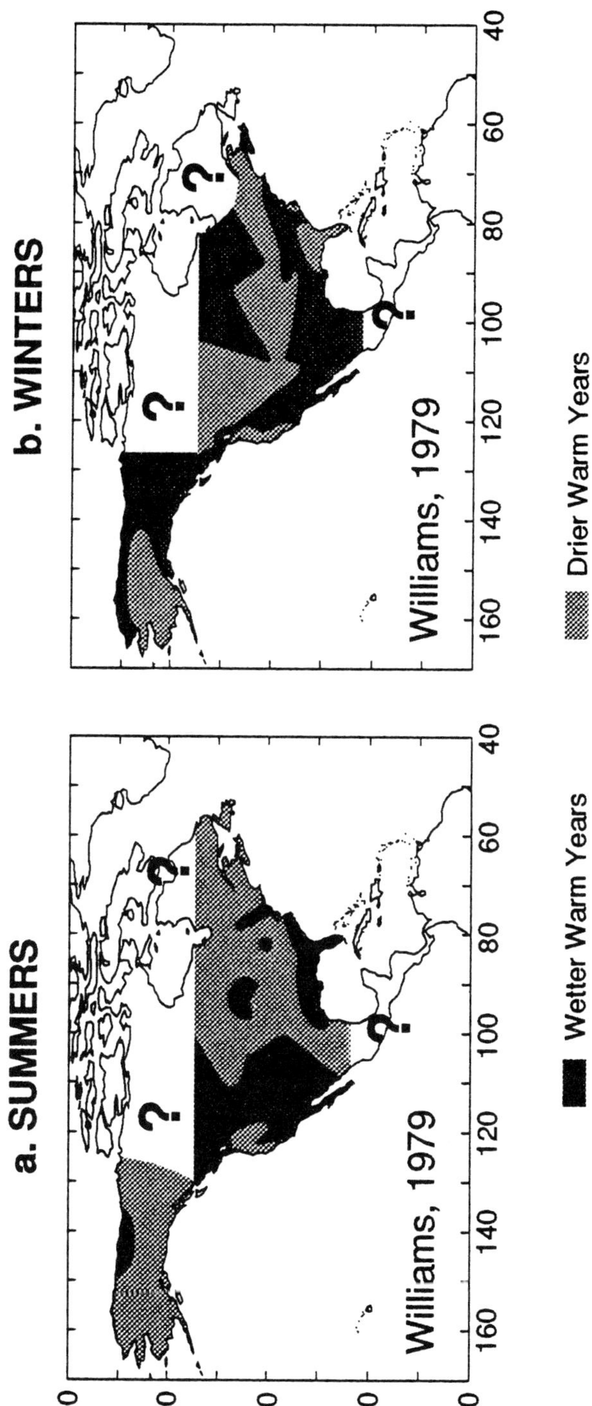

Fig. 3.6 Average deviations from the 70-year-long mean precipitation of (a) 10 warmest Arctic summers, and (b) the 10 warmest Arctic winters (from Williams, 1979).

Butzer (1964) showed that during the last glaciation the storm tracks over this same area were located between 30 and 40°N.

These studies provide a few elements to build a simple reconstruction of the way seasonal patterns of net radiation, evaporation, precipitation, atmospheric surface pressure and surface winds were distributed during remote and recent past climatic conditions. Based on the climate scenario concept, these patterns are used to ascertain how meteorological variables will be distributed over the geography of the region in a warmer Earth. We will then describe what the ocean circulation in the Intra-Americas Sea could be when the equilibrium temperature of the GCS is warmer than today.

Focusing on the environmental conditions over the North Atlantic basin, the reconstructions of remote and recent past climates shown above suggest that in a warmer world:

1) Summer warming over land will be more pronounced than at present, causing lower atmospheric surface pressure, larger net radiation, increased evapotranspiration, and intensified monsoon-type winds over the land. Also, summer warming induces a larger contrast between marine and land seasonally averaged meteorological conditions. Thus, precipitation will increase in a wide belt along the coastal zone and decrease considerably further inland (Palutikof *et al.*, 1984; *q.v.* Fig. 3.4[c]).
2) Winter near-surface thermal and horizontal pressure gradients between land and ocean will be considerably weaker than those developed between the land and sea during summer. In summer, the strong land warming will enhance east–west overturnings, providing kinetic energy to mid-latitude cyclones and anticyclones which jointly transport westerly momentum from the Trade Wind latitudes to the Westerlies. Such an important transport intensifies zonal winds and in turn will boost, via the Coriolis effect, the meridional cell circulation. The kinetic energy that survives dissipation in the thermally driven tropical Hadley cell and in the frictionally driven Ferrel cell, will be transformed into available potential energy which, with the help of the acute differential heating during this season, with a lag of 6–10 weeks (Tucker and Barry, 1984), will drive the east–west overturnings very efficiently. The atmospheric circulation during the summer will be more energetic, and the weather machine will work vigorously and efficiently in the hemispheric redistribution of heat, mass and angular momentum. In autumn, the surface winds thus generated will impart a large portion of their horizontal momentum to the surface layers of the ocean, giving rise to surface currents, mesoscale eddies, coastal upwelling and wind surges.
3) Winter-averaged meridional atmospheric circulation will be drastically different because the stationary (Low and High) sea-level pressure centres will have less contrast. Consequently, the meridional tropical Hadley and the mid-latitude Ferrel cells will scarcely contribute to the available potential energy budget of the GCS during this season. Net radiation during winter will stimulate weaker zonal overturnings between tropical warm air masses and

high-latitude colder air masses, giving to the mid-latitude cyclones and anticyclones only the necessary kinetic energy to maintain a weak zonal wind system over the region. Earth's rotation turns this weakened east–west motion into a correspondingly weakened meridional motion, and during the winter the atmospheric circulation will be slack and the weather machine will be less effective in the redistribution of heat, mass and angular momentum.

Under the atmospheric conditions described above, the circulation of the surface layers of the tropical and subtropical North Atlantic Ocean, including those in the Intra-Americas Sea, will respond with a strong seasonal variability (Gray, Chapter 5) but, due to their relatively large inertia, the response will have a delay of two to four months relative to the corresponding season. In particular, the circulation in the Intra-Americas Sea forced by summer atmospheric conditions will crest in autumn and will not fade before the next spring or summer, depending on the energy input (mainly from the wind stress) during the previous winter. In this way, a mild winter might not give the time necessary to dissipate enough kinetic energy of the surface currents, and the following summer will find favourable conditions to magnify the oceanic motion set by the previous summer. Conversely, a long cold winter may dissipate more kinetic energy than that given to the surface currents during the previous summer. In this case the circulation of the surface layers in the Intra-Americas Sea would weaken.

This line of thought leads to the conclusion that a sequence of hot summers in a warmer Earth will enhance the oceanic surface circulation in the Intra-Americas Sea. Thus, oceanographers will measure swifter surface currents through straits and passages. Analyses of water samples will show clearer evidence of vertical motion in broader areas of upwelling and will detect intense mesoscale motions more often, from the surface down to intermediate water-depths. Conversely, a sequence of cold winters, characterized by relatively weak surface winds, will decrease surface wind stress to the point that it will be insufficient to maintain a moderate surface current system in the Intra-Americas Sea. Concurrently, coastal upwelling, mesoscale motions and coastal wind surges should decrease under these conditions. These two extreme situations of long sequences of hot summers or cold winters influence their own negative and positive feedback processes. An examination of such feedbacks requires the consideration of GCS processes and much longer time-scales, and is outside the scope of the present report.

The point that should be stressed is that in a warmer Earth variability of the oceanic circulation in the Intra-Americas Sea will be basically a result of the summer-winter changes in the circulation of the atmosphere. This will: (1) induce larger fluctuations in the position, speed and direction of surface and subsurface currents; (2) modify the location, areal extent and number of events of coastal and oceanic upwelling; (3) modify the location, areal extent and number of wind surges and erosion–sedimentation episodes; (4) increase the variance of volume, mass and heat transport measurements through passages and straits; and (5) affect the location, size and number of mesoscale eddies (e.g., Gulf of Mexico Loop Current

Rings and the seasonal eddy in the southwestern Caribbean Sea). That is, the expected global climate change will induce a strong annual signal on the circulation in the Intra-Americas Sea. The manifestation of such change will be higher levels of surface current variability, patchiness of surface salinity and temperature, salinity and temperature values in the upper layers, depth of the mixed layer, sea-level variability, erosion–deposition events along the coast, etc.

6 Recommendations

It is of the utmost importance to monitor the general circulation of the Intra-Americas Sea with special emphasis on the annual, interannual and decadal time-scales. Such an effort sets up a few basic demands, particularly on any programme designed for the assessment of climate-change impacts. Regular observations to describe adequately the basin-scale and mesoscale variability of the ocean are necessary to identify basic problems related to sampling and data analyses, and to find solutions relevant to the overall scientific assessment of the environmental impacts that will come from any change of Earth's climate.

In order to assure suitable descriptions of the physical processes of the ocean, no methodology for the systematic sampling of oceanic data should be implemented before answering the following questions:

- What variables must be sampled with greater emphasis?
- What areas of the Intra-Americas Sea require particular attention?
- How long and how frequently should such observations be carried out, and yet remain technically and economically feasible, in order to detect significant changes in, and probable effects of the circulation in this region?

We think that the variables that must be sampled with greater emphasis are:

- Hydrographic (physical, chemical and biological) variables: temperature, salinity, dissolved oxygen, nitrates, phosphates, chlorophyll, ammonia, and primary productivity.
- Sea-surface variables: sea level, evaporation–precipitation, air temperature, atmospheric pressure, back and net radiation, and sea-surface temperature.
- Field kinematical variables: surface, subsurface, intermediate, deep and bottom water currents measured with present-day techniques and modern instrumentation (moored current meters, ADCPs, Pegasus, satellite-tracked buoys, HF radar).
- Upper and lateral boundary variables: surface wind and wind stress, precipitation over the ocean, river discharge and cloudiness.

We suggest that the ocean areas within the Intra-Americas Sea that deserve special attention are:

- Present upwelling zones: the coastal waters off Venezuela and Colombia, the southern shelf and shelf-break waters off Cuba, the shallow waters over the western Yucatan Channel shelf and

Campeche Bank, west and southwest of Dry Tortugas (Florida), and the coastal waters surrounding the Lesser Antilles.
- Areas where strong currents interact with abrupt bottom topography changes around islands and continental shelves: the Yucatan Channel, the passages of the Lesser Antilles, the northern coast of Cuba, the northeastern coast of Venezuela and the banks between the Honduras-Nicaragua border and Jamaica.
- Straits: the Windward, Mona and Anegada-Jungfern passages, the Old Bahamas Channel, the Northwest Providence Channel, the Yucatan Channel, the Straits of Florida between Havana–Key West, and the Lesser Antilles passages.

Current meter moorings should operate for periods of at least three years and hydrographic surveys should be conducted as often as possible, avoiding seasonal bias, to resolve annual, interannual and decadal variability.

Gulf Loop Current Rings are structured water-column formations whose generation, space–time evolution and final dissipation form an outstanding feature of the circulation in the Gulf of Mexico. They have been a central topic of oceanographic research of the Intra-Americas Sea (e.g., Lewis and Kirwan, 1985 and 1987; Muller-Karger *et al.*, 1991). The formation of these mesoscale anticyclonic eddies may be related to fluctuations of the wind stress and wind-stress curl distribution over the Intra-Americas Sea, although their frequency of occurrence seems unrelated to fluctuations in volume transport (Maul and Vukovich, 1993). Therefore, a stronger variability of the surface wind patterns in this region may have an effect on the number of rings per year and on their dimensions, heat content, average salinity, depth and speed of rotation. Such effects should be explored as a problem of climate change.

Sea-level observations are fundamental in any study of climate change. Their fluctuations are correlated to atmospheric pressure changes (inverse barometer effect), thermal expansion of seawater, and surface currents induced by tides and winds. Sea-level gauges are relatively inexpensive and can be placed almost anywhere along the coast or in shallow water. Many countries in the Intra-Americas Sea maintain tide gauges along their coasts, yet there are still long coastal boundaries and islands without sea-level measurements. Any international research project on oceanic circulation related to climate change should fill this observational gap and attempt an integral analysis of all the tidal records available.

Existing historical data sets should be further analysed in order to extract as much information as possible on long-term changes in the general circulation, surface energy balance, and wind-stress patterns over the Intra-Americas Sea. All these oceanographic data should be fed to credible regional ocean circulation models. In the near future accurate predictions of the impact of climate change on the circulation of the region will start to emerge and hopefully, before the turn of the century, the necessary scientific and technical information will be available to governments and supporting agencies committed to ameliorate the adverse impacts of climate change.

7 Conclusions

Oceanic circulation is the open-sea kinematic boundary condition imposed on shelf and coastal current systems, which affects the fresh-water/seawater balance in coastal ecosystems, and modifies the magnitude and location of erosion/deposition processes along the shoreline (Hendry, Chapter 7). Oceanic circulation is also important to the distribution of larvae and eggs and to the migration patterns of economically important species (Alm *et al.*, Chapter 15); it is hence a critical factor in the location of fishing areas. Furthermore, marine oil-spill contingency plans as well as optimum-route programming for marine transportation are designed on the basis of the best information available of the regional average weather conditions, including sea-surface currents. Therefore, it is necessary to be aware of the changes in the oceanic circulation in order to be prepared to ameliorate the expected negative socio-economic impacts, and to manage whatever regional benefits global climate change might bring.

Accordingly, there is an urgent need to describe and understand the variability of the ocean circulation and to investigate trends in long-term climatic records. The characterization of the space-time structure of the ocean climate is crucial to the timely detection of significant changes in oceanic circulation.

Numerical ocean-circulation models together with climate scenarios provide effective tools for integrating remote and recent past climatic data as well as present observations from all possible sources into consistent representations of future climate. Hydrographic surveys and long-term sea level and current measurements in specific areas of the Intra-Americas Sea (passages, straits, zones of upwelling), in addition to marine weather observations and the use of satellite records of oceanic variables, should render sufficient information to augment the skill of specific ocean circulation models of the region and to build more detailed climate scenarios. It will then be possible to give better predictions of the potential impact of climate-induced wind and temperature changes in the ocean circulation of the Intra-Americas Sea.

8 Acknowledgements

We wish to acknowledge valuable criticisms provided by Drs George A. Maul, James K. Lewis and José M. Barberán. Constructive discussions with them were very helpful in formulating the topic presented in this chapter. However, the blemishes, flaws and errors that remain are due to us and are there in spite of their best efforts.

9. Appendix

Definitions of Terms Used in Ocean Climate Studies

(Adapted from the Unites States Committee for the Global Atmospheric Research Program [1975]; *cf.* Maul and Baig, Chapter 17.)

Ocean climate state: The average of a set of oceanic variables over a given period of time in a specific domain of the world ocean. One may speak, for

example, of seasonal, yearly, interannual or decadal ocean climate states.
Ocean climate variation: The difference between ocean climate states of the same kind, as between two winters or between two years.
Ocean climate anomaly: The deviation of a particular climate state from the average of a relatively large number of climate states of the same kind.
Ocean climate variability: The variance among a number of ocean climate states of the same kind.

10 References

Armstrong, R.S., 1967: The subtropical underwater of the eastern Gulf of Mexico. *Comm. Fish. Rev.*, **29(3)**, 46–48.

Armstrong, R.S. and Grady, J.R., 1967: Geronimo cruises on the entire Gulf of Mexico in late winter. *Comm. Fish. Rev.*, **29(10)**, 35–40.

Arrhenius, S., 1896: On the influence of carbonic acid in the air upon the temperature of the ground. *Phil. Mag.*, **41**, 237.

Bolin, B. and Eriksson, E., 1959: Changes of the carbon content of the atmosphere and the sea due to fossil fuel combustion. In: Bolin, B. (ed.), *The atmosphere and the sea in motion*. Rossby Memorial Volume. The Rockefeller Institute Press, New York, pp. 130–142.

Broecker, W.S., 1982: Glacial to interglacial changes in ocean chemistry. *Progress in Oceanography*, **11(2)**, 151–197.

Brucks, J.T., 1971: Currents of the Caribbean and adjacent regions as deduced from drift-bottle studies. *Bull. Mar. Sci.*, **21**, 455–465.

Butzer, K.W., 1964: *Environments and Archaeology*. Aldine Publishing Co., Chicago, 524 pp.

Butzer, K.W., 1980: Adaptation to global environmental change. *Professional Geographer*, **32**, 269–278.

Callendar, G.S., 1938: The artificial production of carbon dioxide and its influence on temperature. *Q.J.R. Meteorol. Soc.*, **64**, 223.

Fairbridge, R.W., 1964: African ice-age aridity. In: *Problems in Paleoclimatology*, London, John Wiley & Sons, Interscience, pp. 356–360.

Febres-Ortega, G.A., 1972: Isanosteric analysis of the eastern Caribbean waters during winter. *Bol. Inst. Oceanogr. Univ. Oriente*, (Cumaná, Venezuela), **11(1)**, 19–34.

Febres-Ortega, G.A., 1978: Circulación y transporte de masa de agua en el Caribe sudoriental. CICAR II Symposium on Progress in Marine Research in the Caribbean and Adjacent Regions. FAO Fisheries Report No. 200, Suppl., pp. 21–23.

Gill, A.E., 1982: *Atmosphere – Ocean Dynamics*. Academic Press, New York, 662 pp.

Gordon, A.L., 1967: Circulation of the Caribbean Sea. *J. Geophys. Res.*, **72(24)**, 6207–6223.

Harris, S.A., and Fairbridge, R.W., 1967: Ice-age climatology. In: *The Encyclopedia of Atmospheric Sciences and Astrogeology*. Reinhold Publishing Corp., London, pp. 454–461.

Hasselmann, K., 1982: An ocean model for climate variability studies. *Progress in Oceanography*, **11(2)**, 69–92.

Hofmann, E.E. and Worley, S.J., 1986: An investigation of the circulation of the Gulf of Mexico. *J. Geophys. Res.*, **91(C12)**, 14221–14236.

Holland, W.R., 1977: The general circulation of the ocean and its modelling. Proc. JOC/SCOR Joint Study Conference on General Circulation Models of the Ocean and Their Relation to Climate (23–27 May 1977, Helsinki), World Meteorol. Organization, Geneva, pp. I–55 to I–114.

Ichiye, T., 1962: Circulation and water mass distribution in the Gulf of Mexico.

Geofis. Int., **2(3)**, 47–76.
IOCARIBE, 1989: Climate change impacts and ocean physical processes: A research proposal of the Task Team of IOCARIBE Experts on Physical Oceanography and Climate. IOCARIBE/ISRP-I/3. Caracas, Venezuela, 32 pp.
Jenne, R.L., 1975: Data Sets for Meteorological Research NCAR Tech. Note 1A-111, Nat'l. Center Atmos. Res., Boulder, CO.
Jones P.D. and Wigley, T.M.L., 1990: Global warming trends. *Sci. Amer.*, **263(2)**, 66–73.
Keehn, P.A., and Rigby, M., (eds), 1968: *Bibliography on Marine Atlases*. Amer. Meteorol. Soc., Washington, DC.
Kellogg, W.W., 1977: Effects of human activities on global climate. WMO Tech. Note No. 156, World Meteorol. Organization, Geneva.
Kellogg, W.W. and Schware, R., 1981: *Climate Change and Society. Consequences of Increasing Atmospheric Carbon Dioxide*. Westview Press, Boulder, CO, p. 178.
Kinder, T.H., Heburn, G.W. and Green, A.W., 1985: Some aspects of the Caribbean circulation. *Mar. Geol.*, **68**, 25–52.
Kraus, E.B., 1982: Oceanic variability on climatic scales. *Progress in Oceanography*, **11(2)**, 61–68.
Lamb, H.H., 1964: The role of atmosphere and oceans in relation to climatic changes and the growth of ice-sheets on land. In: *Problems in Paleoclimatology*. John Wiley & Sons, Interscience, London, pp. 332–348, 361.
Lamb, P.J., 1987: On the development of regional climatic scenarios for policy-oriented climatic-impact assessment. *Bull. Amer. Meteorol. Soc.*, **68(9)**, 1116–1123.
Lewis, J.K. and Kirwan, A.D. Jr., 1985: Some observations of ring-topography and ring-ring interactions in the Gulf of Mexico. *J. Geophys. Res.*, **90(C5)**, 9017–9028.
Lewis, J.K. and Kirwan, A.D. Jr., 1987: Genesis of a Gulf of Mexico ring as determined from kinematical analyses. *J. Geophys. Res.*, **92(C11)**, 727–740.
Maul, G.A., 1978: The 1972–1973 cycle of the Loop Current, Part II: Mass and salt balances of the basin. *Progress in Marine Research in the Caribbean and Adjacent Regions*. CICAR II Symposium, Caracas, Venezuela. FAO Fish. Rep. No. 200, Suppl., 597–619.
Maul, G.A. and Baig, S.R., 1983: A three-year continuous current metre record from the sill of Yucatan Strait. Abstract, EOS, *Trans. Amer. Geophys. Union*, **64(18)**, 233.
Maul, G.A. and Vukovich, F.M., 1993: Aspects of the relationship between the Gulf of Mexico Loop Current and Straits of Florida volume transport. *J. Phys. Oceanogr.* (in press).
Metcalf, W.G., 1976: Caribbean-Atlantic water exchange through the Anegada-Jungfern Passage. *J. Geophys. Res.*, **81(36)**, 6401–6409.
Molinari, R.L., 1978: The relationship of the curl of the local wind stress to the circulation of the Cayman Sea and the Gulf of Mexico. *J. Phys. Oceanogr.*, **8(5)**, 780–784.
Molinari, R.L. and Mayer, D.A., 1982: Current meter observations on the Continental Slope at two sites in the Gulf of Mexico. *J. Phys. Oceanogr.*, **12**, 1480–1492.
Molinari, R.L., Wilson, W.D. and Leaman, K., 1985a: Volume and heat transports of the Florida Current: April 1982 through August 1983. *Science*, **227(4684)**, 295–297.
Molinari, R.L., Maul, G.A., Chew, F., Wilson, W.D., Bushnell, M., Mayer, D., Leaman, K., Schott, F., Lee, T., Zantopp, R., Larsen, J.C. and Sanford, T.B., 1985b: Subtropical atlantic climate studies: Introduction. *Science*, **227(4684)**, 292–311.
Morrison, J.M. and Nowlin, W.D. Jr., 1982: General distribution of water masses within the eastern Caribbean Sea during the winter of 1972 and the fall of 1973. *J. Geophys. Res.*, **87(C6)**, 4207–4229.

Muller-Karger, F.E., Walsh, J.J., Evans, R.H. and Meyers, M.B., 1991: On the seasonal phytoplankton concentration and sea-surface temperature cycles of the Gulf of Mexico as determined by satellites. *J. Geophys. Res.*, **96(C7)**, 12645–12665.

Niiler, P.P., 1976: Observations of low-frequency currents on the west Florida continental shelf. *Mem. Soc. Roy. Sci. Liege., 6 serie tome X*, 331–358.

Niiler, P.P., and Richardson, W.S., 1973: Seasonal variability of the Florida Current. *J. Mar. Res.*, **31(3)**, 144–167.

Nowlin, W.D. Jr., 1971: Water masses and general circulation of the Gulf of Mexico. *Oceanol. Intl.*, 29–33.

Nowlin, W.D., Jr., Paskausky, D.F. and McLellan, H.J., 1969: Recent dissolved oxygen measurements in the Gulf of Mexico deep waters. *J. Mar. Res.*, **27(1)**, 39–44.

Palutikof, J.P., Wigley, T.M.L. and Lough, J.M., 1984: Seasonal Scenarios for Europe and North America in a High-CO_2, Warmer World. US Dept. of Energy, Carbon Dioxide Research Division, Tech. Report TR012, 70 pp.

Parr, A.E., 1935: Report on hydrographic observations in the Gulf of Mexico and adjacent straits made during the Yale Oceanographic Expedition on the *Mabel Taylor* in 1932. *Bull. Bingham Oceanogr. Collect.*, **V(1)**, 1–93.

Perloth, I., 1971: Distribution of mass in the near surface water of the Caribbean. Symposium on Investigations and Resources of the Caribbean Sea and Adjacent Regions. UNESCO, Paris, pp. 147–152.

Revelle, R., and H.E. Suess, 1957: Carbon dioxide exchange between atmosphere and ocean and the question of an increase of atmospheric CO_2 during the past decades. *Tellus*, **9**, 18.

Ropelewski, C.F., Predoehl, M.C. and Platto, M., 1980: A quick reference to selected climate data. Center for Environmental Assessment Services, Washington, DC.

Schneider, S.H., 1989: The changing climate. *Sci. Amer.*, **261(3)**, 38–47.

SCOPE, 1986: The greenhouse effect, climatic change and ecosystems. Bolin, B. (ed.), John Wiley and Sons Publisher, Great Britain, p. 541.

Smith, N.P., 1977: Near-bottom cross shelf currents in the northwestern Gulf of Mexico: A response to wind. *J. Phys. Oceanogr.*, **7**, 615–620.

Stalcup, M.C. and Metcalf, W.G., 1972: Current measurements in the passages of the Lesser Antilles. *J. Geophys. Res.*, **77**, 1032–1049.

Stommel, H. and Fieux, M., 1978: Oceanographic Atlases: A Guide to Their Geographic Coverage and Contents. Woods Oceanographic Institution, Woods Hole, MA.

Sturges, W., 1965: Water characteristics of the Caribbean Sea. *J. Mar. Res.*, **23(2)**, 147–162.

Sturges, W., 1970: Observations of deep-water renewal in the Caribbean Sea. *J. Geophys. Res.*, **75**, 7602–7610.

Sturges, W., 1975: Mixing and renewal of water flowing into the Caribbean Sea. *J. Mar. Res., Suppl.*, **33**, 117–130.

Tucker, G.B. and Barry, R.G., 1984: Climate of the North Atlantic Ocean. In: van Loon, H. (ed.), *Climates of the Oceans*. World Survey of Climatol., Vol. 15, Elsevier, 193–262.

UNEP (United Nations Environment Programme), 1987: Action Plan for the Caribbean Environment Programme. UNEP Regional Coordinating Unit, 14-20 Port Royal Street, Kingston, Jamaica, 24 pp.

Wennekens, M.P., 1959: Water mass properties of the Straits of the Florida and related waters. *Bull. Mar. Sci. Gulf Carib.*, **9(1)**, 1–52.

Williams, J., 1979: Anomalies in temperature and rainfall during warm Artcic seasons as a guide to the formulation of climate scenarios. *Climate Change*, **2**, 249–266.

Williams, J. and Wigley, T.M.L., 1983: A comparison of evidence for late Holocene summer temperature variations in the Northern Hemisphere. *Quaternary Res.*, **20**, 286–307.

WMO/ICSU/UNEP, 1985: Report of the International Conference on the Assessment of the role of Carbon Dioxide and of other Greenhouse Gases in Climate Variations and Associated Impacts, Villach, Austria, 9-15 October 1985. WMO Ref. No. 661, pp. 1–4.

Worthington, L.V., 1955: A new theory of Caribbean bottom-water formation. *Deep-Sea Res.*, **3(1)**, 82–87.

Worthington, L.V., 1966: Recent oceanographic measurements in the Caribbean Sea. *Deep-Sea Res.*, **13**, 731–739.

Worthington, L.V., 1971: Water circulation in the Caribbean Sea and its relationship to North Atlantic circulation. Symposium on Investigations and Resources of the Caribbean Sea and Adjacent Regions. UNESCO, Paris, pp. 181–191.

Wüst, G., 1964: *Stratification and Circulation in the Antillean-Caribbean Basins, Part I., Spreading and Mixing of the Water Types with an Oceanographic Atlas*. Columbia Univ. Press, New York, 20 pp. + appendices.

4

Requirements for Modelling of Future Storm Surge and Ocean Circulation

Aurelio Mercado[1], J. Dana Thompson[2] and James C. Evans[3]

Abstract

The role of numerical models in discerning the implications of climate change on storm surge, storm waves, and ocean circulation is discussed. With respect to storms, the inadequacy of the bathymetric data base is viewed as creating a larger margin of error in predicting the effects of sea level and temperature rise than climate change itself. Modelling of the circulation on the other hand is more prone to uncertainty in the wind stress and (external) thermohaline circulation. In both modelling concerns, existing models and programmes provide the most cost-effective means for case studies of future climate change-induced waves, surges, and currents.

1 Introduction

This chapter summarizes the possible influence of climate change on storm surge and storm waves, and on the circulation of the Gulf/Caribbean waters. The WMO/ICSU/UNEP (1985) scenario of a 1.5°C temperature rise and a 20 cm sea-level rise by the year 2025 is adopted. Numerical models of storm surge and storm waves are used to predict the effect of this scenario on future coastal lands. Then the broader issue of the effect of climate change on circulation is discussed, with emphasis again on the role of numerical models.

2 Two-Dimensional Storm-Surge Modelling

In order to better understand the conclusions reached, it is useful that incorporation of bathymetric and topographic data into into a storm-surge model such as SLOSH be briefly described. Experience has shown that basically the same methodology is used in any two-dimensional computer model where the area of interest is divided into grid cells of a given size and where depth effects are taken into consideration, i.e., wave refraction models, etc.

The available nautical or topographical charts are bound to be decades old. This is the experience in Puerto Rico and it should not be surprising

[1]University of Puerto Rico, Mayaguez, PR 00681–5000, USA
[2]Naval Oceanographic and Atmospheric Research Laboratory, Stennis Space Center, MS 39529, USA (deceased)
[3]University of South Carolina, Columbia, SC 29208, USA

if it is worse in some other areas of the region. Some hydrographic survey charts for Puerto Rico are 90 years old. Topographic maps, being decades old, show sand dunes where there are none, and show the shoreline where it no longer exists. In some places erosion has moved the shoreline inland by a few hundred metres relative to the most recent topographic map. The situation for the rest of the nations of the region is probably similar, although things could change in the near future due to the widespread use nowadays of Geographical Information Systems.

But even if up-to-date topographic data were available the following factors combine to introduce uncertainties which outweigh the effect of a sea-level rise of just a few centimetres: (1) Uncertainties in the measurements (due to instrumental limitations), water-level oscillations due to tides, other long-period waves, and fluctuations in shelf currents, etc. (2) One depth (or elevation) value has to be assigned to a grid cell which might be a few square kilometres in area (for the present Puerto Rico SLOSH model the smallest grid cells are approximately 1.25 × 1.25 miles in size), involving some type of averaging of the dozens (or more) depths (or elevations), which in some cases is done by an 'eyeball' mean. (3) Approximations in the physics of the problem (including boundary and initial conditions, drag coefficients, etc.), and in the meteorological input (for example, errors in the actual central pressure of the storm will introduce errors in the sea-level response due to the inverted barometer effect – the same thing will happen due to errors in the actual track of the storm). (4) Uncertainty in predicting the exact phasing of the astronomical tide and the surge, which introduces another variability in the initial water depths.

2.1 The Regional Situation

The coastal geomorphology of the majority of the region's islands (Hendry, Chapter 7), with the presence of deep water just a few miles offshore and steep land slopes, further diminishes the sensitivity of storm-surge computations due to sea-level rise. In addition to the fact that, as explained above, the final depth assigned to a grid cell has an uncertainty which is greater than the forecasted sea-level rise, the effect of adding 20 cm to a grid cell which already has a large depth will be of no practical consequence. The situation could be different in extensive, relatively flat, low-lying areas (i.e., wetlands), particularly along the north coast of the Gulf of Mexico.

But suppose that the bottom topography is up-to-date and that the depth values assigned to each grid cell are accurate. Will the increase in sea level have any practical effect on the size of the storm surge? From simple physical arguments it is known that the magnitude of the wind setup contribution to the storm surge is inversely proportional to the water depth. Dean and Dalrymple (1984), give a simple equation relating the change in maximum surge height ($\Delta\eta max$) with sea-level rise (S), (uniform) depth of shelf (ho), and wind-stress surge (ηmax):

$$\Delta\eta max = \left(\frac{-\eta max/ho}{1 + \eta max/ho} \right) S \qquad (1)$$

Using as an example a simulation of the 1928 San Felipe hurricane, the strongest on record for Puerto Rico, the maximum surge obtained was 2.4 m. To obtain an estimate of the wind setup contribution we can use the inverted barometer rule of thumb that for each millibar drop below normal conditions sea level will rise by $\simeq$1 cm. This is called the pressure setup. For San Felipe the pressure drop was of the order of 80 mb and hence, the pressure setup is of the order of 0.8 m. Therefore, the wind setup contribution is of the order of 2.4 – 0.8 = 1.6 m. Using this as ηmax, ho = 10 m, and S = 1 m, we get from the above equation $\Delta\eta$max = -0.14 m. The error margin for the SLOSH model is ±20%. For a 2.4 m peak surge this gives an error of ±0.5 m. Hence, the reduction in the peak surge is well within the uncertainty in the SLOSH results. It should be mentioned that the example used is a rare event. More common events will certainly give much smaller results, specially along steeper and deeper shelves. This conclusion is further corroborated by sensitivity runs made with the SLOSH model for Puerto Rico in which depths were increased by up to 1.5 m with at most 2% peak surge reduction in open-coast grid cells and 4% in enclosed and relatively shallow areas like San Juan Bay.

Another factor that is bound to mask the effect of changing water depths on storm surges and storm-surge modelling, at least for the island situation, is the fact that while the narrow and steep shelves are protected from high surges, it makes the islands more vulnerable to wave attack during a hurricane. Under certain circumstances the waves crashing against the coast will further increase the stillwater elevation by the so-called wave setup contribution. For islands this contribution can be as large or larger than the combined wind and pressure setup. But this is a highly complex phenomenon, both in time and space, about which not much is known as far as modelling in complex bottom topographies. A related phenomenon is the ponding effect brought about by waves breaking over offshore reefs. These are examples of physical factors which are extremely important in the determination of the anomalous stillwater elevation produced by a hurricane and which introduce an uncertainty factor that could mask a sea-level rise of 20 cm during the simulation of a given event.

In low-lying areas, such as wetlands, an increase in water depth could have an effect on bottom and wind stresses. But so little is known about the physics of this problem that any inclusion of a sea-level rise of 20 cm in existing theories might not be worthwhile.

Now, as far as sea-level rise is a contributing factor in shoreline erosion, then it could be a factor to take into consideration in the determination of how often bathymetric and/or topographic maps will have to be revised because of the changing position and geomorphology of the shoreline. To the extent that natural or man-made obstructions to the landward penetration of the sea during a storm are affected by sea-level rise, then this is something that should be closely monitored so that these changes are introduced in the input data given to the model.

Given all the uncertainties involved in the determination of the stillwater elevation brought about by hurricanes, the fact is that in many places for coastal planning, it is necessary to estimate the *Nth*-year return period stillwater elevation, and produce maps showing the floodable areas. As argued above, for many islands these *Nth*-year stillwater elevations are relatively small; in fact, they are of the same order as the forecasted sea-level rise. This is something coastal planners should be aware of. Aside from a more frequent revision of flood maps the only other way to account for sea-level rise is to allow, in some way, for X years of relative sea-level rise to be incorporated into the maps. In any case, it should be noted that the simulation of shoreline wave climatology is sensitive to changes in bottom topography and care should be taken in using present conditions to extrapolate to future years.

Hence, it can be concluded that the direct effect of a sea-level rise (of the order of 20 cm) on storm-surge modelling in the region should be of no practical significance. Indirect effects, through continuous displacement of the position of the mean shoreline (specially on sandy, or 'soft' shorelines) and the need it introduces of more frequent updating of coastal maps, is something that should be considered.

2.2 The Effect of a Temperature Increase

Finally, since the problem involves not only sea-level rise but a rise in atmospheric and/or sea temperature, passing mention should be made of the possible effects this might have as far as storm-surge modelling is concerned. The first thing that comes into mind is the potential variation in the atmospheric boundary layer drag coefficient due to variations in the stability of the air column in the marine boundary layer. For many years there have been equations expressing the drag coefficient as a function of the stability of the air column. But the myriad of parametrizations available make the choice important. Much needs to be improved as a recent article shows (Geernaert, 1987). In any case, if the parametrizations used in some planetary boundary layer models take into account the stability of the surface layer then, in theory, changes brought about by a temperature increase should be accounted for. For the same reason, some care will have to be taken in using models in which the drag coefficient is assumed a constant (the SLOSH model being one of them).

There is also the possibility that changes in the atmospheric and oceanic temperatures could bring about changes in the large-scale ocean circulation (Gallegos *et al.*, Chapter 3; Aparicio, Chapter 6). This in turn could bring about changes in nearshore water depths, as discussed below.

Also, there have been suggestions that the changing temperatures might affect the intensity and frequency of occurrence of hurricanes (Gray, Chapter 5). A change in the intensity, towards the high side, might bring about wind speeds outside the range of applicability of existing empirically determined drag coefficient formulas. As for possible variation in the frequency of occurrence of hurricanes, this might be of relevance since almost always a storm-surge study involves the determination of stillwater elevation return periods, and this is going to be highly dependent on the frequency of occurrence of a storm with a given intensity.

2.3 Effect on Hurricane Wave Modelling

The methodology in setting up a grid for numerical modelling of wind waves is similar to the storm-surge case and, therefore, the same uncertainties apply. In the study titled *Responding to Changes in Sea*

Level (NRC, 1987), it is stated that a rising sea level will, due to the increased depths, result in less wave damping and higher wave energy at the shoreline. In addition, the NRC study states that wave generation will be enhanced due to the greater water depths and reduction in bottom friction. For most of the island's situation, since the shelf is narrow and deep water lies close to shore, no changes of substantial engineering significance should be expected for approximately a 20 cm water-depth increase, specially when it is good engineering practice to add a safety factor to all parameters of relevance.

In the same way as for storm-surge generation, consideration will also have to be given to changes in the stratification of the atmospheric boundary layer and its effect on the interchange of momentum and energy between the wind and the ocean.

3 The Role of Dynamical Oceanography and Ocean Modelling

We now turn our attention to circulation modelling. The baseline scenario adopted from the Villach conference (WMO/ICSU/UNEP, 1985) considers a 1.5°C rise in temperature and a 20 cm rise in sea level by the year 2025 due to 'greenhouse' effects. In postulating this scenario two questions related to the physical oceanography of the region arise immediately: (1) What are the 'background' long-term fluctuations in sea level on time-scales of decades to centuries and how can we distinguish this 'noise' from warming-induced rises? (2) How might our knowledge of the physical oceanography of the region be used to improve our capabilities for ocean monitoring and for effective utilization of limited resources?

3.1 Sea-Level Change

In a recent paper, Sturges (1987) discusses the problems associated with detecting the 'carbon dioxide effect' in the sea-level signal. He notes that on the US East Coast the observed rise in sea level is now about 26 cm per 100 years (*cf.* Hanson and Maul, Chapter 9). For the low-latitude rise on the global scale about half is consistent with long-term glacial rebound. The remaining 6 cm per 100 yr is buried in background noise. However, Barnett (1983) found that long-term changes in the internal density structure of selected regions of the world's oceans were not statistically different form zero. Barnett emphasized that explaining secular changes in the density field will require a long time-series of hydrographic observations taken in key locations. In addition, temporal sampling must be 'particularly intense' to filter relatively small expected signals from the background noise.

An apparent sea-level rise in the northern Caribbean Sea on long time-scales need not be due to either an increase in steric height or melting of the polar caps, both usually associated with increased greenhouse gases in the atmosphere. It could also be caused by an increase in the wind field over the interior North Atlantic such that the large anticyclonic 'gyre' intensified, lowering sea level along the margins of the North Atlantic, or it could be due to a contamination of the sea-level record by long-period Rossby waves propagating across the Atlantic from east to west. For example, Sturges (1987) notes that the dominant sea-level signals at periods of 5-10 years and longer are coherent between the US west coast and Hawaii, and on both sides of the Atlantic, with amplitudes of 5-15 cm and Maul and Hanson (1991) show that interannual sea-level signals are spatially coherent and in phase throughout the northern Intra-Americas Sea. At the longest periods detectable in the records, 'signals' at 40-50 year period have amplitudes of 10 cm. Essentially, the longest records are contaminated by ocean wave propagation, perhaps atmospheric forcing, and tectonic motions. Another possible cause for fluctuations in sea level on scales of tens of years is coupled modes of oscillations between basins associated with long-period baroclinic Rossby waves.

The need for long time-series data for sea level at numerous Gulf and Caribbean sites thus becomes critical to sort out the low-frequency wave motions. A key problem then is the difficulty in securing long time-series of sea-level measurements. As Sturges (1987) notes, even the longest records, such as the San Francisco record beginning in the 1850s, are not sufficiently long to resolve long-period variability with statistical reliability. Determination of the causes of sea-level changes is further complicated by geologic activity: sediment accumulation and compaction, and tectonics (Burton *et al.*, 1987). Without technology like the satellite Global Positioning System (GPS), techniques do not exist that can distinguish these geologic causes from each other and from global 'eustatic' sea-level changes.

3.2 Role of Circulation Models

Sturges (1987) asks if it is not possible to reduce the 'noise' in the data by using our knowledge of the physical oceanography. In particular we might use ocean models to simulate sea-level changes of multi-year period generated by wind forcing. If we can understand how the region responds to winds on time-scales ranging from seasons to decades and reproduce the observed sea-level fluctuations on these time-scales, we may have some hope of sorting out the 'greenhouse' signal. Numerical models can also be used to understand the physics of long-term basin coupling, and to estimate the magnitude of such effects. While there are problems with this approach, it is clear that such models are beginning to provide sufficiently realistic results so that we can use them as tools for understanding (Ocean Prediction Workshop, 1986).

Some recent progress in ocean numerical modelling is given in *Oceanography* (Vol. 5(2), 1992) including results for the Atlantic Ocean. Thompson *et al.* (1992) show modelling results for the North Atlantic, and in particular for the Intra-Americas Sea, that are more and more realistic. For example, sea-surface height (SSH) variability in the their model calculations rather well describe the SSH variability seen in satellite altimeter data in the Gulf of Mexico (Jacobs and Leben, 1990) and in the Caribbean Sea (Andrade, 1991). Additionally, Thompson *et al.* show that the thermohaline contribution to the flow in the portion of the Gulf Stream System in the Gulf of Mexico and Caribbean Sea, a part of the global 'conveyor belt', is also important.

Schmitz and Richardson (1991) have estimated that nearly one-half of the volume transport in the Florida Current portion of the 'conveyor belt' has its origin in the South Atlantic Ocean. To illustrate its importance,

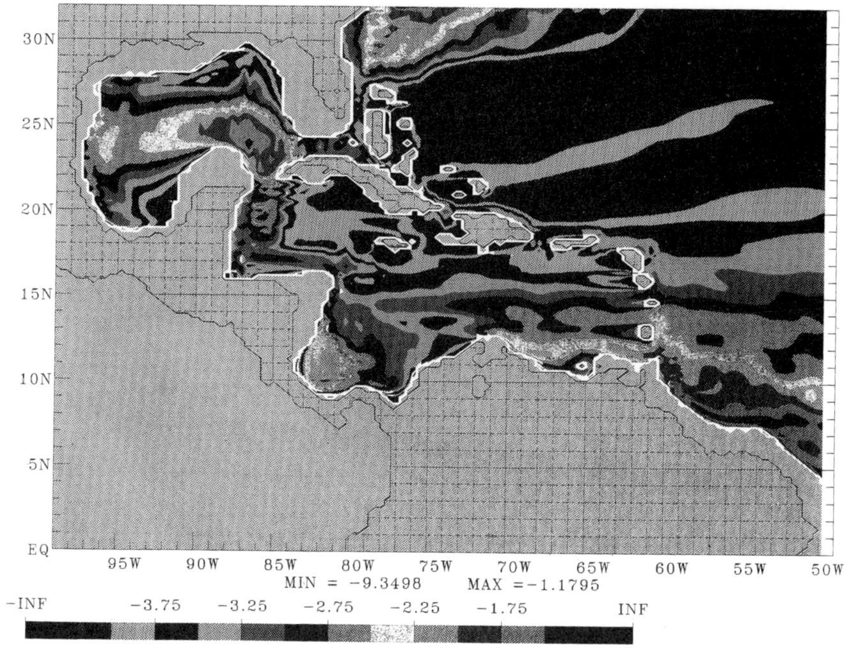

Fig. 4.1a Contours for the log of eddy kinetic energy from a 1.5-layer reduced-gravity model driven by Hellerman and Rosenstein (1983) monthly mean winds to statistical equilibrium in a closed basin. Maximum eddy kinetic energy is 0.07 m^2s^{-2}. See colour plates between pages 210 and 211.

Figs. 4.1a and b show results from two identical 1.5-layer reduced-gravity numerical models driven to statistical equilibrium by the monthly mean winds of Hellerman and Rosenstein (1983). The full extent of the model extends from 20°S to 60°N but only the Intra-Americas Sea is shown. One experiment is with a closed basin (Fig. 4.1[a]) and the other experiment (Fig. 4.1[b]) is with the source being a 20°-wide prescribed inflow at 20°S and the sink occurring at 60°N, also through a 20°-wide open boundary. Note the much larger eddy kinetic energies in the region when South Atlantic inflow is included.

Further, the volume transport through passages of the Gulf of Mexico, Caribbean Sea, and Bahamas are quite correctly modelled when compared to *in situ* observations, revealing a complexity of oceanic flows of interest not only to climate studies, but to problems in surface-current transport as well. Fig. 4.2 shows the annual cycle of volume transports of the Florida Current from observations and from a three-layer numerical model driven by actual winds from the ECMWF (European Centre for Medium-Range Weather Forecasts) operational product. Note the annual cycle of simulated transport is comparable to that observed, with the summertime maximum and the rapid decrease in transport in the autumn well reproduced. Thompson *et al.* (1992) are cautious, however, to note that 'we must [still] determine if our best simulation models are also our

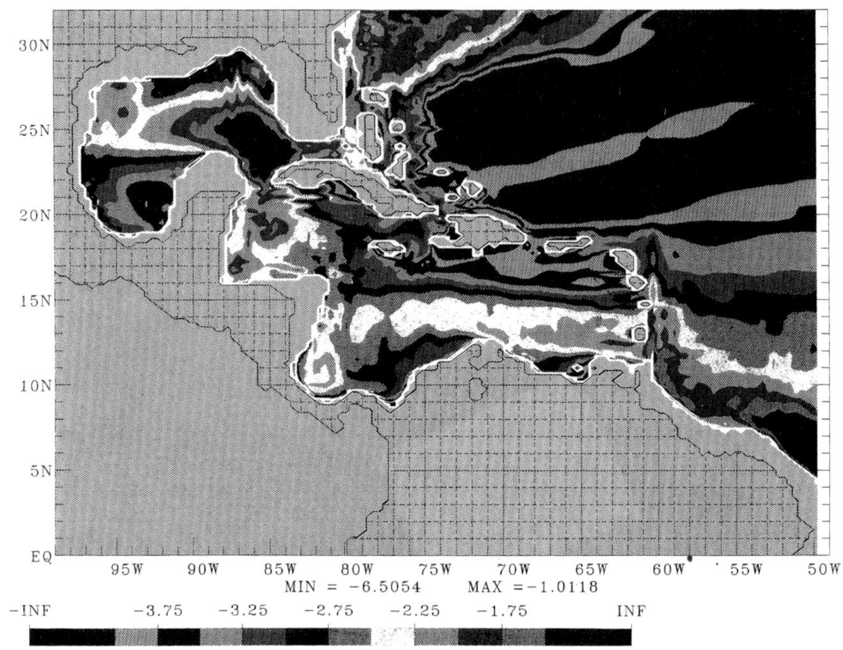

Fig. 4.1b Same as Fig. 4.1a except for an open basin with a 15 Sv South Atlantic surface inflow and a high-latitude surface outflow. Maximum kinetic energy is 0.18 m^2s^{-2}. See colour plates between pages 210 and 211.

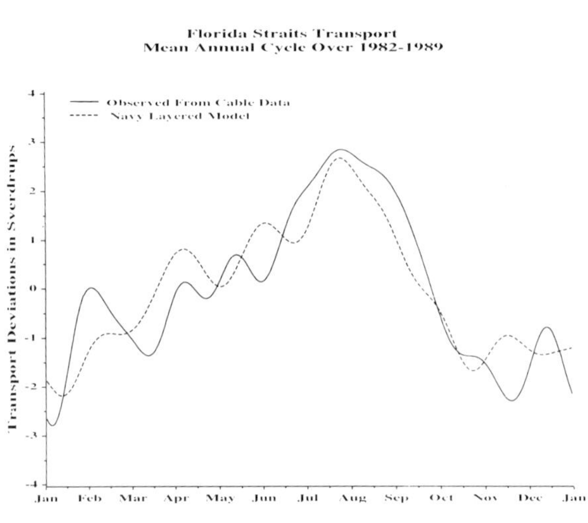

Fig. 4.2 The annual cycle of volume transport at 27°N in the Straits of Florida from observations (Larsen, 1992) and from the numerical model (Thompson *et al.*, 1992).
(1 sverdrup (Sv) = 10^6 m^3s^{-1} = 1 gigaliter per second.)

best prediction models'.

If significant sea-level rises do occur during the next 50 years (*q.v.* Fig. 1.7) it will be even more important that adequate ocean prediction models for short-term ocean response be available. Storm-surge models and wind-wave models which respond on hourly time-scales will be needed more than ever to forecast dangerous conditions to those coastal residents who never experienced serious problems when mean water levels were lower. Basin-scale circulation models will be needed to provide 'ocean weather' forecasts for the short term and climate-scale predictions for decadal time periods. Thus, the entire coastal population of the region are potential beneficiaries of improved warnings from such models. All types of ocean models, from the deep ocean to shallow water, are therefore needed for the problem at hand.

4 Conclusions

It is felt that, as far as most of the region is concerned, a sea-level rise of the order of the one predicted will be of no practical consequence on storm-surge modelling as far as changes that it might introduce in surge heights. There might be indirect effects through possible changes in the position and geomorphology of the coastline. Therefore these conditions will have to be monitored more frequently in order that they are reflected in the relevant maps and charts from which bottom and land topography data values are obtained. Changes in air and/or sea temperatures could have an effect as far as the transfer of momentum and energy from the wind to the ocean is concerned. This might produce changes in the magnitude of the surges. If the warming trend brings about more intense hurricanes then existing equations for the drag coefficient on the high wind-speed side will have to be more carefully defined.

Sea-level rise like the one predicted could introduce more confusion on the concept of return periods since, if not taken into consideration, what is now a 100-year event will be a more frequent event as time goes by, other things being equal. In addition, if the atmospheric temperature increase brings about changes in the hurricane climatology near the area of interest this will further increase the perplexity of this concept. It should also be expected that an increase in the frequency of occurrence of stronger storms will completely wipe out any possible decrease in *Nth*-year return period stillwater elevations due to increased water depths. Increased water depths will allow increased wave heights in the surf zone, in general. This might increase erosion rates and produce greater wave forces. Consequently, the so-called design wave, in the same way as the design stillwater elevation, will be changed.

It is not necessary that each nation of the region possess satellite oceanography and numerical modelling capability to profit from models. Ongoing and planned global geosciences programmes are a resource that can be accessed. The proposed sea-level network in the Intra-Americas Sea in conjunction with altimetric observations from TOPEX/Poseidon and ERS-1 (and to a lesser extent, recently available data from GEOSAT, SEASAT, and GEOS-3), and an expanded *in-situ* observational programme (STACS and WOCE, for example), will provide a valuable baseline for monitoring sea-level changes and understanding the character of that change in the region. Utilization of GPS and similar systems will allow geologic effects to be distinguished from their oceanographic counterparts.

In order to take advantage of such programmes, adequate electronic communications and access to data bases is crucial. United Nations endorsement can strengthen the possibilities for funding of research proposals of importance to the region.

5 References

Andrade, C.A., 1991: Mesoscale variability of the Caribbean Sea from GEOSAT. MS Thesis, US Naval Postgraduate School, Monterey, CA, 55 pp.

Barnett, T.P., 1983: Long-term changes in dynamic height. *J. Geophys. Res.*, **88**, 9547–9552.

Burton, R., Kendall, C.G. St.C. and Lerche, I., 1987: Out of our depth: on the impossibility of fathoming eustasy from the stratigraphic record. *Earth-Sci. Rev.*, **24**, 237–277.

Dean, R.G. and Dalrymple, R.A., 1984: *Water Wave Mechanics for Engineers and Scientists*. Prentice-Hall, Englewood Cliffs, 353 pp.

Greenaert, G.L., 1987: On the importance of the drag coefficient in air-sea interactions. *Dynamics of Atmospheres and Oceans*, **11(1)**, 19–38.

Hellerman, S. and Rosenstein, M., 1983: Normal monthly mean wind stress over the world ocean with error estimates. *J. Phys. Oceanogr.*, **13**, 1093–1104.

Jacobs, G.A., and Leben, R.R., 1990: Loop current eddy shedding estimated using GEOSAT altimeter data. *Geophys. Res. Ltrs.*, **17(12)**, 2385–2388.

Larsen, J.C., 1992: Transport and heat flux of the Florida Current at 27°N derived from cross-stream voltages and profiling data: theory and observations. *Phil. Trans. R. Soc. Lond. A*, **338**, 169–236.

Maul, G.A. and K. Hanson, 1991: Interannual coherence between North Atlantic atmospheric surface pressure and composite southern USA sea level. *Geophys. Res. Lett.*, **18(4)**, 653–656.

NRC (National Research Council) 1987: *Responding to Changes in Sea Level*. National Academy Press, Washington, DC, 148 pp.

Ocean Prediction Workshop, 1986: A status and prospectus report on the scientific basis and the Navy's need. Simmons, W., Tech. Editor. Mooers, C., Robinson, A., Thompson, J.D., Co-convener. (available from the Institute for Naval Oceanography, NSTL, MS., 39529), 494 pp.

Schmitz, W.J. Jr. and Richardson, P.L., 1991: On the sources of the Florida Current. *Deep Sea Res.*, **38**, 5374–5409.

Sturges, W., 1987: Large-scale coherence of sea level at very low frequencies. *J. Phys. Oceanogr.*, **17**, 2084–2094.

Thompson, J.D., Townsend, T.L., Wallcraft, A. and Schmitz, W.J. Jr., 1992: Ocean prediction and the Atlantic Basin: scientific issues and technical challenges. *Oceanography*, **5(1)**, 36–41.

WMO/ICSU/UNEP, 1985: Report of the International Conference on the Assessment of the role of Carbon Dioxide and of other Greenhouse Gases in Climate Variations and Associated Impacts, Villach, Austria, 9–15 October 1985. WMO Ref. No. 661, pp. 1–4, 1986.

Part 3
Physical processes and effects

5

Regional Meteorology and Hurricanes

Calvin R. Gray[1]

Abstract

The meteorological data base for the region, especially the Caribbean Sea, is disorganized and incomplete. Based on what data are available, there are weak indications that while rainfall is decreasing, temperature, wind speed, and evaporation are increasing. Rises in sea-surface temperatures may increase the overall frequency, as well as strength, of tropical storms and hurricanes in the area.

1 Introduction

The climate of the region is tropical (Table 5.1) and rainfall is the most highly variable element both spatially and temporally (Fig. 5.1). Northeasterly Trade Winds associated with large-scale features, such as the Hadley Cell and Inter-Tropical Convergence Zone (ITCZ), provide substantial rainfall, particularly on windward sides of mountain slopes, the amount increasing in general, with rising elevation. Significantly lesser amounts of rainfall occur on leeward slopes due to the rain-shadow effect. Another dominant factor determining the seasonal rainfall and its distribution is the strength and height of the Trade Wind inversion which inhibits convective activity during the winter season when the Bermuda High Pressure cell dominates the Intra-Americas Sea.

The direction from which the prevailing winds approach the coast is also a major determinant of rainfall. Hence, when the prevailing winds are essentially parallel with the coast, this condition diminishes the normal component of orographic uplift of the winds, thus diminishing rainfall. The micro-climate associated with windward/leeward aspects of the wind cannot be overemphasized. In general, more moist conditions exist on slopes facing the northeast and drier conditions on those slopes facing the southwest.

Mesoscale disturbances associated with convergence zones of Tropical Waves are another important rain-producing mechanism in the basin. Local circulations such as the sea-breeze or land-breeze can combine with mean flow conditions and topography to provide localized rainfall and all of the above mechanisms in combination with topography result in extremely variable rainfall conditions on relatively small land masses. The

[1]National Meteorological Service, Kingston 10, Jamaica

Table 5.1 Some climatological normals.

	Period	*Jan.*	*Feb.*	*Mar.*	*Apr.*	*May*	*Jun.*	*Jul.*	*Aug.*	*Sep.*	*Oct.*	*Nov.*	*Dec.*
Piarco Airport, Trinidad and Tobago													
Max. temp. (°C)	1946–75	30	30.3	31.2	31.8	31.6	30.6	30.7	31	31.4	31.3	30.7	30
Ext. max. (°C)	1946–75	31.1	31.6	32.4	33.3	33.2	32.3	32.4	32.7	33.1	32.9	32.4	31.5
Min. temp. (°C)	1946–75	20.4	20.4	20.9	21.9	22.9	23.1	22.7	22.6	22.6	22.5	22.0	21.2
Ext. min. (°C)	1946–75	17.6	17.6	18.3	19.0	21.0	21.3	20.9	20.8	20.9	20.8	20.0	18.5
Rainfall (mm)	1946–80	75	48	35	49	118	265	262	245	181	163	207	152
Sunshine (hours)	1946–75	7.5	7.9	8.1	7.9	7.7	6.3	7.0	6.9	6.8	6.6	6.6	6.7
RH (7am) (%)	1946–75	82	80	78	77	80	85	85	86	85	85	86	85
10 m wind (ms^{-1})	1946–75	4.9	5.6	6.2	6.3	6.5	5.8	4.6	3.6	3.6	3.7	3.7	4.1
Evap. (mm)	1967–84	5.0	6.0	6.8	7.0	6.7	5.3	5.5	5.4	5.4	5.0	4.6	4.5
Manley Airport, Jamaica													
Max. temp. (°C)	1951–80	29.8	29.6	29.8	30.3	30.8	31.2	31.7	31.9	31.7	31.3	31.1	30.5
Ext. max. (°C)	1947–87	32.8	33.3	33.9	34.4	33.9	36	36.7	35.8	36.1	34.4	33.9	33.8
Min. temp. (°C)	1951–80	22.3	22.3	22.9	22.6	24.7	25.3	25.6	25.3	25.3	24.8	24.1	23.1
Ext. min. (°C)	1947–87	15.6	13.9	15.6	18.9	20.0	21.0	20.0	21.7	20.0	20.6	20.6	19.4
Rainfall (mm)	1951–80	18	16	14	27	100	83	40	81	107	167	61	31
Raindays	1951–80	4	4	3	5	5	6	4	6	8	10	6	4
Sunshine (hours)	1951–80	8.3	8.6	8.5	8.7	8.2	7.7	8.2	8.0	7.2	7.4	7.8	7.8
RH (7 am) (%)	1951–80	80	78	77	77	76	73	76	76	78	80	79	78
RH (1 pm) (%)	1951–80	61	62	64	60	66	65	65	68	68	65	65	64
2 m wind (ms^{-1})	1972–86	1.7	1.9	1.9	2.0	2.5	2.9	2.6	2.3	1.8	1.4	1.3	1.5
10 m wind (ms^{-1})	1957–87	3.9	4.4	4.4	4.1	4.8	6.1	5.6	4.8	4.0	3.4	3.0	3.3
Evap. (mm)	1967–87	5.7	5.9	7.0	7.4	7.5	8.0	8.2	7.5	6.6	6.2	5.3	5.4

Abbreviations: Max. temp. = maximum temperature; Ext. max. = extreme maximum; Min. temp. = minimum temperature; Ext. min. = extreme minimum; RH = relative humidity; Evap. = evaporation.

larger landmass of North America is dominated by a monsoonal regime in summer, and by mid-latitude frontal passages in winter.

2 Rainfall

A monthly precipitation data base for the period 1920–1978 was developed for the Dominican Republic, Haiti, Jamaica and Puerto Rico (CEAS, 1979, 1980). A climatic study was made, statistically analysing seasonal rainfall data and radiosonde data observed at these sites. The statistical analysis of seasonal precipitation indicates that since about 1958 rainfall has decreased within the Caribbean Sea (CEAS, 1979).

Tropical cyclones are good rain producers. However, especially since 1970 (W. Gray, 1987) tropical cyclone activity has been significantly suppressed. Gray further notes that this period since 1970 is differentiated from the most active period 1947–1969 by higher surface pressure and 200 mb (12 km or 40,000 ft) westerly zonal winds. Figs. 5.2 and 5.3 indicate the annual frequency of all storms (hurricanes, tropical storms and tropical depression) entering the Caribbean Sea and the annual frequency of all storms originating within the Caribbean Sea between 1886 and 1986, respectively (Molina and Gray, 1986; C. Gray, 1987). Fig. 5.4 hints at a trend in the occurrence of tropical cyclones in the region.

The cumulative annual rainfall departure from the median is examined for annual rainfall at Manley and Piarco (Henry, 1987). Fig. 5.5 therefore gives some hints of possible trends and cycles in the rainfall. The word 'hint' is purposely chosen given the small data base. However, since the mid-1960s a downward trend in annual rainfall is evident and this agrees with the analysis in the CEAS (1979) report.

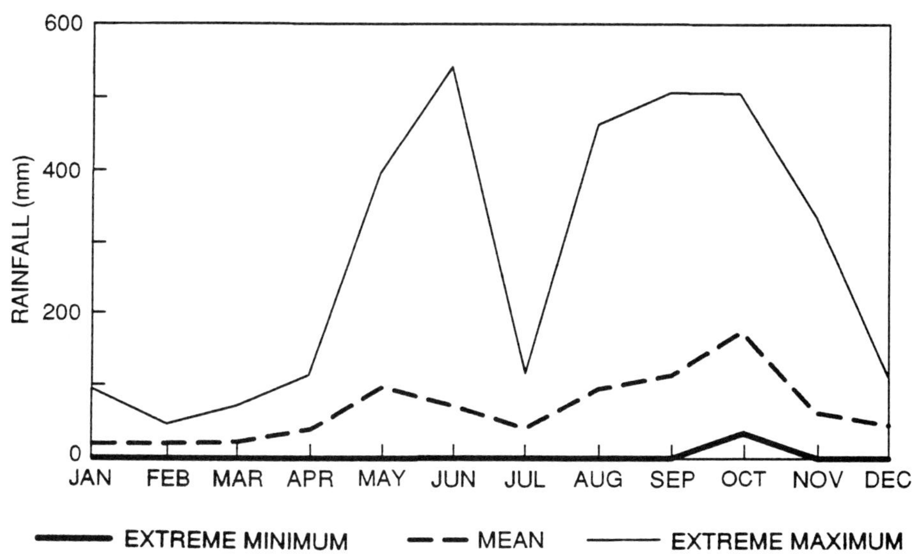

Fig. 5.1 Extreme maximum, mean, and extreme minimum rainfall at Manley Airport, Kingston, Jamaica based on data from 1951 to 1980.

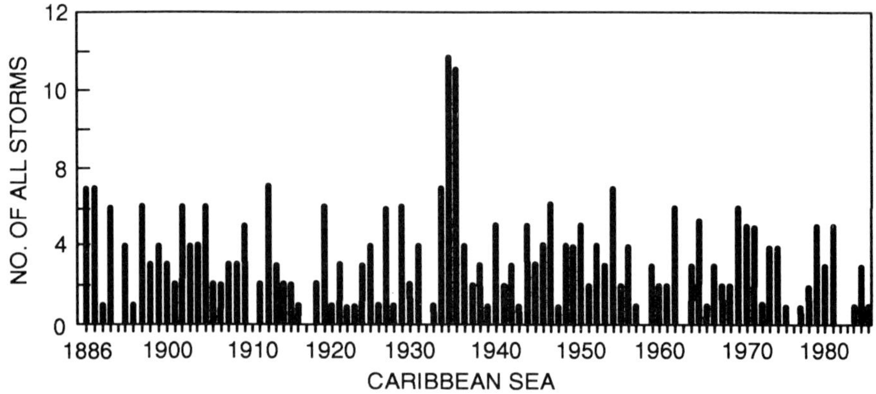

Fig. 5.2 Number of tropical storms observed in the Caribbean Sea between 1886 and 1986.

3 Wind, Temperature and Evaporation

One effect of upper-level winds as they relate to tropical cyclone activity has been mentioned earlier. On the surface, however, a look at small samples (Table 5.2), especially for the northeastern Caribbean Sea (Chen *et al.*, 1985), suggests a generally similar trend between Jamaica in the north-central Caribbean Sea, and Antigua and Barbados in the Leeward and Windward Island Chain from 1970 to 1980 (Fig. 5.6). Unfortunately, since 1980 only data from Jamaica and Trinidad and Tobago have been available, but they strongly suggest a continued trend toward increasing surface wind speeds.

The increasing wind speeds at the surface and decreasing levels of rainfall that would again be consistent with the findings from the CEAS (1979) report which states dry conditions during the summer season are associated with abnormally low values of temperatures at the 850 and 700 mb pressure levels, increased stability between these pressure surfaces, a tendency for a more southerly component in the Trades, marginal evidence for increased low-level divergence, stronger flow in the low-level Trades, increased sea-level pressure and decreased 700 mb

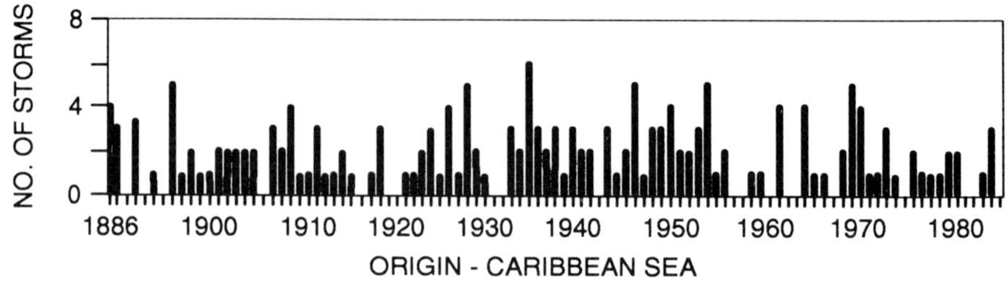

Fig. 5.3 Number of tropical storms originating in the Caribbean Sea between 1886 and 1986.

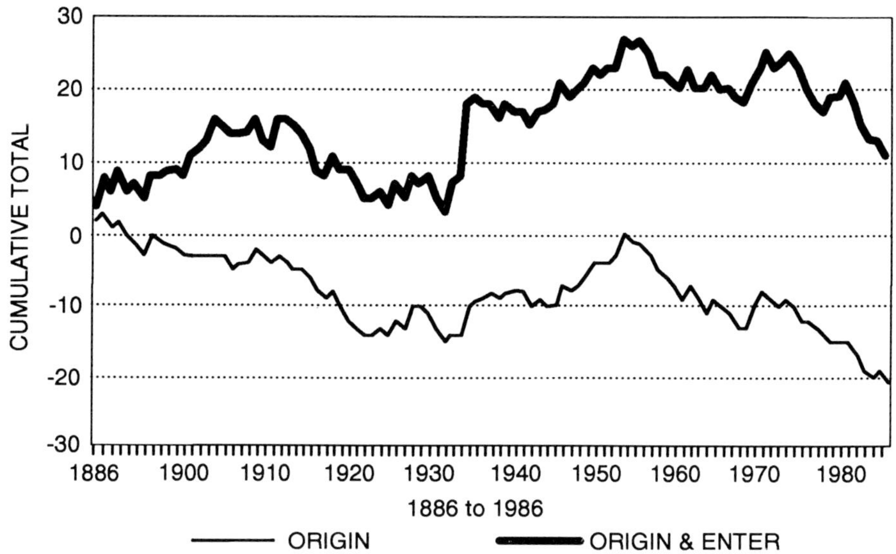

Fig. 5.4 Cumulative annual departure from the median of tropical cyclones (a) originating in the Caribbean Sea, and (b) total observed in the Caribbean Sea.

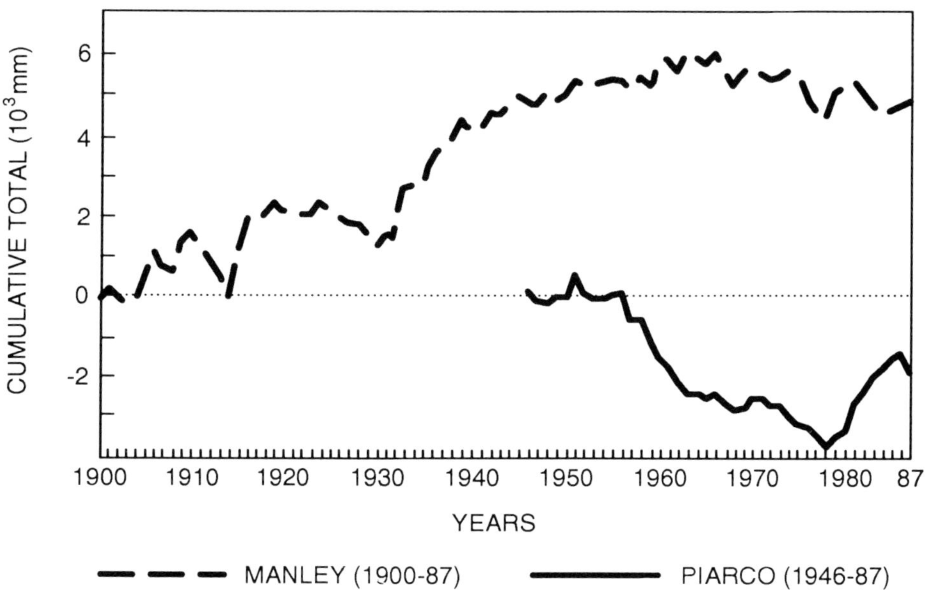

Fig. 5.5 Cumulative annual departure from the median of rainfall (in 10^3 mm) at Manley Airport, Kingston, Jamaica, and at Piarco Airport, Trinidad and Tobago.

Table 5.2 Seasonal variation in wind speed.

	Period	Jan.	Feb.	Mar.	Apr.	May	Jun.	Jul.	Aug.	Sep.	Oct.	Nov.	Dec.
Grantley Adams International Airport, Barbados.													
10 m wind (ms^{-1})	1969–78	5.8	5.6	6.0	5.8	5.9	6.5	6.2	5.7	4.5	4.8	5.0	5.4
Ext. gust (ms^{-1})		12.9	12.9	14.4	12.4	12.4	14.4	14.4	14.9	14.4	15.5	14.4	12.9
Vere Bird (Coolidge) International Airport, Antigua.													
10 m wind (ms^{-1})	1971–80	6.6	6.4	6.4	6.3	6.3	7.2	7.2	6.8	5.4	5.0	5.3	6.1
Ext. gust (ms^{-1})		21.1	19.1	19.6	18.5	18.0	22.7	22.1	21.6	22.7	21.6	18.5	24.2

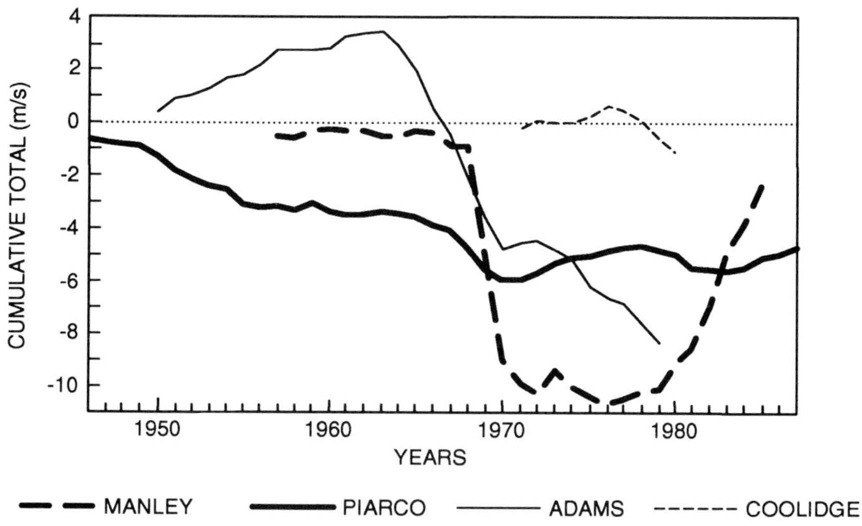

Fig. 5.6 Cumulative annual departure from the median of surface wind speed (ms^{-1}) from Manley Airport, Jamaica; Piarco Airport, Trinidad and Tobago; Adams Airport, Barbados; and Coolidge Airport, Antigua.

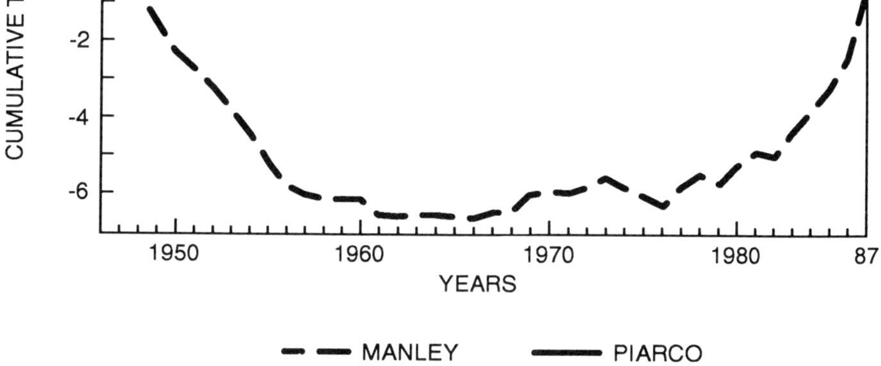

Fig. 5.7 Cumulative annual departure from the median of air temperature (°C) at Manley Airport, Jamaica and at Piarco Airport, Trinidad and Tobago.

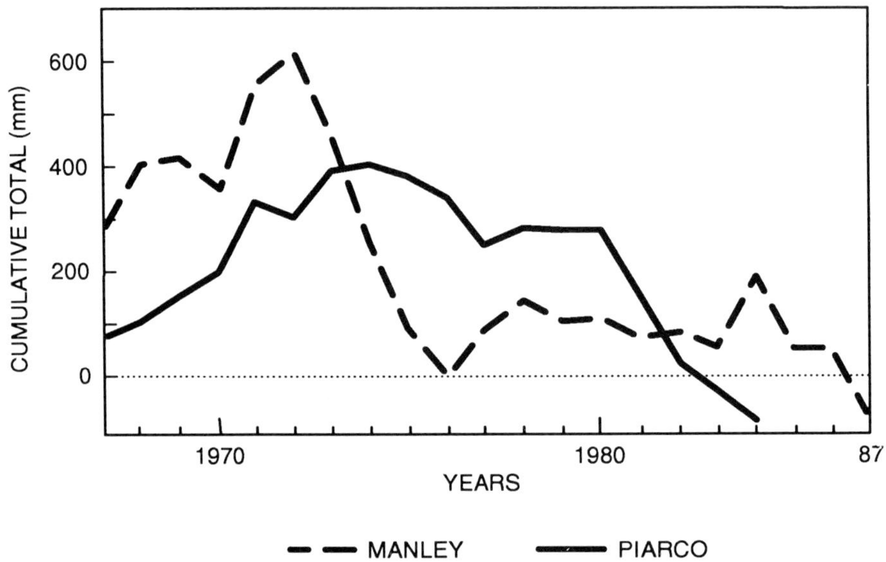

Fig. 5.8 Cumulative annual departure from the median of evaporation (mm) at Manley Airport, Jamaica and at Piarco Airport, Trinidad and Tobago.

height (Lamming, 1982; Chen *et al.*, 1985). Prior to 1970, wind data analysed for Adams, Piarco and Manley Airports indicate that the trends for the surface wind regime are generally similar (Fig. 5.6). It is quite possible therefore that when the post-1980 wind data for Adams and Coolidge are analysed, we will find that the surface wind speeds at these locations are also increasing.

Temperature and evaporation data were available for analysis only from Jamaica and Trinidad and Tobago at this time. The mean annual maximum temperature has increased distinctly since the mid-1950s (Fig. 5.7) but evaporation has decreased since the early 1970s (Fig. 5.8). The interpretation of this analysis is difficult since in general if wind and temperature are increasing (Figs. 5.6 and 5.7), one would normally expect increased evaporation (Hare, 1985; WCDP, 1987).

What can be stated at this time, in the absence of adequate data, is that the rate of evaporation from a free water surface depends on the moisture deficit, atmospheric pressure, wind speed, temperature, shape, size, and, in the case of seawater, salinity of the evaporating surface and other factors. Likewise, the amount of evaporation from the surface of soil and plants is profoundly influenced by meteorological conditions, primarily, moisture deficit, wind, air temperature, solar radiation, soil temperature, air pressure and factors such as soil colour and relief.

These factors can operate simultaneously and with mutual interaction, each depending on the presence and intensity of the other. Nevertheless, an increase in evaporation is anticipated in the Caribbean Sea based solely on the expected increase in surface wind, air temperature and surface pressure. Furthermore, increased evaporation would be consistent with the decreasing rainfall at the Manley Airport station (Fig. 5.5) and in the region in general (CEAS, 1979, 1980).

4 Some Likely Effects of Temperature Rise

In developing some likely effects it is assumed that:
1) rainfall will continue to decrease;
2) temperatures will continue to increase;
3) surface wind speed will continue to increase; and
4) evaporation will increase.

4.1 Rainfall

Decreasing levels of rainfall will increase the likelihood of droughts, resulting in reduced crop yields or even the loss of perennial plants, especially shrubs and trees. The loss of vegetation could increase the reflective capacity (albedo) of the surface for solar radiation. Over time and with continuing deterioration of the vegetative cover, desertification will result, given human pressure on the land aggravated by climatic stress.

Decreasing levels of rainfall also will result in the lowering of water tables. The pumping of water from deep below the surface can result in subsidence which could lead to a local rise in the relative sea level. Further, the lowering of the water table near an estuary, can cause seawater to seep inland.

Decreasing rainfall also will mean an increase in irrigation in areas of rain-fed agriculture. If 'water management' in these areas is imperfect, then mineral salts could become concentrated in the surface soil by the evaporation of saline soil solutions.

4.2 Wind

Increasing surface winds will result in considerable soil erosion as the wind removes the fertile top soil and exposes the subsoil. Soil erosion results in reduced water-holding capacity of the soil thus reducing potential soil-moisture reserves, thereby increasing the vulnerability of crops to short-term dry spells.

Increase in surface winds will result in decreased crop yields, resulting from the most obvious influence which the wind has on plants, that is, mechanical damage. In the region, there is considerable risk from nematodes which feed on roots, and physically weaken the roots to the extent that they become incapable of resisting even moderate winds. Increased surface winds would intensify its negative influence by increasing evaporation from soil, plant and water surfaces, thereby drying plants and increasing transpiration. Increased surface winds will also intensify the movement of dust into the region and/or increase the flight of insects which often affects the spread of fungi and other pathogens (*cf.* de Sylva, Chapter 14).

In the area of renewable sources of energy, the increase in surface wind speeds will be a boost for the generation of electricity, but mostly over the Leeward and Windward Islands, provided that the gustiness of the wind does not increase significantly. In the vicinity of the wind energy flux maxima (800 W/m^2) in the Caribbean Sea about 14°N, 71°W, the gustiness of the wind has already limited the generation of electricity from the wind, hence any further increase in wind speeds should add to the existing problems.

4.3 Evaporation

Increased levels of evaporation from soil surfaces will result in the reduction of soil-moisture reserves leading to drought, reduced crop yields, decreased animal production, increased irrigation and in general, impoverishment. With respect to the hydrological cycle, increasing levels of evaporation will result in more water vapour being available for the formation of clouds and subsequent rainfall. This, however, assumes that all other necessary and sufficient conditions have been satisfied.

5 Impact of Climate Change on Hurricanes

During the present century, there have been on average about four hurricanes in the Atlantic, Caribbean Sea, and Gulf of Mexico during the year. Based on historical data, an increase in sea-surface temperature (SST) of 1.5°C is found to be associated with an increase in hurricane frequency of 1.6 ± 1.2 per year (Shapiro, 1982). This predicted increase is about 40% above normal, but is subject to substantial uncertainty. The greater the SST, the more moisture is evaporated from the ocean, so that more latent heat is available for release in convection. All else being equal, more active convection will tend to lead to more frequent intensification of tropical systems to tropical storm and hurricane strength, and to more intense storms. Hurricane Gilbert (Fig. 5.9) may be a harbinger of such future storms.

5.1 Increased SST

The hurricane may be thought of as a thermodynamic engine, with heat input from latent heat of condensation. Based on a simple model of the hurricane in this context, Emanuel (1987) has used a measure of thermodynamic efficiency to evaluate the expected increase in the potential strength of hurricanes in an environment with increased SST, and correspondingly increased surface air temperature. For typical environmental conditions, Emanuel finds that an increase in SST of 1.5°C would increase the potential maximum hurricane wind by about 8%, or 6 ms^{-1}. This increase, if realized, could have a substantial impact on the destructive potential of the storms.

Neither the estimates of increased hurricane frequency nor strength may be realized, however, due to other factors that are difficult to evaluate. Changes in the atmospheric circulation associated with future SST increases may not be the same as the historical variations on which the estimates of hurricane frequency are based. Since these changes modify the tropical storms' environment, they could prevent the storms from reaching the potential increased strength estimated from thermodynamic considerations. For example, an increase in vertical shear of the zonal wind over the tropical Atlantic would tend to decrease the frequency of tropical storm formation, thereby moderating an increase in frequency from higher

Fig. 5.9 Colourized visible image of Hurricane Gilbert observed from the Geostationary Operational Environmental Satellite (GOES-7) on 13 September 1988. Until 1992 with the advent of Hurricane Andrew, Hurricane Gilbert was the most intense Atlantic tropical storm on record, causing 318 deaths (202 in Mexico alone) and US$ 5,000,000,000 in estimated damages. At NOAA aircraft flight altitude (10,000 ft or 3000 m) maximum wind gusts of 173 knots (89 m/s) were recorded, and the central pressure was 888 mb. See colour plates between pages 210 and 211.

SSTs, while a decrease in vertical shear would tend to have the opposite effect (see review by W. Gray, 1979).

5.2 Modelling

Numerical modelling simulations of the atmospheric changes that would be produced by increased CO_2 (see Wigley and Santer, Chapter 2, as well as reviews by Hansen *et al.*, 1984; and Schlesinger and Mitchell, 1987) have all been with global-scale models that are incapable of simulating or predicting hurricane genesis. The model analyses concentrate on changes in parameters, such as air temperature and precipitation, that are only indirectly related to tropical cyclogenesis. Thus, the actual changes in hurricane climate that would occur in an altered environment is unknown.

Moreover, from a regional point of view, changes in the location of hurricane formation and changes in track may be more important than overall changes in strength or basin-wide frequency of development. Prediction of this regional impact is beyond the present state of knowledge and modelling ability.

6 Conclusions

The greatest potential concern to the region if there is a climate-induced temperature rise is associated with tropical storms. While the estimates of increased hurricane frequency and potential strength over the Atlantic, Caribbean Sea and Gulf of Mexico may be useful guides of the direction and general magnitude of the changes that might be possible due to increased SSTs, the likely impact on any particular region is quite uncertain (Harrison, 1984).

7 Acknowledgements

Some of the material in this chapter was drawn from L.J. Shapiro as well as other efforts in *Implications of Climate Change in the Wider Caribbean Region* (UNEP (OCA)/CAR WG.1/INF.3, 183 pp.). The figures were redrawn by D.A. Senn, and the satellite image of Hurricane Gilbert was provided by Paul Hebert, US National Weather Service.

8 References

CEAS (Center for Environmental Assessment Services), May 1979: A study of the Caribbean Basin Drought/Food Production Problem, Progress Report.

CEAS (Center for Environmental Assessment Services), December 1980: Drought/Subsistence Food Shortages in the Caribbean Basin and Sub-Saharan Africa.

Chen, A. A., Daniel, A., Gray, C.R., Chambers, W., Archibald, A., 1985: Wind Power in Jamaica based on Climatic Data. Univ. of the West Indies, Mona Campus.

Emanuel, K.A., 1987: The dependence of hurricane intensity on climate. *Nature*, **326(2)**, 483–485.

Gray, C.R., 1987: History of Tropical Cyclones in the Caribbean 1886 to 1986. National Meteorological Service, Jamaica.

Gray, W.M., 1979: Hurricanes: their formation, structure likely role in the tropical circulation. In: Shaw, D.B. (ed), *Meteorology Over the Tropical Oceans*. Roy. Meteorol. Soc., pp. 155–218.

Gray, W.M., 1987: Cyclical Trends in Hurricane Destruction Potential and Discussion of Factors Which Will Be Used in the Seasonal Forecast, Ninth Annual Hurricane Conference, 03 April 1987, Orlando, FL.

Hansen, J.A., Lacis, A., Rind, D., Russell, G., Stone, P., Fung, I., Ruedy, P. and Lerner, J., 1984: Climate sensitivity: analysis of feed-back mechanisms. In: *Climate Processes and Climate Sensitivity* (Maurice Ewing Series, No. 5), American Geophysical Union, Washington, DC, pp. 130–163.

Hare, F.K., 1985: World Meteorological Organization – Report No. 653, Climate Variations, drought and desertification.

Harrison, P., 1984: Population, Climate and Future Food Supply. *Ambio*, **13**.

Henry, E.B., 1987: Rainfall Climatology/Flood Run-Off, Caroni Watershed, Trinidad and Tobago. Oklahoma Climatological Survey.

Lamming, S.D., 1982: Wind Resource Assessment Project, Barbados. Caribbean Meteorological Institute.

Molina, M. and Gray, C.R., 1986: Frequency distribution of Hurricanes and Tropical Storms in Jamaica 1900–1980. National Meteorological Service, Jamaica.

Schlesinger, M.E. and Mitchell, J.F.B., 1987: Climate model simulations of the equilibrium climatic response to increased carbon dioxide. *Rev. Geophys.*, **25**, 760–798.

Shapiro, L.J., 1982: Hurricane climatic fluctuations. Part II: Relation to large-scale circulation. *Mon. Wea. Rev.*, **110**, 1014–1023.

WCDP (World Climate Data Programme), 1987: The Global Climate System, Autumn 1984 – Spring 1986. CSM R84/86, World Meteorological Organization, Geneva.

6

Meteorological and Oceanographic Conditions along the Southern Coastal Boundary of the Caribbean Sea, 1951–1986

Ruben Aparicio[1]

Abstract

An attempt is made to define local climatic conditions along the southern coastal boundary of the Caribbean Sea based on data collected at coastal weather stations during the period 1951-1986. Examination of the records has been addressed to point out interannual variability that could serve as evidence of climatic change during the most recent decades. Some basic considerations about the physical characterization of the upwelling regime and the structure of the water column stability pattern for the region are also presented. In addition, a brief discussion of sea-surface temperature data and the relative sea-level variability from tide-gauge records for the area of concern has been also included.

1 Introduction

Potential effects of global climatic changes, typified by relative sea-level fluctuations and an expected atmospheric and sea-surface warming on the coastal ecosystems of the Caribbean Sea and adjacent areas, are actually relevant aspects of the Caribbean Environment Programme of the United Nations Environment Programme (UNEP), as well as other bodies, in particular IOC and IOCARIBE.

The main goal of this chapter is to present the available time histories of key climatological and oceanographical parameters for the southern coastal boundary of the Caribbean Sea in order to delineate regional evidence for long-term (multi-decadal) climatic change. Assuming that the regional coastal upwelling regime is one of the most important local oceanographic features susceptible to be altered by any global warming, an attempt is made to summarize its more general characteristics.

2 Regional Climatology

General climatic controls of the region have already been discussed by Gray (Chapter 5). However, the enormous spatial variability typical of the region forces us to mention some particular factors characteristic of the

[1] Universidad de Oriente, Cumana, Venezuela, SA

southern section of the Caribbean Sea:

1) Among other processes, the seasonal fluctuation of the intertropical convergence zone (ITCZ) is usually thought to be the most important factor controlling the regional rainfall pattern.
2) On a shorter spatial scale, the coastal sea-surface temperature variability seems to have a very strong influence on the local weather (Aparicio, 1986).
3) Tropical cyclone activity is not significant. During this century, according to Depradine *et al.* (1973), only three tropical storms over the southern coastal boundary of the Caribbean Sea have reached hurricane intensity. In the period 1961-1977, tropical storms south of 13°N have been observed only in 1961, 1963, 1969, 1971 and 1974 (Herrera and Febres, 1981).

A 35-year record (1951-1986) for three selected exposed locations along the southeastern coastal boundary of the Caribbean Sea, supposed to be regionally representative, at an average latitudinal location of 11°N and covering the longitudinal range 62–72°W, reveals the following:

2.1 Precipitation

In contrast with the trend toward decreasing level of rainfall within the Caribbean Sea, reported previously for high latitudes (CEAS, 1979) no meaningful trend in precipitation is observed during the period of time under consideration (Fig. 6.1[a]). Low spatial variability is also evident.

2.2 Evaporation

In opposition to the rainfall regime, the local evaporation pattern shows appreciable spatial variations (Fig. 6.1[b]). An abrupt westward intensification of the evaporation rate seems to occur approximately in the eastern Gulf of Venezuela (around 70°W). A small trend of increasing evaporation since the early 1960s can be detected.

2.3 Wind Stress

The spatial distribution of the alongshore component of the surface wind stress, estimated by using the drag coefficient approximation given by Large and Pond (1981), shows a remarkable westward intensification (Fig. 6.2). Also shown in the records is a pronounced decreasing trend of the zonal component of the surface wind stress since the early 1960s.

2.4 Air Temperature

Air-temperature data also show manifest longitudinal spatial variability, with the values increasing westward. A regional coherent pattern of air-temperature change seems to exist over the period recorded. This pattern is associated with a linear increase in air temperature at the rate of 0.1°C per decade. The records (not shown) exhibit considerably larger interannual variations. Due to the short length of the records, this finding should be considered with caution (see Hanson and Maul, Chapter 9).

An attempt to summarize qualitatively the general findings is presented in Table 6.1. The regional spatial variability observed could be easily

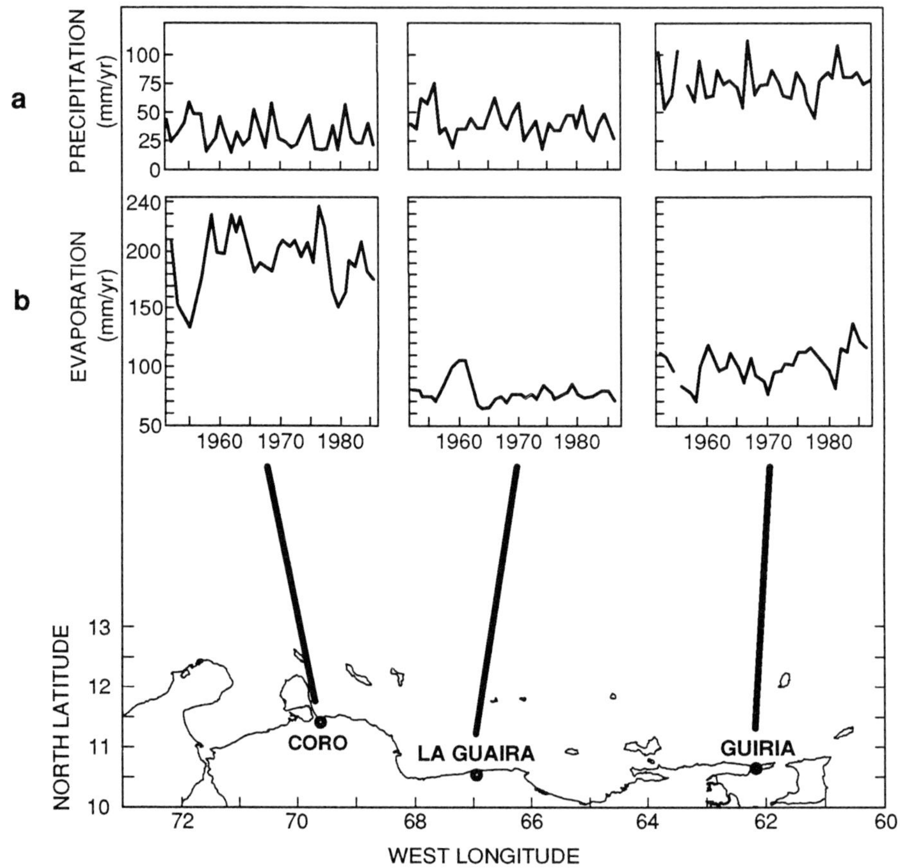

Fig. 6.1 Time-series of annual mean values of (a) precipitation (in mm/yr), and (b) evaporation (in mm/yr), for three selected locations along the southeastern coastal boundary of the Caribbean Sea, during the period 1953–1986.

Table 6.1 Qualitative summary of regional climatology and long-term trends for the southern Caribbean Sea.

Parameter	*Spatial variability*	*Long-term trend*
Evaporation	High	Increase
Precipitation	Low	Not significant
Wind stress (alongshore component)	High	Decrease
Air temperature	High	Increase

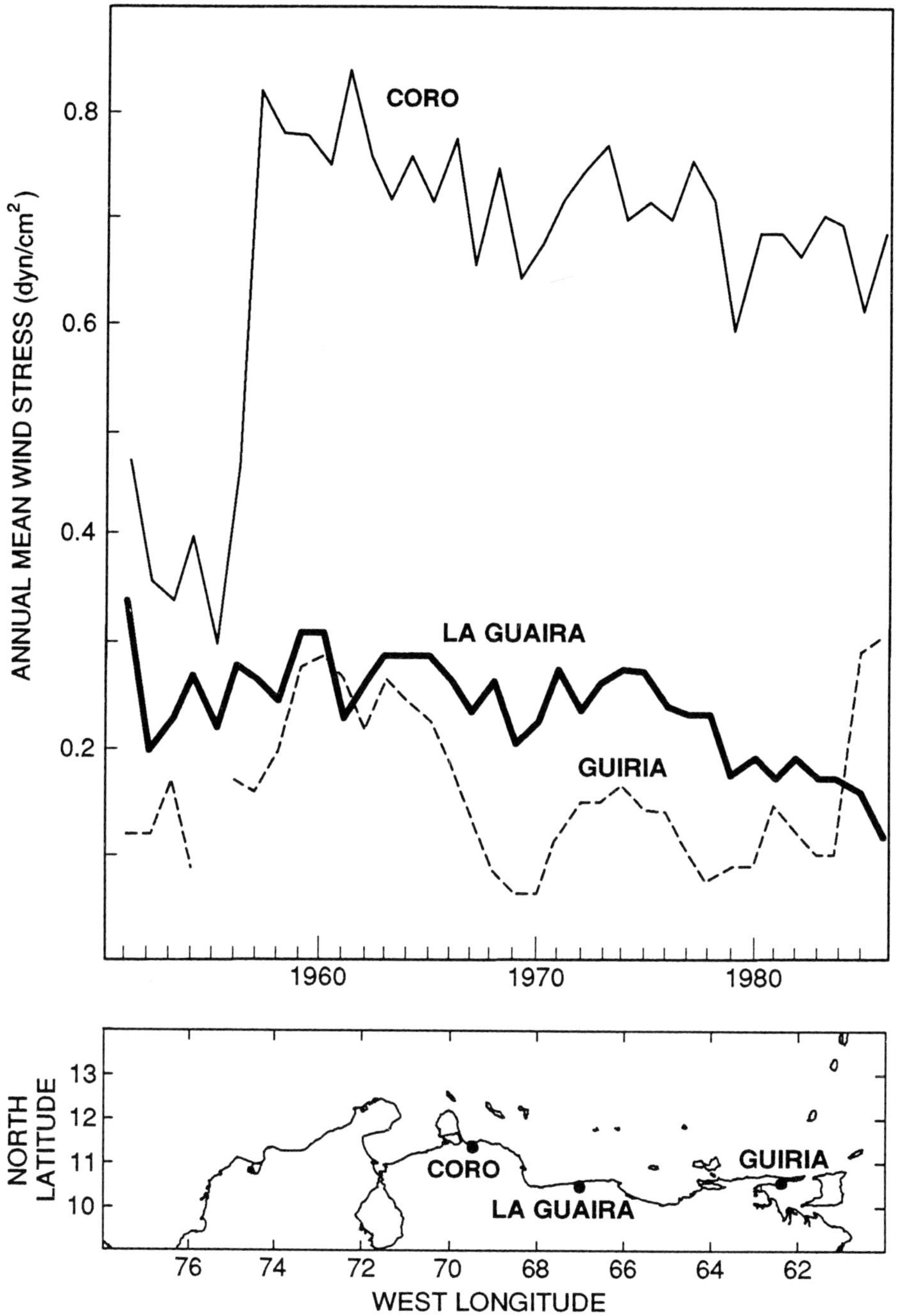

Fig. 6.2 Time-series of annual mean values of the alongshore component of the surface wind stress, (in dyn/cm^2) for three selected locations along the southeastern coastal boundary of the Caribbean Sea, during the period 1951–1986.

interpreted in terms of an intensification of the wind field from east to west, which in combination with higher values of air temperature at the west of the area under concern, would accordingly enhance the local rate of evaporation.

3 Regional Oceanography

According to previous descriptive oceanographical studies for the coastal area of the southern Caribbean Sea (see Richards, 1960; Gade, 1961; Ljöen and Herrera, 1965; Griffiths and Simpson, 1972; Febres and Herrera, 1976; Maul and Kavanagh, 1988; Muller-Karger *et al.*, 1989) the following two general regional oceanographic features may be outlined: (1) Occurrence of seasonal coastal upwelling, thought to be a local wind-induced phenomenon. (2) Appreciable seasonal differences in the vertical distribution of the water masses reflecting the strong influence of fresh-water discharge from south American rivers (Amazon, Orinoco and Magdalena) and the seasonal fluctuation of the local heating pattern. These two features indicate consideration of the regional wind field and the density gradient field as primary driving mechanisms controlling the local pattern of coastal surface flow.

3.1 Coastal Upwelling

Coastal upwelling occurs where surface water moves along a coast with an offshore component, creating a horizontal surface divergence and inducing water of deeper layers to ascend toward the surface. It is well known that this upwelled water transports large quantities of nutrients to the surface layer. The combination of high nutrient content, oxygen supply due to wave action, and incident solar radiation results in enhanced biological productivity that leads to important consequences for fishery activities. Coastal upwelling also affects the deposition of sediments rich in organic matter, which may lead to formation of phosphorites. Further, coastal upwelling critically influences local weather and climate. In the region, coastal upwelling appears to be confined to its southern area. Due to its biological, climatological and geological regional implications, one is obligated to pay attention to the general physical characteristics of that boundary process:

1) Negative correlations, found to be statistically highly significant between long-term time-series of monthly mean values of the alongshore wind stress component and coastal sea-surface temperature, indicate that the local regime of wind stress is a primary forcing mechanism for coastal upwelling on the southern Caribbean Sea (Aparicio, 1986).
2) The local surface wind stress field presents a remarkable month-to-month persistency in its mean directional distribution (Fig. 6.3), reflecting the regional predominance of the northeast Trade Winds; it implies favourable conditions for coastal upwelling along the entire shoreline of interest throughout the entire year.

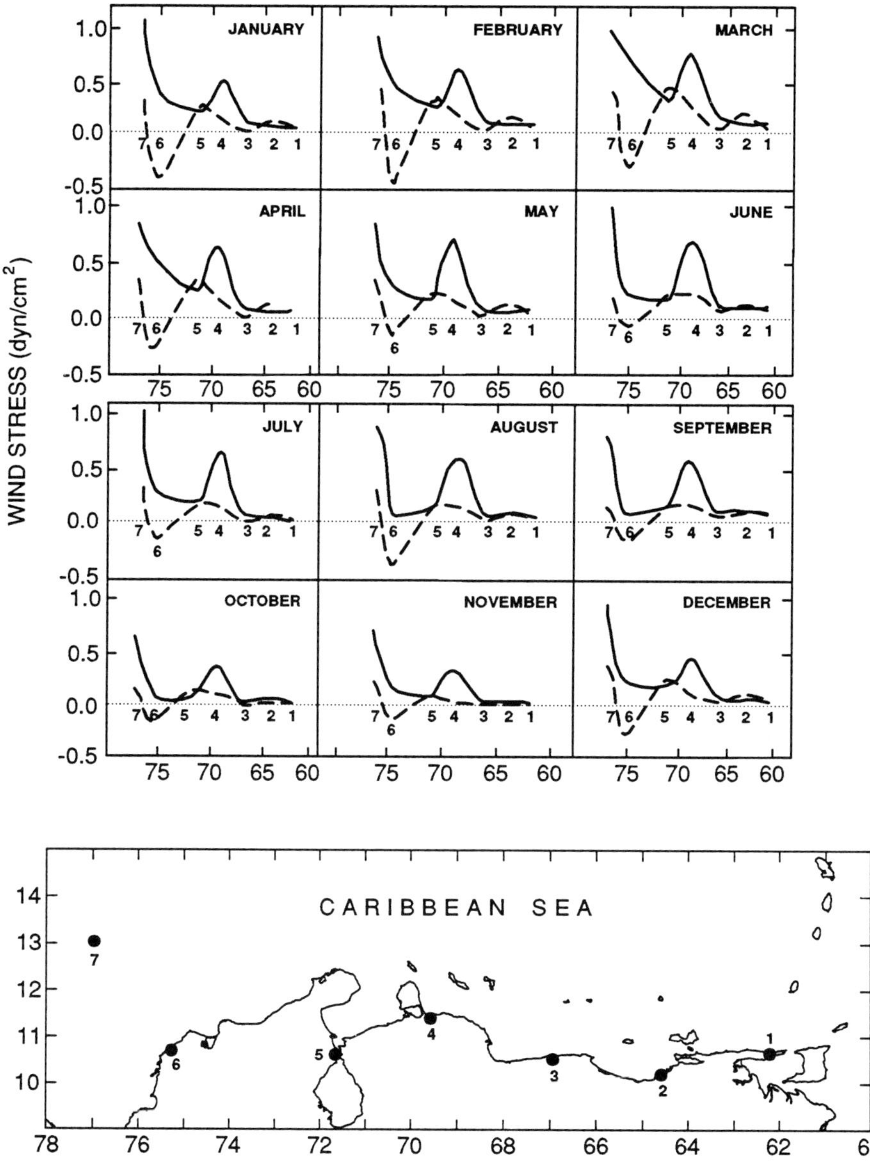

Fig. 6.3 Seasonal cycle of the spatial variability of the alongshore (solid line) and cross-shore (dashed line) components of the surface wind stress (dyn/cm^2) for the southern coastal boundary of the Caribbean Sea, during the period 1961–1971. The remarkable directional persistency of the local surface wind stress regime implies favourable conditions for a wind-induced coastal upwelling occurrence throughout the whole year.

3) With regard to the longitudinal spatial extension, several studies (Gordon, 1965; Perloth, 1971; Febres, 1972), based largely on historical hydrographical data analysis for the entire Caribbean Sea, suggest the existence of a non-continuous spatial regime of upwelling along its southern boundary. In that scheme, two clear zones of coastal upwelling are evident. The first one affecting the Venezuelan coast between 60°W and 67°W, and the second one affecting the Colombian coast, from 72°W to 75°W. However, continuity in the alongshore spatial extension regime could be inferred from similarity among the local coastal sea-surface temperature anomaly patterns given for that area by Aparicio (1986; Fig. 6.4). From Fig. 6.4 is also evident a degradation east–west in the intensity of that process, such as postulated by Richards (1960).
4) The vertical velocity at the bottom of the Ekman layer on the south-eastern section of the Caribbean Sea has been estimated to be about 80 cm/day (Aparicio, 1986). Similar estimates have been given by Ljöen and Herrera (1965) and Fajardo (1979).

3.2 Vertical Stratification

Vertical stratification in the ocean is quantitatively described by the vertical stability of an oceanic water column, which depends on the vertical distribution of density, which in turn is a function of temperature, salinity and pressure. Other things being equal, a water column is stably stratified if the density increases with depth. A commonly used index of the static stability of the upper 100 m in the water column is the Hesselberg-Sverdrup stability parameter, E, which is proportional to the vertical gradient of sigma-t, σ_t (i.e., the density parameter at atmospheric pressure). The total expression for stability, however, contains additional small terms related to the vertical gradient of temperature change due to compressibility (see, for example, Pond and Pickard, 1978). Another factor used as indicator of stratification is the Brunt-Väisälä frequency, N, which is of prime importance in studies of various types of wave motion in the ocean.

An examination of historical hydrographic surveys for the area enclosed between 12–15°N and 16–68°W (southeastern Caribbean Sea) reveals mainly a close relationship between the vertical distribution of the

Table 6.2 Vertical stability characteristics of the Caribbean Sea.

Approximate depth range (m)	*Approximate temp. range (°C)*	*Identifying feature*	*Average Brunt-Väisälä frequency (cycles/hour)*
$0 < Z < 200$	27–20	Shallow salinity maximum	54
$200 < Z < 1200$	20–5	Salinity minimum at $600 < Z < 800$	18
$1200 < Z < 2000$	< 5	Deep salinity maximum	< 11

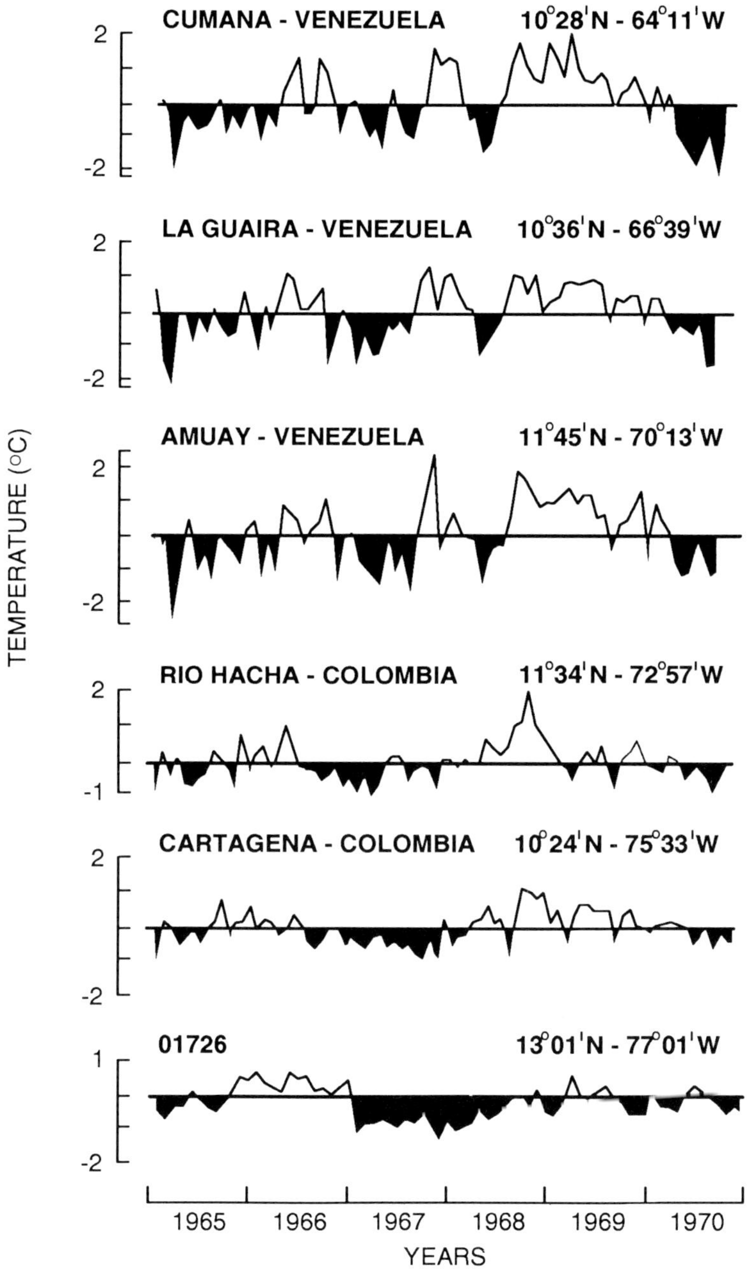

Fig. 6.4 Time-series of monthly mean values of coastal sea-surface temperature anomalies (in °C), for six selected locations along the southern boundary of the Caribbean Sea, during the period 1965–1970. Negative anomalies associated with cold periods have been darkened. Similarity among the anomaly patterns suggest continuity in the regional alongshore spatial extension regime of coastal upwelling occurrence.

static stability and that of salinity. It is also evident that there are conditions of neutral stability (E = 0) below 1200 m. The vertical profile of the Brunt-Väisälä frequency, given for a hypothetical average regional water column, discriminates three stability layers. Table 6.2 lists these layers by depth, temperature range, main identifiable feature and the value of N in cycles/hour. Although water type is not included, it is not difficult to identify the various water layers found in the Caribbean Sea (see, for example, Metcalf, 1976; Morrison and Nowlin, 1982). Values of N reported in Table 6.2 compare quite well with values for the Intra-Americas Sea extracted from the horizontal world distribution of annual mean Brunt-Väisälä frequency given by Levitus (1982) for selected standard depth levels.

3.3 Sea-Surface Temperature Variability

An abrupt alongshore east–west intensification of the sea-surface temperature (SST) between 71°W and 72°W separates the southern coastal boundary of the Caribbean Sea into two different zones. Higher temperatures are always observed in the western zone, i.e., the Colombian Guajira coastal area. That can be clearly seen in the mean seasonal variation of SST for the whole coastal boundary (Fig. 6.5). A comparison with the mean seasonal cycle for the entire area of the Caribbean Sea, as reported by Colon (1963), reveals a large-scale spatial homogeneity in the thermal characteristics of the Caribbean Sea near-surface layer.

A 24-year SST record (1953-1976) for three selected locations along the Venezuela shoreline (Fig. 6.6) reveals no significant long-term trend with large interannual variability dominating the whole signal. Even though time histories of SST from the world's oceans show a linear trend rise (Kukla *et al.*, 1977), that 'global' signal is not apparent in the region. There seems to exist no connection between a regional atmospheric warming and the thermal behaviour of the coastal surface layer. Indeed, while the regional range for long-term air temperature annual variation at any location is no larger than 2°C, the range of SST averages 4.8°C.

3.4 Relative Sea-Level Variability

In order to examine the local pattern of the long-term relative sea level (RSL) variability, mean annual sea-level data from four tide-gauge stations, with record length ranging from 20 to 34 years and located on the eastern and central part of the southern Caribbean Sea coastal boundary, have been inspected. It is assumed that a linear trend estimate for the mean annual sea level at any given station can indicate the general rise or lowering of RSL with some assurance. So, least-squares best-fit regression lines, correlation coefficients, standard errors and confidence levels for the lines were calculated for each station.

Tabulated results (Fig. 6.7) show the geographical location and the data record length for each tide-gauge station. Coastal locations No. 2, 3 and 4 exhibit a general rise of RSL. However, the most eastern tide-gauge station (Carupano, No. 1) shows a lowering trend and the maximum rate of

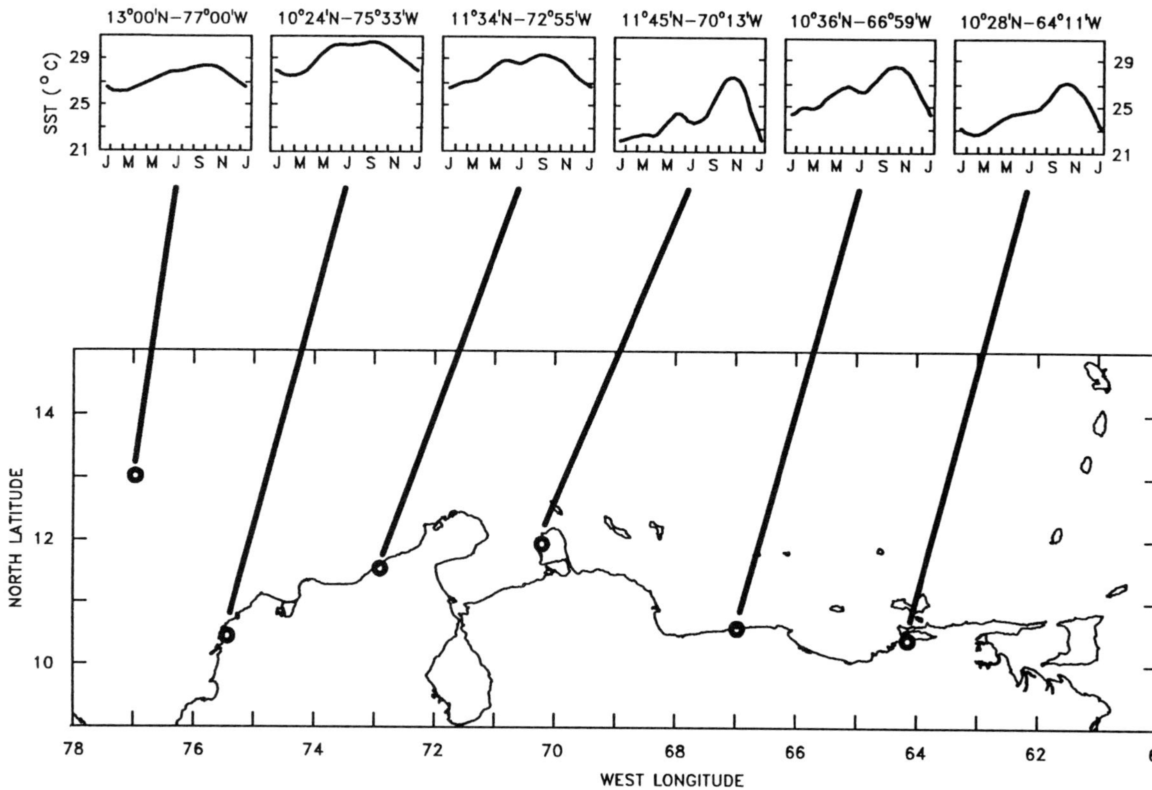

Fig. 6.5 Seasonal pattern of the sea-surface temperature (°C) at six locations along the southern coastal boundary of the Caribbean Sea during the period 1965–1970.

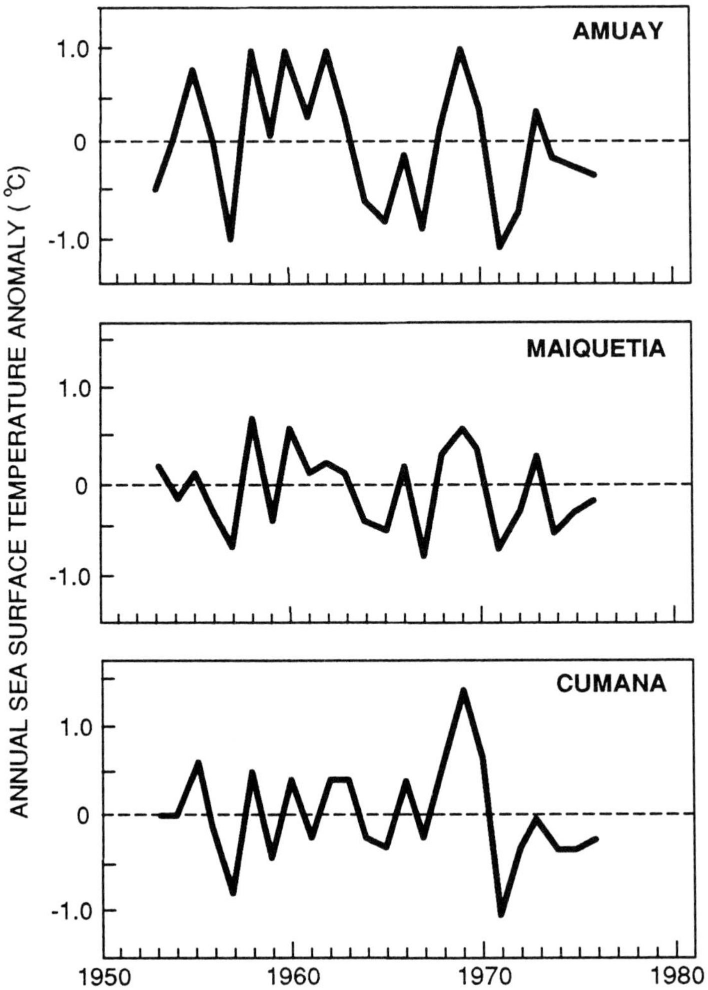

Fig. 6.6 Time-series of annual mean values of sea-surface temperature anomaly (°C) for three selected locations along the southeastern coastal boundary of the Caribbean Sea during the period 1953–1976.

change. Except for the change trend at station No. 1, these estimates seem to confirm previous regional findings (*q.v.* Hanson and Maul, Chapter 9). Discrepancy at station No. 1 could arise from having made use of a very different data record length: local data from 1976 to 1986 were not available to those authors. A rapid apparent decrease in RSL

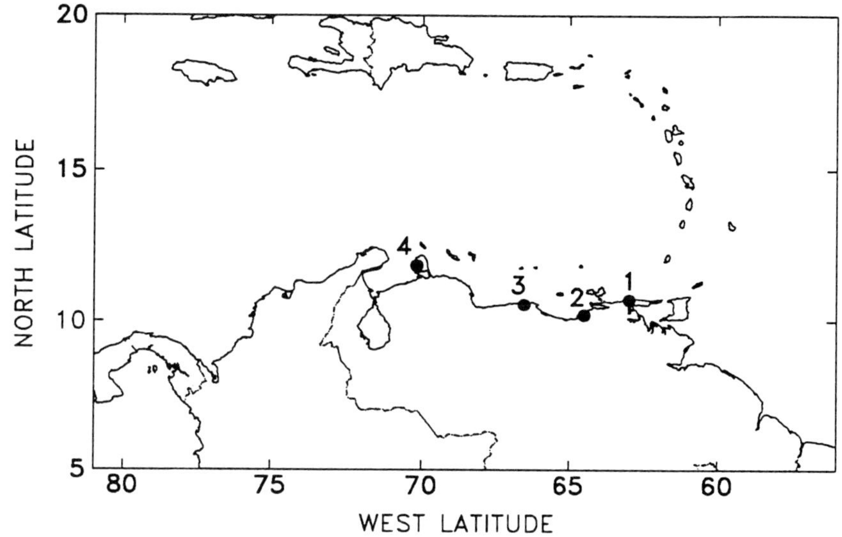

STATION CODE	LOCATION	LATITUDE	LONGITUDE	RECORD LENGTH
1	CARUPANO	10°40' N	63°15' W	1967 – 1986
2	CUMANA	10°28' N	64°12' W	1953 – 1976
3	LA GUAIRA	10°18' N	66°56' W	1953 – 1987
4	AMUAY	11°45' N	70°13' W	1953 – 1979

STATION CODE	n	RSL CHANGE (mm/year)	S.E. (±)	r	CONFIDENCE LEVEL
1	20	−2.46	0.09	−0.47	≈96.50%
2	24	1.96	0.05	0.68	>99.95%
3	34	2.10	0.07	0.69	>99.95%
4	27	0.90	0.08	0.30	≈87.00%

Fig. 6.7 Location of the tide-gauge stations distributed on the eastern and central section of the southern coastal boundary. Table shows RSL changes (mm/yr) from linear-trend estimates, correlation coefficients, standard errors, and confidence levels.

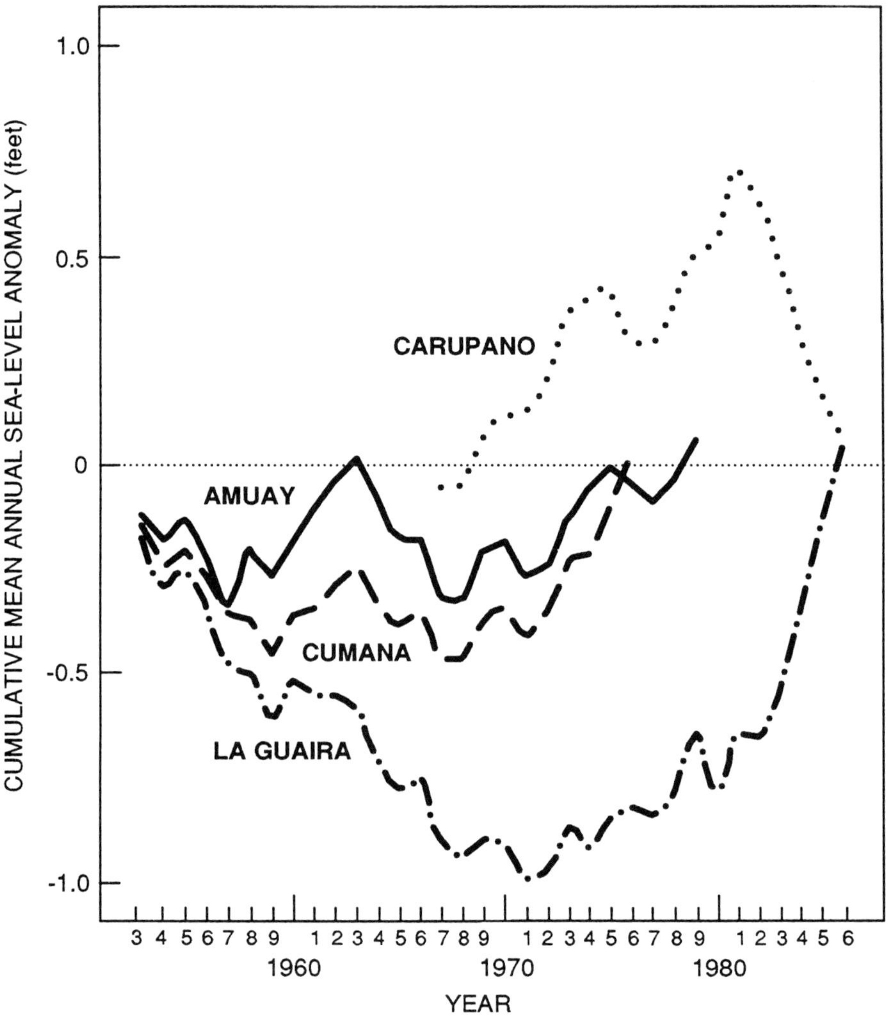

Fig. 6.8 Cumulative mean annual anomalies of sea level (ft) for tide-gauge stations typifying RSL variability along the southeastern coastal boundary of the Caribbean Sea.

during the last five years of that period of time seems to be responsible for that disagreement. It can be seen in Fig. 6.8 where cumulative mean annual anomalies of sea level (ft) are plotted for each tide-gauge station. Proximity of Carupano to the boundary between the Caribbean Tectonic Plate and the South American Tectonic Plate may explain the local RSL change in terms of tectonically induced uplift of the land mass. This location emerges as a very interesting region for attempting a reliable separation of a suspected high 'tectonic signal' from the local sea-level record. New geophysical measurement systems to determine crustal movements, e.g., Very Long Baseline Interferometry, could be used to do that.

4 Summary

1) A conclusive statement emerges from the comparison between information in Table 6.1 and that given in previous publications (CEAS, 1979; Gray, Chapter 5): Meteorological conditions at the surface for the northern section of the Caribbean Sea (at latitudes higher than 16°30′N) differ considerably from those characterizing the most southern coastal boundary. Those differences apply basically to the trends observed in the wind field, precipitation, and evaporation regimes.
2) Existing local records seem to corroborate the 'global' signal of atmospheric warming at a rate of 0.1°C/decade.
3) No regional evidence could be found for a 'global' signal of sea-surface temperature increase.
4) The region seems to exhibit a coherent pattern of RSL increase, at an average rate of 0.17 cm/yr. However, the most eastern tide-gauge station analysed in this study (Carupano; 10°40′N, 63°15′W) shows a falling RSL at a maximum regional rate (−0.25 cm/yr).

5 References

Aparicio, R., 1986: Upwelling along the southern coastal boundary of the Caribbean Sea: Physical characterization, variability and regional implications. Master's thesis. Florida Institute of Technology, Melbourne, FL, 183 pp.

CEAS (Center for Environmental Assessment Services), 1979: A study of the Caribbean Basin Drought/Food Production Problem, Progress Report.

Colon, J., 1963: Seasonal variations in heat flux from the sea surface to the atmosphere over the Caribbean Sea. *J. Geophys. Res.*, **68(5)**, 1421–1430.

Depradine, C.A., Rudder, G. and Lamming, S., 1973: Some characteristics of Hurricanes in the Eastern Caribbean. Final report of The Caribbean Meteorological Institute to the Regional Beach Erosion Control Programme Barbados.

Fajardo, G., 1979: Contribucas as estuds das condicoes oceanograficas das aguas adjacentes a costa Colombiana no mar do Caribe. Dissertacao. Instituto Oceanografico da Universidade de São Paulo.

Febres-Ortega, G.A., 1972: Isanosteric analysis of the eastern Caribbean waters during winter. *Bol. Inst. Oceanogr., Univ. Oriente*, **11(1)**, 19–34.

Febres-Ortega, G.A. and Herrera, L.E., 1976: Circulacion y Transporte de masas de agua en el Caribe suroriental. In: *Symposium on Progress in Marine Research in the Caribbean and Adjacent Regions.* FAO Fisheries Report No. 200 Supplement, Caracas, Venezuela.

Gade, H., 1961: On some oceanographic observations in the southeastern Caribbean Sea and the adjacent Atlantic Ocean with special reference to the influence of the Orinoco River: *Bol. Inst. Oceanogr., Univ. Oriente,* **1(2)**, 287–342.

Gordon, A., 1965: Quantitative Study of the Dynamics of the Caribbean Sea. Ph.D. Dissertation, Columbia Univ., New York, NY.

Griffiths, R.C. and Simpson, J., 1972: Upwelling and other oceanographic features of the coastal waters of northeastern Venezuela. MAC Series *Recursos y Explotacion Pesqueros*, **2(4)**, 1–72.

Herrera, L.E. and Febres-Ortega, G.A., 1981: Principales acciones meteorólogicas y procesos ciclogenéticos que afectan las aguas costa afuera de Venezuela. *Revista Técnica Intevep*, **1(2)**, 119–133.

Kukla, G., Angell, J.K., Korshover, J., Dronia, H., Hoshiai, M., Namias, J., Rodenwald, M., Yamamoto, R., and Iwshima, I., 1977: New data on climatic trends. *Nature*, **270**, 573–580.

Large, W. and Pond, S., 1981: Open Ocean momentum flux measurements in moderate to strong winds. *J. Phys. Oceanogr.*, **11**, 324–336.

Levitus, S., 1982: Climatological Atlas of the World Ocean. NOAA Professional Paper 13.

Ljöen, R. and Herrera, L., 1965: Some oceanographic conditions of the coastal waters of Eastern Venezuela. *Bol. Inst. Oceanog., Univ. Oriente*, **4(1)**, 7–50.

Maul, G.A. and Kavanagh, N.M., 1988: *English-Spanish Bibliography in Physical Oceanography and Ocean Climate for the Caribbean Sea and Adjacent Regions.* Intergovernmental Oceanographic Commission Document IOC/INF-783, Paris, 49 pp.

Metcalf, W., 1976: Caribbean-Atlantic water exchange through the Anegada-Jungfern Passage. *J. Geophys. Res.*, **81(36)**, 6401–6409.

Morrison, J. and Nowlin, W. Jr., 1982: General distribution of water masses within the eastern Caribbean Sea during the winter of 1972 and fall of 1973. *J. Geophys. Res.*, **87(C6)**, 4207–4229.

Muller-Karger, F.E., McClain, C.R., Fisher, T.R., Esaias, W.E. and Varela, R., 1989: Pigment distribution in the Caribbean Sea: Observations from space. *Progress in Oceanogr.*, **23**, 23–69.

Perloth, I., 1971: Distribution of mass in the near surface water of the Caribbean Sea. In: *Symposium on Investigation and Resources of the Caribbean Sea and Adjacent Regions*. UNESCO, Paris.

Pond, S. and Pickard, G., 1978: *Introductory Dynamic Oceanography*. Pergammon Press, London, 241 pp.

Richards, F.A., 1960: Some chemical and hydrographic observations along the north coast of South America. *Deep-Sea Res.*, **7(3)**, 163–182.

7

Sea-Level Movements and Shoreline Changes

Malcolm Hendry[1]

Abstract

Relative change in sea level has provided underlying control on a general picture of shoreline retreat during the Holocene epoch within the Intra-Americas Sea. Retreat has been caused by landward migration of coastal lowlands, including wetlands, in response to rising water levels assisted by sediment transfer through physical processes. Land loss has persisted despite a decrease in rate of sea-level rise over the last 5000 years, though locally at river mouths and in sheltered wetland locations recent progradation has occurred. Inadequate beach-sand management practices, coastal construction and hurricanes have contributed significantly to recent coastal changes.

Regionally relative sea level is rising at an average of 3 mm/yr, but considerable variability is evident. Locally this may not be due to vertical tectonic motion because extrapolated rates of land movement, while subject to error in interpretation, are often an order of magnitude or more lower than historical rates of sea-level rise. The cause(s) of variation in sea-level records across the region remain to be identified.

All countries of this region stand to suffer the consequences of future sea-level rise impacts. Especially at risk are low-lying island chains – including the Bahamas, Turks and Caicos and Caymans – and continental areas with coastal plains supporting dense populations near to or below present sea level, primarily the Guianas, Central America (Belize) and the heavily developed barrier islands of the Gulf of Mexico and Florida. Tourism-based economies of the Antillean islands may be badly affected by loss of remaining pocket beaches on already narrow and sediment-starved shelves.

Potential problems can be addressed through greater understanding of regional sea-level variability, especially by strengthening the Global Sea Level Observing System; territorial studies to define the local extent of risk from sea-level change and cost-benefit analysis of response options identified; development of vulnerability indices for sea level and hurricane-related coastal impacts; environmental-impact assessment on coastal development, incorporating climate-change considerations and institutional strengthening in relation to decision-making on coastal land-loss issues. These improvements are necessary to combat existing problems whether or not a greenhouse-related pulse of increased sea-level rise is superimposed on present trends.

---

[1]University of the West Indies, Cave Hill Campus, Barbados

1 Introduction

As part of the UNEP Regional Task Team contribution on climate change this study reviews coastal changes and the processes responsible for them in the Intra-Americas Sea (see Fig. 1.1; *cf.* Fig. 10.2) in order to identify areas at greatest risk under the scenario of a 20 cm sea-level rise by the year 2025 (UNEP, 1989a). This value is referred to here as UNEP Sea Level Rise (USLR). The assessment recognizes new programme directions and weaknesses in existing programmes that need strengthening if the problems of sea-level rise are to be adequately addressed.

In this chapter the term 'eustatic' refers to changes in ocean water level, whether by deglaciation, thermal expansion or other factors (Mercado *et al.*, Chapter 4). Tectonic changes are those solely due to land movements regardless of any sea-surface changes that may have taken place during the interval under consideration. Relative changes in sea level include any change in elevation of the sea surface whether caused by movement of land relative to the sea or changes in sea surface relative to the land.

2 Theoretical Basis for Coastal Change Related to Sea-Level Rise

Acting singly or in combination, changes in sea level, sediment supply, wave climate and nearshore currents, storm activity and anthropogenic intervention may cause changes in shoreline position or land loss (Pilkey *et al.*, 1989). With respect to sea-level rise alone, elevation of mean sea level will subject higher levels of the shoreface to wave and current action. In theory the shore profile adjusts to this through displacement of shoreface material in the offshore area, resulting in net translation of the sediment/water interface in a landward direction. This process is referred to as the Bruun Rule (Bruun, 1962) and is demonstrated graphically in Fig. 7.1. Note that the original Bruun Rule is concerned only with the impacts of sea-level rise, all other variables being considered stable. Galvin (1983) noted that the relative importance of this factor diminishes in situations where there is a strong alongshore energy flux.

In addition to physical displacement, land loss also can be caused by submergence. In theory shoreline displacement due to submergence can be calculated relatively simply and is dependant on the slope of the coastal area, although caution must be applied where topographic breaks (like barrier islands or scarps) occur on otherwise low-gradient plains (Galvin, 1983). Large portions of most territories under consideration in this study are below, close to or within a metre or two of present sea level and likely to be affected by relatively small changes in this level whether due to submergence or sediment transfer.

3 Relative Sea-Level Change in the Intra-Americas Sea

An attempt to separate signals that contribute to relative sea-level change is performed here by analysing Holocene sea-level records, regional tectonic movements and historical data on trends of sea-surface fluctuations as recorded by tide gauges. Emphasis is placed on the component of change

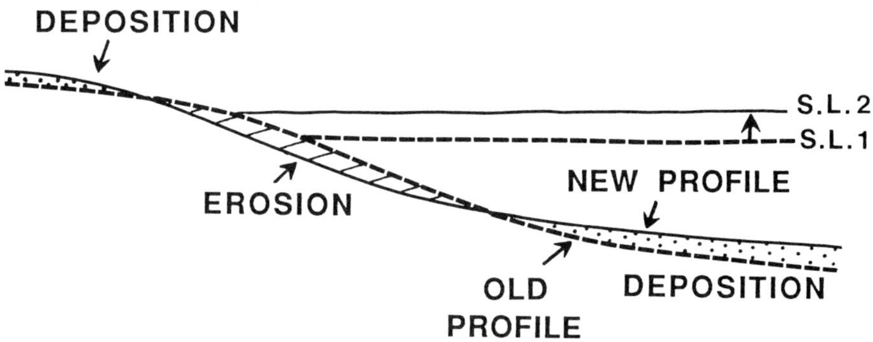

Fig. 7.1 The effect of sea-level rise on sandy substrates. Changes in other parameters, for example, sediment budget, are not considered. Elevation of sea level by an increment creates a 'gap' at the sediment surface, which represents disturbance in the profile of 'dynamic equilibrium' that had been established under previous sea-level conditions. The gap is filled by removal of foreshore sediment to offshore, resulting in shoreface retreat and landward advance of the water line. Overtopping of the beach during storms transfers sediment landwards, contributing to the process of shore migration.

due to tectonic factors, as this element is often invoked to account for variability in regional sea-level records. Variations in relative sea-level movements that may be caused by tectonic overprint are explained in Fig. 7.2.

3.1 Holocene Sea-Level Changes

A longer-term picture of sea-surface changes is available from the construction of sea-level curves for the Holocene epoch (the last 10,000 years; Fig. 7.3). Authors of these curves have generally discounted tectonic overprint or adjusted for it. If tectonic factors can be discounted in interpreting these curves then other reasons must be invoked for differences. Digerfeldt and Hendry (1987) speculated that geoidal deformation might explain regional variability, but also noted that many more such curves were required before the details of deformation could be properly defined or explained.

Variation between these curves supports the suggestion by Digerfeldt and Hendry (1987) that extra-regional curves should not be used for interpretation of local shoreline history. Moreover, intra-regional variability is great enough to merit caution in applying these curves to other territories within the Intra-Americas Sea. Evaluating causes of differences between these curves will be important in further work as they indicate that future sea-surface changes across the region will not be necessarily uniform, an added problem for prediction and modelling. In this context we must bear in mind that several international projects including IGCP (International Geological Correlation Program) Projects 61 and 270 have already determined that no single 'eustatic' curve can be regarded as representative of regional or global sea-surface changes (Pirazzoli and Grant, 1987).

The general signal emerging for Holocene curves from the insular Caribbean Sea is of a relatively rapid rise in sea level to about 5000 years before present (ybp) at rates up to *c.* 10 mm/yr, with a substantial slackening in the rate of change to between 1–3 mm/yr in historical times (Fig. 7.3). The most marked deviation from this general Holocene picture is found on the Atlantic borderland of northern South America. The curve for Surinam indicates that present sea level was reached at 5000 ybp, remaining relatively stable till the present, though there have been suggestions that littoral sediments of the latest Holocene record a higher-than-present phase (+2 or +3 m) (Prost, 1990). Others argue strongly against this suggestion (e.g., Roeleveld and Van Loon, 1979). Farther south, in Brazil,

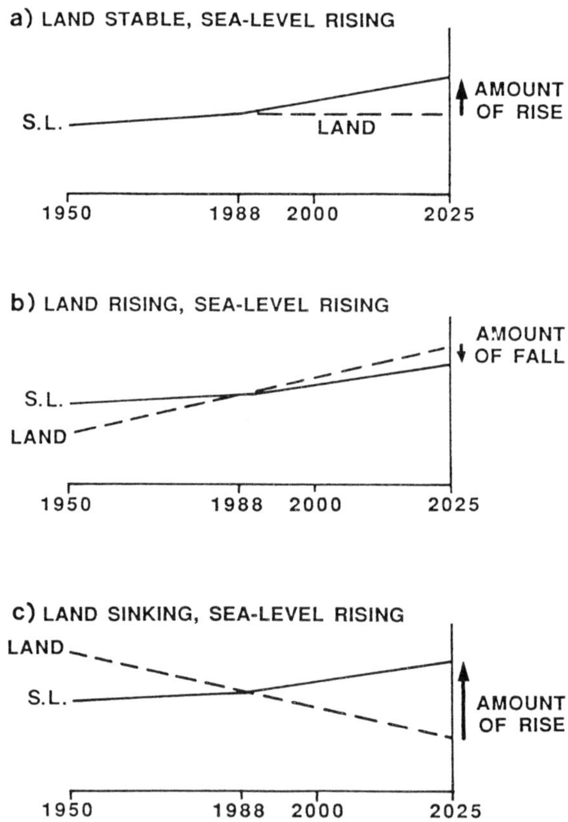

Fig. 7.2 Examples of variability in local sea-level rise. The assumptions are that sea level (S.L.) will rise at a faster rate in the future than in historical times and that land movement (positive or negative) is continuous. The second assumption may not hold everywhere because of the episodic fashion in which land displacement generally takes place, as in the Costa Rican earthquake of 1991. This type of effect is more important for locations at the tectonically active boundary of the Caribbean Plate (Fig. 7.4) and where fluid withdrawal has caused land subsidence as in the Mississippi Delta and Lake Maracaibo (c). In these cases the relative rise in sea level is much greater than would be expected for a tectonically stable setting (a). Vertical axis not drawn to scale.

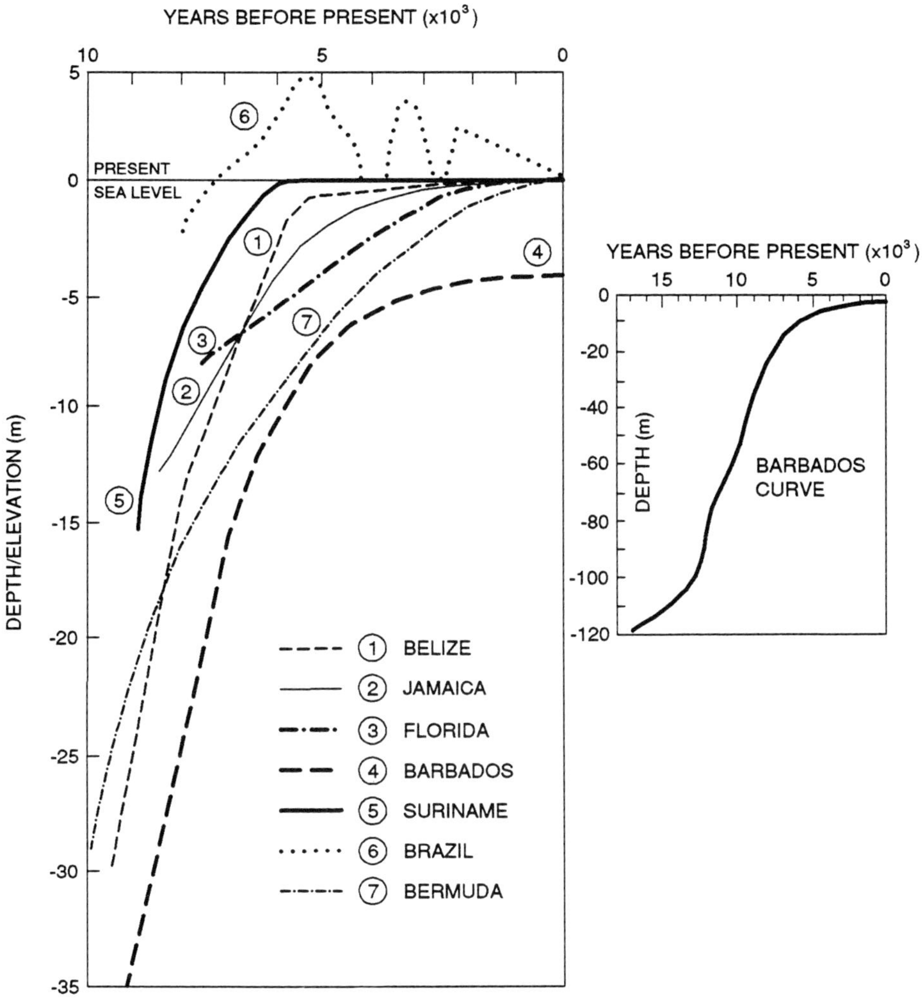

Fig. 7.3 Local sea-level curves for the Intra-Americas Sea. 1 = Belize (Westphal, 1986); 2 = Jamaica (Digerfeldt and Hendry, 1987); 3 = Florida (Shinn *et al.*, 1989); 4 = Barbados (Fairbanks, 1989); 5 = Surinam (Prost, 1990). A curve for Brazil (6; Suguio *et al.*, 1988) is also included for comparison as is the Bermuda curve (7; Neumann, 1971), which is frequently used to assist interpretation of coastal history in the rest of the region. The overall signal from the insular Caribbean Sea is of a relatively rapid rise to about 5000 ybp, followed by a slower rate. The Barbados curve is based on radiometrically dated corals, which are not as good for definition of former sea-level position as some peat deposits, which are used in most other curves. Also, the upper part of the Barbados curve is not accurately constrained by local data because in the upper 15 m it incorporates data points from other island locations. The full de-glacial sea-level record, as defined for Barbados, is shown in the smaller diagram.

there are well-established departures from the record where Holocene sea level has risen above the present at least three times in the last 6000 years (Suguio *et al.*, 1988).

These differences suggest that sea-level forcing factors also vary between the insular Caribbean Sea and the open-ocean margins of South America. Variation in open-ocean dynamic circulation, perhaps related to changes in Gulf Stream activity, is one possible cause for these (Morner, 1990) in addition to changes in the geoid along the Atlantic coast of South America over time-scales of centuries (Martin *et al.*, 1985). Interpretation of coastal and sea-level history and predictions for the future must bear in mind these differences between the Guianas and the rest of the region.

A central question to all concerns of possible greenhouse-induced sea-level changes and their impact on shorelines is whether the apparent peaking of sea-level curves represents the zenith of an interglacial event and whether under normal conditions we might anticipate a decline in sea level as climate shifts back to a colder phase. Greenhouse projections, including the one adopted by the UNEP Task Team, do not allow for peaking followed by decline: they project sharply increased rates of rise over those of recent historical times as the greenhouse signal makes its presence felt (see Fig. 1.7).

3.2 Historical Sea-Level Records

Three major recent reviews of historical sea-surface changes from tide-gauge records have been undertaken for the Caribbean Sea (Pirazzoli, 1986; Aubrey *et al.*, 1988; Hanson and Maul, Chapter 9). In addition, Maul and Hanson (1990) studied data points from US stations on the Gulf of Mexico and east Florida. Aubrey *et al.* (1988) targeted a different area for review: major areas of overlap within the UNEP study region are in islands of the Caribbean Sea and also data points from Pacific and Caribbean Central America (including Mexico) and northern South America. Pirazzoli (1986) incorporated a number of the same sites in a global review, although the analysis was less exhaustive than in the other two articles. Both Aubrey *et al.* (1988) and Hanson and Maul (Chapter 9) advise caution because tide-gauge records for the region in many cases are very short, making extrapolation difficult.

Results emerging from each analysis are similar: most local trends are positive, though some are negative (see, for example, Table 9.2 and Fig. 6.7). Due to sparsity of tide stations, Aubrey *et al.* (1988) were unable to account for changes that were systematic with latitude or region on the basis of oceanographic and atmospheric causes. Nor was there any apparently uniform link between ENSO events and sea-level responses through the Caribbean Sea in the absence of direct oceanographic teleconnection between the region and the Pacific. At Key West, however (a site not included by Aubrey *et al.*, 1988), a statistically significant relationship is found between lower sea levels in the year prior to, and higher sea levels during the year of ENSO events (Hanson and Maul, Chapter 9). Disparity between these interpretations is striking and clearly requires further work.

In the light of such variability, provision of average figures for regional change may appear spurious, but at least they give an idea of the overall picture. Hanson and Maul (Chapter 9) use records since 1930 of longer than 25 years to calculate an average relative sea-level rise of 3.6 ± 2.5 mm/yr, a value that includes some rapidly subsiding regions of the northern Gulf of Mexico. This figure is similar to the Pirazzoli (1986) calculation of +3.0 mm/yr, but the data in this case were not selected on the basis of lengths of record. Aubrey and Emery (1988) deselected records with low confidence levels and calculated a value +2.37 mm/yr. All these calculations for the region are higher than the world average of +1.2 mm/yr (Gornitz and Lebedeff, 1987). The USLR scenario provides for an increase, regionally, to *c.* 5.3 mm/yr over the regional historical rate of *c.* 3.0 mm/yr.

The general diversity of values suggests that in the latest Holocene, when the overall rate of relative sea-level rise decreased (Fig. 7.3), the importance of regional and local forcing factors increased in relation to sea-level changes caused by global glacio-eustatic forcing. The main problem is to identify the source of these signals. As a first attempt at this, we can look at the published information on tectonic changes for the region.

3.3 Regional Tectonic Factors

A substantial body of literature exists on the structure and tectonics of the Caribbean Sea and adjacent areas (see Mann *et al.*, 1984; and Pindell and Barrett, 1990, for general geological and tectonic overviews). Of interest here are studies that provide evidence of uplift or subsidence and the rates at which they proceed in coastal and shelf areas.

The general geological framework for the region (Fig. 7.4) is dominated by the Caribbean Plate, bounded to the east and west by zones of active convergence and to the north and south by extensive strike-slip fault systems. Active underthrusting associated with plate subduction occurs in Central America, where the Cocos Plate underrides the Caribbean Plate, and in the eastern Caribbean where the North American Plate is subducted beneath islands of the Lesser Antilles.

The boundary of the Caribbean Plate with the South American Plate is less clearly defined because in many places direct evidence of contact has been buried beneath thick sediment piles derived from the South American craton. However, Mann *et al.* (1990) record that Neogene strike-slip motion in the northern and southern plate boundary zones of the Caribbean Plate has been accompanied by extensive vertical components of movement. Vertical changes must not be interpreted as simple epeirogenic motion: the structural style that accompanies this motion is highly complex (Mann *et al.*, 1990).

Uplift and subsidence rates quoted below are based on published data from emergent reefs, elevated bathyal sediments, age-depth-facies relationships in cores and geodetic relevelling, amongst other indicators. Except where stated the data refer to land movements.

The basis for grouping this information takes account not only of broad geological provinces (Fig. 7.4) but also aspects of sedimentary processes including sediment supply, allowing for discussion of these processes following this section.

1) ***Northern South America***

   The Guianas (Guyana, Surinam, French Guiana): Overall subsidence appears to characterize the borderland region of the continental

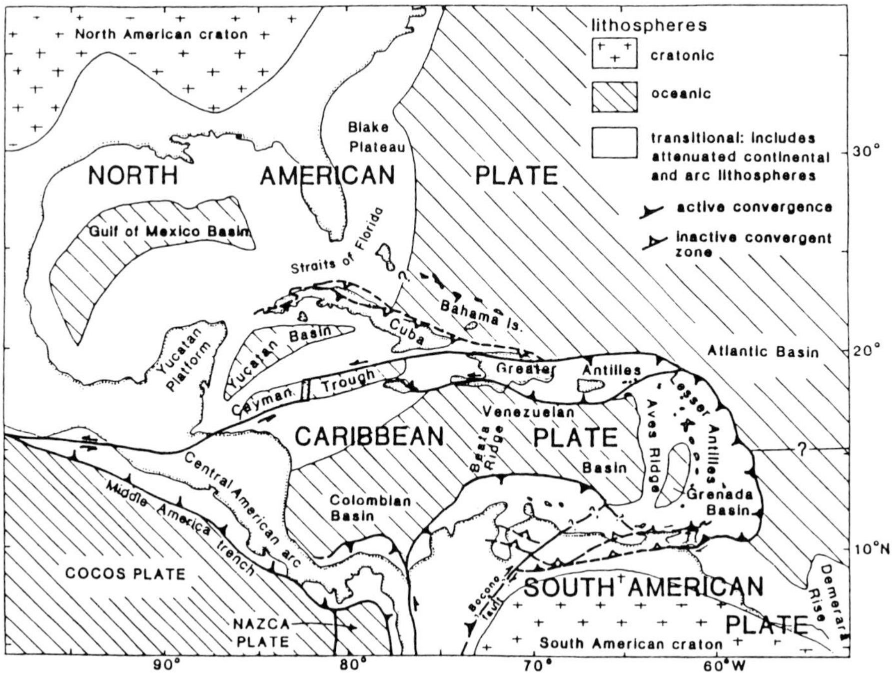

Fig. 7.4 Geological structure of the region. From Speed (1987). Cratonic areas are ancient parts of the North and South American continents, bordered by younger 'transitional' crust that separates them from oceanic crust. Lithosphere is essentially the outer surface or crust of the earth. Relative motion along plate or fault boundaries is indicated by half-arrows. Vertical tectonic motion is most pronounced at the boundaries of the Caribbean Tectonic Plate. Coastal systems in cratonic areas are characterized by large volumes of sediment input through major rivers, as in the Gulf of Mexico (e.g., Mississippi) or the Guianas (Amazon, Orinoco and Magdalena). The Greater Antillean islands are able to support some large river systems though shelves are narrow and sediment easily lost to deep water. Lesser Antillean islands, many still volcanically active, generally have limited sediment budgets in coastal areas due to small, seasonal river input, narrow shelves and compartmented shorelines. Carbonate supply is particularly abundant on bank areas such as the Bahamas and on sections of the Central American and Antillean island shelves. (1 = Mississippi Delta; 2 = Magdalena Delta; 3 = Lake Maracaibo; 4 = Orinoco Delta; refer to Fig. 10.2 for broader geography of territories bordering the Intra-Americas Sea.)

shield, at a rate between 0.15–0.20 mm/yr (Kumar *et al.*, [1977] in Prost, 1990).
Colombia: There is reported Holocene uplift of 2–15 mm/yr between Cedro and Covenas while between Covenas and Berrugas there is 0.7 mm/yr subsidence (Page, 1983). This coastline appears to have a complex deformation history, with both mud diapir intrusion and tectonic displacement.
Curaçao: Schubert and Szabo (1978) estimated uplift of 0.03 mm/yr since the last interglacial (120 ka ybp) based on elevated reef displacement.
Venezuela: Between 1925 and 1972 there was subsidence of 5 m (87 mm/yr) on the eastern side of Lake Maracaibo due to oil extraction (Henneburg, 1983). On the north central coast, beachrock has been uplifted at between 0.03–0.6 mm/yr over the last 2–3 ka (Schubert *et al.*, 1977). In the Venezuelan islands movement is also variable. Aves Ridge is subsiding between 0.038 and 0.06 mm/yr (Schubert and Laredo, 1984); La Blanquilla is rising at between 0.003 and 0.03 mm/yr since 130 ka ybp (Schubert and Laredo, 1984), and La Orchillo has subsided at a rate between 0.02 and 0.04 mm/yr over the same time interval (Schubert and Valastro, 1976).

2) ***Central America/Costa Rica***
Hare and Gardner (1985) report differential uplift for the Nicoya Peninsula on the Pacific coast, while Lundberg (1982) proposed general uplift at a rate of +0.05 mm/yr since the early Pliocene. Along the mouth of the Rio Terrabe, Alt *et al.* (1980) suggested a rapid uplift rate of 25 mm/yr, though this is based on a single carbon-14 date. In the Pacific Osa Peninsula, Corrigan *et al.* (1990) report an uplift rate of 0.2 mm/yr. For the Pacific Burica Peninsula in Panama, Corrigan *et al.* (1990) calculated an uplift rate of +0.94 mm/yr through the Quaternary period. These rates of uplift are amongst the highest for the region under review.

3) ***Gulf of Mexico/Florida***
There are very limited tectonic data for the Gulf states of the USA and the Gulf coast of Mexico (Amos Salvador, pers. comm.). Primary controls on subsidence are the position of depocentres from river outlets draining the continental interior that have caused sediment loading and crustal downwarping. These depocentres have changed position on several occasions, the present focus being at the mouth of the Mississippi River though even here there have been five changes in the position of delta lobes during the last 7500 years (Coleman, 1988). Thus, the rate of subsidence has varied through space and time.

Maximum rates of relative sea-level rise are presently found in the Eugene Island coastal area (+11.9 mm/yr), decreasing west to Galveston (+6.3 mm/yr) and Port Isabel (+3.1 mm/yr) on the border with Mexico. Relative rates of rise decrease rapidly east of the Mississippi to +2.4 mm/yr at Pensacola (Hanson and Maul, Table 9.2). Rates of rise on the west Florida coast are consistently close to 2.0 mm/yr (Penland and Ramsay, 1990). Within the Louisiana coastal zone the main factor driving relative sea-level rise is compaction of

Holocene deltaic sediments (Penland and Ramsay, 1990). The west Florida platform appears to have remained relatively stable through the late Pleistocene and Holocene (Fairbridge, 1974; Evans *et al.*, 1989) though there is some evidence, reviewed by Shinn *et al.*, (1989) and Lidz and Shinn (1991), for a gradual westerly subsidence of southwest Florida relative to the east. Fairbridge (1974) suggested a subsidence rate of *c.* 0.6 mm/yr for this area.

4) ***Antillean Islands***

Barbados: Bender *et al.* (1979) estimate uplift of Pleistocene reef tracts at a rate of between 0.2–0.45 mm/yr. Barbados is unique in the Antilles, being located on top of the accretionary prism of the Lesser Antilles subduction zone. Relatively rapid uplift is due primarily to intrusion of mud diapirs from depths of over 1000 m between the Miocene and the present (Speed, 1988).

Barbuda: Brasier and Mather (1975) suggest the island was relatively stable during the Neogene, but no data are available on younger time intervals.

Cayman Islands: Stable since 120 ka ybp (Woodroffe *et al.*, 1983).

Cuba: Southeast, next to the Cayman trough, geodetic relevelling has generated displacement values between ±2.8 mm/yr (Liliyenburg *et al.*, quoted in Mann *et al.*, 1990). Horsfield (1975) described uplift in the southeast as part of a domal subregional pattern consistent with uplift characteristics in neighbouring northwest Haiti (Dodge *et al.*, 1983). On this basis, uplift of the order of 0.2–0.3 mm/yr may be representative over longer time intervals.

Dominican Republic: Mann *et al.* (1984) and Taylor *et al.* (1985) report displacement of mid-Holocene reef corals in the Enriquillo Valley up to 2 m above their assumed level of formation in a valley bounded by thrust faults. This gives a displacement of +0.4 mm/yr (Holocene, uncertain). Reefs of probable last-interglacial-age adjacent to the valley have not been deformed (Schubert and Cowart, 1980).

Guadeloupe (Grande-Terre): Garrabe and Andreieff (1985) report progressive uplift with an overprint from faulting, and Bonneton (1990) suggests uplift rates of between 0.03 and 0.07 mm/yr since 120,000 ybp.

Haiti: Coral reef terraces in the northwest have been uplifted at an average of 0.3 mm/yr over the last 130 ka (Dodge *et al.*, 1983). This figure is amongst the highest uplift rates for the Greater Antilles (Horsfield, 1975).

Jamaica: Block faulting has caused differential displacement of Pleistocene terraces around the island. At several locations on the north coast for example uplift and subsidence rates of +0.14 mm/yr, +0.05 mm/yr and −0.03 mm/yr have been calculated (Horsfield, 1972; Cant, 1973). In the southwest Hendry and Head (1985) report relative stability since 120 ka, while Horsfield (1974) suggested relative southwards tilting in the south-central part of the island. On the west coast Hendry (1987) calculated subsidence of a fault block at a rate of 0.2 mm/yr through the late Pleistocene. Cant (1971) estimated uplift of an adjacent block at +0.05 mm/yr. In the east the Blue Mountain Block has risen at a rate of up to 0.22 mm/yr since the late Middle Miocene (Horsfield, 1973).

Marie-Galante: The island emerged during the Pleistocene and has been tilted southwest and extensively faulted. Quaternary tectonic history of the island appears similar to Grande-Terre (Garrabe and Andreieff, 1985).

Martinique: Based on the disposition of various terrace levels, the island, together with southern and eastern shelves, is subject to significant tectonic movement, but unfortunately no quantitative information is provided (Froidefond *et al.*, 1985).

St. Eustatius: Late Pleistocene (*c.* 32 ka) limestones display steep dips to the south, due to expansion and uplift on the flank of the Quill volcano. Rates of movement are unknown (Maury *et al.*, 1990).

St. Kitts: Pleistocene limestones have been uplifted on the western flank of Mt. Misery, a stratovolcano (Maury *et al.*, 1990), but again amounts or rates of displacement are not given.

Tobago: Variable movements in the southwest are separated from the remainder of the island by faults. One Pleistocene reef location shows 8 m of uplift over the last 125,000 years (a rate of +0.06 mm/yr [Wadge and Hudson, 1985]).

Virgin Islands: Holmes and Kindinger (1985) provide evidence for recent tectonism of the central Virgin Islands shelf. Vertical movements are differential, but no values of relative displacement are given. This evidence is supported by Lewis and Draper (1990) who indicate that the interrelationships of these movements are not yet understood.

5) ***Platform Islands***
Bahamas Platform: Subsidence of 0.03 mm/yr over several million years (Goodell and Garman 1969).

3.4 The Importance of Tectonics in Interpreting Regional Relative Sea-Level Movements

Comparison between these average tectonic displacement rates and historical sea-level change rates (Chapter 9) indicates that recent sea-surface movements are generally one or two orders of magnitude greater than the tectonic displacement values, with a few exceptions, mainly in Central America, Lake Maracaibo and Louisiana (due to petroleum extraction) and along the Colombian coast (due to mud diapirism). At Port au Prince, Haiti, the high rate of relative sea-level rise may reflect a high rate of tectonic subsidence. On first analysis therefore the tectonic component in the bulk of historical sea-level records appears limited. However, two factors need to be born in mind:

1) The assumption of linear rates of change in land level associated with average tectonic displacement figures.
2) The lack of site-specific correspondence between location of tide gauges and more general measurements of tectonic displacement. It is probably necessary to provide greater geodetic control of land movements close to tide gauges, perhaps through relevelling.

3.5 Detection and Trend of Future Sea-Level Changes

A recent analysis of the tide-gauge data by Maul (1990) suggests that as yet there is no indication of a greenhouse-induced sea-level signal in the

Intra-Americas Sea. Should there be such an increase it may in some instances reverse the trend, where this is currently negative, to one of submergence. Where presently positive, the rate of rise will accelerate (Fig. 7.2). If we know how much of the historical signal is due to various causative processes, it should be possible to model the anticipated change at each site by superimposing a 20 cm rise on the data.

In the case of tectonic deformation, for example, if all the negative change of −3.97 mm/yr at Santo Tomas de Castilla in Guatemala (Hanson and Maul, Table 9.2) were due to tectonic uplift, then the expected net change between 1990 and 2025 (35 years) would be (−3.97 x 35) + 20 cm = +6.1 cm. This would reverse the current trend, but the total change would be relatively insignificant when compared to other sites. Using the same assumption for the Port au Prince, Haiti data, if the present rate of relative sea-level rise of +12.36 mm/yr were solely due to tectonic factors (for example, land subsidence) the expected change by 2025 would be a relative rise of +63.3 cm. Because much of downtown Port au Prince is presently very close to sea level such a change could have substantial impacts.

These kinds of calculations are necessary for modelling of future potential impacts at site-specific scales, especially in those areas where variability is demonstrably due to tectonics or fluid withdrawal. As the preceding analysis indicates, however, the number of sites where these factors are likely to be important over the next 30–40 years will be relatively few except in locations where rapid vertical motion accompanies tectonic movements, as in the 1991 Costa Rican earthquake. How significant are the other, less well understood factors that have influenced historical sea-level variability and created more systematic regional differences in the Intra-Americas Sea (Fig. 7.3), and what will their impact be on the 'global' USLR value of 5.3 mm/yr rise to 2025? Will their importance be 'drowned' by the predicted eustatic pulse? There is clearly a substantial amount of work still to be done on interpretation of factors influencing sea-surface movements.

4 Regional Patterns of Shoreline Change

Quantitative information on shoreline changes for substantial sections of the Intra-Americas Sea is hard to find. Only in the last few years have systematic efforts at shoreline monitoring started, most notably in the eastern Caribbean Sea, Cuba and the Guianas. The rest of the region appears to lag behind the USA in measurement and interpretation of shoreline fluctuations in heavily populated and economically critical coastal areas. The following section summarizes the literature obtained for this study.

1) ***Northern South America***
   French Guiana: Sedimentation is strongly influenced by Amazon River input. Prost (1990) and Prost *et al*. (1990) demonstrated migration of the Kourou-Sinnamary mud bank 12 km to the northwest between 1979 and 1989. As a result, the coast adjacent to the town of Kourou eastward to Cayenne changed from accretional to erosional as the tail-end of the

mud bank passed. Over the same time interval, the front of the bank passed from Plage Degonde to almost the outlet of the Sinnamary River, causing shore progradation of between 1–2 km (Figs. 7.5 and 7.6)

Surinam: The shoreline has been generally progradational within the later Holocene, punctuated by occasional phases of erosion or non-deposition (Prost, 1990). These young coastal plain deposits expanded significantly at *c.* 6500 ybp, at the end of the early Holocene sea-level rise in this area (Roeleveld and Van Loon, 1979; Wong, 1990; Fig. 7.3). At this time supplies of sediment from the Amazon and some smaller rivers especially those in French Guiana were restored to the newly submerged shelf (Krook, 1990a,b). Inland penetration of marine-influenced vegetation also appears to have reached a maximum at 6500 ybp as recorded in the presence of *Rhizophora sp.* peat upstream in riverine deposits of both Surinam and Guyana (Roeleveld and Van Loon, 1979).

Westerly migration of submerged mud banks up to 5 m high, 50–60 km long, and 10–20 km wide along the continental shelf under the influence of waves and currents has been documented in detail for the Surinam coast by Rine and Ginsburg (1985; Fig. 7.5). Mud-bank spacing is of the order of 30–60 km, with longshore migration averaging 1.5 km/yr. The net effect of this movement is shoreline advance: progradation occurs where mud banks ('sling mud') are attached to the coastline, while interbank areas are characterized by erosion and reworking of coarser sediments into fine sand beach ridges, which are ultimately incorporated in the chenier plain. Wells and Coleman (1978) estimate that 100×10^6 m^3/yr of sediment are transported by these mud banks along the coast of the Guianas, with over twice that volume being transported annually in suspension.

Augustinus *et al.* (1989), through zero-baseline referenced time-series analysis from 1947–1981, demonstrated migration of mud banks westwards between the Surinam and the Coppename rivers. For example, 15 km east of the Coppename, the shoreline eroded up to 200 m between 1947 and 1957, then prograded by over 300 m between approximately 1957 and 1981 as the mud bank moved alongshore. These predictable forcing mechanisms and coastal responses will assist modelling of longer-term predictions for these coastlines.

Guyana: Earle (1987) reports cyclical erosion and accretion along the coastline and attributes this to the alongshore passage of mud and sand waves northwest from the primary source area, the Amazon River. This suggestion is supported by empirical studies of shelf-sediment transport along coastlines of the Guianas (Prost, 1990). Earle (1987) indicates this process has had serious impacts on the stability of the country's sea-defence works. Much of the infrastructure and population is at or below sea level and requires dykes, dams and drainage canals as protection against inundation (Jones, 1986).

Venezuela: In the northeast, including the northern Gulf of Paria, numerous archaeological sites dated between 6000 and 2300 ybp are many tens of kilometres from the present coastline. These are thought to identify former estuarine shorelines, dominated by mangroves (Sanoja, 1989). Aparicio-Castro *et al.* (1989) summarized

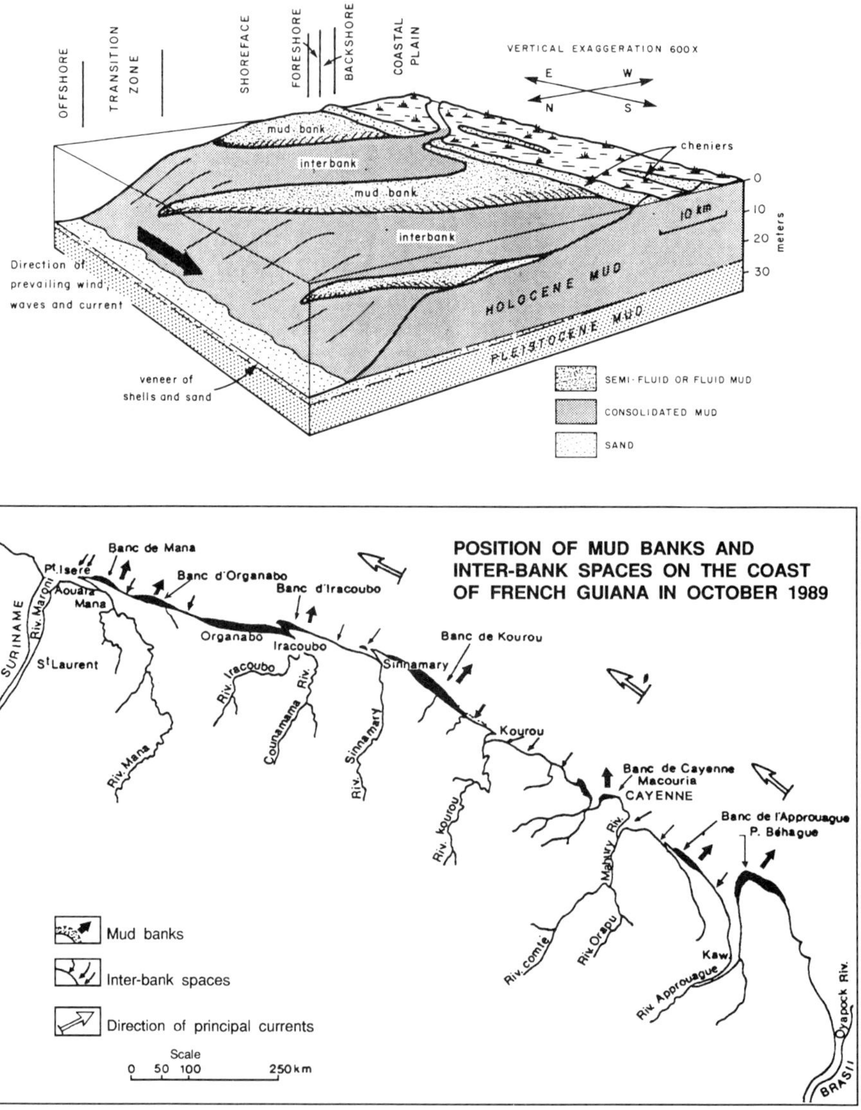

Fig. 7.5 Coastal sedimentary processes in the Guianas. Dominated by large volumes of fine sediment discharge, primarily from the Amazon River, phases of coastal erosion and accretion are controlled by juxtaposition of mud waves during their migration alongshore. Mud-bank accretion (see Fig. 7.6) allows for shoreline advance and colonization by wetland vegetation. When interbank areas pass, the coast erosion from waves and currents winnows out finer material leaving fine sand cheniers. This cumulative process results in net long-term shoreline advance. Top diagram modelled on Surinam is from Rine and Ginsgurg (1985). The lower diagram from Prost (1990) shows conditions on the coast of French Guiana, the location of bank and interbank areas clearly defined as zones of short-term accretion and erosion, respectively. Sea level in the Guianas may have reached the present stage up to 6000 years ago.

Fig.7.6 Mud bank on the coastline of French Guiana. The gently sloping mud surface extends as far as the eye can see, its landward margin terminating against a narrow sand beach backed by wetland. The mud bank dampens wave energy across the shelf. In other locations the landward edge of the mud bank borders directly against the wetland system (*cf.* Fig. 7.5). See colour plates between pages 210 and 211. Photograph by Malcolm Hendry.

areas of coastal land loss and gain. Progradation was noted mainly in those areas with river input (Orinoco, Mitare Delta) or in relatively sheltered, swampy locations (Gulf of Paria, southern shore of Lake Maracaibo). Seven other sites display active erosion: northern shore, Araya Peninsula; Unare Bay; Tucacas; east shore, Medanus Isthmus; east coast, Paraguana Peninsula; south side, Guajira Peninsula; and western coast, Gulf of Venezuela. Graf (1972) suggested that erosion might have been associated with tectonic motion in several areas, including the north-central coastal peninsulas.

Aruba: Phases of erosion and accretion on Aruba are linked to seasonal swell wave events generated by distant storms and hurricanes, which move sediment eastward from the only locations of substantial beach build-up on the west-facing coast. The dominant wind-wave conditions generated by Trade Winds return sediment lost in this way to the west, and overall there appears to be net transport in that direction. Some engineering stability measures have been proposed to protect beaches and adjacent properties, especially hotels, from this transient sediment movement (Kohsiek *et al.*, 1987). The summer/winter swell-dominated process described here is common throughout the Antillean islands.

Colombia: Available evidence suggests a generally dynamic coastline. For example, at Barranquilla, Cartagena and the Gulfs of Morrosquillo and Urabe, erosion and accretion of as much as ±20 m/yr have been recorded (Vernette *et al.*, 1984; Correa, 1990).

2) ***Central America***

Belize: Repeated flooding is a major problem, related in many instances to surge associated with hurricane passage. According to Gentry (1971, reported in Kjerve and Dinnel, 1983) the Belize coastline is subject to a tropical storm or hurricane on average 112 times in every 100 years. These storms have had major impacts on the geomorphology and vegetation of the cays along the Belize Barrier Reef, the second largest barrier reef in the world (e.g., Stoddart, 1963, 1969, 1974) and are a cause of great concern in areas of population concentration such as Belize City (Jones, 1986).

3) ***Gulf of Mexico/Florida***

Mexico Padre Island (north Mexico), a microtidal barrier system with low population density, shows recent evidence of susceptibility to storm damage (Schwartz and Anderson, 1986), and is contiguous with a section of the Texas barrier island system that is presently eroding (Dolan *et al.*, 1988). In general the whole coast is recessive, except at river mouths and locally at shoreline structures (Bird, 1985).

USA – Gulf Coast and Florida: Compilations of the *Living with the Shore* series (Morton *et al.*, 1983; Doyle *et al.*, 1984; Kelley *et al.*, 1984; Pilkey *et al.*, 1984; Canis *et al.*, 1985) and the *National Atlas of Coastal Erosion and Accretion* (Dolan *et al.*, 1988) provide a comprehensive data base for these coastlines. The Atlas compiles data for time-series that in some instances extend back over 100 years, though 50 years is about the average, with careful filtering based on assessment of the accuracy and reliability of information. Within the Gulf of Mexico, some 63% of shorelines are recorded as eroding, the average erosion rate being 1.8 m/yr (Dolan *et al.*, 1988). This figure masks the variability in data sets, however. For example, the coast of Louisiana is receding at the rate of 4.2 m/yr, for an overall land loss of approximately 124 km²/yr (EPA, 1987). The EPA (1987) considered this rapid change was primarily due to subsidence in the deltaic regions of the Mississippi River as a result of compaction, fluid withdrawal (oil and gas exploration) and canalization that directs inorganic sediments away from the wetlands and into the sea, a viewpoint supported by Penland and Ramsey (1990). In this area are some of the best examples of impacts from a high rate of relative sea-level rise combined with anthropogenic influences on lowland shorelines. Experience here will help to guide planning for other sites bordering the Intra-Americas Sea in the event of rapid future greenhouse-induced sea-level change. The barrier islands of the Texas coastline are also nearly all retreating, at rates between 1.0 and 2.9 m/yr (Dolan *et al.*, 1988).

On the Florida west coast the picture is generally erosional, with some segments showing progradational trends. Maximum erosion rates occur just south of Tampa Bay where recession between 3.0–4.9 m/yr is recorded. Other sections, north and south of this area, are retreating at less than 1 m/yr. Prograding barrier islands in northern Pinellas County are accounted for by late Holocene decrease in rate of sea-level rise coupled with increased sediment supply as the barrier islands have migrated inland over a relict sediment source (Evans *et al.*, 1985). The Dolan *et al.* (1988) map shows no data for the extreme southwestern coast of Florida, an area known as 'The Ten Thousand Islands'. Parkinson (1989) suggests that progradation has occurred there in the latest Holocene. In this case the change was accounted for by decreased late-Holocene sea-level rise, which in this lower energy environment led to wetland expansion.

No recent data exist for the Florida Keys and Florida Bay area, up to Miami. Lidz and Shinn (1990) reconstructed the palaeogeography of this area for four time windows (8,6,4 and 2 ka ybp) to demonstrate the sequence of platform flooding, shoreward reef displacement and progressive drowning and separation of the Florida Keys. Lidz and Shinn (1990) projected that if the present 3.8 mm/yr rate of rise was sustained, most of the Keys would flood within 260–520 years, though an increase in this rate would significantly shorten the time.

The east Florida coast shows short sections of accretion (less than 1 m/yr) between long segments of coast that are retreating at less than 1 m/yr. Three short (<20 km) segments south of Cape Canaveral, north of West Palm Beach and at Fort Lauderdale are retreating faster, between 1.0–2.9 m/yr, though it should be noted that the most recent data cited for east Florida are from 1965 (Dolan *et al.*, 1988). Most of the 10 beach renourishment projects mentioned by Leonard *et al.* (1990) for this coast commenced after this date.

4) ***Antillean Islands***

Antigua: Deane *et al.* (1973) recorded erosion of 0.3 m/yr between 1942 and 1968 at Fort James on the northwest of the island, requiring construction of a revetment to protect the road. Cambers (1985) reports severe erosion on the northwest coast at Dickinson Bay, where hotel development dominates the coastline. Foundations of the Blue Heron Hotel on the southwest coast were exposed after Hurricane Klaus in 1984. Sand mining in Antigua, like many of the other eastern Caribbean Sea islands, is a major cause of concern for beach instability.

Barbados: Cambers (1988) includes data on erosion for the period 1954–1982 on the south and west coasts of the island. Fifteen of 22 monitored sites displayed erosion, for an average rate of loss of 0.3 m/yr over the 28-year period. Further work between 1983 and 1986 confirmed the continuation of these trends, though Cambers (1988) points out that spatial variation is masked by average values. Twelve out of 19 beaches along the south coast accreted over the period, though this appears usually in association with groyne structures (Cambers, 1988). Narrow beaches in front of coastal property have caused considerable government concern in the light of continuing erosion, itself a partial function of reef pollution and degradation (Proctor and Redfern, 1983, 1984). A major government coastal conservation project is now under way, with financial assistance from the Inter-American Development Bank (Hendry and Nurse, 1991) to improve coastal conservation measures.

Cuba: Juanes *et al.* (1985) and Juanes *et al.* (1986) described inter- and intra-annual variability of beaches on the 20 km-long Varadero Peninsula in the north of the island. Processes responsible for this include shoreline orientation in respect of winter storms (northers) which cause erosion on northwest-facing beaches, and summer Trades that return sediment from the wider shelf to the east. Concern about long-term stability and management of this coast has led to development of a major beach renourishment programme using borrow from the offshore shelf to improve beaches in this major tourist area.

Dominica: West coast beaches are generally narrow, and Cambers (1985) reports that the coastal road, a critical link in the island's communication and transport system, has been the focus of substantial concern and protection work. There are few buildings between the road and the sea. Lasting impacts from Hurricane David in 1979 were reported at Massacre, where a coastal road was destroyed, and between Mahaut and Coulibistri the road required gabion protection after the storm. Gabions protecting the road at Gueule Lion Pointe were destroyed by Hurricane Klaus (Cambers, 1985).

Dominican Republic: The Country Environmental Profile (JRB Associates, 1981) only records shoreline stability problems with respect to aggregate mining. Severe land loss was experienced on the delta of the Nigua River due to loss of coastal protection from natural sand, a process that exposed adjacent property to storm surge during the passage of Hurricane Allen.

Grenada: The major tourist beach at Grande Anse on the southwest coast eroded 0.7 m/yr between 1970 and 1984, accelerating from

0.4 m/yr between 1951 and 1970. This resulted in removal of one hotel to another location; extrapolation of these trends provides gloomy predictions for other resort sites. Erosion rates at Beausejour Bay on the west coast and Levera on the north are greater, averaging 2.9 and 3.6 m/yr, respectively. Possible contributory causes are reef health problems from pollution plus sand mining (Cambers, 1985, 1988). Cambers (1985) reports damage from Hurricane Klaus in 1984 on the north and west coasts and serious ongoing erosion for the northern island of Carriacou.

Jamaica: Wood (1976) recorded 11 locations of accretion mainly along the south coast of the island in association with river systems draining the interior in this direction. Relative tectonic tilting in the south of the island compared to the north was invoked to account for these differences in sedimentation patterns. The most dramatic erosional phases are associated with storm events, although locally storms may encourage downdrift accumulation (Bacon, 1990; Hendry and Bacon, 1990). Earthquakes also have had a significant impact on coastal land loss due to liquefaction and slumping (Hendry, 1979a, b).

The Country Environmental Profile for Jamaica (GOJ *et al.*, 1987) identifies 17 locations of 'critical erosion', distributed fairly evenly around the island, both north and south coasts included. They were described in this way due to impending threats to infrastructure (houses, schools, roads, hotels, power stations, etc.) by erosion. The report indicates that in five cases illegal removal of beach sand is the primary cause, with, ironically, some of the material being sold to government departments. Loss of beach, exposure of roadways and evacuation of residents are all results of these activities, with loss of foreshore varying between approximately 7 and 25 m at some of these locations over about a 15-year period up to 1987. The GOJ *et al.* (1987) report also highlights coastal structures such as seawalls and groynes as contributing to erosion and beach narrowing. For example, a groyne has been implicated in the loss of 90% of Sunset Beach in Montego Bay.

There are marked differences in long-term behavioural patterns of some Jamaican shorelines. Hendry (1987) and Hendry and Digerfeldt (1989) demonstrated landward migration of the carbonate barrier beach along the west coast of the island during Holocene sea-level rise (Figs. 7.7 and 7.8). The reverse response has been described along the Palisadoes tombolo on the south coast, where abundant river-derived sediment coupled with decreasing rate of sea-level rise caused avulsion of the depositional axis to seawards on at least four occasions in the last 6000 years (Hendry, 1990a).

Montserrat: There is very little information for this island, but reports of accelerated erosion due to the impacts of hurricanes and heavy seas in recent years have been conveyed to the author (J. Jeffers, pers. comm.). Cambers (1981a) reported 18 m of erosion on reclaimed land at Plymouth, near the jetty, as a result of Hurricane David in 1979, though hurricanes Frederick in 1979 and Allen in 1980 caused no further damage.

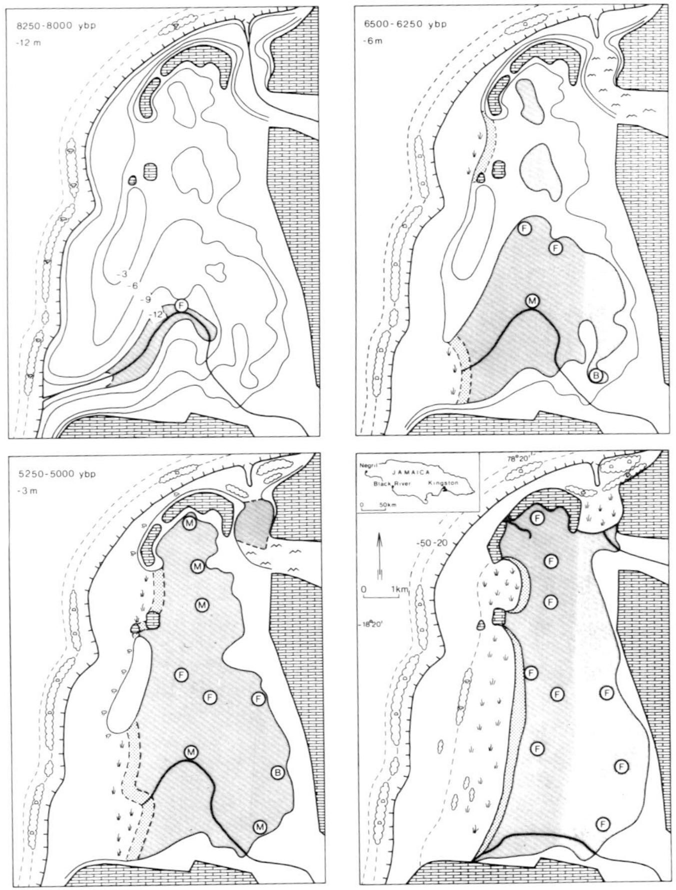

Fig. 7.7 Shoreline retreat under rising sea-level conditions. Contours (and bottom topography) in metres relative to present sea level. Located on the western (leeward) side of Jamaica, this wetland at Negril developed behind a retreating beach which was nourished by carbonate supply from seagrass and reef environments (see Fig. 7.8). Sea level rose steadily though at decreasing rates during the late Holocene in Jamaica (Fig. 7.3) and longshore processes in this area are subdued. Landward transfer of beach sediments may largely have been a function of storm activity. Four time windows are shown representing sea-level conditions at −12 m, −6 m, −3 m and present sea level (from Hendry and Digerfeldt, 1989).

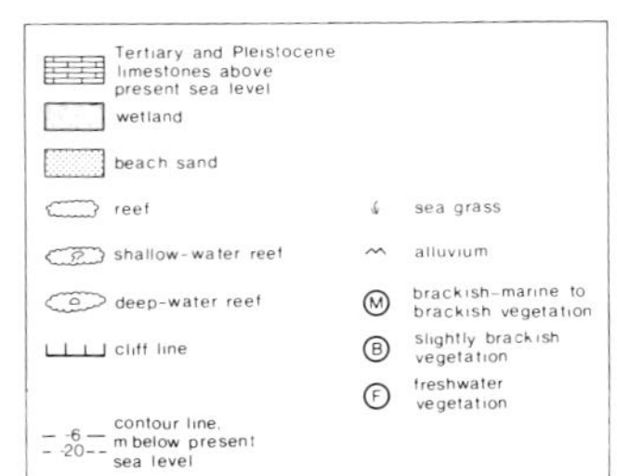

Fig. 7.8 The beach-wetland system at Negril, western Jamaica (see Fig. 7.7). In a setting typical of many islands this wetland is confined in a coastal basin, its landward limit defined here by a fault scarp. This boundary restricts potential for further wetland expansion under a rising sea-level scenario. Dense seagrass beds offshore are the primary source area for carbonate sediments of which the beach is composed. This proximate, general source type is quite distinct from the point sources along shorelines dominated by river discharge where longshore transport is required to nourish the coast. See colour plates between pages 210 and 211. Photograph by Malcolm Hendry.

Puerto Rico: Morelock (1984) reported work on 10 sites of severe coastal erosion that were monitored until 1977. Five sites were along the northeast coast, while four were on the south-central coast and one at the northern end of the west coast. At several of the sites erosion was accounted for by shoreline adjustment to wave regime, but at others, including Punta Salinas and Arecibo, coastal changes are accounted for by anthropogenic modification of the bay areas through dredging and construction. At Boca de Cangrejos, Pozuela and Las Mareas, erosion was already underway before shoreline construction and has continued since that time. There have been some attempts at reclamation and stabilization where property was threatened.

Thieler (1990) reports that most north-coast beaches are presently eroding at varying rates. Storm losses followed by recovery are common on the narrow beaches near San Juan, but 16 km east the beach at Punta Uvero has been retreating by over 3.0 m/yr. At Arecibo the 0.55 m/yr rate of retreat between 1950–1977 was halted only by construction of a seawall, but there is now no beach for recreation or storm protection. An updrift seawall and jetties have caused sand starvation at Boca de Cangrejos, where the beach is eroding at 4 m/yr. Jetties and channels constructed on the Rio de Bayamon also interfered with sediment supply to Levittown, where 30 m of beach eroded between 1950–1987.

St. Kitts/Nevis: In St. Kitts, coastal cliffs close to Basseterre on the southwest-facing coast are eroding and further north erosion of 0.7 m/yr from 1968–1983 was measured for a cliffed area. The coast road in the area of Brimstone Hill was re-routed in 1970 due to erosion. Erosion of a spit near Dieppe Bay Town at a rate of 1–3 m/yr since Hurricane David in 1979 is attributed to storm-related reef damage. The northeast coast is used extensively for sand mining, with resultant removal of dune vegetation cover. Hurricane Klaus in 1984 is also reported to have caused major damage to coastal structures including seawalls and roads, with beach erosion to the north of Halfway Tree. A beach in Frigate Bay South completely disappeared during the storm. In general, sand mining and pollution of reefs, as well as storms, are the major worries for an island with limited beach resources (Cambers, 1983a, 1985).

Severe erosion problems are also reported in Nevis for the period 1968–1983 on the west-central coast between Fort Charles and the northern end of Pinneys Beach. On the north coast between Newcastle and Camps, erosion rates in excess of 1 m/yr were measured over the same time interval, though other parts of the north and west coasts are eroding at slower rates (Cambers, 1983b). Natural, not anthropogenic causes, are believed to be responsible for erosion. Sand mining takes place on the southeast coast of Nevis, and poor management of the site at Indian Castle Estate has caused concern over dune and general coastal stability (Cambers, 1983b).

St. Lucia: In an excellent article on the causes and effects of beach stability, Williams (1986) described the history of beach sand mining, which appears to be the major cause for coastal land loss. Mining affected all the island's major beaches and led to the commissioning of numerous studies on the nature of the problem and potential solutions, which included opening pumice quarries for aggregate crushing. At Vigie, Choc and Reduit Beaches, erosion rates of 0.8 m/yr, 0.4 m/yr and 0.7 m/yr, respectively, were measured for the period 1941–1970. These changes resulted from beach mining (Deane, 1973). Erosion of a spit at Marigot Bay followed hotel construction in 1979, and at Anse La Raye, further south, beach mining also has caused severe erosion, threatening a school (Cambers, 1985). Nesting turtles were also under threat due to mining. Towle (1985) provides a detailed account of how badly planned coastal development, including sand mining, caused substantial changes in wave climate, littoral sediment transport and ultimately beach instability at Gros Islet Bay in the northwest of the island. This in turn created major environmental and socio-economic problems, especially in local fishing communities.

St. Vincent and the Grenadines: Cambers (1981b, 1985) reports widespread erosion at bays around the island of St. Vincent. Replenishment from natural sources is restricted due to the extremely episodic nature of river discharge on this high, volcanic island. Legislation has been being enacted to control beach mining, but success has yet to be evaluated. The primary cause of beach erosion is thought to be aggregate mining, and small block-making factories were found close to a number of mined beaches. Several hotels, houses and roads were found to be under threat from beach loss. Blasting for channels, wrongly spaced groynes and hurricanes also figure in the problems (Cambers, 1981b).

Trinidad and Tobago: O'Brien and Lawson (1988) studied coastal processes and shoreline change at two bays on the southeast-facing coast of Tobago. Applying sediment transport calculations to wave parameters, they calculated net losses of 58,880 m^3/yr over a beach front of some 0.6 km at Queens Bay and 9430 m^3/yr over 0.45 km at Richmond Beach. Aerial photographs were unavailable to confirm the erosion, though observations of cliff retreat and exposure of roots of coastal vegetation supported the calculations for Queens Bay. Georges *et al.* (1988) attributed fluctuations in plan shape of Erin Spit to cyclicity in the supply side from nearby cliffs and to erosion during higher water levels. The Institute of Marine Affairs of Trinidad and Tobago is currently compiling an inventory of coastal erosion problems for both islands.

5) ***Platform Islands***

Bahamas: Bahamas coastlines are dynamic and exhibit substantial late Holocene and recent changes in many locations, both erosional and depositional. An interesting component of the Holocene record is the generally abundant carbonate sediment production and supply from extensive shallow bank areas (Hine *et al.*, 1981; Strasser and Devaud, 1986; Wilber, 1986). There are no known historical studies of shoreline change and no monitoring programmes. Critical future problems in addition to shoreline changes will be submergence leading to expansion of wetlands, territorial loss and the intrusion of saline water in shallow groundwater lenses on these low-lying carbonate

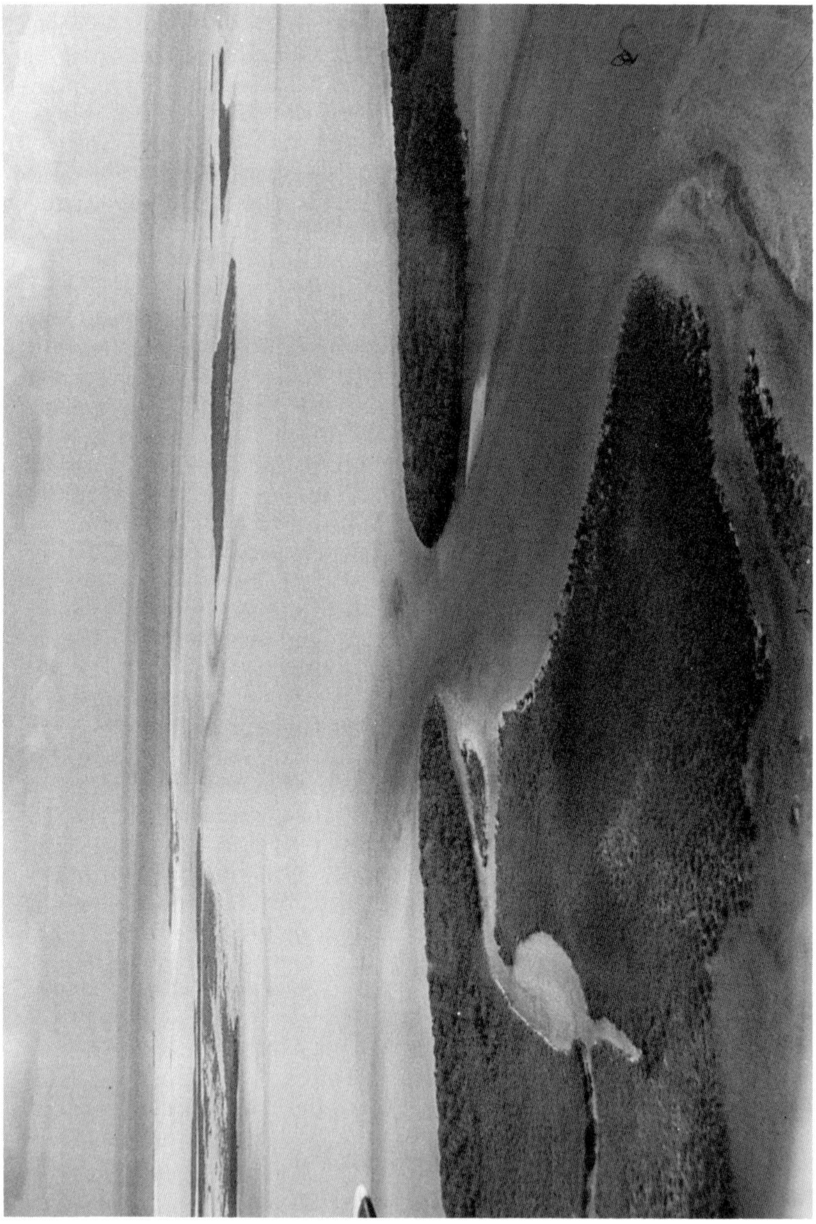

Fig. 7.9 Carbonate platform islands, the Bahamas. Dissected by tidal channels, the islands are typically at elevations of 1 m or less above sea level. Bank areas produce abundant carbonate sediment but supra-tidal areas are extremely susceptible to ocean processes and storms, and shorelines in exposed areas are mobile. Developed and undeveloped islands in platform settings are susceptible to flooding and submergence as well as shoreline instability in a rising sea-level scenario, and are key locations for further impact-assessment work. See colour plates between pages 210 and 211. Photograph by Malcolm Hendry.

islands (Fig. 7.9; Cant and Weech, 1986; R.V. Cant, pers. comm.).

Turks and Caicos Islands: On West Caicos there are several onshore late-Pleistocene to Holocene beach-ridge packages cross-cutting each other, with erosion of one submerged set on the east coast providing evidence of very active shoreline fluctuations (Hendry, 1990b). For Grand Turk, Cambers (1980) suggests that between Cockburn Town, located centrally on the west coast, and English Point, approximately 2 km south, there was net long-term erosion for the years 1906–1969 at rates between 0.3 and 0.45 m/yr. In contrast, the southern tip of the island shows accretion, accounted for by southerly movement of sediment from the central coast. Some evidence of apparently cyclical erosion/accretion at headlands, some erosion at seawalls, local erosion/accretion at groynes and beachrock exposure on certain beaches also provide local evidence of shoreline changes on the west coast. There is concern over stability at several sites of economic value, including telecommunications and hotel facilities, plus concern that relatively stable beaches be maintained for tourism-related activities. Cambers (1980) suggests long-term west-coast erosion is forced primarily by winter storms and hurricanes. One beach site on Grand Turk has been designated for sand mining (Williams, 1988).

4.1 Summary Characteristics of Regional Shoreline Changes

1) ***Northern South America:*** Coastal sedimentation is dominated by massive inputs of material derived from the continental interior and transported to the shelf by major river systems such as the Amazon (Brazil), Orinoco (Venezuela) and Magdalena (Colombia). The sand shoals, deltas and mud banks generated by river input and longshore transport are continually moving, causing repeated phases of accretion and erosion. On the coastline of the Guianas, for example, cyclical shoreline advance and retreat accompanies passage of huge mud banks northwestwards along the continental shelf. Major population centres occupy land at or below sea level from French Guiana to Colombia, on extensive, low-lying coastal plains.

2) ***Central America:*** The coastline of Belize is flanked by the second largest barrier reef in the world. In the shelter behind the reef are numerous keys and sand shoals on a shallow platform. In contrast to northern South America, sediment transport is limited, but the mangrove systems, which dominate the coastline, are sensitive to sea-level variations. The low-lying coastal plain and its inhabitants are continually exposed to the impacts of sea-level variations on several scales, including storm surge. The Belize coastline has been subjected to tropical storm or hurricane effects on average 112 times every 100 years.

3) ***Gulf of Mexico/Florida:*** Dominated in the north by Mississippi River sediments, shorelines are characterized by barrier island–lagoon–wetland systems, which have retreated during the last few

thousand years except in one or two locations such as the Ten Thousand Islands area of southwest Florida. In that area, a relatively sheltered location and a slowing in the rate of sea-level rise in the last 2-3000 years have allowed mangrove communities to expand seawards. Gulf coast barrier islands are preferred sites for residential development, continually requiring initiation of response strategies to prevent or restrict land loss due to sand movement.

4) ***Antillean Islands:*** Coastlines in this island chain are typified by small, compartmental beaches with restricted sediment-supply systems, except in the Greater Antilles where the bigger river systems debouch to the shelf. Shorelines are generally close to the edge of the shelf, over which much sediment is lost to deeper water, especially during storms. Many of these sediment-starved beaches are backed by old cliff lines, with short distances between the beach face and upland area (Fig. 7.10). These beaches are critical sites: the present trend to erosion, exacerbated by poor planning, inadequate sand management and other anthropogenic impacts described above, may remove even these vestiges of beach from economically essential locations.

5) ***Platform (low elevation) Islands*** (e.g., Bahamas, Turks and Caicos, Cayman Islands): These types of islands are characterized by low elevations, with the corresponding threat of submergence and major loss of habitable land area due to even small changes in this level. While the extensive bank areas of the Bahamas, Turks and Caicos and Nicaraguan Rise produce large volumes of carbonate sediment – the basis for island formation over geological time-scales – their exposure to oceanic processes and elevations relative to sea level make them extremely susceptible to changes in shoreline configuration.

Long-term shoreline recession is generally evident during the period of Holocene sea-level rise in the Intra-Americas Sea, though local progradation has been facilitated by sediment input at river mouths, or reduced energy conditions on shallow offshore gradients during the late-Holocene decrease in rate of sea-level rise. These kinds of examples provide indications of the relative importance of sea-level rise compared to other processes in the longer-term picture of shoreline change. In contrast, fluctuations in longshore transport rates can have an effect considerably larger than rapid sea-level rise on shoreline stability sometimes over relatively short time-scales, a process very clearly demonstrated along the coastline of the Guianas.

Three other factors are clearly of major importance in forcing coastal change over historical time-scales. Not surprisingly they are beach sand mining (especially in the Antilles), hurricanes, and coastal structures such as seawalls and groynes. Anthropogenic activities in a few instances may have triggered beach instability, while in others it has been superimposed on existing shoreline changes due to natural events. The importance of hurricane impacts compared to sea-level change in the region cannot be overstressed. It is these 'crescendo events', as Fairbridge (1989) calls them, which in the short term (seasonal) present a greater risk for major

Fig. 7.10 Pocket beaches on the southeast coast of Barbados. These beach systems are characteristic of many Antillean islands: physically separated processes influencing the beach are largely confined to each bay. Beach stability in such cases is controlled to a great extent by onshore-offshore transport in a conservative sedimentary regime. An old sea-cliff defines the landward beach boundary. The long-term integrity of these beaches is under threat in rising sea-level conditions due to several factors including the upland area against which they are anchored: this may in time function as a vertical seawall, inducing wave reflection and sediment re-suspension, thus inhibiting deposition. See colour plates between pages 210 and 211. Photograph by Malcolm Hendry.

impacts in coastal areas over the impacts of sea-level rise, bearing in mind they generally do not occur south of latitude 10°N.

Awareness of the frequency of storms in prehistorical and historical times is increasing; so too is appreciation of the economic impacts of these events (e.g., McGowen and Scott, 1975; Davis and Andronaco, 1987; Davis *et al.*, 1989; Thieler *et al.*, 1989; Bacon, 1990; Hall *et al.*, 1990; Reading, 1990; Hendry and Bacon, 1990). Although impacts are local, not regional, the frequency of occurrence of major storms throughout the region opens the possibility that most coastlines may suffer the impact of one or more storm events over the period considered in the UNEP scenario for sea-level rise. Future research and planning must take this important factor into account.

5 Sea-Level Movement and Shoreline Change: Implications for Management

Cramped into the narrow, urbanized, overpressured coastal zone of lands surrounding the Intra-Americas Sea are the bulk of population, infrastructure, tourist facilities, fishing and manufacturing industry, seaports and airports, much of it on reclaimed ground and within 1–2 m of present sea level. Any damage that results in loss of tourism revenue will have substantial impacts on the balance of payments, foreign currency reserves, employment and social stability for smaller island states that have become almost entirely dependant on tourism as a primary source of revenue (McElroy and de Albuquerque, 1989; Nurse, 1989).

There has been much discussion of general implications of climate change and sea-level rise for coastal areas of the world (e.g., EPA, 1986; Titus, 1986; Titus *et al.*, 1987; Hekstra, 1989; Vellinga and Leatherman, 1989; Leggett, 1990) and for the region (Hendry, 1988; Gable *et al.*, 1989). These papers describe situations that typify many of the world's shorelines with accelerated sea-level rise. Included in these overview assessments are loss of wetland where suitable intertidal areas above present levels are unavailable; land loss due to flooding (permanent elevation of the coastal water table); changes in coastal sedimentation patterns (including erosion and accretion) following profile adjustment to changing inshore water depths; further cliff retreat especially in weaker lithologies; increased flood hazard from storms on elevated water levels and increased saline intrusion in rivers, estuaries and coastal aquifers.

It is necessary to remember that sea level has been rising for the last 17,000 years (Fig. 7.3) and is still generally rising. Our responses must recognize this underlying factor in addition to the possibility of greenhouse-related effects that may accelerate the rate of rise (Fig. 1.7).

Unfortunately there is very little research dealing with potential physical impacts in this region. Papers by Leatherman (1984) on Galveston Bay and by Lidz and Shinn (1990) on the Florida Keys provide examples of what can be achieved in this regard. Similarly, economic forecasting using various risk scenarios could be attempted, such as for Galveston, Texas (Gibbs, 1984), allowing for model improvements as the data base broadens. Relatively simple models like this will have substantial value in generating public and political awareness. Detailed economic valuation of particular coastal resources will also be useful for this purpose (e.g., Bell and Leeworthy, 1986).

The WMO/UNEP (1990) Intergovernmental Panel on Climate Change (IPCC) concentrates on three major response alternatives to sea-level rise and coastal land loss problems. These are the so-called RAP (retreat, accommodation, protection) options:

1) *Retreat:* this includes prevention of coastal development, permitted development on the understanding that it may need to be abandoned, and no direct government role other than withdrawal of subsidies and provision of information on risks.
2) *Accommodation:* strategies here include improvement in drainage, storm preparedness planning, elevation of housing on stilts, adaptation of crops to types compatible with saline environments, regulation of groundwater pumping, etc.
3) *Protection:* these include the many hard or soft engineering options available for defense against flooding, inundation, erosion, saline intrusion and loss of natural resources.

The following discussion seeks to identify some of the key coastal issues that need to be addressed in planning for these options in the Intra-Americas Sea:

5.1 Ecological Issues Related to Coastal Land Loss

Islands of the region presently support as much as 50% of the land area under wetland cover, as in Grand Cayman (Ellison and Stoddart, 1991). The regional average is over 9.0%, a figure second only to the Pacific islands (19.0%) and far in excess of the next highest proportion of 3.4% for southeast Asia in the global subdivision of 20 areas adopted by the IPCC (WMO/UNEP, 1990). Wetlands are prime development sites in the region, occupying low land next to the sea. They are frequently regarded as nuisance areas and in general are poorly understood in the context of their critical ecological function in the island systems (Bacon, 1989; Bossi and Cintron, 1990). In addition, the Atlantic borderlands of Guyana, Surinam and French Guiana support swaths of swamp forest many kilometres in width along hundreds of kilometres of coastline, while there is close association of mangroves and reefal communities along the barrier reef system in Belize.

Coastal wetlands of western Jamaica had been able to sustain a vertical accumulation rate that kept pace with the mid-Holocene sea-level rise of *c.* 3.8 mm/yr, at a time when vegetation was dominated by mangrove communities (Hendry and Digerfeldt, 1989). In a review of global mangrove response to sea-level rise Ellison and Stoddart (1991) indicate that mangroves could keep pace with a sea-level rise of up to 9 mm/yr, would be stressed by rates of 9–12 mm/yr, and may not persist in expansive mode beyond this. At the rate of rise adopted by UNEP for this study (*c.* 5.3 mm/yr) wetlands should not be directly stressed from this source. Snedaker (Chapter 12) indicates that the most likely source of stress that could result in land loss would be reduced precipitation, which would push the salt-water/fresh-water interface further inland, along with the mangroves.

Limits for this type of wetland migration will be existing coastal-plain infrastructure plus boundaries with upland areas. Many wetlands in islands are accommodated by coastal basins whose limits are well-defined breaks-of-slope along fault scarps or old sea-cliffs. In these locations, we should expect overall decrease in wetland area as a consequence of mangrove retreat.

Continued loss of beach front by longshore movement or permanent sand loss to offshore evidently will provide for significant ecological changes. Not only will nesting habitats for endangered species such as turtles be impacted, but so too will the natural protection beaches afford to adjacent habitats such as lagoons and wetlands.

5.2 Shoreline Change and Economic Issues

Costs associated with adaptive responses to climate change problems cannot be generalized as they depend on the nature of the particular problem and the choice of response to deal with it. Most states of the region are presently spending money to manage and protect coastal areas. The magnitude and purpose of this expenditure can be expected to change as awareness of existing problems increases. As Alm *et al.* (Chapter 15) point out, however, there will be beneficiaries of this expenditure, including the construction industry. The issue may be how to identify which part of the existing national budget can be allocated to response mechanisms, or how to raise additional revenue where budget adjustments are inadequate. The possibility of cost-sharing options with coastal property owners and users will have to be addressed.

In a general sense it might be expected that intense and continuing development of the region's coastal areas would demand provision of a greater portion of resources for protection and conservation measures. This will require an increase in awareness and especially action on the part of decision-makers. A primary objective of UNEP, IPCC and IOC (Hendry and Loucks, 1991) type programmes is generation of this awareness and provision of direction on where the major future costs may lie.

The economic cost of retreating from an eroding shoreline will be high, involving complete relocation of infrastructure and population. In island settings it would be especially difficult due to constraints on space in already crowded coastal plains and because of resource-use conflicts, for example, with agriculture. In addition, where beaches and other resources are allowed to degenerate, the value of the resource to the critical tourism sector also will decline, a change that will have major impacts on foreign-currency inflows and employment. Each decision on whether to move inland from the coast or opt for protection, stabilization and other ecosystem support measures should be accompanied by cost-benefit analysis, as in the ongoing Feasibility Study on Coastal Conservation being conducted by the Government of Barbados.

For tourism-dependant economies, protection and stabilization may be the only realistic option. On very small islands the extreme case would be reached if complete submergence of the island required the entire population to be moved. This scenario is not out of the question, for example, in islands of the Bahamas and Turks and Caicos, or in the fishing communities of the Pedro and Morant Cays in Jamaica, where severe local socio-economic problems will result. The critical problems faced by low islands in response to rising sea levels have already been brought into sharp focus in the Maldives (Wells and Edwards, 1989).

In Guyana, the major city of Georgetown is situated on land an average of 1 m below sea level and has been protected since the time of Dutch administration by dykes and other sea defenses. Of the 750,000 population 90% live within 5 km of the coast (Jones, 1986). Earle (1987) reports that the costs for new sea-defence construction between 1980–1984 totalled US$ 12,774,050 and in the same time period a total of US$ 88,652,309 was spent on new drainage and irrigation works. Needs for new construction and maintenance of old defenses and drainage place additional burdens on the Guyanese economy. Protection of this enormous economic, cultural and social investment may be very high, and a full economic projection of the options and their costs should be undertaken.

WMO/UNEP (1990) have produced the first comprehensive, if general, cost estimates for protection of the world's shorelines from land loss in the face of sea-level rise for a rise of 1 m in 100 years, about double the USLR scenario. Using various categories and formulae for different shorelines and infrastructure, they calculated costs for protection of low coasts, cities, harbours, island elevation and beach nourishment. An extract of the data covering territories of the Intra-Americas Sea is shown in Table 7.1.

WMO/UNEP (1990) point out that the accuracy of figures given to two decimal places may be misleading given the assumptions, uncertainties and constraints involved in their analysis. The estimates do not include costs of unprotected dry lands or ecosystems that would be lost, nor the impacts of dealing with saline intrusion or increased storm frequency. Nonetheless these numbers do provide a good sense of the relative scale of the problem even though substantial refinement will be necessary based on more detailed local modelling of expected impacts and in the light of options finally adopted by each government. No general response model can be applied. For all the island territories the cost of new construction alone would be US$ 11.1 billion, at an average of 0.20% of GDP, the latter figure only exceeded by the small Pacific islands and small Indian Ocean islands in the IPCC subdivision (WMO/UNEP, 1990). At present, there is a major discussion in the international scientific literature on relative merits and cost-effectiveness of protection measures on eroding shorelines using hard and soft engineering options (e.g., Kraus and Pilkey, 1988; Dean, 1988; Bottin, 1990; Bruun, 1990; Hemsley, 1990; Leonard *et al.*, 1990; Pilkey, 1990; Schwartz and Pilkey, 1990). This debate and evaluation promises to improve techniques for beach protection and restoration. For a small initial investment, tide gauges should be installed in all the region's harbours; most of the seawalls, roads and piers in these critical facilities are within a metre or so of present sea level, and even increases at the USLR rate will create flood risk. Early warning of increased sea level is necessary, and the critical value of improved sea-surface monitoring is central to nearly all recommendations of this paper.

5.3 Political and Sociological Issues

Vellinga and Leatherman (1989) observe that a substantial lag time may exist between the recognition of particular problems, their analysis, and decision-making at the local level to deal with them. They suggest 10–20 years may elapse before strategies are developed to combat the problems recognized. During this time sea level may have changed by an additional number of centimetres, requiring further revision of problem analysis. Clearly, forward modelling and attempts to encourage early acceptance of potential problems must accompany expanded scientific efforts to detect

Table 7.1 Cost of coastal protection by country for countries of the Intra-Americas Sea (from data supplied by the Coastal Zone Management Subgroup to the IPCC [IPCC, 1990], summary of which appears in WMO/UNEP, (1990).

Country	*Total costs (US$ millions)*	*Annual costs (US$ thousands) per capita*	*As % of GDP*
Anguilla	83	11,786	10.31
Antigua/Barbuda	152	1870	1.01
Aruba	646	9447	1.85
Bahamas	2565	10,915	2.67
Barbados	297	1172	0.27
Belize	527	3085	2.93
British Virgin Islands	93	7725	1.24
Cayman Islands	228	10,350	1.04
Colombia	784	30	0.02
Costa Rica	232	92	0.08
Cuba	2030	199	0.13
Dominica	14	161	0.17
Dominican Republic	238	37	0.04
El Salvador	27	5	0.01
French Guiana	533	6197	2.96
Grenada	54	554	0.67
Guadeloupe	236	708	0.20
Guatemala	1007	123	0.11
Guyana	995	1250	2.12
Haiti	124	23	0.07
Honduras	32	8	0.01
Jamaica	462	197	0.19
Martinique	192	585	0.14
Mexico	9889	123	0.06
Montserrat	3	225	0.09
Netherlands Antilles	908	5159	0.66
Nicaragua	956	282	0.35
Panama	263	118	0.06
Puerto Rico	334	102	0.02
St. Kitts/Nevis	140	3033	2.33
St. Lucia	123	879	0.82
St. Vincent and the Grenadines	55	497	0.55
Surinam	2622	6638	1.94
Trinidad and Tobago	1720	1431	0.21
Turks and Caicos	223	24,739	8.10
USA*	86,819	360	0.02
Venezuela	3155	177	0.05
US Virgin Islands	230	2018	0.26

*This figure includes all coasts, not just those bordering the Intra-Americas Sea.

and monitor problem sites (Hendry and Loucks, 1991).

The overall consequences of even small increases in sea level such as the 20 cm USLR value will have major impacts on the political identity of several states, by threatening to reduce land area. Related problems may arise with regard to territorial claims in respect of the Exclusive Economic Zone (EEZ). At present, for example, substantial areas of the eastern Caribbean Sea fall under Venezuelan territorial control. Aves Island, located on the Aves Ridge in the eastern Caribbean Sea, is only 600 m long but it entitles Venezuela to claim 135,000 km^2 of its total 630,000 km^2 EEZ (Aparicio-Castro *et al.*, 1989). There is evidence of island erosion in the last two decades, a process that may have been aided by mining for guano (Schubert and Laredo, 1984). With a maximum elevation of only 3.7 m above sea level, erosional processes combined with known subsidence of the Aves Ridge may result in loss of the island in the event of a greenhouse-pulsed sea-level rise and lead to forfeiture of territorial claims.

Similar considerations need to be given to other low-lying island areas and territories where land loss may impinge upon the baseline from which EEZ rights are extended. These areas include Cuba, the Bahamas, Turks and Caicos, Netherlands Antilles, Jamaica (Pedro and Morant Cays) and Colombia. The continental borderlands may also have these problems if there is loss of coastal lowlands and a continuation of barrier island migration. Shoreline changes are accompanied by many small-scale, localized problems relating to boundary definition (Gustadt, 1990).

Impacts of various options on coastal communities must be carefully evaluated and as with all proposed changes affecting these people, they must be consulted and incorporated in the decision-making process. Resettlement is the most drastic response to flooding or erosion and will have the most severe sociological implications, requiring substantial adjustment to a new environment and loss of valued infrastructure which may have cultural significance. Complete separation of communities from their original area is the greatest source of dislocation and should be avoided (WMO/UNEP, 1990). The same report points out that public health and safety will be the main concerns if attempts are made to accommodate existing lifestyles to rising water levels. Given existing coastal pollution conditions throughout the region (UNEP, 1989b), the problem of drainage and sewage treatment must be given a high priority.

Hard engineering structures, should they be adopted for protective purposes, are known to have severe effects on nearshore sedimentation patterns while not necessarily changing the volume of sand in the littoral system. Seawalls, for example, while protecting property, may accelerate beach sand losses and may prevent sand from returning. The amenity and aesthetic value of the beach may thus be substituted for a bulwark to protect coastal property. This aesthetic factor is most important in the tourist-oriented coastal zone of the region and must be carefully weighed in developing appropriate responses to regional land-instability problems.

6 Recommendations

The analysis attempted in this paper highlights major gaps in our understanding of sea-level movements and potential future impacts for the

Intra-Americas Sea. To address these weaknesses will require approaches across many discipline areas. The recommendations made here have commonality with strategies proposed by the IPCC (WMO/UNEP, 1990). IPCC conclusions are general in nature and recognize the need for development of specific requirements to improve management at local and regional levels depending on particular characteristics of the problem. The following suggestions are designed to fulfill this regional requirement for the territories of the Intra-Americas Sea.

The need for three types of actions is recognized: territorial studies, scientific data base improvement and institutional strengthening.

6.1 Territorial Studies

These studies are required to more clearly define the potential problems in a rising sea-level situation. The areas at most risk fall into three categories:

Category 1

1) States comprising entirely low islands where nearly all political boundaries, population and infrastructure are at some degree of risk. These include the Bahamas, Turks and Caicos Islands and Cayman Islands.
2) Continental states where significant portions of the population and infrastructure are situated below or close to present sea level. These include the Guianas and Central America, especially Belize.

Category 2

This incorporates the Antillean islands where narrow, overpressured coastal plains and their associated marine resources are threatened by even relatively small additional losses of beach and other shoreline areas.

Category 3

Includes coastal areas with inhabited barrier islands, spits or reclaimed lowlands, at or within a few metres above present sea-level. Some examples of these are west Florida and the other USA Gulf States, Jamaica (Kingston) and Colombia (Cartagena). Many harbours have moorings and facilities located on reclaimed land.

In each of the identified categories, estimates should be made of areas likely to be lost by both inundation (submergence) and physical relocation of substrates. Various options for mitigation must be examined at the local level, and this work should be integrated with socio-economic studies of the implications of possible options.

6.2 Scientific Data Base Improvement

Extensive gaps in our knowledge of causes and potential effects of changing climatic and environmental conditions on coastal stability require that urgent attention be given to improving the information base.

1) *Sea–Surface Studies*
   a) Immediately improve the network of regional sea-surface monitoring stations. The framework for these activities already has been outlined by the Intergovernmental Oceanographic Commission (IOCARIBE, 1986; IOC, 1988) through programmes linked to

GLOSS, the Global Sea Level Observing System.

b) At sites with historically rapid relative sea-level changes (see Fig. 9.8) geodetic relevelling will identify ground movement effects and their impact on the local sea-level record.

2) *Vulnerability Indices*

a) Establish a vulnerability index for hurricane impacts, with advice for coastal landowners and prospective landowners (private and government) on risk areas and types of risk, plus what to expect from a storm and how best to prepare for one. Some efforts have already been made for Antigua, Barbados, Belize, Cayman Islands, Guyana, Jamaica, St. Lucia and the Turks and Caicos Islands (Jones, 1986), and recommendations for extending the nature of this work have followed (Hendry and Bacon, 1990).

b) Parallel with but linked to 2(a), develop a detailed vulnerability index for coastal problems in relation to sea-level rise. This index should build upon approaches already under way that are yielding useful results (Gornitz and Kanciruk, 1989). The output from each index will reflect different priorities, though many of the basic data requirements will be common to both.

3) *Modelling*

a) Extend efforts to obtain sea-level curves for the period since the last glacial maximum (*c.* 18,000 ybp) for as many locations as possible. This will assist in: (i) the interpretation of causes for differences detected between those curves that already exist for the region; (ii) provide an empirical base for local modelling of postglacial shelf and coastal history. Some potential target sites are coastal wetlands in territories that have no such curves, for example, Venezuela, Guyana, Mexico, Cuba and other Antillean islands.

b) Further modelling of Quaternary shoreline responses to sea-level change will aid prediction of future responses to such change. The framework for conducting such studies exists within International Geological Correlation Program, Project 274, on Coastal Evolution in the Quaternary (IGCP, 1989) and the International Union for Quaternary Research (INQUA, 1990a,b). The Caribbean and Adjacent Regions Secretariat of the Intergovernmental Oceanographic Commission are also pursuing plans for regional studies of sea-level change impacts and their management (IOCARIBE, 1989; Hendry and Loucks, 1991). The importance of cooperation in regional research initiatives must be stressed to avoid unnecessary wastage and/or duplication of limited resources.

c) Work on the causes of relative sea-level change and in particular its variability across the region.

6.3 Institutional Strengthening

1) Improve government capability to apply appropriate coastal management techniques. This strengthening must include: (a) in-house coastal monitoring to identify sites under erosional stress; and (b) immediate improvement of coastal-management practices related to sand and wetland management, especially in island locations.

2) Persuade governments of the need to encourage and support activities which contribute to systematic mapping of coastal-zone resources, thus providing the data needed for identification of stressed and high-risk sites (e.g., Butler and LeBlanc, 1990).
3) Plan for a series of seminars and workshops aimed at government planners, policy-makers and coastal-zone user groups (Hendry and Loucks, 1991) alerting participants to existing coastal-zone management requirements and the potential for exacerbated stress in a climate- change situation.
4) Encourage governments to continue promotion of environmental impact assessment for all coastal development, and encourage incorporation of climate-change considerations in these assessments.

7 Summary

Shorelines of the Intra-Americas Sea have generally retreated through historical times, continuing a trend that developed during the earliest periods of Holocene shelf submergence. Retreat has been a function of landward migration by coastal lowlands in response to rising water levels, coupled in many instances with physical relocation of coastal substrates due to erosion and sediment transfer. Land loss has persisted despite a decrease in rate of relative sea-level rise over the last 5000 years. Progradation has occurred in a few areas over this interval, including shorelines adjacent to river systems discharging large volumes of sediment (e.g., the Guianas), and also in some sheltered swampy locations (e.g., west Florida).

Relative sea-level change provides underlying control in this overall retreat pattern over a scale of millennia, but other factors have contributed to coastal instability in recent times. These include poor beach sand management, particularly illicit sediment removal, and coastal construction. Hurricanes have been important agents of geomorphological change over geological time-scales and are an increasingly damaging factor for coastal infrastructure due to expansion of the built environment.

The duration and distribution of tide-gauge observations limits our ability to interpret sea-level forcing mechanisms. Geological data suggest that in most areas the component of relative sea-level change caused by tectonic factors is an order of magnitude or more lower than the historical rate of sea-level rise. However, tectonic measurements do not necessarily coincide with locations of sea-surface observations, and errors may accompany the assumption of linear tectonic deformation. As yet undefined oceanic forcing factors appear to play a primary role in the variability of historical and prehistorical sea-surface movements within the Intra-Americas Sea. On the decadal time-scale under consideration for planning purposes, relative sea-level movements caused by land motion therefore may only be important for shoreline changes at a few locations. These include Lake Maracaibo and Louisiana, where land subsidence has been accelerated by hydrocarbon withdrawal and by channelization of rivers. Anomalously high relative sea-level rise in western Haiti may incorporate a tectonic subsidence component, while tectonics and diapirism also may be influencing land uplift rates in central America and Colombia.

At the average regional rate of rise of about 3.0 mm/yr, or under the projected lower-case greenhouse scenario of about 5.3 mm/yr adopted by the UNEP/IOC Joint Task Team, there will be continued stress on coastal stability and also therefore on human occupation of the coastal zone in the Intra-Americas Sea. Additional sand loss may eliminate many beaches, particularly the compartmented beaches that characterize island locations. Coastal wetlands will continue to retreat landwards, but this natural accommodation to sea-level rise will be restricted by existing coastal development and also by topographic breaks, resulting in further reduction of wetland area.

All countries adjacent to the Intra-Americas Sea stand to suffer consequences from future relative sea-level change. Some areas may be at relatively greater risk than others where territory is almost entirely within a few metres of sea level. For these states substantial loss of land area will likely accompany even small increases in sea level. Loss of territorial claims on the Exclusive Economic Zone plus flooding will accompany land loss due to permanent submergence and erosion. Costs projected by the Intergovernmental Panel on Climate Change for the coastal protection option in Caribbean Sea islands exceed US$ 11.0 billion (0.20% of GDP) for an assumed sea-level rise of 10 mm/yr over 100 years, about twice the UNEP rate. Several territories have reached a critical stage of coastal land loss and are involved in development of expensive protection procedures. For island states, which experience land-use conflicts due to limited space on narrow coastal plains, options for management may be limited to coastal protection aimed at preservation of remaining beach areas for tourism, recreation and inherent protective functions.

Recommendations for strategies to address these problems fall into three areas. The first identifies categories for Territorial Studies, including low island areas of the Bahamas, Turks and Caicos, and Cayman Islands and continental coastal areas with concentrations of population and infrastructure close to or below sea level, for example, the Guianas and Belize. Category 2 includes the Antillean islands where coastal erosion is threatening the integrity of remaining pocket beaches and where coastal developments cannot be relocated further inland. Category 3 includes countries with population concentration in coastal areas such as barrier islands, spits or low-lying coastal plains. Examples of this type include coastlines bordering the Gulf of Mexico, plus Jamaica (west Kingston) and Colombia (Cartagena), among others.

Scientific data base improvements require expansion of the Global Sea Level Observing System; development of vulnerability indices for (1) hurricane and (2) long-term sea-level change impacts in coastal areas, plus further modelling of prehistorical sea surface and associated shoreline changes as a basis for predicting future behaviour. Additional efforts must be made to understand the nature of oceanic forcing of relative sea-level movements.

Institutional strengthening measures call for improvement of in-house shoreline monitoring capability (especially in relation to coastal erosion and ecological stress) and of skills in shoreline management; seminars for government planners, policy-makers, and user groups, on coastal management and climate-change issues, plus further local promotion of Environmental Impact Assessment that includes climate-change considerations.

Sceptics may argue that mobilization of significant resources to deal with possible impacts of greenhouse-induced climate change is wasteful, particularly as some doubt whether these changes will occur. However, many countries in the region are trying to come to grips with existing problems of shoreline migration and land loss due to past and present sea-level change and anthropogenic impacts in addition to a host of other coastal-resource management and environmental problems (UNEP, 1989b). The majority of countries are doing so with minimal institutional, technical and financial resources and in some cases none at all.

The recommendations made in this paper are made with a view to an overall strengthening of the region's coastal zone management capability, an objective that would be essential even without a greenhouse effect. The overall improvement in capability of countries bordering the Intra-Americas Sea to manage their own critical coastal resources, which underpin a substantial portion of economic activity, will be enormously beneficial.

8 Acknowledgements

A review of this nature cannot be undertaken without assistance from professionals active in the field. Special thanks to the following persons who sent reprints or corresponded on various matters, and apologies to any whose name has been inadvertently omitted: Richard Cant (Bahamas), Leonard Nurse (Barbados), Gillian Cambers (British Virgin Islands), Captain Fernando Londomo (Colombia), Alejandro Gutierrez, Alejandro Echeverria, Carlos Brenes (Costa Rica), Jose Juanes, Jorge Foyo (Cuba), Patrick Nunn (Fiji), Marie-Therese Prost (French Guyana), Marta Pilon de Pacheco (Guatemala), Peter Bacon (Jamaica), Gerardo Gold (Mexico), John Jeffers (Montserrat), Jeff Sybesma, Tom van't Hoff (Netherlands Antilles), Bruce Taggart, Robert Webb (Puerto Rico), Glen Gemerts (Surinam), Oswald Williams (Turks and Caicos), Eric Blommestein (UNECLAC, Trinidad), Barbara Lidz, Gren Draper, Richard A. Davis, Cyril Galvin, Tighe Geoghegan, Robert Ginsburg, Vivien Gornitz, Al Hine, Stephen Leatherman, John Lewis, Paul Mann, George Maul, Ian Macintyre, Orrin Pilkey, Amos Salvador, Maurice Schwartz, Gene Shinn, E. Robert Thieler, James Titus, Peter Wilcoxen (USA), Ruben Aparicio Castro, Carlos Schubert, Carlos Suarez (Venezuela). The author is also grateful to George Maul, John Milliman and Gren Draper who reviewed earlier versions of the manuscript and George Maul and Linda Pikula who organized a data base search. The work was supported by the United Nations Environment Program, through the Regional Coordinating Unit in Kingston, Jamaica and also forms a contribution to IGCP Project 274 *Coastal Evolution in the Quaternary*.

9 References

Alt, J.N., Harpster, R.E. and Schwartz, D.P. 1980: Late Quaternary deformation and differential uplift along the Pacific coast of Costa Rica. Geol. Soc. Amer., Abstracts with Prog., 12, 378–379.

Aparicio-Castro, R., Castenada, J.and Perdomo, M.1989: Regional implications of relative sea-level rise and global climate change along the marine boundaries of Venezuela. Int. Workshop on Adaptive Options and Policy Implications of Sea-Level Rise and Other Aspects of Global Climate Change, IPCC Working Group 111, Miami, p. 16.

Aubrey, D.G., Emery, K.O. and Uchupi, E., 1988: Changing coastal levels of South America and the Caribbean from tide-gauge records. *Tectonophysics*, **154**, 269–284.

Augustinus, P.G.E.F., Hazelhoff, L. and Kroon, A. 1989: The chenier coast of Surinam: modern geological development. *Mar. Geol.*, **90**, 269–281.

Bacon, P.R., 1989: Ecology and management of swamp forests in the Guianas and Caribbean region. In: Lugo, A.E., Brinson, M. and Brown, S. (eds), *Ecosystems of the World, Vol. 15, Forested Wetlands*, Elsevier, pp. 213–250.

Bacon, P.R., 1990: Assessment of economic impacts of Hurricane Gilbert on coastal and marine resources in Jamaica: UNEP Regional Seas Reports and Studies, No.110, 78 pp.

Bell, F.W., and Leeworthy, V.R. 1986: An economic analysis of the importance of saltwater beaches in Florida. Report No. 82, Florida Sea Grant College, 166 pp.

Bender, M.L., Fairbanks, R.G., Taylor, F.W., Matthews, R.K., Goddard, J.G. and Broecker, W.S., 1979: Uranium series dating of the Pleistocene reef tracts of Barbados, West Indies. *Geol. Soc. Amer. Bull.*, **90**, 577–594.

Bird, E.C.F., 1985: *Coastline Changes: A Global Review*. John Wiley and Sons, 219 pp.

Bonneton, J.G., 1990: Marine terraces and Quaternary crustal movements of Guadeloupe island (Lesser Antilles). Symposium International Sur L'evolution des Littoraux des Guyanes et de la Zone Caraibe Meridionale Pendant le Quaternaire. ORSTOM/IGCP 274, Cayenne, French Guiana, Abstracts, pp. 7–11.

Bossi, R., and Cintron, G. 1990: Mangroves of the Wider Caribbean: towards sustainable development. UNEP, Caribbean Conservation Association and Panos Institute, Washington, 33 pp.

Bottin, R.R. Jr., 1990: Case study of a successful beach restoration project. *J. Coast. Res.*, **6**, 1–14.

Brasier, M., and Mather, J. 1975: The stratigraphy of Barbuda, West Indies. *Geol. Mag.*, **112**, 271–282.

Bruun, P., 1962: Sea level rise as a cause of shore erosion. *J. Waterways and Harbours Divns.*, **88**, 117–130.

Bruun, P., 1990: Beach nourishment – improved economy through better profiling and backpassing from offshore sources. *J. Coast. Res.*, **6**, 265–277.

Butler, M.J.A. and LeBlanc, C., 1990: Study to evaluate the need for a Caribbean habitat mapping programme. Champlain Institute/CERMES Report to International Center for Ocean Development, Halifax, Nova Scotia, 78 pp.

Cambers, G., 1980: An assessment of the coastal erosion on the west coast of Grand Turk Island, West Indies. Caribbean Planning and Development Company, Barbados, 65 pp.

Cambers, G., 1981a: Hurricane protection at the port of Plymouth, Montserrat. Caribbean Planning and Development Co., Barbados, pp. 1.1–7.1.

Cambers, G., 1981b: Coastal erosion in St. Vincent and the Grenadines. Caribbean Planning and Development Co., for World Tourism Organisation, 97 pp.

Cambers, G., 1983a: Coastal erosion in St. Kitts – Nevis: Vol. 1, St. Kitts. Caribbean Oceanographic Consulting Company, Barbados, 114 pp.

Cambers, G., 1983b: Coastal erosion in St. Kitts – Nevis: Vol 2, Nevis. Caribbean Oceanographic Consulting Company, Barbados, 114 pp.

Cambers, G., 1985: An overview of coastal zone management in six east Caribbean islands. UNESCO, Regional Office for Science and Technology for Latin America and the Caribbean, 69 pp.

Cambers, G., 1988: Coastal erosion and conservation in two eastern Caribbean islands, Barbados and Grenada. Transactions, 9th Caribbean Geological Conference, Barbados, pp. 9.1–10.4.

Cant, R.V., 1971: Aspects of the geology of the Pleistocene marine terraces of Jamaica, West Indies. Ph.D. Dissertation, Univ. West Indies, Jamaica, 372 pp.

Cant, R.V., 1973: Jamaica's Pleistocene reef terraces. *Geol. en Mijnbouw*, **52**, 152–160.

Cant, R.V. and Weech, P.S. 1986: A review of the factors affecting the development of Ghyben-Hertzberg lenses in the Bahamas. *J. Hydrol.*, **84**, 333–343.

Canis, W.F., Neal, W.J., Pilkey, O.H. Sr., and Pilkey, O.H. Jr., 1985: *Living with the Alabama–Mississippi shore*. Duke University Press, Durham, NC, 215 pp.

Coleman, J.M., 1988: Dynamic changes and processes in the Mississippi River delta. *Geol. Soc. Amer. Bull.*, **100**, 999–1015.

Correa, I.D., 1990: Registros historicos de cambios de linea de costa en los litorales Caribe y Pacifico de Colombia. Symposium International Sur L'evolution des Littoraux des Guyanes et de la Zone Caraibe Meridionale Pendant le Quaternaire. ORSTOM/IGCP 274, Cayenne, French Guiana, Abstracts, p. 37.

Corrigan, J., Mann, P. and Ingle, J.C. Jr., 1990: Forearc response to subduction. *Geol. Soc. Amer. Bull.*, **102**, 628–652.

Davis, R.A. and Andronaco, M. 1987: Hurricane effects and post-storm recovery, Pinellas County, Florida (1985–1986). *Coastal Sediments '87*, Amer. Soc. Civil Eng., pp. 1023–1036.

Davis, R.A., Knowles, S.C. and Bland, M.J., 1989: Role of hurricanes in the Holocene stratigraphy of estuaries: examples from the Gulf coast of Florida. *J. Sed. Pet.*, **59**, 1052–1061.

Dean, R.G., 1988: Managing sand and preserving shorelines. *Oceanus*, **31**, 49–55.

Deane, C., Thom, M. and Edmunds, H., 1973: Eastern Caribbean coastal investigations 1970–73. Regional Beach Erosion Control Program, Faculty of Engineering, Univ. West Indies, Trinidad.

Digerfeldt, G. and Hendry, M.D., 1987: An 8000-year Holocene sea-level record from Jamaica: implications for interpretation of Caribbean reef and coastal history. *Coral Reefs*, **5**, 165–169.

Dodge, R.E., Fairbanks, R.G., Benninger, L.K. and Maurasse, F., 1983: Pleistocene sea levels from the raised reefs of Haiti. *Science*, **219**, 1423–1425.

Dolan, R., Anders, F. and Kimball, S., 1988: *National Atlas of coastal erosion and accretion*. US Geological Survey, Denver, Colorado.

Doyle, L.J., Sharma, D.C., Hine, A.C., Pilkey, O.H. Jr., Neal, W.J., Pilkey, O.H. Sr., Martin, D. and Belknap, D.F., 1984: *Living with the west Florida shore*. Duke University Press, Durham, NC, 222 pp.

Earle, T.A, 1987: The development and management of the Atlantic coast of Guyana in Coastal Zone Management of the Caribbean Region: a Status Report. CSC Tech. Rept. No. 227, Commonwealth Science Council, London, pp. 1–8.

Ellison, J.C., and Stoddart, D.R., 1991: Mangrove ecosystem collapse during predicted sea-level rise: Holocene analogues and implications. *J. Coast. Res.*, **7**, 151–166.

EPA, 1986: *Effects of changes in stratospheric ozone and global climate, Volume 4: sea-level rise*. US Environmental Protection Agency/United Nations Environment Programme, 193 pp.

EPA, 1987: Saving Louisiana's coastal wetlands: the need for a long-term plan of action. Report of the Louisiana Wetland Protection Panel, Louisiana Geological Survey. US Environmental Protection Agency, 102 pp.

Evans, M.W., Hine, A.C., Belknap, D.F. and Davis, R.A., 1985: Bedrock controls on barrier island coastal development: west-central Florida coast. *Mar. Geol.*, **63**, 263–283.

Evans, M.W., Hine, A.C. and Belknap, D.F., 1989: Quaternary stratigraphy of the Charlotte harbor estuarine-lagoon system, southwest Florida: implications of the carbonate-siliciclastic transition. *Mar. Geol.*, **88**, 319–348.

Fairbanks, R.G., 1989: A 17,000-year glacio-eustatic sea-level record: influence of glacial melting rates on the Younger Dryas event and deep-ocean circulation. *Nature*, **342**, 637–642.

Fairbridge, R.W., 1974: The Holocene sea-level record in south Florida. In: Gleason, P.J. (ed.), *Environments of South Florida: Present and Past*. Miami Geol. Soc., Memoir No.2, 223–232.

Fairbridge, R.W., 1989: Crescendo events in sea-level changes. *J. Coast. Res.*, **5**, ii–v.

Froidefond, J.M., Berthois, L., Griboulard, R., Julius, C. and Pons, J.C., 1985: Terrases submergees d'origine recifale variations du niveau marin at activite neotectonique sur le plateau sud et est de la Martinique. Symposium, Geodynamique des Caraibes, Editions Technip, Paris, pp. 143–154.

Gable, F., Gentile, J. and Aubrey, D., 1989: Global environmental change and its related effects (sea-level rise) on the coastal Caribbean. Carib. Cons. Assoc., Ann. Gen. Meeting, Curacao, Netherlands Antilles, p. 21.

Galvin, C., 1983: Sea level rise and shoreline recession. Coastal Zone '83, *Amer. Soc. Civil Eng.*, **3**, 2684–2705.

Garrabe, F., and Andreieff, P., 1985: Sedimentation et tectonique Plio-Quaternaires comparees de Marie-Galante et de Grande-Terre (Guadeloupe). Symposium, Geodynamique des Caribes, Editions Technip, Paris, pp. 155–160.

Georges, C., Hudson, D. and Bachew, S., 1988: The dynamics of Erin Spit, south Trinidad. Transactions, 11th Caribbean Geological Conference, Barbados, pp. 11.1–11.13.

Gibbs, M.J., 1984: Economic analysis of sea-level rise: methods and results. In: Barth, M.C., Titus, J.G., and Ruckelshaus W.D. (eds), *Greenhouse Effect and Sea Level Rise*, Van Norstrand Rheinhold Co., pp. 215–252.

GOJ and Ralph Fields Associates, 1987: Jamaica: country environmental profile. Natural Resources Conservation Division, Government of Jamaica and Ralph Fields Associates, Inc., for International Institute for Environment and Development, 362 pp.

Goodell H.G. and Garman, R.K., 1969: Carbonate Geochemistry of Superior Deep Test Well, Andros Island, Bahamas. *Assoc. Pet. Geol. Bull.*, **53**, 513–536.

Gornitz, V. and Lebedeff, S., 1987: Global sea-level changes during the past century. *Soc. Econ. Pal. Mineral. Spec. Publ. No. 41*, 3–16.

Gornitz, V., and Kanciruk, P., 1989: Assessment of global coastal hazards from sea-level rise. Proc. 6th. Symp. Coastal and Ocean Managt., Charleston, SC, Amer. Soc. Civil Eng., 1345–1359.

Graf, C.H., 1972: Relaciones entre tectonismo y sedimentacion en el Holoceno del noroeste de Venezuela. Bolotin de Geologia, Memoria IV Congreso Geologica Venezolano, Tomo II, Caracas, 1126–1136.

Gustadt, A.M., 1990: Coastal processes and legal boundaries. *J. Geol. Eductn.*, **38**, 94–100.

Hall, M.J., Young, R.S., Thieler, E.R., Priddy, R.D. and Pilkey, O.H. Jr., 1990: Shoreline response to Hurricane Hugo. *J. Coast. Res.*, **6**, 211–221.

Hare, P.W., and Gardner, T.W., 1985: Geomorphic indicators of vertical tectonism along convergent plate margins, Nicoya Peninsula, Costa Rica. In: Hack J. and Morisawa M. (eds), Tectonic Geomorphology. Proc. 15th Geomorph. Symp. Ser., Binghampton, New York, pp. 76–104.

Hekstra, G.P., 1989: Global warming and rising sea levels: the policy implications. *The Ecologist*, **19**, 4–15.

Hemsley, J.M., 1990: Monitored completed coastal projects: status of a programme. *J. Coast. Res.*, **6**, 253–263.

Hendry, M.D., 1979a: Historical evidence of shoreline evolution for the Palisadoes, Kingston, Jamaica. *J. Geol. Soc. Jamaica*, **17**, 39–48.

Hendry, M.D., 1979b: A study of coastline evolution and sedimentology: the Palisadoes, Jamaica. Ph.D. Dissertation, Univ. of the West Indies, Kingston, 232 pp.

Hendry, M.D., 1987: Tectonic and eustatic control on late Cenozoic sedimentation within an active plate boundary zone, west coast margin, Jamaica. *Geol. Soc. Amer. Bull.*, **99**, 718–728.

Hendry, M.D., 1988: Geology: physical processes, coastal impacts. In: Maul, G.A. (ed.), *Implications of Climatic Changes in the Wider Caribbean Region.* UNEP(OCA)/CAR WG.1/INF. 3, Regional Coordinating Unit, Kingston, Jamaica, pp. 53–64.

Hendry, M.D., 1990a: Geology and beach management. In: Bacon, P.R., *Assessment of the Environmental Impact of Resort and Residential Development on West Caicos.* West Caicos Development and Construction Company, p. 115.

Hendry, M.D., 1990b: 6000 years of shoreline migration and cannibalization, south Jamaica. Int. Geol. Corrltn. Prog. Project 200/ORSTOM, Int. Symp. Quaternary Coast. Changes in South Amer. and the Caribbean, Cayenne, French Guiana, Abstract.

Hendry, M.D., and Bacon, P.R., 1990: Hurricane impacts on Caribbean beaches: the development of a data base and guidelines for planning and management. Proc. 12th Carib. Geol Conf., St. Croix, USVI, Miami Geological Society, pp. 5–9.

Hendry, M.D. and Digerfeldt, G., 1989: Paleogeography and paleoenvironments of a tropical coastal wetland and adjacent shelf during Holocene submergence, Jamaica: *Paleogeography, Paleoclimatology, Paleoecology*, **73**, 1–10.

Hendry, M.D. and Head, S.M., 1985: Late Quaternary sea-level changes and the development of the raised reef/dune sequence at Great Pedro Bluff, southwest Jamaica. Proc. 5th Int. Coral Reef Cong., Tahiti, 3, 119–124.

Hendry, M.D. and Loucks, R.A., 1991: Global change and coastal land loss: management and decision-making in support of sustainable coastal area management within the Caribbean and adjacent regions. Program Proposal prepared for the Intergovernmental Oceanographic Commission, Sub-Commission for the Caribbean and Adjacent Regions (IOCARIBE), Cartagena, Colombia, p. 56.

Hendry, M.D. and Nurse, L.A., 1991: Shoreline erosion and accretion: a key factor for sustainable development of Caribbean coastal resources. Proc. of the CIDA/UWI Workshop on Sustainable Development in the Caribbean, Univ. of the West Indies, Barbados.

Henneberg, H.G., 1983: Geodetic control of neotectonics in Venezuela. *Tectonophysics*, **97**, 1–15.

Hine, A.C., Wilbur, R.J. and Neumann, A.C., 1981: Carbonate sand bodies along contrasting shallow bank margins facing open seaways in Northern Bahamas. *Amer. Assoc. Pet. Geol. Bull.*, **65**, 261–290.

Holmes, C.W. and Kindinger, J.L., 1985: Late Pleistocene-Holocene geology of the central Virgin Island platform. *Mar. Geol.*, **64**, 41–64.

Horsfield, W.T., 1972: A late Pleistocene sea-level notch and its relation to block faulting on the north coast of Jamaica. *J. Geol. Soc. Jamaica*, **12**, 18–22.

Horsfield, W.T., 1973: Late Tertiary and Quaternary crustal movements in Jamaica. *J. Geol. Soc. Jamaica*, **13**, 6–13.

Horsfield, W.T., 1974: Major faults in Jamaica. *J. Geol. Soc. Jamaica*, **14**, 1–14.

Horsfield, W.T., 1975: Quaternary vertical movements in the Greater Antilles. *Geol. Soc. Amer. Bull.*, **86**, 933–938.

IGCP, 1989: Report of the International Geological Correlation Program. *Geol.* Correlation, No. 17, IGCP/IUGS/UNESCO, Paris, p. 138.

INQUA, 1990a: Americas Subcommission of INQUA Commission on Quaternary Shorelines, Newsletter No. 3, International Union for Quaternary Research, ORSTOM, Cayenne, French Guiana, p. 21.

INQUA, 1990b: Americas Subcommission of INQUA Commission on Quaternary Shorelines, Newsletter No. 4, International Union for Quaternary Research, ORSTOM, Cayenne, French Guiana, p. 29.

IOC (Intergovernmental Oceanographic Commission), 1988: IOC Task Team on the Global Sea-Level Observing System (GLOSS), Report of the Second Session, Honolulu, Hawaii, Intergovernmental Oceanographic Commission, Paris, 23 pp. + Annexes.

IOCARIBE (IOC Sub-Commission for the Caribbean and Adjacent Regions), 1986: IOCARIBE Workshop on Physical Oceanography and Climate. IOC Workshop Report No. 45, Intergovernmental Oceanographic Commission, Paris.

IOCARIBE, 1989: Interdisciplinary seminar on research problems in the IOCARIBE region. IOC Report No. 67, IOCARIBE, Cartagena, Colombia, 37 pp. + Annexes.

IPCC (Intergovernmental Panel on Climate Change) 1990: Strategies for adaptation to sea-level rise. Report of the Coastal Zone Management Subgroup, Response Strategies Working Group, Ministry of Transport and Public Works, The Hague, The Netherlands, p. 122.

Jones, E.B., 1986: Coastline vulnerability study of the Caribbean. Report to Pan-Caribbean Disaster Planning and Preparedness Project, Caritech Associates Ltd., Jamaica, 5 vols.

J.R.B. Associates, 1981: The Dominican Republic: Country Environmental Profile. United States Agency for International Development, 118 pp.

Juanes, J.L., Ramirez, E. and Medvediev, V.S. 1985: Dinamica de los sedimentos en la Peninsula de Hicacos, Cuba, I: variaciones morfologicas del perfil de playa. *Ciencias de la Tierra y del Espacio*, **10**, 69–84.

Juanes, J.L., Ramirez, E., Caballero, M., Medvediev, V.S. and Yurkevitch, M.G., 1986: Dinamica de los sedimentos en la Peninsula de Hicacos, Cuba, II: Efecto de las olas de viento en la zona costera. *Ciencias de la Tierra y del Espacio*, **11**, 93–102.

Kelley, J.T., Kelley, A.R., Pilkey, O.H. Sr. and Clark, A.A., 1984: *Living with the Louisiana shore*. Duke Univ. Press, Durham, NC, 164 pp.

Kjerve, B., and Dinnel, S.P., 1983: Hindcast hurricane characteristics on the Belize barrier reef. *Coral Reefs*, **1**, 203–207.

Kohsiek, L.H.M., Hulsbergen, C.H. and Terwindt, J.H.J., 1987: Beach erosion along the west coast of Aruba, Netherlands Antilles. *J. Coast. Res.*, **3**, 37–53.

Kraus, N.C. and Pilkey, O.H. (eds), 1988: The effects of seawalls on beaches. *J. Coast. Res.*, Spec. Issue No.4.

Krook, L., 1990a: Evidence of Amazon provenance of part of sandy sediments in the coastal and shelf areas of the Guianas. Symposium International Sur L'evolution des Littoraux des Guyanes et de la Zone Caraibe Meridionale Pendant le Quaternaire. ORSTOM/IGCP 274, Cayenne, French Guyana, Abstracts, 92–95.

Krook, L., 1990b: Geological and morphological evolution in the hinge zone of northern Surinam. Symposium International Sur L'evolution des Littoraux des Guyanes et de la Zone Caraibe Meridionale Pendant le Quaternaire. ORSTOM/IGCP 274, Cayenne, French Guyana, Abstracts, 97–101.

Leatherman, S.P., 1984: Coastal responses to sea-level rise: Galveston Bay, Texas. In: Barth, M.C. and Titus J.G. (eds), *Greenhouse Effect and Sea Level Rise: a Challenge for this Generation*. Van Norstrand Rheinhold Co., pp. 151–178.

Leggett, J. (ed.), 1990: *Global warming: the Greenpeace report*. Oxford Univ. Press, 554 pp.

Leonard, L., Clayton, T. and Pilkey, O.H., 1990: An analysis of replenished beach design parameters on US east coast barrier islands. *J. Coast. Res.*, **6**, 15–36.

Lewis, J.F. and Draper, G., 1990: Geology and tectonic evolution of the northern Caribbean region. In: Dengo, G. and Case, J.E. (eds), *The Geology of North America, Vol. H, The Caribbean Region*. Geol. Soc. Amer., pp. 77–140.

Lidz, B.H., and Shinn, E.A., 1991: Paleoshorelines, reefs and a rising sea: south Florida, USA. *J. Coast. Res.*, **7**, 203–230.

Lundberg, N., 1982: Evolution of the slope landward of the Middle America Trench, Nicoya Peninsula, Costa Rica. In: Leggett, J.K. (ed.), *Trench-Forearc Geology*. Geol. Soc. Amer., Spec. Publ. 10, 133–147.

Mann, P., Schubert, C. and Burke, K., 1990: Review of Caribbean neotectonics. In: Dengo, G., and Case, J.E. (eds), *The Geology of North America, Vol. H., The Caribbean Region*, Geol. Soc. Amer., pp. 307–338.

Mann, P., Taylor, F.W., Burke, K. and Kulstad, R., 1984: Subaerially exposed coral reef, Enriquillo Valley, Dominican Republic. *Geol. Soc. Amer. Bull.*, **95**, 1084–1092.

Martin, L., Flexor, J.M., Blitsgow, D. and Suguio, K., 1985: Geoid indications along the Brazilian coast during the last 7000 years. Proc. 5th. Int. Coral Reef Cong., Tahiti, 3, 85–90.

Maul, G.A., 1990: Climate changes and physical oceanography in the Caribbean region. 2nd Congress on Marine Sciences, Havana, Cuba, Abstract.

Maul, G.A. and Hanson, K., 1990: A Century of Southeastern United States Climate Change Observations: Temperature, precipitation and sea-level. *Global Change: A Southern Perspective*, Southeast Regional Climate Center, Columbia, SC, pp. 139–155.

Maury, R.C., Westbrook, G.K., Baker, P.E., Bouysse, Ph. and Westercamp, D., 1990: Geology of the Lesser Antilles. In: Dengo G. and Case, J.E. (eds), *The Geology of North America, Vol. H, The Caribbean Region*; Geol. Soc. Amer., pp. 141–166.

McElroy, J. and de Albuquerque, K., 1989: Tourism styles and policy responses in the open economy-closed market context. Conf. on Economics and the Environment, Carib. Cons. Assoc., Bridgetown, Barbados, p. 18.

McGowen, J.H., and Scott, A.J., 1975: Hurricanes as geologic agents on the Texas coast. *Estuarine Res.*, **2**, 23–46.

Morelock, J., 1984: Coastal erosion in Puerto Rico. *Shore and Beach*, January 1984, 18–27.

Morner, N.A., 1990: Sea-level changes and paleoclimate in view of ocean circulation changes and the key position of data from the Brazilian-Guiana region. Symposium International Sur L'evolution des Littoraux des Guyanes et de la Zone Caraibe Meridionale Pendant le Quaternaire. ORSTOM/IGCP 274, Cayenne, French Guyana, Abstracts, 136–139.

Morton, R.A., Pilkey, O.H. Jr., Pilkey, O.H. Sr. and Neal, W.J., 1983: *Living with the Texas shore*. Duke Univ. Press, Durham, NC, 190 pp.

Neumann, A.C., 1971: Quaternary sea-level data from Bermuda. *Quaternaria*, **14**, 41–43.

Nurse, L.A., 1989: Global sea-level rise: a statement from the island of Barbados. Small States Conference on Sea-Level Rise, Ministry of Planning and Environment, Mali, p. 9.

O'Brien, C.B. and Lawson, D., 1988: Nearshore processes and sedimentation at Queens and Richmond Bays, Tobago, West Indies. Trans. 11th Caribbean Geological Conference, Barbados, pp. 14.1–14.25.

Page, W.D., 1983: Holocene deformation of the Caribbean coast, northwestern Colombia. 10th Carib. Geol. Conf., Cartagena, Colombia, Abstract.

Parkinson, R.W., 1989: Decelerating Holocene sea-level rise and its influence on southwest Florida coastal evolution: a transgressive/regressive stratigraphy. *J.*

Sed. Pet., **59**, 960–972.

Penland, S. and Ramsey, K.E., 1990: Relative sea-level rise in Louisiana and the Gulf of Mexico: 1908–1988. *J. Coast. Res.*, **6**, 253–502.

Pilkey, O.H., 1990: A time to look back at beach replenishment. Editorial, *J. Coast. Res.*, **6**, iii–vii.

Pilkey, O.H., Morton, R.A., Kelley, J.T. and Penland, S., 1989: *Coastal land loss*. Amer. Geophys. Union, Washington, DC, 73 pp.

Pilkey, O.H., Sharma, D.C., Wanless, H.R., Doyle, L.J., Pilkey, O.H. Sr., Neal, W.J. and Gruver, B.L., 1984: *Living with the east Florida shore*. Duke Univ. Press, Durham, NC, 259 pp.

Pindell, J.L., and Barrett, S.F., 1990: Geological evolution of the Caribbean region; a plate-tectonic perspective. In: Dengo, G. and Case, J.E. (ed.), *The Geology of North America, Vol. H, The Caribbean Region*. Geol. Soc. Amer., pp. 405–432.

Pirazzoli, P., 1986: Secular trends of relative sea-level (RSL) changes indicated by tide-gauge records. *J. Coast. Res.*, Special Issue No.1, 1–26.

Pirazzoli, P. and Grant, D.R., 1987: Lithospheric deformation deduced from ancient shorelines. In: Kasahara, K. (ed.), *Recent Plate Movements and Deformation*. Geodynamics Series, Vol. 20, Amer. Geophys. Union and Geol. Soc. Amer., pp. 67–72.

Procter and Redfern, 1983: Coastal Conservation Study, Technical Reports. Procter and Redfern International Ltd., for Ministry of Housing and Lands, Govt. of Barbados, 8 vols.

Procter and Redfern, 1984: Coastal Conservation Study, Diagnostic Report. Procter and Redfern International Ltd., for Ministry of Housing and Lands, Govt. of Barbados.

Prost, M.T., 1990: *Les cotes des Guyanes*. ORSTOM, Cayenne, French Guiana, p. 213.

Prost, M.T., Lointier, M. and Charron, C., 1990: La zone de Sinnamary: donnees geomorphologiques, hydrologiques et teledetection. In: Excursion Guide B, Les Plaines de Kourou et de Sinnamary, Symposium internationale sur l'evolution des littoraux des Guyanes et de la zone Caraibe meridionale pendant le Quaternaire, ORSTOM, Cayenne, French Guiana, pp. 12–36.

Reading, A.J., 1990: Caribbean tropical storm activity over the past four centuries. *Int. J. Climatol.*, **10**, 365–376.

Rine, J.M. and Ginsburg, R.N., 1985: Depositional facies of a mud shoreface in Surinam, South America – a mud analogue to sandy, shallow-marine deposits. *J. Sed. Pet.*, **55**, 633–652.

Roeleveld, W. and Van Loon, A.J., 1979: The Holocene development of the Young Coastal Plain of Surinam. *Geol. en Mijnbouw*, **58**, 21–28.

Sanoja, M., 1989: Origins of cultivation around the Gulf of Paria, northeastern Venezuela. *National Geog. Res.*, **5**, 446–458.

Schubert, C. and Cowart, J.B., 1980: Terrazas marinas del Pleistoceno a lo largo de la costa suroriental de la Republicana Dominica: cronologia priliminar. Trans. 9th. Carib. Geol. Conf., Santo Domingo, Dominican Republic, pp. 681–688.

Schubert, C. and Laredo, M., 1984: Geology of Aves Island (Venezuela) and subsidence of Aves Ridge, Caribbean Sea. *Mar. Geol.*, **59**, 305–318.

Schubert, C. and Szabo, B.J., 1978: Uranium-series ages of Pleistocene marine deposits on the islands of Curacao and La Blanquilla, Caribbean Sea. *Geol. en Mijnbouw*, **57**, 325–332.

Schubert, C. and Valastro, S., 1976: Quaternary geology of La Orchilla Island, central Venezuelan offshore, Caribbean Sea. *Geol. Soc. Amer. Bull.*, **87**, 1131–1142.

Schubert, C., Valastro, S. and Coward, J.B., 1977: Evidencias de levantamiento reciente de la norte-central (Cordillera de la Costa), Venezuela. *Acta Scientifica Venezolana*, **28**, 363–372.

Schwartz, M.L. and Anderson, B.D., 1986: Coastal geomorphology of Padre Island, Mexico. *Shore and Beach*, **54**, 22–29.
Schwartz, M.L. and Pilkey, O.H. (eds), 1990: Artificial beaches. *J. Coast. Res.*, Spec. Issue No. 6, p. 140.
Shinn, E.A., Lidz, B.H., Kindinger, J.L., Hudson, J.H. and Halley, R.B., 1989: *Reefs of Florida and the Dry Tortugas: A guide to the modern carbonate environments of the Florida Keys and Dry Tortugas*. US Geol. Survey, St. Petersburg, Florida, 53 pp.
Speed, R.C., 1987: Caribbean geological evolution. Report to Joint Oceanographic Inst., US Science Adv. Comm., US Geol. Survey and Nat. Sci. Foun., Dept. Geol. Sciences, Northwestern University, p. 211.
Speed, R.C., 1988: Geologic history of Barbados: a preliminary synthesis. Trans. 11th Carib. Geol. Conf., Barbados, pp. 29.1–29.11.
Stoddart, D.R., 1963: Effects of Hurricane Hattie on the British Honduras reefs and cays, 30–31 October 1961. *Atoll Res. Bull.*, **95**, 1–142.
Stoddart, D.R., 1969: Post-hurricane changes on the British Honduras reefs and cays: Re-survey of 1965. *Atoll Res. Bull.*, **131**, 1–25.
Stoddart, D.R., 1974: Post-hurricane changes on the British Honduras reefs and cays: Re-survey of 1972. Proc. 2nd. Int. Coral Reef Symp., 2, 473–483.
Strasser, A. and Davaud, E., 1986: Formation of Holocene limestone sequences by progradation, cementation and erosion. *J. Sed Pet.*, **56**, 422–428.
Suguio, K., Martin, L. and Flexor, J.M., 1988: Quaternary sea-levels of the Brazilian coast: recent progress. *Episodes*, **11**, 203–208.
Taylor, F.W., Mann, P., Valastro S. Jr. and Burke, K., 1985: Stratigraphy and radiocarbon chronology of a subaerially exposed Holocene coral reef, Dominican Republic. *J. Geol.*, **93**, 311–332.
Thieler, E.R., 1990: Coastal erosion in Puerto Rico: the need for multiple solutions. The Coastal Society, 12th Ann. Conf., Abstract.
Thieler, E.R., Bush, D.M. and Pilkey, O.H., 1989: Shoreline response to Hurricane Gilbert: lessons for coastal zone management. Proc. Sixth Coast. and Ocean Mangt., Amer. Soc. Civil Eng., Charleston, SC, pp. 765–775.
Titus, J.G., 1986: Greenhouse effect, sea-level rise and coastal zone management. *Coastal Zone Manag. J.*, **14**, 147–172.
Titus, J.G., Chin, Y.K., Gibbs, M.J., LaRoche, T.B., Webb, M.K. and Waddell, J.O., 1987: Greenhouse effect, sea-level rise and coastal drainage systems. *J. Water Res. Planning Mgt.*, **113**, 216–227.
Towle, E.L., 1985: The island microcosm. In: Clarke, J.R. (ed.), *Coastal Resources Management: Development Case Studies*. Research Planning Institute, Columbia, SC, pp. 589–738.
UNEP (United Nations Environment Program), 1989a: Implications of climatic changes in the Wider Caribbean Region: preliminary conclusions of the Task Team of Experts. UNEP, Carib. Envt. Prog., Tech. Rept. No. 3, 39 pp.
UNEP (United Nations Environment Program), 1989b: Regional overview of environmental problems and priorities affecting the coastal and marine resources of the Wider Caribbean. UNEP, Carib. Envt. Prog., Tech. Rept. No. 2, 39 pp.
Vellinga, P. and Leatherman, S.P., 1989: Sea-level rise, consequences and policies. *Climatic Change*, **15**, 175–189.
Vernette, G., Leseur, P. and Klingebiel, A., 1984: Evolucion morfologia y sedimentologica de la flecha litoral del laguito (Bahia de Cartagena). *Centro de Investigaciones Oceanograficas e Hidrograficas, Boletin Cientifico, No. 5*, 3–23.
Wadge, G. and Hudson, D., 1985: Neotectonics of southern Tobago. In: Rodrigues, K.R. (ed.), Trans. First Geol. Conf., Geol. Soc. Trinidad and Tobago, pp. 7–20.
Wells, J.T. and Coleman, J.M., 1978: Longshore transport of mud by waves: northeastern coast of South America. *Geol. en Mijnbouw*, **57**, 353–359.

Wells, S. and Edwards, A., 1989: Gone with the Waves: *New Scientist*, November, pp. 47–51.

Westphal, M.J., 1986: Anatomy and history of a ringed-reef complex, Belize, central America. MS Thesis, Univ. Miami, 135 pp.

Wilber, J.R., 1986: Carbonate island response to sea-level rise: Little San Salvador and West Plana Cay. 3rd Symposium on the Geology of the Bahamas, CCFL Bahamian Field Station, Abstracts and Program.

Williams, B.P., 1988: Grand Turk coastal study, Little Bluff Point to English Point, stage 1 report. W.A. Dawson Ltd. and Posford Duvivier, Report to Govt. of Turks and Caicos Islands, 38 pp. + appendices.

Williams, M., 1986: Beach sand mining in St. Lucia. In: Geoghegan, T. (ed.), *The Caribbean Seminar on Environmental Impact Assessment, Case Studies*. CERMES, Univ. of the West Indies/Carib. Cons. Assoc./IRES, Dalhousie University, pp. 1–22.

WMO/UNEP (World Meteorological Organisation/United Nations Environment Programme), 1990: *Climate Change: the IPCC response strategies*. Intergovernmental Panel on Climate Change, 270 pp.

Wong, T.E., 1990: Quaternary stratigraphy of Surinam. Symposium International sur L'evolution des Littoraux des Guyanes et de la Zone Caraibe Meridionale Pendant le Quaternaire, ORSTOM/IGCP 274, Cayenne, French Guiana, Abstracts, p. 197.

Wood, P.A., 1976: Beaches of accretion and progradation in Jamaica. *J. Geol. Soc. Jamaica*, **15**, 24–31.

Woodroffe, C.D., Stoddart, D.R., Harmon, R.S. and Spencer, T., 1983: Coastal morphology and Late Quaternary history, Cayman Islands, West Indies. *Quat. Res.*, **19**, 64–84.

8

River Discharge Variability Including Satellite-Observed Plume-Dispersal Patterns

Frank E. Muller-Karger[1]

Abstract

Time-series (1978–1985) of Coastal Zone Colour Scanner (CZCS) images of the Western Atlantic were examined to determine the dispersal pattern of the Amazon (Brazil), Orinoco (Venezuela), Magdalena (Colombia), Rio Grande (Mexico/United States), and Mississippi (United States) river water in the region. These data were examined in the context of long time-series of annual mean river discharge and rainfall to characterize these rivers, detect secular trends in discharge, and determine present-day surface patterns of plume dispersal.

The discharge record length varied from 171 years (Mississippi, 1817–1989) to four years (Magdalena, 1969–1972). Thus, many of these data were not useful for addressing climate-change questions. While rainfall did not show secular trends, the discharge for both the Amazon (84 years) and Orinoco (23 years) showed small but significant positive linear trends. The Mississippi's record (171 years) did not show a trend. Neither the Orinoco nor the Amazon discharge tracked annual mean rainfall well. There was a strong correlation between mean annual rainfall in southern Texas and discharge of the Rio Grande, and between rainfall along the Mississippi and its discharge, showing that discharge trends can be inferred by pooling a large number of meteorological stations. Hydrographs suggest a general increase in interannual variability as a result of human impact. Small changes in rainfall can lead to significant variations in the flow of a river. These results point to the importance of increasing the geographical density of meteorological observations in South America.

The CZCS satellite images show that plume dispersal depends primarily on the seasonal variation of the wind and current system. The Amazon has a reduced direct effect on the Caribbean Sea between June and January, when it is carried by the North Equatorial Counter Current (NECC) toward Africa. During this period, the Orinoco River plume engulfs Trinidad, Tobago, Grenada, St. Vincent and St. Lucia, but not Barbados, and reaches Puerto Rico. From about February to May, the NECC weakens and Amazon water drifts into the Caribbean Sea, also affecting Barbados. Both Amazon and Orinoco water tend to remain south of 14°N during this time due to a larger influx of North Atlantic water into the basin forced by strong Trade Winds. Interannual variability was not quantifiable with this short series. The Mississippi dispersed primarily along the coast to the west of its delta, without showing dramatic change between seasons. However, there were enormous differences in the size of the plume between years, consistent

[1]University of South Florida, St. Petersburg, FL 33701, USA

with variations in discharge. Typically, the Magdalena also extended west along the coast.

1 Introduction

Within the last decade, international concern about a potential change in climate due to anthropogenic generation of greenhouse gases has grown, leading to a hunt for evidence of secular temperature trends on regional and global scales. The concern has spurned the development of models that can predict global temperature increase and sea-level rise. As envisioned by the World Meteorological Organization (WMO), the International Council of Scientific Unions (ICSU), and the United Nations Environment Programme (UNEP), an increase of 1.5°C in temperature and 20 cm rise in sea level may be a realistic *global* climate-change scenario by the year 2025 (see Maul, Chapter 1). Such changes will be accompanied by important geographical, geophysical, and social changes in the Caribbean Sea and Gulf of Mexico region. As pointed out by Maul, however, the local response to global change is not predictable at present. Therefore, it is imperative that we begin reconstructing the environmental history of the region and establish a series of environmental monitoring programmes so that changes can be detected if and when they occur.

This study is the result of an effort by UNEP, in cooperation with the Intergovernmental Oceanographic Commission (IOC) and in consultation with the University of South Florida, to provide background information on the variability in discharge of the largest rivers affecting the region (Fig. 8.1). Stage height or proxy discharge data for the Amazon (Brazil), Orinoco (Venezuela), Magdalena (Colombia), Rio Grande (Mexico/United States) and Mississippi (United States) rivers were assembled and are presented with general rainfall statistics over the respective catchment areas. Unfortunately the series for the Magdalena and the Rio Grande were too short to address climate-change issues, and the Rio Grande discharge data are influenced by damming. This study also includes an overview of the patterns of river plume dispersal as observed from space using the Coastal Zone Colour Scanner (CZCS). This information will be used to determine if significant trends in river discharge have a noticeable impact on the region in the near future. Finally, recommendations are presented for river studies that are needed to understand better the repercussions of climate change in the region.

2 Objectives

The information presented in this report is intended to document the availability and limitations of river discharge and rainfall data for the largest rivers of South, Central and North America affecting the region: the Caribbean Sea, the northeastern coast of South America, and the Gulf of Mexico (Fig. 8.1). CZCS satellite data for this area are presented to document the near-surface circulation patterns and oceanographic phenomena that affect the colour of the water. It is expected that the discharge, rainfall and satellite data series will help in the development of conceptual models of regional biological oceanographic processes, in addition to help planning of physical oceanographic expeditions and the layout of meteorological stations in the catchment area of the Amazon and Orinoco rivers.

The specific goals of this study were to:

1) Obtain long time-series of stage height or proxy river discharge data to define the character and recent history of the Amazon, Orinoco, Magdalena, Rio Grande and Mississippi rivers and determine if these rivers are experiencing secular trends in discharge.
2) Obtain long time-series of mean precipitation values at locations representative of the catchment area for the rivers listed above to determine if rainfall is experiencing secular trends.
3) Determine any relationship between the rainfall estimates obtained from available meteorological stations and river stage height or proxy

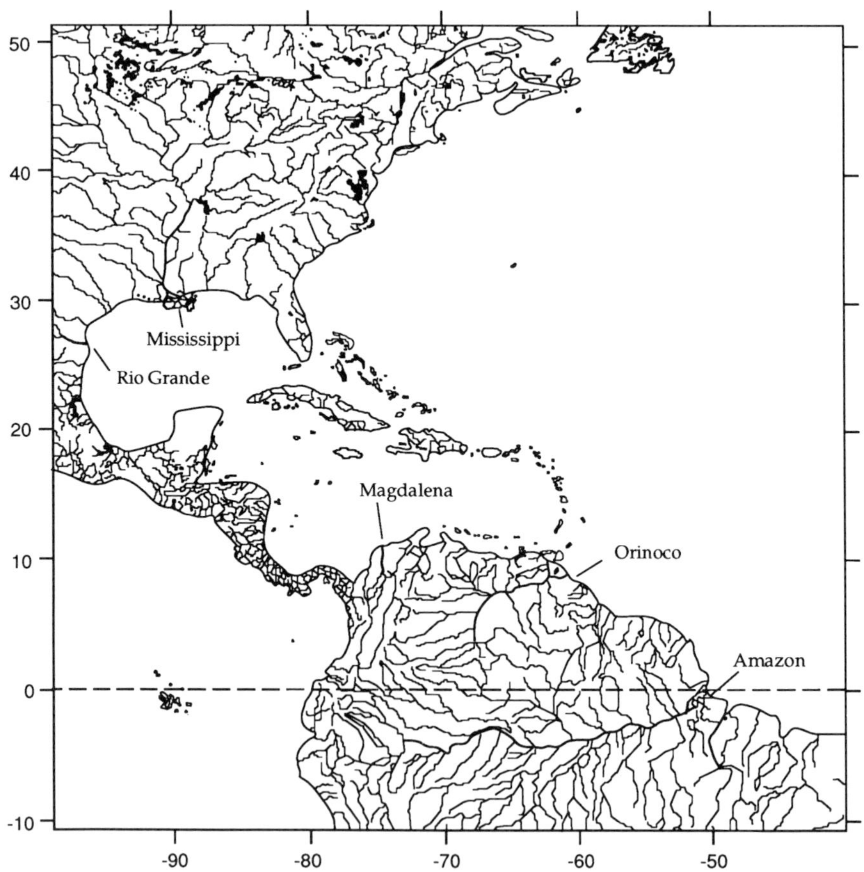

Fig. 8.1 Outline map of the region: the Caribbean Sea, the northeastern coast of South America, the Gulf of Mexico, and the Florida–Bahamas region. Rivers that affect the region were outlined using the World Database II files. The rivers of interest for this study are the Amazon (Brazil), Orinoco (Venezuela), Magdalena (Colombia), Rio Grande (Mexico/United States), and Mississippi (United States).

discharge rates.

4) Obtain a time-series describing the two-dimensional, historical surface patterns of dispersal of the plumes of the Amazon, Orinoco and Mississippi rivers using the Coastal Zone Colour Scanner data set to assess the possible impact of changes in discharge on the region.

3 Methods

3.1 River Discharge Data

Annual mean stage height data (cm) at Manaus on the Rio Negro (Manaus Harbor Ltd-Portobras) were used to estimate flow within the Amazon River (Brazil). The data were obtained from several sources: The period 1903–1985 was read from a time-series plot presented in Sternberg (1987; see also Richey *et al.*, 1989a). The period 1957–1988 was obtained from R. Mahon (pers. comm.). The data for the overlapping period (1957–1985) were the same, so the tabulated data from Mahon were used for the post-1956 period for statistical analyses. The stage height data were converted to flow units (m^3s^{-1}) by calibrating a monthly climatology of stage height against a monthly hydrology obtained from the data in UNESCO (1979).

Annual mean stage height data (cm) at Ciudad Bolivar (1965–1988) were used to document changes in the flow of the Orinoco River (Venezuela). Data are courtesy of W. Lewis and R. Mahon. The stage height data were converted to flow units (m^3s^{-1}) by calibrating a monthly climatology of stage height against a monthly hydrology obtained from the data using a formulation provided by W. Lewis (pers. comm.).

Monthly mean discharge data for the Magdalena River (Colombia) were obtained from the UNESCO river-discharge data base maintained at the NASA Climate Data Service (NCDS), Goddard Space Flight Center, Maryland. Annual means were derived for the brief period covered in the data base (1969–1972). The history of these data including geographical location of the gauging station were not available. These data were the least useful for studying recent trends in discharge affecting the region.

Daily discharge values (cubic feet per second) for the Rio Grande at Brownsville, Texas (United States) were obtained from the United States Geological Survey (Mrs Mary Bell Peters, USGS National Center, National Water Data Exchange, Reston, VA 22092). The data were converted to m^3s^{-1} and annual means were derived from the daily values. These data represented the second shortest record available for this study (1966–1983). The Rio Grande is dammed at a couple of locations (Amistad Reservoir and Falcon Lake), and thus is not a good choice for monitoring changes in climate.

Discharge values (cubic feet per second) for the Mississippi River (United States) were obtained from the Waterways Experiment Center, Mississippi (contact Alan Teeter [601-634-2820], Jim Tuttle [601-634-5888] or Charles Elliott [601-634-5692]). Stations downstream of Vicksburg, Mississippi were probably contaminated around 1963 when the Old River control structure allowed about 30% of the Mississippi discharge into the Atchafalaya River. Only annual mean discharge estimates were available for the period 1817–1929; daily values were available thereafter. Values were converted to m^3s^{-1} and annual means were derived from the daily values for the post-1929 period.

3.2 Annual Mean Rainfall Estimates (mm/month)

As a simple index of rainfall in the catchment areas of interest, data from all available meteorological stations within each basin were used to derive annual mean (mm/month) rainfall estimates. Means are based on 12-monthly values per year, pooled for all basin stations. The geographical density of meteorological stations in South America is extremely low, particularly within the catchment area of the Amazon and Orinoco rivers. In contrast, extremely high geographical data density was available for the Rio Grande and Mississippi rivers. Two data bases were used:

1) For the Amazon, Orinoco and Magdalena rivers, the World Monthly Surface Station Climatology was used, which is based on the World Meteorological Organization (WMO) station summaries. The following stations were extracted:
   a) Amazon River: Manaus (period covered 1910–1987);
   b) Orinoco River: San Fernando, Ciudad Bolivar, Santa Elena de Uairen (period covered 1941–1987);
   c) Magdalena River: Barranquilla, Medellin, Bogota (El Dorado and Meteorological Center) (period covered 1866–1987).

2) For the Rio Grande and the Mississippi rivers, monthly mean precipitation values from the Climatic Division data base of the National Climatic Data Center (NCDC) were used to obtain annual mean precipitation estimates over large areas representative of each catchment area:
   a) Rio Grande: included all the stations within southern Texas (period covered 1895–1988).
   b) Mississippi River (period covered 1895–1988): southwest and southeast Illinois (IL), western Kentucky (KY), Bootheel division in Missouri (MO), east-central and southeast Arkansas (AR), western Tennessee (TN), northwest, central, and southeast Louisiana (LA), and upper delta division and central Mississippi (MS).

3.3 Coastal Zone Colour Scanner Data

A total of 660 CZCS images of the western tropical Atlantic and eastern Caribbean Sea were obtained for the period November 1978 to December 1985. An additional 733 CZCS images covering the western Caribbean Sea and Gulf of Mexico were obtained covering the period November 1978 through May 1980. The individual calibrated raw CZCS radiance images were subsampled every fourth pixel and every fourth line at the Goddard Space Flight Center (NASA, Greenbelt, MD) by Dr Gene Feldman. After correcting for atmospheric radiance components, pigment concentrations (approximately chlorophyll-a plus phaeopigment and coloured dissolved organic matter) were computed on the basis of ratios of the blue (443 nm) or blue-green (520 nm) water-leaving radiances relative to the green radiance (550 nm) (Gordon *et al.*, 1983a; Gordon *et al.*, 1983b).

These radiances are a measure of the visible radiance backscattered from the upper optical depth of the ocean (Clark, 1981), or the upper one-fifth of the euphotic zone.

Images of pigment concentration were geographically mapped to a cylindrical equidistant projection covering the region of interest. Monthly composite images were derived based on the arithmetic mean of congruent pixels showing concentrations between 0.04 and 20 mg m^{-3}. The average concentrations were colour coded, using purple and blue for low concentrations, and increasingly warmer colours for higher concentrations. Locations with no valid pixels due to the presence of clouds or due to lack of programmemed coverage were assigned a black colour. An unquantified and uncorrected source of error in the pigment data is due to colour changes of the water when dissolved organic matter is present, as occurs in river plumes. Possible sources of error in this region are described in detail by Muller-Karger *et al.* (1989). Using a mass-balance approach, Muller-Karger *et al.* estimate that at least 50% of the pigment-concentration values derived with the CZCS within a river plume (i.e., the Orinoco plume) represents viable phytoplankton biomass, even at distances of 800–1000 km from the mouth of the river. This estimate has recently been substantiated by direct ship observations in the Orinoco plume (R. Bidigare, 1990, pers. comm.).

The discolouration of river plumes due to the presence of phytoplankton and dissolved organic matter can be used below as a visual tracer of the circulation. Inferences about the circulation are made based on the distribution of pigments. This approach is justified since, when the source terms for physical properties (i.e., sources of salinity, temperature, nutrients, etc.) are larger than the loss terms due to turbulence and advection, patches of these properties remain coherent.

4 Results

River Discharge and Annual Mean Rainfall Estimates

The time-series of daily river discharge data were used to derive a monthly mean hydrology (Fig. 8.2). There is a clear difference in the phasing of the peak discharge of the Amazon, Orinoco, and Mississippi rivers. The discharge of tropical rivers depends primarily on the position of the Intertropical Convergence Zone (ITCZ), while the Mississippi derives its flood waters from melting snow during the Northern Hemisphere spring.

Fig. 8.3 shows the series of annual mean stage height values (m) and mean rainfall estimates (m/month) for the Amazon at Manaus, and Fig. 8.4 shows a scatter plot of these two variables. The summary statistics for the period 1903–1987 (last year of rainfall data) are given in Table 8.1. The serial correlation statistics between stage height and rainfall data were derived based on annual mean values of these quantities, with no time lag.

Similarly, Fig. 8.5 shows the series of annual mean stage height values (m) for the Orinoco at Ciudad Bolivar and mean rainfall estimates (m/month) for the Orinoco catchment area over the same period. Fig. 8.6 shows a scatter plot of the two variables. The summary statistics for

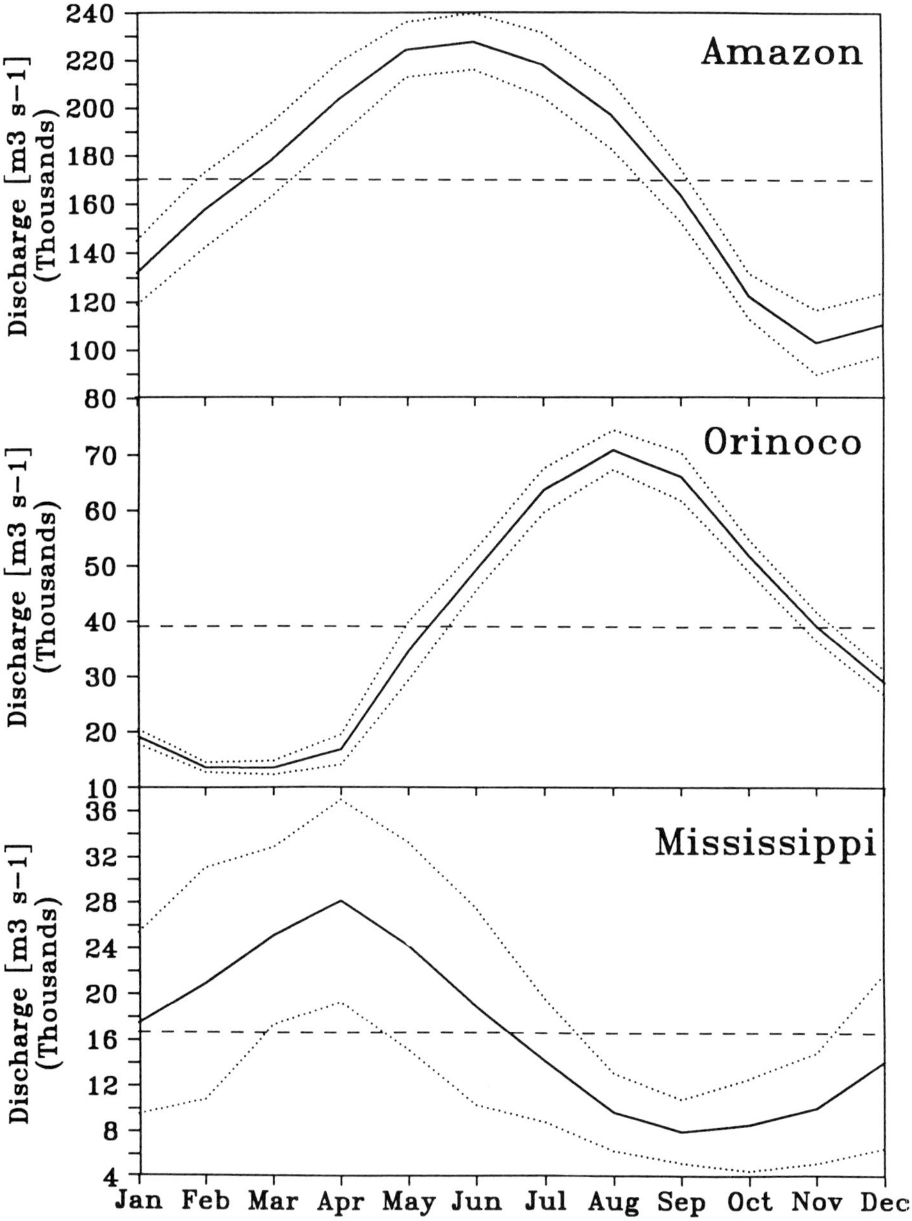

Fig. 8.2 Mean (solid lines) plus or minus one standard deviation (stippled line) of the monthly mean discharge (m^3s^{-1}) of the Amazon River at Manaus, Brazil (1969–1978), Orinoco River at Ciudad Bolivar, Venezuela, (1979–1983; includes a correction for the Caroni River), and Mississippi River at Vicksburg, United States (1929–1989). The broken line indicates the long-term annual mean discharge value.

the period 1965–1988 (Table 8.1) also were based on annual mean values of these quantities with no time lag.

Fig. 8.7 shows the series of annual mean discharge rates (m^3s^{-1}) for the Rio Grande at Brownsville and mean rainfall estimates (mm/month) for the south Texas region, and Fig. 8.8 shows the scatter between the two

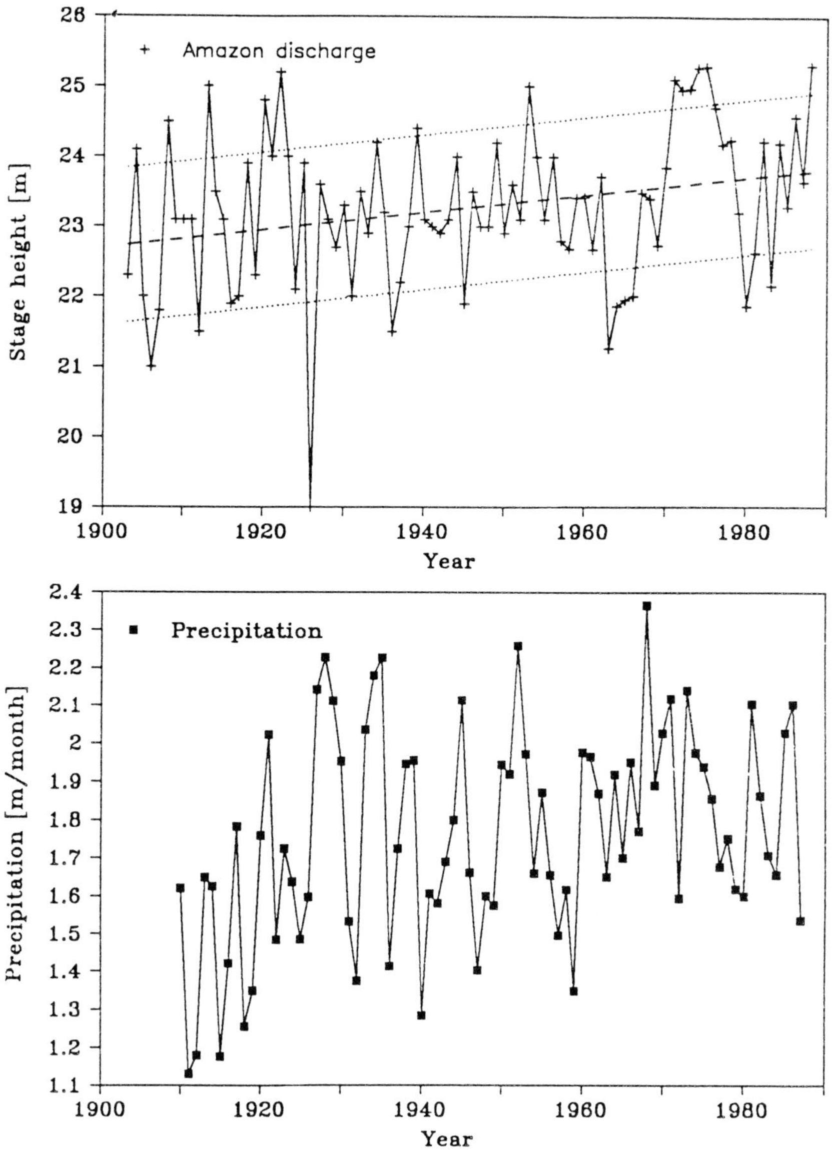

Fig. 8.3 Time-series of annual mean stage height values (m) (1903–1988) and annual mean precipitation estimates (m/month) (1910–1987) at Manaus on the Amazon. The linear trend (± SE of the slope) shown for the Amazon stage height was obtained by least-squares regression (see Table 8.1).

variables. The summary statistics for the period 1966–1983 are also given in Table 8.1.

Finally, Fig. 8.9 shows the series of annual mean discharge rate values (m^3s^{-1}) for the Mississippi at Vicksburg and mean rainfall estimates (mm/month) for the Mississippi's catchment area over the same period, with Fig. 8.10 showing the scatter between these two variables. The summary statistics for the period 1817–1989 are also presented in Table 8.1.

4.2 Coastal Zone Colour Scanner Images

Since plumes of the larger rivers typically extend several hundred to thousands of kilometres from the delta, their dispersal pattern is impossible to define by ship surveys alone. With the advent of satellite instruments capable of measuring slight variations in the colour of ocean water, however, this task became feasible. Between November 1978 and July 1986, the experimental CZCS on NASA's Nimbus-7 satellite recorded minute changes in water colour related to the distribution of phytoplankton pigments, other suspended matter, and dissolved organic matter associated with river plumes drifting near the surface. Figs. 8.11 and 8.12 show monthly composites of CZCS-derived pigment concentrations over the region. Few data were collected over the western tropical Atlantic and eastern Caribbean Sea over the 7.5

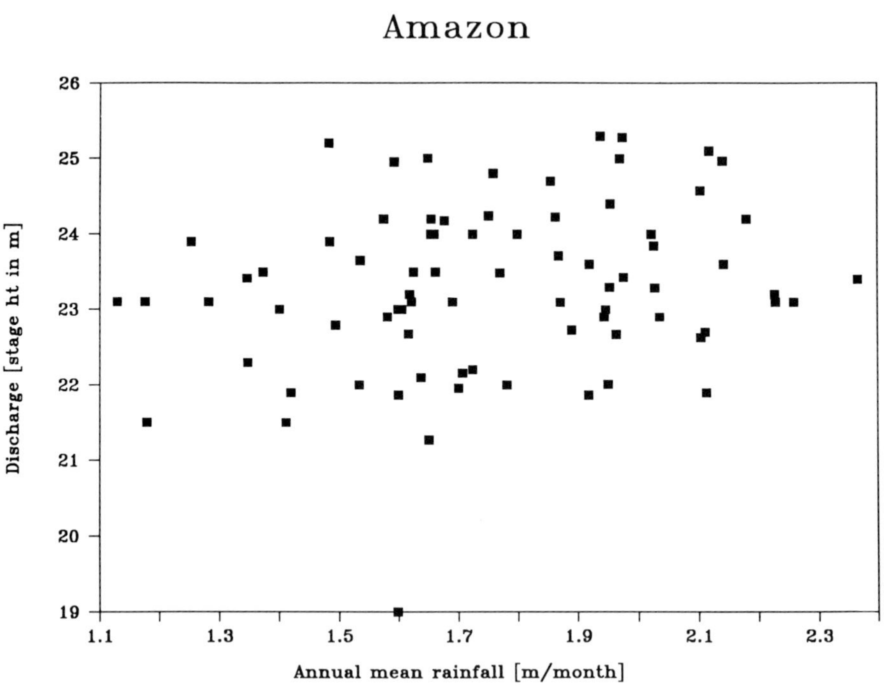

Fig. 8.4 Amazon. Scatter plot of the annual mean precipitation estimates (m/month) against the annual mean stage height values (m) at Manaus on the Amazon (1910–1987).

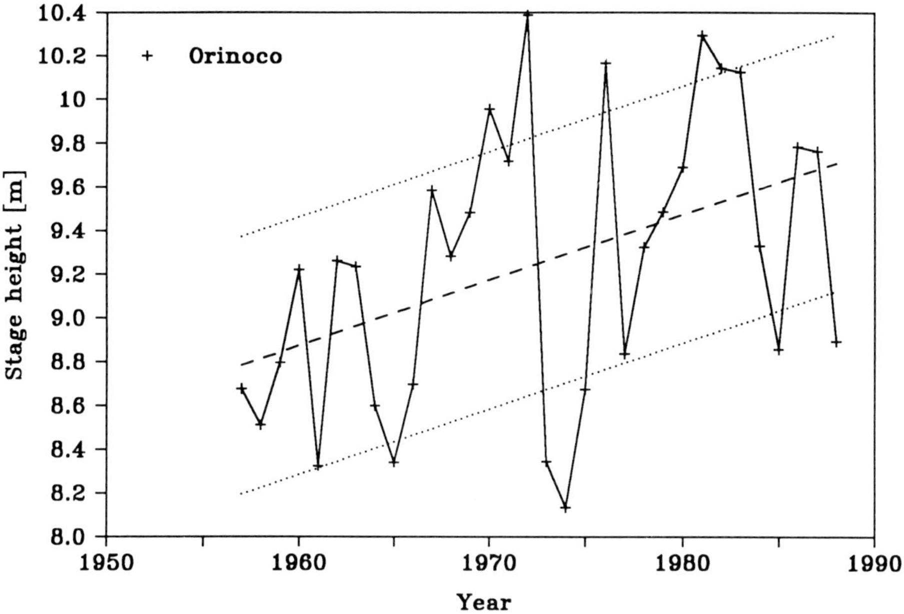

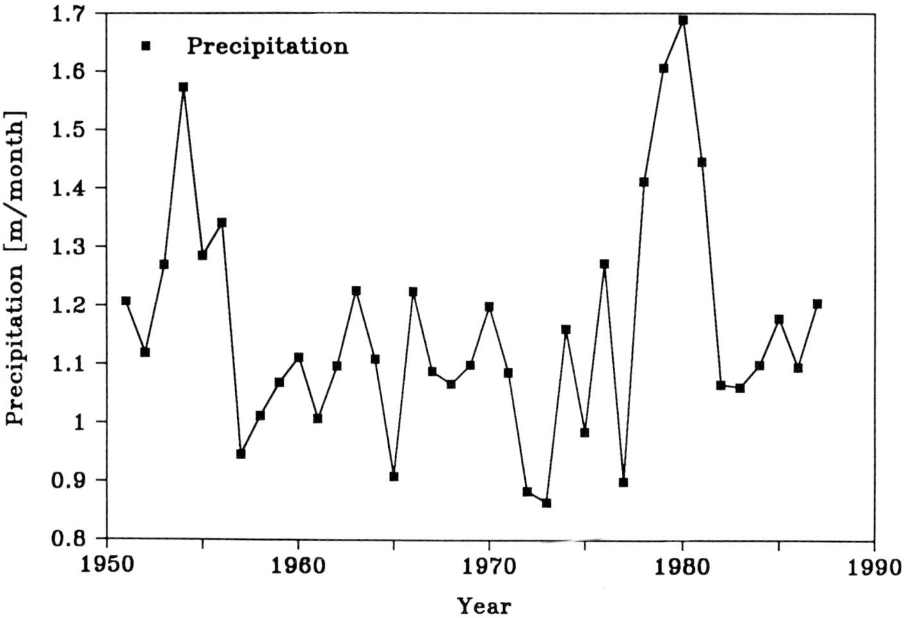

Fig. 8.5 Orinoco. Time-series of annual mean stage height values (m) (1965–1988) at Ciudad Bolivar on the Orinoco and annual mean precipitation estimates (m/month) (1941–1987) at several stations within the Orinoco's catchment area. The linear trend (± SE of the slope) shown for the Orinoco stage height was obtained by least-squares regression (see Table 8.1).

Table 8.1 Average, minimum, maximum and (year) for time-series of discharge of the Amazon, Orinoco, Mississippi, and Rio Grande rivers, and estimated rainfall in the respective catchment area. Summary correlation statistics for these parameters are presented in the bottom half of the table. The short time-series of discharge values for the Magdalena River are given at the bottom of the table.

	Amazon		**Orinoco**		**Mississippi**		**Rio Grande**	
	Stage (cm)	*Rain (mm/month)*	*Stage (cm)*	*Rain (mm/month)*	*Stage (m^3/s)*	*Rain (mm/month)*	*Stage (m^3/s)*	*Rain mm/month*
Average	2327.1	1758.4	924.8	1119.4	16,455.3	107.5	48.0	49.0
Minimum	1900.0 (1926)	1129.6 (1911)	813.3 (1974)	864.8 (1973)	8182.4 (1895)	77.4 (1963)	5.9 (1970)	25.0 (1919)
Maximum	2531.0 (1988)	2366.4 (1968)	1038.3 (1972)	1879.2 (1943)	26,887.9 (1973)	143.6 (1957)	102.3 (1976)	77.0 (1956)

Summary correlation statistics

Amazon River and catchment area
Stage height (cm) = Year · 1.25–103.365
Slope significant at $P<0.02$ (Dof = 84, SE = 0.48, $r^2 = 0.07$)
Precip. (mm/month) = Year · 4.18–6392.18
Slope significant at $P<0.01$ (Dof = 76, SE = 1.33, $r^2 = 0.11$)
Orinoco River and catchment area
Stage height (cm) = Year · 2.98–4961.08
Slope significant at $P<0.02$ (Dof = 30, SE = 1.12, $r^2 = 0.19$)
Precipitation slope not significantly different from 0 (Dof = 29)

Rio Grande and catchment area
Stage height: No regression was obtained (short records)
Precipitation: No regression was obtained
Mississippi River and catchment area
Stage height is negative but is not significantly different from 0 (Dof = 171)
Precipitation: No regression was obtained
Magdalena River
Annual discharge estimates (m^3s^{-1})
1969–2295.5
1970–2568.9
1971–3250.4
1972–2348.2

years of CZCS operation, the best coverage being between November 1978 and March 1980 and in late 1984–1985. This presents a serious aliasing problem and makes interpretation of time-series of regional CZCS images difficult.

Due to the size of the discharge, the Amazon and Orinoco plumes were the easiest to trace with the CZCS. Only a few images showed the Magdalena (Fig. 8.13) or Mississippi River (Fig. 8.14) discharge moving offshore. Most of the images examined suggest that both the Magdalena and the Mississippi plumes turn west and remain close to the coast. Muller-Karger *et al.* (1991) examine a time-series of CZCS images of the Gulf of Mexico that shows occasional and short-lived entrainments of the Mississippi plume along the edge of the Loop Current (*cf.* Maul, 1977). Fig. 8.13 shows an extreme case in which the Magdalena plume was carried offshore as a thin jet across the Caribbean Sea almost to Jamaica. Typically, the smaller plumes of these rivers are hard to differentiate in the CZCS pigment products from sediment resuspended nearshore, since the algorithm used was designed to quantify offshore oceanic phytoplankton concentrations (chlorophyll plus phaeopigment). The colour of coastal waters is typically the result of many processes such as sediment input, sediment resuspension, and input of yellow dissolved organic matter

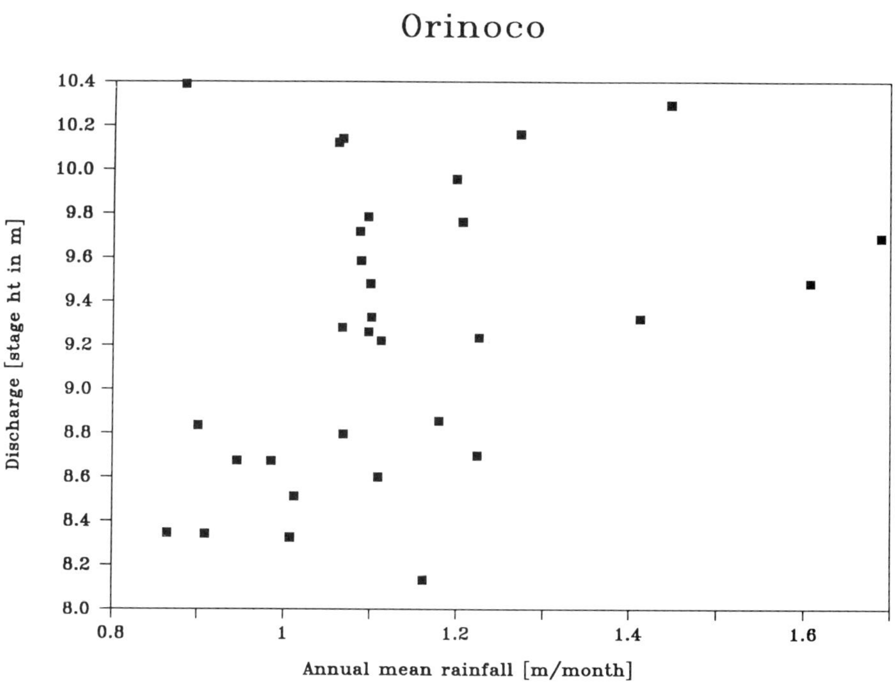

Fig. 8.6 Orinoco. Scatter plot of the annual mean precipitation estimates (m/month) at several stations within the Orinoco's catchment area against the annual mean stage height values (m) at Ciudad Bolivar (1965–1987).

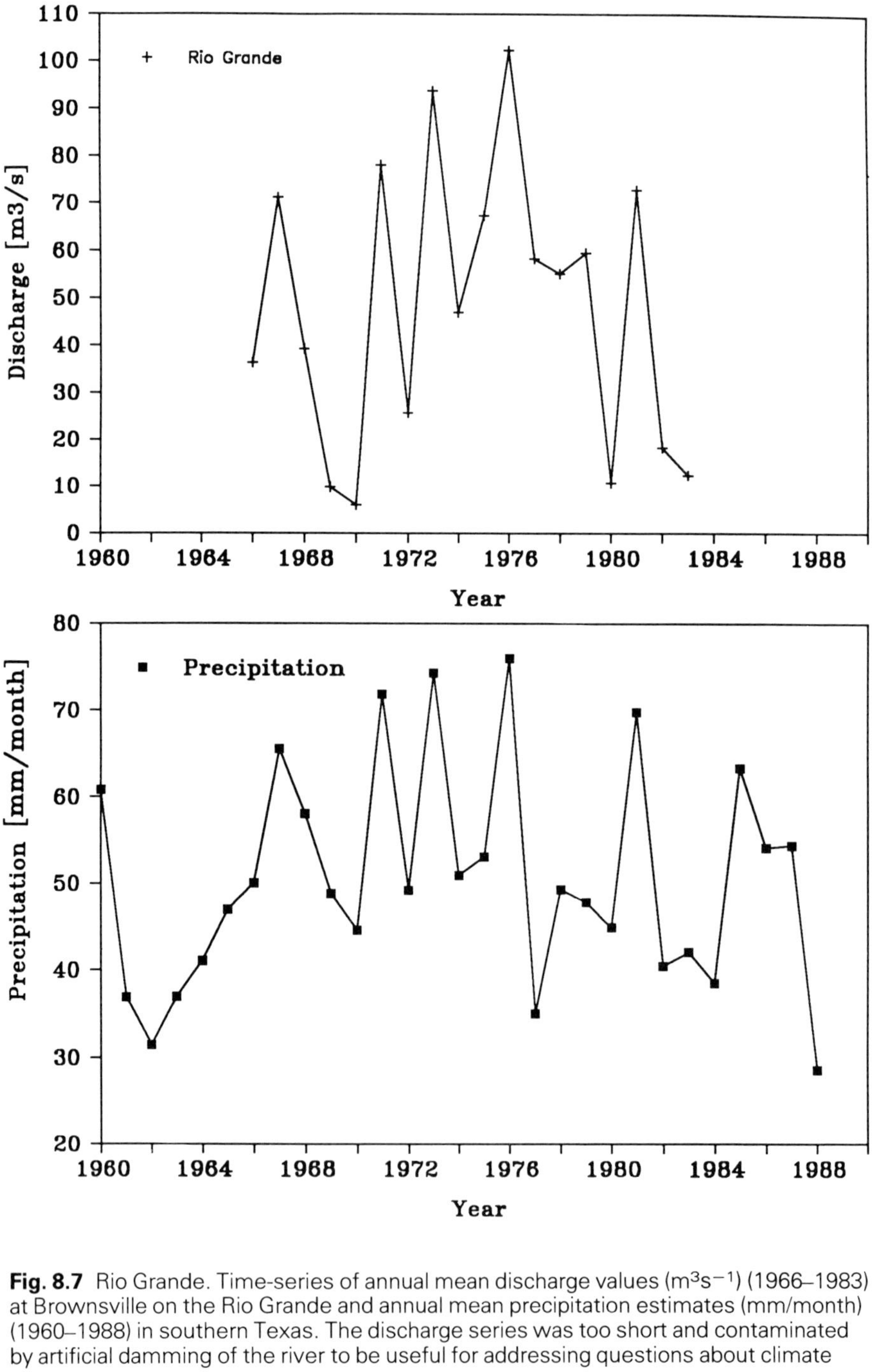

Fig. 8.7 Rio Grande. Time-series of annual mean discharge values (m^3s^{-1}) (1966–1983) at Brownsville on the Rio Grande and annual mean precipitation estimates (mm/month) (1960–1988) in southern Texas. The discharge series was too short and contaminated by artificial damming of the river to be useful for addressing questions about climate change.

('Gelbstoff') in addition to phytoplankton.

5 Discussion

5.1 River Discharge and Rainfall

Rivers integrate the effects of weather over their catchment area, making discharge an efficient variable for monitoring changes in global climate. Variability in the discharge of the Amazon may be coupled to El Niño-Southern Oscillation (ENSO) phenomena, with lower than normal discharge rates occurring during El Niño years (Richey *et al.*, 1989a). It is also tempting to draw conclusions on teleconnections between the strong drought in the African Sahel and the increased stage heights for the Amazon seen between the mid-1960s and mid-1970s (Fig. 8.3). However, mechanistic connections between these phenomena are ill-defined and difficult to prove. Therefore, the discussion is limited to the long-term behaviour of the discharge of the largest rivers affecting the region, including comparisons with long-term records of rainfall, but without exploring the causes for low or high discharge on any particular year. The fate of the discharge, as inferred from the seasonal dispersal patterns revealed in the CZCS data, is also discussed in the context of climate change.

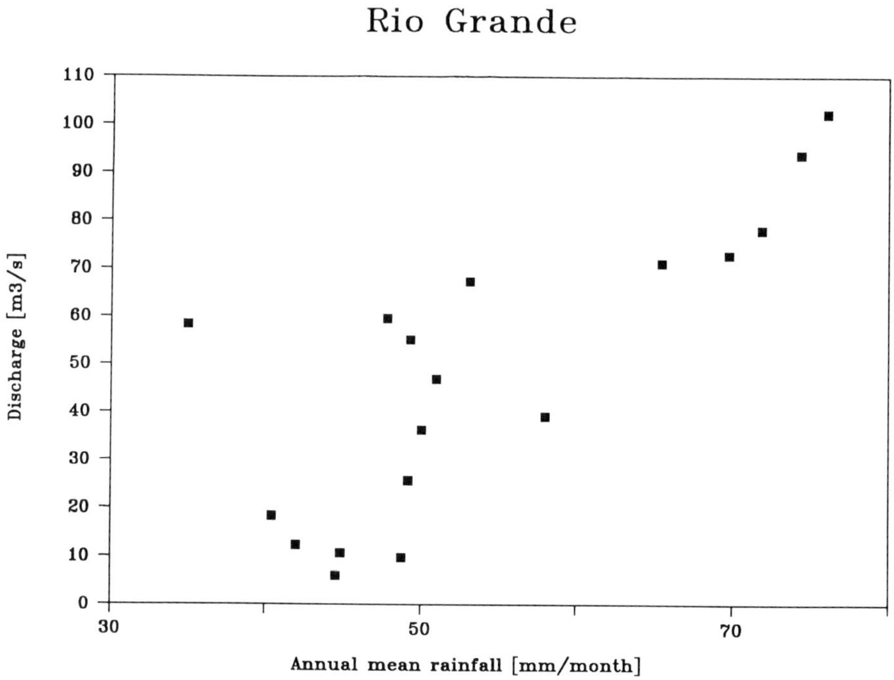

Fig. 8.8 Rio Grande. Scatter plot of the annual mean precipitation estimates (mm/month) in southern Texas against the mean discharge values (m^3s^{-1}) at Brownsville on the Rio Grande (1966–1983).

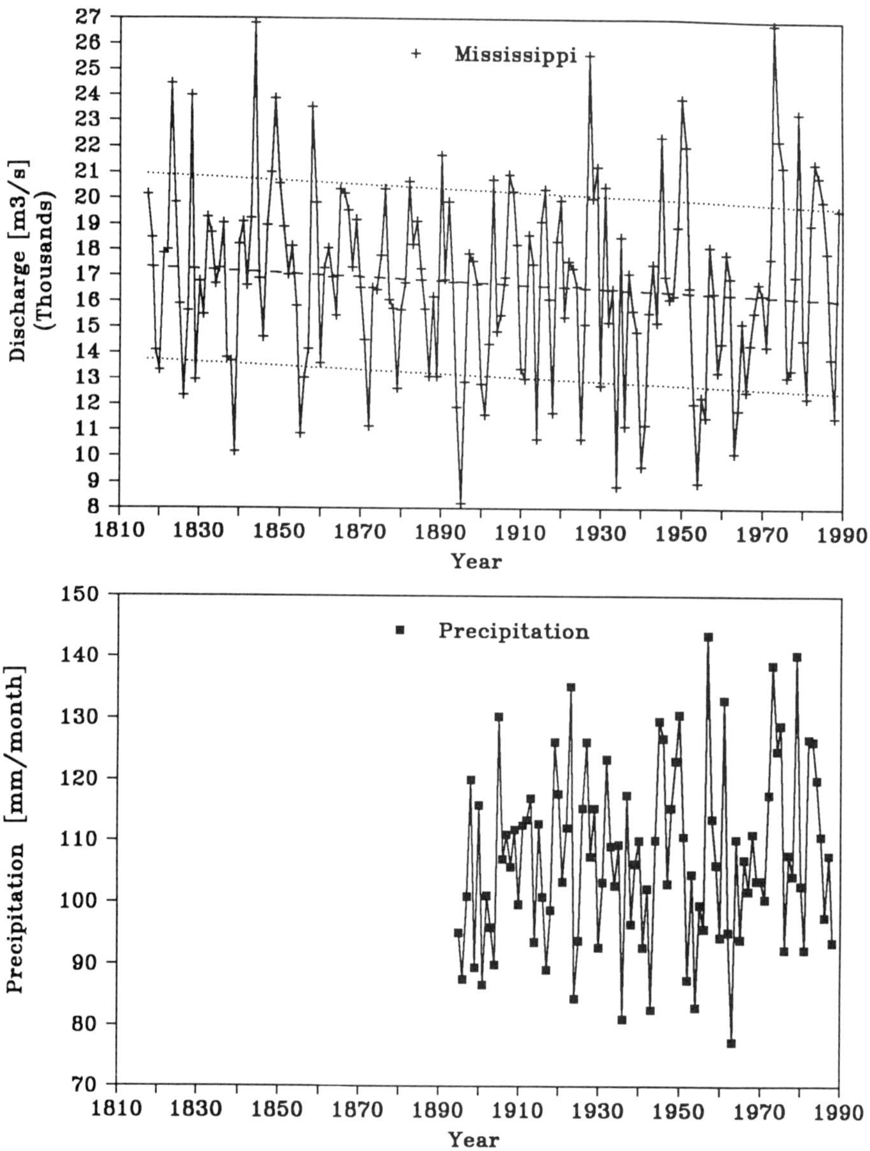

Fig. 8.9 Mississippi. Time-series of annual mean discharge values (m^3s^{-1}) at Vicksburg on the Mississippi (1817–1989) and annual mean precipitation estimates (mm/month) (1895–1988) along the Mississippi valley. The discharge did not show a trend in spite of massive human development along the Mississippi banks.

Both the Amazon and Orinoco river stage heights show weakly positive trends over the past 50–100 years (Figs. 8.3 and 8.5). On average, the Amazon river stage has increased 1.25 cm per year since 1903, and the Orinoco 3 cm per year since 1965 (both trends significant at the 2% level). Assuming that the average minimum stage height observed during low

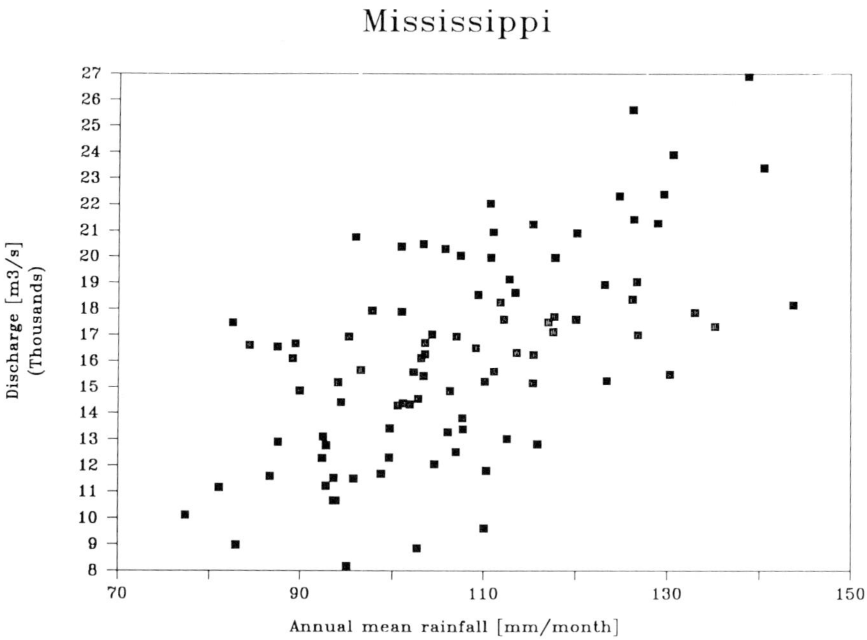

Fig. 8.10 Mississippi. Scatter plot of the annual mean precipitation estimates (mm/month) along the Mississippi valley against the mean discharge values (m^3s^{-1}) at Vicksburg on the Mississippi (1895–1988).

discharge in the Amazon (*c.* 17.5 m in November) is equivalent to a flow rate of 10×10^4 m^3s^{-1}, that the average maximum stage height observed during high discharge (approximately 27.5 m in June) is equivalent to a flow rate of 23×10^4 m^3s^{-1} (UNESCO, 1979), and that the relationship between discharge and stage height is linear, the observed trend indicates that the Amazon discharge is increasing at a rate of about 163 $m^3s^{-1}yr^{-1}$, or 5.1×10^9 m^3yr^{-1}. Note that the relationship between stage height and discharge is usually non-linear; thus, the calculated increase may be too low. On the other hand, if the gauge sites are sinking, the discharge change many not be occurring at all.

Only about 45–50% of the precipitation over the catchment area of the Amazon is channeled into the river, the rest being lost as evapotranspiration (Vorosmarty *et al.*, 1989). Also, drainage of rain water occurs over varying time-scales and follows complicated pathways to the main stream (Richey *et al.*, 1989b). Furthermore, water artificially diverted will impact downstream flow rate measurements, in particular in smaller rivers. Thus, a straightforward connection between incident rainfall and river discharge rates is not easy to establish. The problem is exacerbated in the tropics, where the Intertropical Convergence Zone (ITCZ) migrates about the equator on a seasonal basis, local microclimate effects are very strong, and meteorological stations are sparse. The Amazon River is an extreme case: its catchment area is large (> 3×10^6 km^2) and straddles the equator. While either the northern or southern half of the drainage basin experiences a 'dry' season, the other half receives enough rainfall to dampen variations in the amplitude of seasonal discharge relative to rivers at higher latitudes.

Since the catchment area of the Amazon is large, only 0.14 mm/month increase in monthly mean precipitation rate (assuming 100% of precipitation goes to the river) over this area would be sufficient for the 163 $m^3s^{-1}yr^{-1}$ increase in discharge. Thus, even the smallest change in precipitation has a large impact on the hydrological cycle if applicable over a large area. Widespread deforestation and development and climate change may push precipitation in either direction. In particular, deforestation will increase runoff by decreasing evapotranspiration. Discharge rates will be directly affected by these changes.

Current estimates for rainfall variability over the region associated with climate change are large and inconsistent between models (Maul, Chapter 1; Wigley and Santer, Chapter 2). Annual-average changes range from −12 mm/month to about +12 mm/month over the northern portion of South America. Clearly, changes of this magnitude would lead to noticeable changes in the discharge rate of tropical rivers. A change of +12 mm/month over the entire Amazonian catchment area (3×10^6 km^2) would cause an increase of over 7000 m^3s^{-1} in the discharge of the Amazon if 50% of the rainfall is recycled by evapotranspiration. While the trend for rainfall at Manaus was, in fact, positive (4.2 mm/month yr^{-1}), it was much larger than that needed to balance the weak trend in the increase of the Amazon's discharge.

In contrast, rainfall at stations near the Orinoco River did not show a trend. As with the Amazon, visual inspection of the Orinoco time-series (Fig. 8.5) suggests a possible relationship between stage height and precipitation. However, the scatter plots (Figs. 8.4 and 8.6) show no apparent relationship between these variables for either the Amazon or the Orinoco. Neither is there a clear connection between years of minimum discharge and years of minimum precipitation. This may be an artifact, however, because rain water may be stored in the basin for some time, in particular in floodplain lakes or as groundwater recharge. Thus, there may be a lag of months to years between rainfall and discharge in the river system. Three-year and four-year running means on the rainfall data for the Orinoco and Amazon catchment, plus a one to two year lag, did not increase the correlation in scatterplots against stage height. Examination of such lags is necessary but only will be possible using longer time-series and more complete rainfall records.

The discussion above clearly shows why single or a few stations cannot be used in a water budget designed to project discharge rates of large tropical rivers. A sparse network of meteorological stations within the catchment area will bias the rainfall records to peculiarities of local climate. The best relationships between rainfall and river discharge were obtained when a large number of stations were pooled to obtain an estimate of rainfall over the catchment area. Both the Rio Grande and the Mississippi showed improved relationships between these variables (Figs. 8.8 and 8.10).

The largest rivers affecting the Gulf of Mexico are the Rio Grande and the Mississippi. The Rio Grande has been tapped for development needs, and thus the record at Brownsville (Fig. 8.7) is not useful to detect changes induced by climate variability.

The Mississippi discharges on the average about 1.7×10^4 m^{3}s^{-1} (average range = 0.81×10^4 m^{3}s^{-1} in September to 2.81×10^4 m^{3}s^{-1} in April), and shows large variability in its monthly climatology compared with the Amazon and Orinoco rivers (Fig. 8.2). Even though development along the banks of the Mississippi and its tributaries since 1817 has been enormous, no significant change in discharge is seen. While discharge has shown interannual variability since the beginning of the record, there appears to be a tendency for higher frequency and larger amplitude variations after about 1900. Rainfall over the Mississippi valley has not shown a significant trend in the last 90 years either, and there is a good relationship between rainfall and discharge. Thus, the Mississippi's discharge may be a valuable index for the effects of climate.

Clearly, a good understanding of the hydrological cycle is necessary before good models of climate can be implemented. Modern models that address climate change are still simplistic in their formulation (*q.v.* Wigley and Santer, Chapter 2). Also, while attempts are being made to model the relationship between water balance and river transport on continental scales (Vorosmarty *et al.*, 1989) or to model the routeing of water into particular rivers based on the input of tributaries and storage of water along a valley (Meade *et al.*, 1983; Richey *et al.*, 1989b), these models have not been linked to global ocean or atmospheric circulation models. The reverse is true as well: general global circulation and climate-change models, such as those discussed by Wigley and Santer, do not include the effects of forest clearing, urbanization, erosion, etc.

It is not easy to identify and quantify the anthropogenic factors that affect the water budget within a catchment area. As Sternberg (1987) points out, the effects of deforestation vary with the size and environment of the area cleared. For example, deforestation of mountainous areas eliminates fog drip, which is a significant water source, while also accelerating erosion of the soil. In contrast, clear-felling of forest areas in lower-lying areas typically increases water flow and subsurface storage. In both cases, evapotranspiration is reduced, but the consequences for the resupply of water to the atmosphere are not understood.

These problems indicate that (1) a dense, representative, network of meteorological stations should be installed around the larger rivers so that long-term trends can be examined with confidence, and (2) general circulation models should include the microclimate effects of continents, forests, the hydrological cycle and the human impact on them.

5.2 Where Does the River Water Go?

5.2.1 Amazon River discharge

The dispersal of the Amazon's discharge affects surface salinity, concentration of phytoplankton, and phytoplankton species composition around Barbados and the western tropical Atlantic (e.g., Ryther *et al.*, 1967; Borstad, 1982a, b; Muller-Karger *et al.*, 1988). While the flow of the Amazon ranges from about 10×10^4 m^{3}s^{-1} in November to about 23×10^4 m^{3}s^{-1} in June (mean is approximately 17.3×10^4 m^{3}s^{-1}; Fig. 8.2), these changes in the flow seem to play a minor role in the dispersal pattern of the discharge. The path followed by the Amazon water during February–May is very different from that traced during June–January. The change is consistent with patterns in the circulation inferred from historical ship drift (Richardson and Walsh, 1986), buoy drift (Richardson and Reverdin, 1987; Muller-Karger *et al.*, 1988), current meters (Flagg *et al.*, 1986; Johns *et al.*, 1990) and numerical models (Philander and Pacanowski, 1986), and consistent with patterns seen in the distribution of sea-surface salinity by Perlroth (1966) and Cochrane (1969).

During February–May, the bulk of the Amazon's water flows into the Caribbean Sea in a broad (150–200 km), continuous band of high pigment (> 2 mg m^{-3}) and sediment concentrations (*q.v.* Fig. 8.11). Offshore in the tropical Atlantic, pigment concentrations are relatively low, typically << 0.1 mg m^{-3}. This is the period during which Amazon water most consistently and clearly affects the Barbados region, as Amazon water drifted to the northwest, and water displaced earlier into the interior Atlantic during June–January drifts westward with the North Equatorial Current (Richardson and Reverdin, 1987; Muller-Karger *et al.*, 1988).

During June–January, the discoloured water originating at the mouth of the Amazon flows first to the north and northwest very close to the coast, but then veers offshore near 5°N, 50°W (Fig. 8.12). Subsequently it flows north to approximately 11°N, 53°W, then it retroflects southward and then east, forming a southern meander trough near 4°30′N, 46°W. A similar pattern was observed in June–January buoys, clearly indicating that the North Brazil Current turns east in a sharp loop off the Guianas (Richardson and Reverdin, 1987; Muller-Karger *et al.*, 1988). Phytoplankton concentrations are highest near the continental margin where the current separates (> 2 mg m^{-3}) and decreases in the plume offshore to values of about 0.5 mg m^{-3}. If all of the Amazon water discharged during June–January moves eastward, about 60% of the annual volume discharged by the Amazon would disperse offshore (Muller-Karger *et al.*, 1988). This is the period during which the clearest waters are observed around Barbados.

5.2.2 Orinoco River discharge

The other river that has a strong influence on the Caribbean Sea is the Orinoco. It discharges an average of 3.9×10^4 m^3s^{-1} (range = 1×10^4 m^3s^{-1} in March to 7×10^4 m^3s^{-1} in August; Fig. 8.2). In this case, the large amplitude of the seasonal variation in the flow plays a large role in the fate of the plume, along with variation in the wind intensity and changes in the flux of water through the Caribbean Sea (see Muller-Karger *et al.*, 1989). These changes also are tied to the seasonal migration of the ITCZ about the equator.

The images show that the Orinoco's impact on the southern Windward Islands is directly related to the seasonally varying discharge. During minimum discharge in February–April, most of the Orinoco's outflow enters the Gulf of Paria, between Trinidad and Venezuela. A fraction flows around eastern Trinidad but typically remains between Trinidad and Tobago. The discharge then enters the Caribbean Sea through Dragon's Mouth and Grenada Passage and continues on a west–northwestward drift, remaining within the southern Caribbean Sea.

The effect of the Orinoco River on the western tropical Atlantic east of the Antilles Islands is accentuated during this period by the formation, or migration into this region, of large ephemeral (1–2 months) anticyclonic eddies that transport pigment-rich water far offshore into the tropical Atlantic, east or northeast of Trinidad. CZCS images show such eddies in November 1978, late January–March 1980, and in January 1982. The 1980 event engulfed Barbados for a period of over a month in pigment concentrations between 0.2 and 0.5 mg m^{-3}.

Even though temporal resolution of the sequence of images was poor, during the period of maximum discharge (July–November), the Orinoco's outflow completely engulfed Trinidad, Tobago, Grenada and St. Vincent. While during the first half of the year pigments near Tobago are low (< 0.5 mg m^{-3}), during July–November concentrations increase to over 1 mg m^{-3}. Such dramatic effects are seen at Grenada and St. Vincent as well, but St. Lucia seems to be the northernmost of the Lesser Antilles regularly affected by the Orinoco. Any increase or decrease in the maximum discharge rate would affect the extent to which these islands are affected by the river plume.

5.2.3 Mississippi River discharge

Compared to the Amazon and the Orinoco rivers, the Mississippi's discharge has a large standard deviation (Fig. 8.2), with fluctuations at times reaching over a factor of 3 between consecutive years. Note the large difference in the discharge between 1979 and 1980: in 1979 the flow rate was about 2.35×10^4 m^3s^{-1}, while in 1980 the rate dropped to about 1.46×10^4 m^3s^{-1}, about a factor of 1.6 times less. This situation is not unusual for the Mississippi, but such variability is more rare in the Orinoco and Amazon rivers. In both of the latter rivers, extreme variation of a factor of 1.2 occurred only once in the record (Figs. 8.3 and 8.5). Of course, small variations in the discharge of these rivers can represent a large volume of water. This difference is clearly visible in the time-series of CZCS images for the northern Gulf of Mexico (Muller-Karger *et al.*, 1991).

Fig. 8.14 shows two CZCS images in which the plume of the Mississippi can be clearly seen extending to the west of the delta along the coast of Louisiana and Texas. The series of images suggests that most of the plume spreads to the west along the coast, and that only a fraction, which rarely exceeds 30% of the area of the plume, extends eastward along the coasts of Mississippi, Alabama, and Florida. Much of the Mississippi's water is released into the Gulf of Mexico on the western side of the delta through its distributary the Atchafalaya River, which carries about 30% of the Mississippi's volume.

The 1979 satellite images suggest that the Mississippi's plume can extend along the coast all the way to Mexico, but it is difficult to separate the plume from phytoplankton blooms and resuspended sediment material. In contrast, in 1980 the portion of the plume that went west was much narrower and did not appear to reach as far along the coast. The issue of plume dispersal in the northern Gulf of Mexico will need to be addressed by examining the full CZCS record (November 1978–June 1986), using the water-leaving radiance CZCS images in addition to the CZCS pigment products, and using estimates of the coastal wind and current velocities during the period of observation.

The series of images of the Mississippi suggest that the Mississippi plume rarely extends seaward of the shelf break, except directly off the birdfoot delta due to the proximity of the break to the region where the Loop Current, or anticyclonic eddies shed from it, reach their northernmost extension. Occasionally, streamers may be pulled off into the interior of the Gulf from the birdfoot delta by these current systems. The streamers can follow the cyclonic front of the Loop Current along the Florida shelf for a few hundred kilometres. However, the satellite data for 1979 and 1980 suggest that the streamers are typically very narrow, implying that little of the discharge is dispersed offshore via this mechanism (Muller-Karger *et al.*, 1991). In any case, Maul (1978) has shown that the salinity of the surfacial Gulf of Mexico waters is a balance between evaporation, precipitation, river runoff, and the anticyclonic eddies of the Gulf Loop Current.

5.3 Interannual Variation in Plume-Dispersal Patterns

The CZCS, buoy-drift and ship-drift data suggest that there is a strong seasonal cycle in the circulation of the western tropical Atlantic and the Caribbean Sea. But the CZCS data also show conspicuous interannual differences in the dispersal patterns of the river plumes. Patterns associated with the Amazon plume vary both during first and second semester periods. During January–May 1979 there did not seem to be a regular pattern in the distribution of pigments in the western tropical Atlantic southeast of Barbados. Rather, pigments sometimes traced the remnants of a retroflection (e.g., January–February 1979, March 1983, February 1984), and at times they showed relatively uniform concentrations over the region (January–March 1980, January–May 1985).

The second semester dispersal patterns also show large differences between years. While there are few images for this period, in 1978, 1979 and 1982 most of the discoloured water originating from the Amazon seems to have been entrained in the retroflection. However, it is not clear if this is the most common pattern during this period. The sequence of CZCS images shows patches of high pigment concentrations near Barbados between June and August of 1980 and 1981 which may have been the result of Amazon water lost to the northwest away from that retroflecting into the NECC, as was clearly seen in 1985 and perhaps in 1984 (the patches in 1984 seem to be more directly related to the Orinoco's plume as described below). Unfortunately no images were collected south of Barbados in the second semester of 1980 and 1981 to delineate the origin of these patches. Such high concentrations near Barbados were not apparent in the second semesters of 1978, 1979 and 1982–1985, and were only seen briefly in 1980, suggesting that these patches are not a common occurrence around Barbados during the second semester.

Under anomalous wind conditions, Orinoco water can reach far into the tropical Atlantic and affect Barbados. This is an anomalous situation due to extreme changes in the wind or the circulation. An example of this situation was recorded in November 1984, when hurricane winds caused the Orinoco plume to extend about 1000 km to the northeast of the delta into the Atlantic.

The ephemeral anticyclonic eddies observed east of Trinidad during the first few months of the year also seem to vary in intensity. The CZCS images show that the eddies seen in January–March 1980 and January 1982 affected Barbados more dramatically compared to the eddy seen in November 1978. Clearly, spatial variability is so large in this region that coarse sampling from ships could lead to misconceptions about the circulation.

5.4 Erosional Effects of Sea-Level Rise on Deltas

The balance between a possible rise of sea level and accretion of land by entrapment of sediment discharged by rivers along the coast is a complicated and yet unresolved issue. The modern deltas of the great rivers affecting the region are all geologically well developed, in great part because they drain trailing continental edges. However, because of the gentle gradient of trailing margins relative to collision margins, small changes in sea level can affect large geographical areas. In particular, a rise in sea level of 1–3 m foreseen over the next century (e.g., Hansen *et al.*, 1981; Walsh, 1990) is likely to flood large portions of the modern deltas, increasing erosional problems both on the shoreface and on the shelf (Suter *et al.*, 1987). The problem may be aggravated by oil drilling activity since subsidence is accelerated as oil is retrieved from the ground, but may be compensated in part by increased sediment supply from erosion of upstream land.

Currently it is not clear if the Amazon Delta is an area of subsidence or emergence. Much of the continental shelf off the Amazon is covered with relict sands, while sediments from the Amazon are dispersed in the nearshore environment along the entire coast of northeastern South America, reaching the Caribbean Sea (van Andel and Postma, 1954; Betzer *et al.*, 1977; Milliman *et al.*, 1982). About 50% of the particles $> 2\ \mu m$ and over 80% of particles finer than $2\ \mu m$ off the Orinoco Delta are of Amazonian origin (Eisma *et al.*, 1978; Meade *et al.*, 1983). These observations suggest that as much as 40% of the Amazon's annual sediment output to the Atlantic moves toward the Orinoco every year (Milliman and Meade, 1983).

The Amazonian sediments dispersed along the coast to the northeast clearly influence both Amazon and Orinoco delta morphology and dynamics. The enormous flux of Amazonian sediment may help alleviate the impact of sea-level rise in the area. Most of the sediment appears to move along Brazil and the Guyanas inshore of the 10 m isobath (*cf.* Nittrouer *et al.*, 1986), forming a series of sediment waves about 5 m high, 50–60 km long, and 10–20 km wide, that transit the coast at about 1 km yr^{-1}. Little of the sediment seems to move offshore (Rine and Ginsburg, 1985). The movement of these waves appears to be due to resuspension by strong tides and the seasonal intensification of wave action by changes in wind intensity. Higher transport rates occur in February–May relative to July–November (Delft Hydr. Lab., 1962; Wells and Coleman, 1981; Rine and Ginsburg, 1985). The residual flow in coastal regions is about 0.15–0.30 ms^{-1} (Delft Hydr. Lab., 1962), compared with about 0.90 ms^{-1} seaward of the shelf break (Muller-Karger *et al.*, 1988). It is still not clear how the currents in the area affect sediment transport. However, since it is hard to envision large changes in the discharge of water or sediment from the Amazon in the short term, the main factors that would affect sediment dispersal are changes in the seasonal wind and current pattern.

Currently, only a portion of the sediments derived from the Amazon are stored in the Orinoco Delta region. Some sediment continues to the Caribbean Sea where they are stored on the continental shelf off northeastern Venezuela or buried on the continental slope (Milliman *et al.*, 1982). While a large fraction of the Orinoco's sediments are stored in the inner delta (van Andel and Postma, 1954; Meade *et al.*, 1983), enough leaves the Orinoco to accrete mud flats in the northern portion of the delta, combined with Amazonian mud. While some authors consider the Orinoco Delta as being sediment-starved and a zone of subsidence (Royo y Gomez, 1956), others find that enough sediments are discharged to fill the nearby Gulf of Paria in the next 5000–10,000 years (van Andel, 1967).

The Mississippi Delta is one of the largest and best known delta formations in the world (Shepard, 1973). While the Mississippi and the Atchafalaya rivers discharge large amounts of sediment, the Mississippi Delta seems to be retreating and subsiding. The reasons for this behaviour are not entirely clear, but large-scale human intervention has reshaped the dispersal of the Mississippi's sediments. In the past, the Mississippi formed various distinct sediment lobes, all much larger than the present birdfoot delta, indicating that sediment is currently being lost from the delta. However, the fate of this sediment is not clear. Large but yet unquantifiable changes may be expected in the next 50–100 years in the shape and dynamics of the Mississippi Delta. A concerted effort using remote sensing techniques needs to be implemented to trace the Mississippi's sediment and help construct a sediment budget.

The effect of a sea-level rise over the next 50–100 years therefore may vary with location, and will depend as much on changes in the dispersal patterns as on changes in the sediment loads. However, currently there is not enough information at hand to predict with certainty what changes may be expected for the large deltas present in this region. While a sea-level increase is most likely to lead to delta erosion and subsidence, other changes associated with climate variability may aggravate or compensate for this effect. If rainfall increases, sediment load may arrest erosional effects. However, if rainfall decreases, heavy erosional damage to a delta is almost certain. Again, we are in a situation in which we know little about the current status of deltas, and can only speculate about the possible effects of climate change.

As mentioned above, a change in climate is likely to lead to a marked change in the wind and current regime affecting the region (Gallegos *et al.*, Chapter 3). An increase in the heat content of the tropical ocean is likely to increase wind activity, in particular the frequency of storms. If these physical oceanographic processes dominate fluvial processes and sediment flux, the morphology of the deltas will change. Fisher *et al.* (1969) show how delta morphology changes from elongate or lobate to cuspate with increased domination of physical oceanographic processes. It is not clear, however, how long it may take for such processes to have an effect due to the size of the deltas of these large rivers.

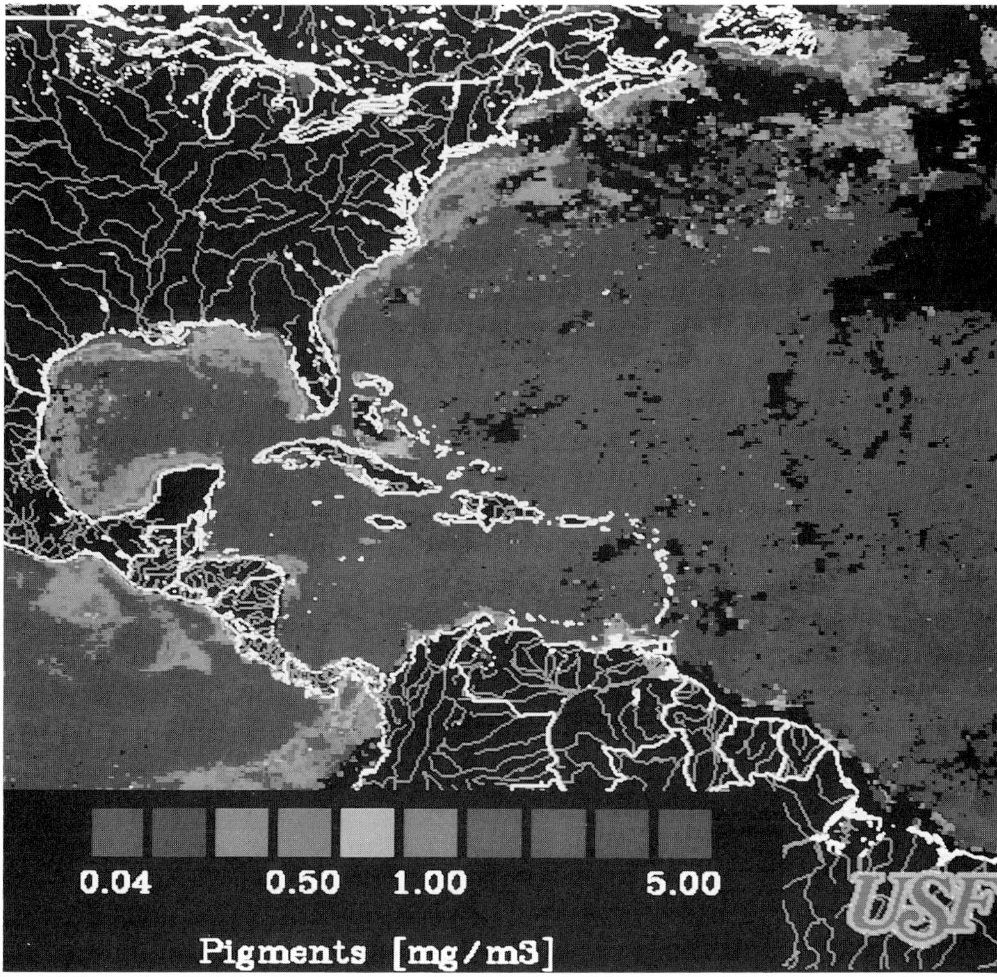

Fig. 8.11 February 1980. Monthly mean pigment concentration field obtained from CZCS satellite imagery showing the western North Atlantic Ocean in February 1980. Concentrations were colour coded, with warm colours (yellow, orange, and red) representing high phytoplankton pigment concentrations or other coloured materials in river plumes, and cooler shades (purple, blue, and green) representing low concentrations. Black represents no data, due either to cloud cover or lack of programmed coverage. Since the CZCS was an experimental instrument, rather than an operational one, data coverage was limited to specific requests and short periods of intensive data collection. This created many data gaps. Land was masked brown. Rivers that affect the region have been outlined using the World Database II files. See colour plates between pages 210 and 211.

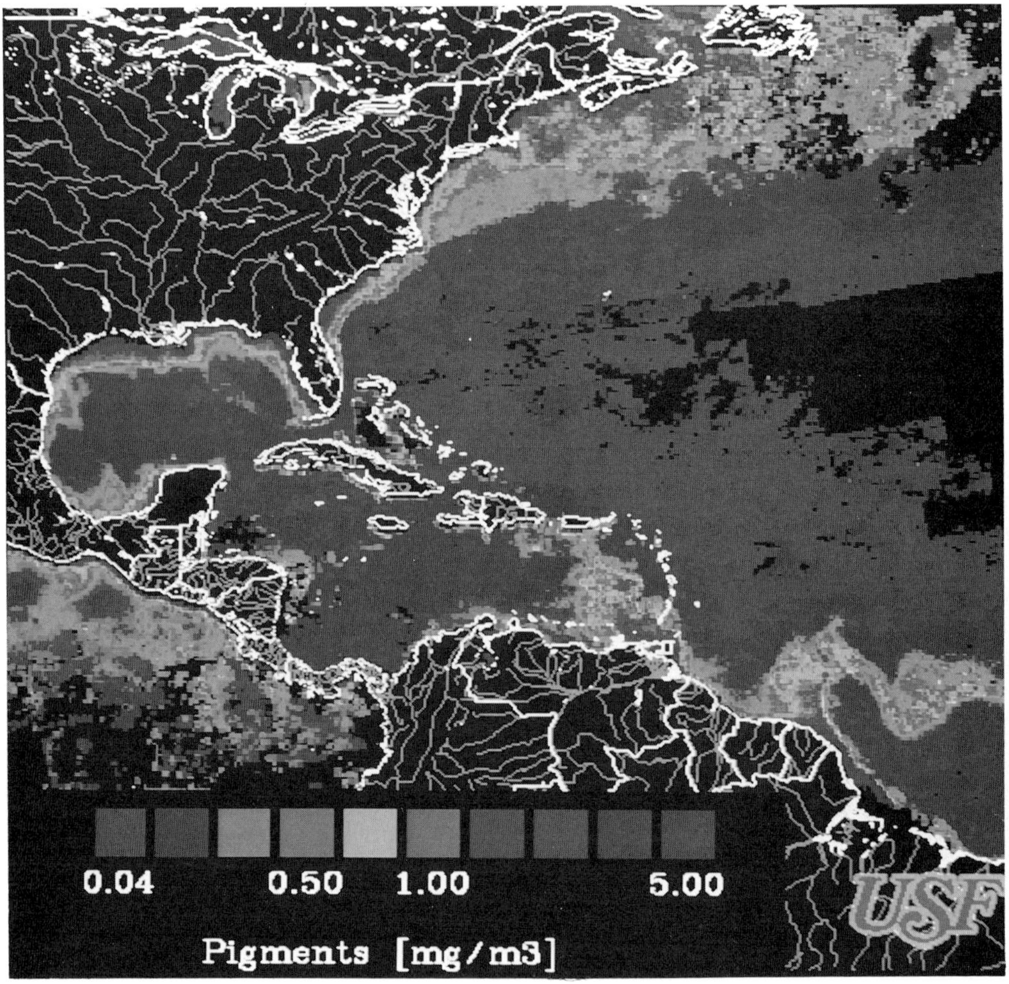

Fig. 8.12 October 1979. Monthly mean pigment concentration field obtained from CZCS satellite imagery showing the western North Atlantic Ocean in October 1980. Concentrations were coded as in Fig. 8.11. See colour plates between pages 210 and 211.

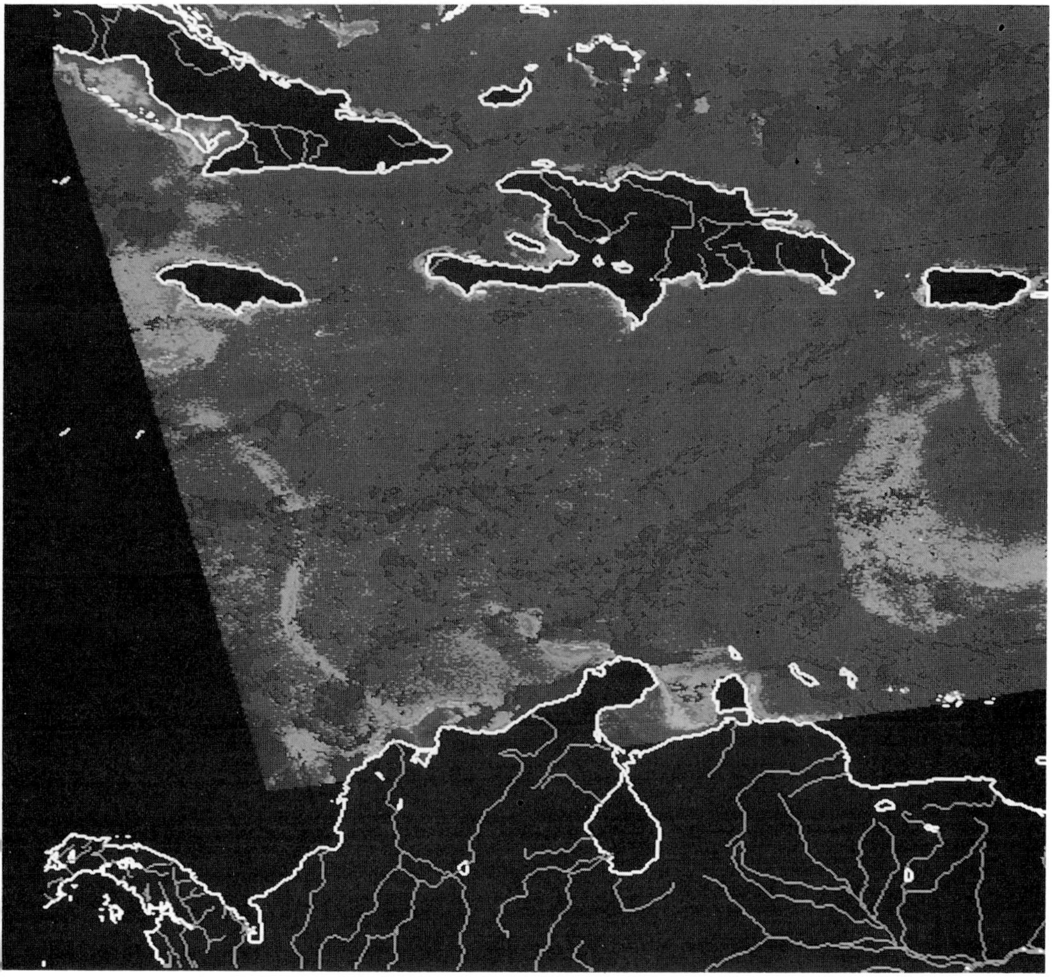

Fig. 8.13 11 October 1982. CZCS image of the western Caribbean Sea showing an extreme case of the dispersal of the Magdalena River where the plume extends across the western Caribbean Sea to the northwest and almost reaches Jamaica. Typically, the plume flows along the coast of Colombia to the west. Clouds are masked grey and areas with no data, masked black. Rivers that affect the area are outlined. See colour plates between pages 210 and 211.

a)

b)

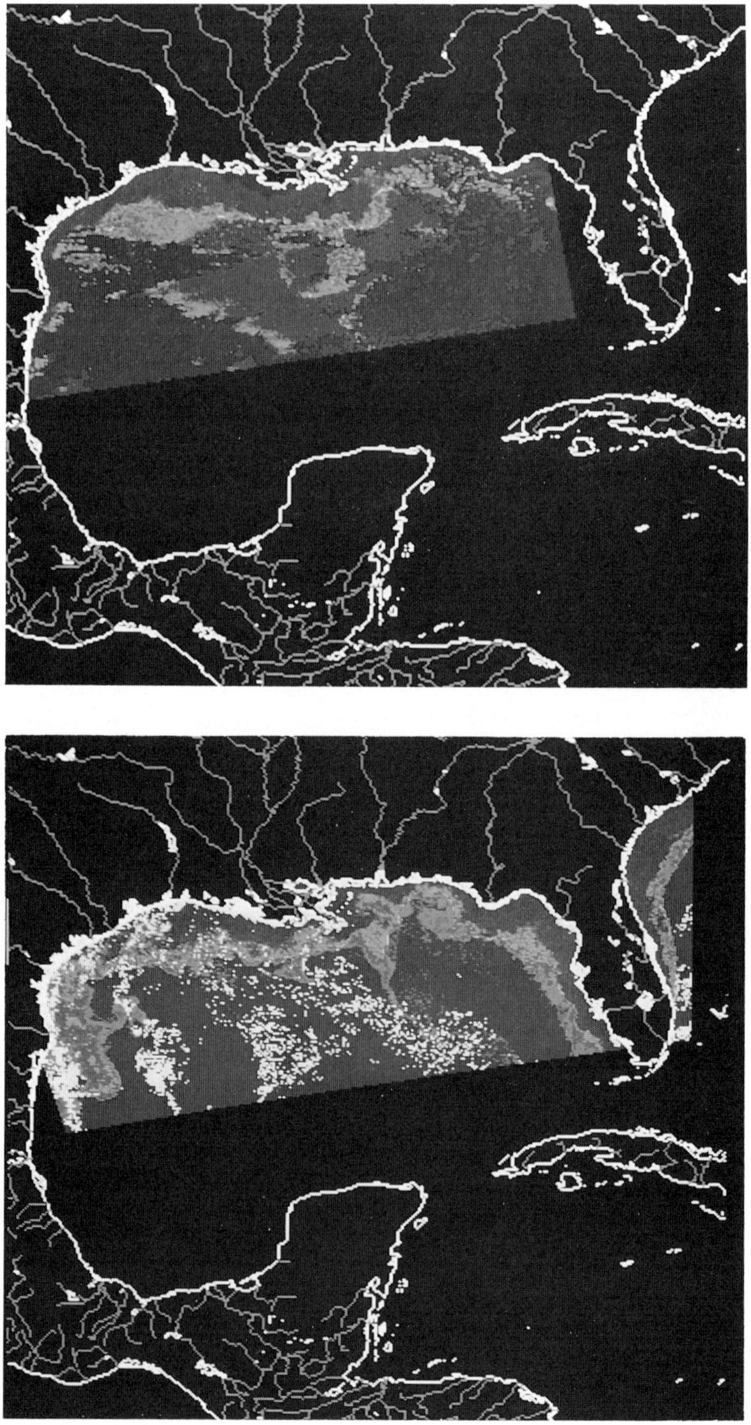

Fig. 8.14 CZCS images of the Gulf of Mexico showing the Mississippi River plume. Clouds are masked grey and areas with no data, masked black. Grey shades near the mouth of the Mississippi indicate high sediment concentrations. (upper) High discharge in 1979 (24 March); (lower) Low discharge in 1979 (8 October). Rivers that affect the area are outlined. See colour plates between pages 210 and 211.

Clearly, the impact of sea-level change on delta regions is a key issue for the economy of the countries around the region. Sedimentation rates, patterns and rates of sea-level variability must be carefully monitored. Without accurate estimates of the direct effect of human development and of global warming on regional evaporation–precipitation balances, it is hard to predict what will happen to river discharge under a climate-change scenario. It is important to begin to address interactions between the general circulation and water-budget models, so that we may begin to understand the consequences of clearing forests like the Amazon, estimated to reach upwards of 200×10^3 km^2 deforested by the year 2000. Also, research addressing delta dynamics needs to include a remote sensing component to accurately determine the pattern of sediment dispersal and accumulation. Otherwise, the local response to global change is not predictable at this time.

6 Recommendations and Conclusions

Several issues need to be addressed to assess the impact of climate change on the region:

1) A representative grid of meteorological stations along the major rivers of Central and South America needs to be installed as soon as possible. This grid is necessary to understand the regional global hydrological cycles, and in particular for separating how changes in climate or in land use may affect the contribution of different parts of a catchment area to the discharge of rivers. The grid would become an essential component of other programmes which address global climate, in particular the Tropical Rain Measuring Mission (TRMM) and Earth Observation System (EOS).
2) Climate changes are likely to have a larger, more immediate effect on discharge of smaller rivers, rather than on the larger rivers affecting the region. In order to detect changes, a careful record of river stage height and calibrated discharge needs to be maintained for the major rivers and their tributaries; particular attention to geodetic levelling of gauging sites on a regular basis must be given. Thus, tributaries should be identified where discharge is not being estimated, and gauges installed and maintained there. It is also necessary to document human factors which have an impact on discharge, in particular widespread clear-felling and deforestation. This effort will allow the construction of dynamic water budgets, adjustable for varying flow in tributaries, to understand and model flow along the main stream.
3) Because of their great hydrographic, oceanographic and sedimentologic influence on the southeast Caribbean Sea, the Amazon forest and the catchment area of the Orinoco need to be studied using a time-series of satellite images to determine patterns and changes in deforestation, land use, and their effect on river discharge. For this purpose, historical and current Landsat visible data and Advanced Very High Resolution Radiometer (AVHRR) infrared data can be used to map vegetation changes in the region. To map sediment movement in the delta, near the river mouth and in the dispersal region along

the coast, Landsat, AVHRR and historical CZCS satellite data can be combined for analysis. The launch of a new ocean colour scanner (Sea-viewing Wide Field-of-view Sensor – SeaWiFS) in the mid-1990s will provide a tool for real-time monitoring of the dispersal of sediment and river water. This will provide the basis for sediment thematic maps and sediment budgets necessary to address questions of delta subsidence or emergence, and the effect of sediments or pollutants carried downstream to other communities along the coast or nearby island nations.

4) Basic airborne surveys of the length of the Amazon, Orinoco and Mississippi rivers, including their deltas, are needed using high-resolution spectral imaging technology. The surveys should be accompanied by an in-depth *in situ* sampling programme focusing on sediment load, sediment composition, and water-quality measurements, including estimates of the concentration of phytoplankton and dissolved organic matter. This comprehensive initial survey is needed to understand future changes in sediment input and water quality as affected by different tributaries, understanding the impact of different patterns of land use, and to lay the basis for understanding sediment deposition and dispersal in the delta and adjacent regions under changing climatic conditions.

7 Acknowledgements

G. A. Maul (NOAA/AOML, Miami, Florida), J. J. Walsh (University of South Florida, St. Petersburg, Florida), and C. R. McClain (NASA Goddard Space Flight Center) provided helpful comments on the CZCS images, the river discharge series, and on the validity of using short series for studying global climate change. T. R. Fisher (Horn Point Environmental Laboratories, University of Maryland) provided helpful comments on the manuscript. Processing of the satellite data was largely carried out on the software environment 'dsp', developed at RSMAS, University of Miami, and implemented at the University of South Florida. Post-processing and final analyses were carried out using software developed at the University of South Florida. This work was supported by United Nations Environmental Programme contract No. 9-26-08020.

References

Betzer, P.R., Eggiman, D.W., Carder, K.L., Kester, D.R. and Betzer, S.B., 1977: Seasonal patterns in suspended calcium carbonate concentration during the dry and wet seasons in the eastern Caribbean. In: Andersen, N.R. and Malahoff, A. (eds), *The fate of fossil fuel* CO_2 *in the oceans*. Plenum Press, New York, pp. 63–79.

Borstad, G.A., 1982a: The influence of the meandering Guiana Current and Amazon River discharge on surface salinity near Barbados. *J. Mar. Res.*, **40**, 421–434.

Borstad, G.A., 1982b: The influence of the meandering Guiana Current on surface conditions near Barbados – temporal variations of *Trichodesmium* (Cyanophyta) and other plankton. *J. Mar. Res.*, **40**, 435–452.

Clark, D.K., 1981: Phytoplankton pigment algorithms for the Nimbus-7 CZCS. In: Gower, J.F. (ed.), *Oceanography from Space*. Plenum Press, New York, pp. 227–237.

Cochrane, J.D., 1969: Low sea-surface salinity off northeastern South America in summer 1964. *J. Mar. Res.*, **27**, 327–334.

Delft Hydraulics Laboratory, 1962: Demerara coastal investigation: Report on siltation of Demerara Bar Channel and coastal erosion in British Guiana. Delft, The Netherlands, Vol. 1–3, 240 pp.

Eisma, D., van der Gaast, S.J., Martin, J.M. and Thomas, A.J., 1978: Suspended matter and bottom deposits of the Orinoco delta: Turbidity, mineralogy, and elementary composition. *Netherlands J. Sea Res.*, **12(2)**, 224–251.

Fisher, W.L., Brown, L.F. Jr., Scott. A.J. and McGowen, J.H., 1969: Delta systems in the exploration of oil and gas – a research colloquium. Univ. of Texas, Bur. Econ. Geol., Austin, 102 pp.

Flagg, C.N., Gordon, R.L. and McDowell, S., 1986: Hydrographic and current observations on the continental slope and shelf of the western equatorial Atlantic. *J. Phys. Oceanog.*, **16**, 1412–1429.

Gordon, H.R., Clark, D.K., Brown, J.W., Brown, O.B., Evans, R.H. and Broenkow, W.W., 1983a: Phytoplankton pigment concentrations in the Middle Atlantic Bight: Comparison of ship determinations and CZCS estimates. *Applied Optics*, **22**, 20–35.

Gordon, H.R., Brown, J.W., Brown, O.B., Evans R.H. and Clark, D.K., 1983b: Nimbus 7 CZCS: reduction of its radiometric sensitivity with time. *Applied Optics*, **22(24)**, 3929–3931.

Hansen, J., Johnson, D., Lacis, A., Lebedeff, S., Lee, P., Rind, D. and Russell, G., 1981: Climate impact of increased atmospheric carbon dioxide. *Science*, **213**, 957–966.

Johns, W.E., Lee, T.N., Schott, F.A., Zantopp, R.J. and Evans, R., 1990: The north Brazil Current retroflection: seasonal structure and eddy variability. *J. Geophy. Res.*, **95(C12)**, 22103–22120.

Maul, G.A., 1977: The annual cycle of the Gulf Loop Current, Part I: Observations during a one-year time series. *J. Mar. Res.*, **35(1)**, 29–47.

Maul, G.A., 1978: The 1972–1973 Cycle of the Gulf Loop current, Part II: Mass and Salt Balances of the Basin. In: FAO Fisheries Report No. 200 Suppl., 587–619.

Meade, R.H., Nordin, C.F. Jr., Perez Hernandez, D., A. Mejia B. and Perez Godoy, J.M., 1983: Sediment and water discharge in Rio Orinoco, Venezuela and Colombia. Proceedings of the Second International Symposium on River Sedimentation. 11–16 October 1983. Nanjing, China. Water Resources and Electric Power Press, Beijing, 1134–1144.

Milliman, J.D. and Meade, R.H., 1983: World-delivery of river sediment to the oceans. *J. Geol.*, **91(1)**, 1–21.

Milliman, J.D., Butenko, J., Barbot, J.P. and Hedberg, J., 1982: Depositional patterns of modern Orinoco/Amazon muds on the northern Venezuelan shelf. *J. Mar. Res.*, **40(30)**, 643–657.

Muller-Karger, F.E., Walsh, J.J., Evans, R.H. and Meyers, M.B., 1991: On the seasonal phytoplankton concentration and sea-surface temperature cycles of the Gulf of Mexico as determined by satellites. *J. Geophy. Res.*, **96(C7)**, 12645–12665.

Muller-Karger, F.E., McClain, C.R., Fisher, T.R., Esaias, W.E. and Varela, R., 1989: Pigment distribution in the Caribbean Sea: observations from space. *Progress in Oceanog.*, **23**, 23–69.

Muller-Karger, F.E., McClain, C.R. and Richardson, P.L., 1988: The dispersal of the Amazon's water. *Nature*, **333**, 56–59.

Nittrouer, C.A., Curtin, T.B. and Demaster, D.J., 1986: Concentration and flux of suspended sediment on the Amazon continental shelf. *Continental Shelf Res.*, **6**, 151–174.

Perlroth, I., 1966: The distribution of water type structure in the first 300 feet of the equatorial Atlantic. In: Proc. of the Symposium on Oceanography and

Fisheries Resources of the Tropical Atlantic. Abidjan, Ivory Coast, UNESCO, pp. 185–191.

Philander, S.G.H. and Pacanowski, R.C., 1986: A model of the seasonal cycle in the tropical Atlantic. *J. Geophys. Res.*, **91(C12)**, 14192–14206.

Richardson, P.L. and Walsh, D., 1986: Mapping climatological seasonal variations of surface currents in the tropical Atlantic using ship drifts. *J. Geophys. Res.*, **91(C9)**, 10537–10550.

Richardson, P.L., and Reverdin, G., 1987: Seasonal cycle of velocity in the Atlantic North Equatorial Countercurrent as measured by surface drifters, current meters and ship drifts. *J. Geophys. Res.*, **92(C4)**, 3691–3708.

Richey, J.E., Mertes, L.A.K., Dunne, T., Victoria, R.L., Forsberg, B.R., Tancredi, A.C.N.S. and Oliveira, E., 1989b: Sources and routing of the Amazon River flood wave. *Global Biogeochemical Cycles*, **3**, 191–204.

Richey, J.E., Nobre, C. and Deser, C., 1989a: Amazon River discharge and climate variability: 1903 to 1985. *Science*, **246**, 101–103.

Rine, J.M. and Ginsburg, R.N., 1985: Depositional facies of a mud shoreface in Surinam, South America – A mud analogue to sandy, shallow-marine deposits. *J. Sed. Pet.*, **55(5)**, 633–652.

Royo y Gomez, J., 1956: Quaternary in Venezuela. In: Stratigraphical lexicon of Venezuela (English Edition). Boletin de Geologia. Special Publication No. 1. Ministerio de Minas e Hidrocarburos, Venezuela, 468–478.

Ryther, J.H., Menzel, D.W. and Corwin, N., 1967: Influence of the Amazon River outflow on the ecology of the western tropical Atlantic. I. Hydrography and nutrient chemistry. *J. Mar. Res.*, **25(1)**, 69–83.

Shepard, F.P., 1973: *Submarine Geology*. Third Edition, Harper and Row, 517 pp.

Sternberg, H.O., 1987: Aggravation of floods in the Amazon River as a consequence of deforestation? *Geografiska Annaler.*, **69(A1)**, 201–219.

Suter, J.R., Berryhill, H.L. Jr. and Penland, S., 1987: Late quaternary sea-level fluctuations and depositional sequences, southwest Louisiana continental shelf. In: Nummedal, D., Pilkey, O.H. and Howard, J.D. (eds)., *Sea-Level Fluctuation and Coastal Evolution*. Society of Economic Paleontologists and Mineralogists, Special Publication No. 41. Tulsa, Oklahoma, pp. 199–219.

UNESCO, 1979: Discharge of selected rivers of the world. Volume III (part III). Mean monthly and extreme discharges (1972–1975), pp. 54–70.

van Andel, T.H., 1967: The Orinoco delta. *J. Sed. Pet.*, **37(2)**, 297–310.

van Andel, T.H. and Postma, H., 1954: Recent sediments of the Gulf of Paria. Vehr. der K. Nederl. Adad. Van Wetensch., *Afd. Natuurkunde, Eerste Reeks*, **20**, 1–244.

Vorosmarty, C.J., Moore, B. III, Grace, A.L. and Gildea, M.P., 1989: Continental scale models of water balance and fluvial transport: an application to South America. *Global Biogeochemical Cycles*, **3**, 241–265.

Walsh, J.J., 1990: Arctic carbon sinks: present and future. *Global Biogeochemical Cycles*, **3**, 393–411.

Wells, J.T. and J.M. Coleman, 1981: Physical processes and fine-grained sediment dynamics, coast of Surinam, South America. *J. Sed. Pet.*, **51**, 1053–1063.

9

Analysis of Temperature, Precipitation, and Sea-Level Variability with Concentration on Key West, Florida, for Evidence of Trace-Gas-Induced Climate Change

Kirby Hanson[1] and George A. Maul[2]

Abstract

Meteorological and sea-level data for Key West, Florida, have been examined for evidence of changes during recent decades that may be attributed to increasing trace gases in the atmosphere. The 136-year air-temperature record (1851–1986) gives the evidence that a slight warming has occurred, but there has been no appreciable change since 1950. However, there are questions of the reality of the warming because of the varied temperature-observing conditions over the period of record. The 101-year precipitation record (1886–1986) gives evidence that no significant change in precipitation has occurred during the period of record. In addition to Key West weather, sea-level records from all 62 stations on file with the Permanent Service for Mean Sea Level that cover the Caribbean Sea, Gulf of Mexico, the Bahamas, and Bermuda have been examined for linear trends. Average (±1 standard deviation), sea-level rise is 0.4 cm/yr (±0.6 cm/yr) for all stations (mean record length 20 years), and 0.3 cm/yr (±0.4 cm/yr) for those stations with records of ≥10 years in length. A regional maximum of ≥+1.0 cm/yr is centred in the northwestern Gulf of Mexico, an area of subsidence. Regional minima (≃–0.3 cm/yr) occur in the southwestern Gulf and in the Lesser Antilles, where there is diastrophism. Average sea-level rise at Key West, a site of tectonic stability, is 0.22 cm/yr (±0.01 cm/yr) for the period 1913–1986. Key West sea level seems unrelated to local air temperature, barometric pressure, precipitation and records of coral growth, but is significantly lower than normal during the year preceding a strong El Niño – Southern Oscillation event, and higher than normal during the event itself. There is no evidence for accelerated sea-level rise at this site.

1 Introduction

The United Nations Environment Programme (UNEP) expressed concern over possible future climatic changes resulting from the continuing increase of trace gases in the atmosphere. For CO_2-doubling, climate-change predictions range from temperature warmings of 1.5 to 4.5°C

[1]Correspondence address: 11412 SW 82 Ave., Ocala, FL 32676, USA
[2]National Oceanic and Atmospheric Administration, Miami, FL 33149, USA

and sea-level rises of 15 to 50 cm (Charney, 1979; Smagorinsky, 1982; and Hoffman *et al.*, 1983; IPCC, 1990). As a result, the UNEP has invited studies in several regions of the possible impact of these changes on the marine and coastal environment, in most cases in cooperation with other bodies.

In this chapter we will refer to the region (i.e., the northeast coast of South America, the Caribbean Sea, the Gulf of Mexico, and the Bahama – Florida area of the Atlantic Ocean) as the Intra-Americas Sea. Meteorologically, the region is dominated by easterly Trade Winds punctuated by monsoonal events, all of which contribute to the interannual variability of air temperature, precipitation and sea level by variation in wind stress and wind-stress curl. Oceanographically, it is a region of complex currents including the formation area of the Gulf Stream System. Geologically, three tectonic plates, i.e., the North American, Caribbean and South American, have varying amounts of vertical movement that influence relative sea-level change (Cartwright *et al.*, 1985).

The purpose of the study reported in this chapter is two-fold: (1) To examine the historical meteorological data for Key West, Florida, USA, for information about previous climatic changes at that location. (2) To study all the available records on file with the Permanent Service for Mean Sea Level (PSMSL; Pugh *et al.*, 1987) to assess: (a) What is the mean rise of relative sea level in the region based on historical observations? (b) Is there significant sea-level change spatial variability within the region? (c) Are there any inter-relationships between meteorological variables, sea-level change, and other physically meaningful variables? In particular, we wish to define any temperature and/or sea-level rise during the most recent portion of the record. In addition, we have examined the precipitation record for Key West to further assess climate change.

The meteorological and sea-level data for Key West have been selected for this study because of the length of the record: 136 years (1851–1986) for weather and 74 years (1913–1986) for sea level. We also have evaluated how well air temperature at Key West serves as a proxy for sea-surface temperature.

2 The History of Meteorological and Sea-Level Observations

The chronology of annual temperature for Key West (Fig. 9.1), has been calculated from monthly temperatures from two sources. Data for the period 1851–1970 were obtained from the World Monthly Surface Station Climatology (Spangler and Jenne, 1985), and for the period 1971–1986 from the National Climatic Data Center (NOAA, 1987). A photograph of one of the earliest sites for weather observation, the Sand Key Lighthouse off Key West, is shown in Fig. 9.2.

Meteorological observations for Key West have been taken at a number of locations, both in the city and at nearby airports, over the 136 years of record (Table 9.1). The observing location until 1870, is unknown to us. From 1870 to 1931, 10 different observing locations were used within the city. In 1931 observations were initiated at the airport, and have continued at various airport locations, except for the period 1953–1975,

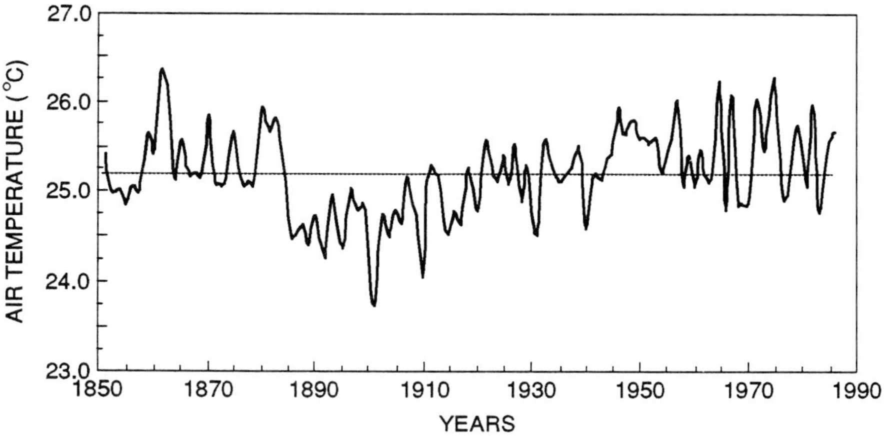

Fig. 9.1 Annual average air temperature, Key West, Florida, 1851–1986.

Fig. 9.2 A 1903 photograph showing on the right, the original Sand Key weather station. The Sand Key Lighthouse (shown on the left) was established in the 19th century and still serves to guide ships in the Straits of Florida. The weather station is now at Key West Airport, its 17th location in a century.

when observations were taken at the Post Office building in the city. Elevation of thermometers above ground ranged from 1–15 m at the latter 15 observing locations (Table 9.1). Observing conditions are not known for the earliest two locations.

Cotton-Region Shelters, containing the thermometers, were used continuously since first employed at the Louvre Hotel in 1872. Prior to that time 'window units' were used, usually on the north side of the building. From this heterogeneous mix of temperature-observing conditions, we cannot assume consistent observing conditions over this long period of record, and therefore cannot assume secular trends in the data represent real conditions.

Sea-level observations at Key West date back to the 1840s, but were not continuous until 1913. Full one-year series started July 1846, and were fairly complete until May 1852 and for June 1857–May 1858, after which sporadic measurements of several months duration are available for 1882, 1896, 1898, 1899, and 1903. The data prior to 1913 were observed at Fort Taylor; from 1913–1926 at Curry's Wharf; and from 1926 to date at the Naval Base. These locations are shown in Fig. 9.3.

Tide gauges are always referenced to several survey markers, known as tidal benchmarks, by conducting levelling (vertical) surveys. Thus, if a gauge is destroyed or moved, the measurements can be related in a simple arithmetic way to earlier measurements. These precise levelling surveys are also conducted once or twice a year to ensure that the gauge hasn't been moved by accident or shifted due to settling. Sea-level records are

Table 9.1 History of locations and heights of the maximum-minimum thermometers at Key West, Florida, for the period 1830–1974 (from NOAA, 1987)

Location	*Beginning year*	*Ending year*	*Elevation of thermometers above ground (m)*
1. US Army Surgeons (location unknown)	1830	1870	?
2. Russell House	1870	1870	?
3. Tift & Co. Building	1870	1871	4.9
4. Duval Street	1871	1872	5.5
5. Louvre Hotel	1872	1882	13.1
6. Wall Building	1882	1886	6.1
7. US Naval Depot	1886	1886	14.6
8. Waite Building	1887	1903	12.8
9. Weather Bureau Building	1903	1911	3.4
10. Island City Bank Building	1911	1913	12.5
11. Weather Bureau Building	1913	1931	3.0
12. Airways Station Building	1931	1942	1.2
13. Boca Chica Airport	1942	1944	1.5
14. Boca Chica Airport	1944	1953	5.8
15. US Post Office Building	1953	1957	1.5
16. Key West International Airport	1957	1958	7.3
17. Key West International Airport*	1958	1974	5.5

*(Max.–Min. removed in 1974)

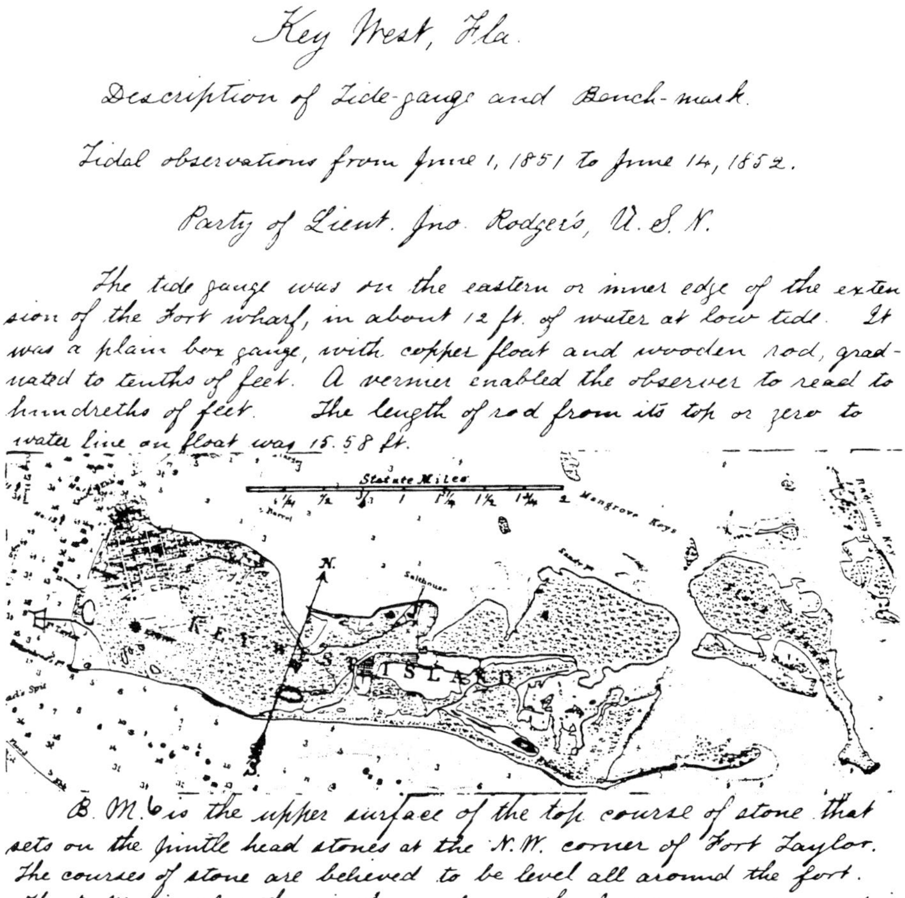

Key West, Fla.

Description of Tide-gauge and Bench-mark.

Tidal observations from June 1, 1851 to June 14, 1852.

Party of Lieut. Jno. Rodgers, U.S.N.

The tide gauge was on the eastern or inner edge of the extension of the Fort wharf, in about 12 ft. of water at low tide. It was a plain box gauge, with copper float and wooden rod, graduated to tenths of feet. A vernier enabled the observer to read to hundredths of feet. The length of rod from its top or zero to water line on float was 15.58 ft.

B.M. 6 is the upper surface of the top course of stone that sets on the pintle head stones at the N.W. corner of Fort Taylor. The courses of stone are believed to be level all around the fort. The B.M. is also the ninth one from the bottom course, and is 0.18 ft. above the reading point of tide gauge.

Fig. 9.3 Map of the tide-gauge location at Key West dating from the 1840s at Fort Taylor. The map is from a survey (c. 1860) and the note describing the tide-gauge benchmark location is from 1888. Many more such records are in the archives and it should be a high priority to locate them for analysis.

thus referenced to benchmarks that provide continuity in the records, a feature unfortunately not also true of the weather records (*q.v.* Table 9.1).

3 The Key West Temperature Record

With this limitation in mind, we low-pass filtered the temperature time-series (Fig. 9.1) using a 13-year 'box-car' filter. The filtered record (Fig. 9.4) indicates that the temperature averaged about +0.2°C during 1860–1880, similar to averages after the mid-1940s (expressed as departures from the mean for the entire period). The low-pass filtered time-series shows that temperatures averaged below the mean, about −0.6°C, during the period 1890–1910.

The low-pass filtered time-series indicates an upward trend for the 1890s to about the 1950s, with no apparent trend after. Thus, concerning evidence in the data for a trace-gas-induced warming, it could be argued that the average temperature has been above the mean during the last 30 years. But it is also important to note that (1) although the trend has been upward from the 1890s, it has been relatively constant during the immediate past 30 years, and (2) an explanation is needed for the above-the-mean temperatures of the last century (1860–1880).

While the high-frequency (interannual) variability may be real, it is also possible that the low-frequency trends in the data are a result of changing observational conditions, and therefore, are not real. That is, they may result from a changing sampling bias over the years. Yet, it is also possible that the low-frequency variability of the data is real. Encouraging in this respect is the remarkable similarity between the Key West low-pass filtered temperature and the average for the Northern Hemisphere, especially from about 1890–1950 (Jones *et al.*, 1986).

We also prepared a time-series of air temperatures for the Straits of Florida for the period 1888–1979, based on ship data (COADS; Woodruff

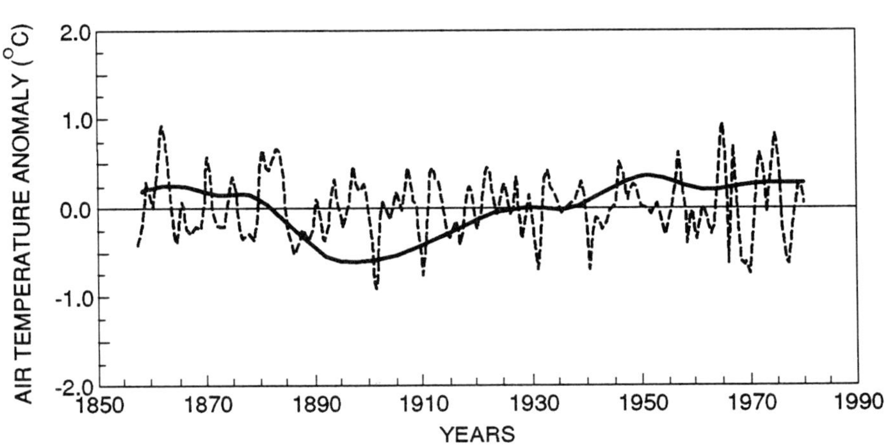

Fig. 9.4 Low-pass filtered and residual annual temperature, Key West, Florida, 1857–1980.

et al., 1987) only (not shown). These data are not supportive of the low temperatures reported for Key West (Fig. 9.1) from about 1885–1920. The ship data indicated below normal temperatures from 1905–1910, and above normal temperatures during the last few years of the 1800s, and from the early 1940s to about 1950. Otherwise, air temperatures as determined by ship observations in that area are close to normal. There is no apparent long-term trend in these COADS data.

In the future, it will be important to compare Key West temperatures with those of other nearby locations in order to determine the degree to which the Key West historical record represents real conditions. For the present, the record neither can be accepted nor rejected as representative of lower tropospheric conditions at Key West. Thus, we cannot use the record for the detection of trace-gas-induced climatic change during the past few decades (*cf.* Gray, Chapter 5).

4 The Key West Precipitation Record

The precipitation data for Key West began in 1886 and are complete for the 101-year period through 1986. The history of precipitation observations for Key West is the same as for temperature, except that tipping-bucket gauges were used during the entire period of record. Precipitation is easily measured, and the stability of a precipitation time-series is likely to be influenced mainly by changing exposure conditions. Over the 101-year period of observations at Key West, the observation sites have varied locally. Thus, we decided to use this time-series to look for secular trends at Key West. The annual precipitation chronology is given in Fig. 9.5.

The annual precipitation record has been low-pass filtered, as described for the temperature record. The low-pass filtered precipitation time-series and the residual of the original precipitation time-series are illustrated in Fig. 9.6. This shows that much of the variability in annual precipitation is in the high-frequencies rather than low. There is no apparent trend in the low-pass filtered time-series (Fig. 9.6).

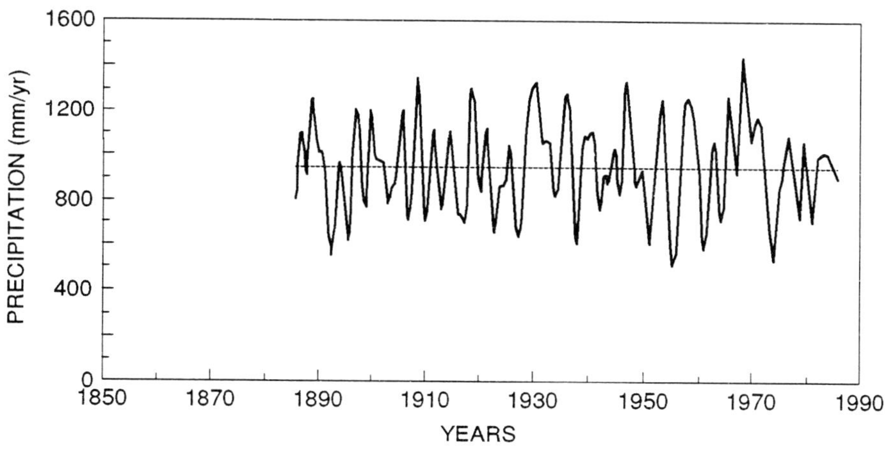

Fig. 9.5 Annual precipitation, Key West, Florida, 1886–1986.

In a recent study, Hanson and Maul (1991) analyzed the precipitation record of Florida for evidence of El Niño effects. They found five- to six-year variability in rainfall in particular during winter and spring, and statistically significant evidence of rainfall being timed to Pacific El Niños. In particular, there is below normal rainfall during winter and spring of the year prior to an El Niño and above normal during the same seasons for the second year of a major El Niño. B. Hernandez (pers. comm.) has shown similar effects on frontal passages at Havana, Cuba. These interannual events often dominate the Key West rainfall record and mask climatic change.

5 Summary of Meteorological Analysis

Analysis of the Key West temperature record for the 136-year period (1851–1986) shows: (1) relative warmth for 1851–1880; (2) strong cooling from 1880–1890; (3) an apparent increase in temperature from 1890–1950; and (4) temperatures above normal (by 0.2 to 0.3°C) from 1950 to the present. This temperature chronology has some features in rough agreement with that of the Northern Hemisphere average (Jones *et al.*, 1986). It is obvious, of course, that the warming from 1890 to about 1950 is in the direction needed to support the argument of trace-gas-induced warming. However, there are two serious questions: (1) Why is the warming mainly in the period 1890–1925, and no warming subsequent to 1950? (2) How can the warm temperatures be explained during the mid- to late 19th century?

Because of changing location and exposure in temperature sampling conditions at Key West over this 136-year period, we conclude that it is neither possible to accept nor totally reject the reality of these temperature observations. If the Key West temperature chronology is real, then an additional explanation must be sought to account for the warm temperatures at Key West during the last century (1850–1880).

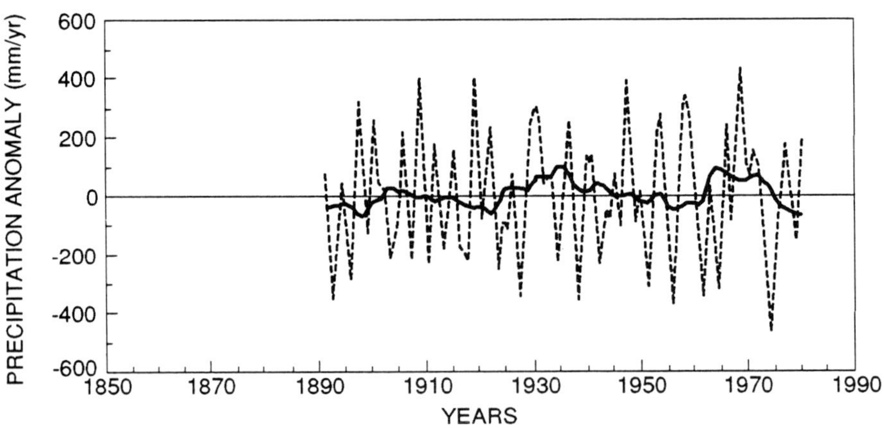

Fig. 9.6 Low-pass filtered and residual annual precipitation, Key West, Florida, 1892–1980.

Analysis of the 101-year precipitation data set for Key West indicates there is no significant trend in the data, and that precipitation during the last 10 to 20 years is not significantly different than during the previous 80 to 90 years.

Confirmation of the Key West low-frequency temperature chronology, given here, is a problem that should be investigated using other temperature or proxy indicators. We plan to continue this confirmation effort, and encourage others in this work, as well.

Interested readers will find two excellent summary articles on this subject. Ellsaesser *et al.* (1986) have examined the question of global climate trends in recorded data, particularly land-station observed data. They concluded: 'that while we are witnessing a warming of terrestrial climate, we cannot now identify its cause.' The other is an examination of global surface temperature COADS data by Oort *et al.* (1987). They point out some differences between ocean- and land-station temperature trends, and suggest the uncertainty in the data of the late 19th century is a significant problem to be overcome.

6 Linear Trends in Sea Level

We now turn our attention to sea-level change. At monthly and longer periods, sea level in the Intra-Americas Sea is dominated by an annual signal with an amplitude of approximately 20 cm. Interannual variability has an amplitude of approximately 5 cm superimposed on a linear trend of varying amount. In the sections that follow, the linear trends in the region are discussed, followed by a more thorough study of the 74-year record of annual sea level at Key West, Florida (Fig. 9.7). It will be shown that the 20 cm rise in global sea level by 2025 scenario in UNEP (1987) and other sources (e.g., Hoffman *et al.*, 1983; Cartwright *et al.*, 1985; Titus, 1986; IPCC, 1990) is wholly inadequate for detailed climate-change impact planning at specific sites because of the influence of other effects, such as

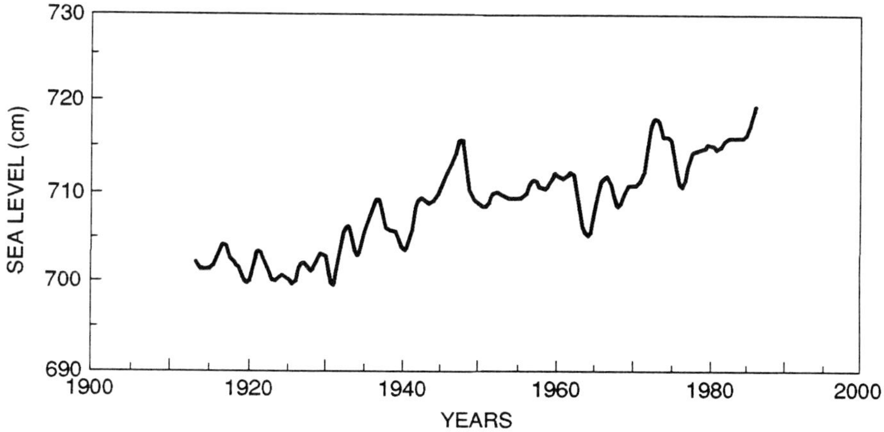

Fig. 9.7 Time-series of annual mean sea level at Key West, Florida, from the Revised Local Reference data on file with the Permanent Service for Mean Sea Level.

Table 9.2 Summary of data availability and linear trends from records at the permanent service for mean sea level, 1990.

Country/City	*Latitude*	*Longitude*	*Years*	*n*	*Change (cm/yr)*	*SE (±cm/yr)*	*r*
Bermuda							
[1]St. George*	32°22.0′N	64°42.0′W	1932–1986	40	0.247	0.056	0.58
Colombia							
[2]Riohacha*	11°33.0′N	75°55.0′W	1953–1969	15	0.197	0.180	0.29
[2]Cartagena*	10°24.0′N	75°33.0′W	1949–1984	32	0.513	0.025	0.97
Costa Rica							
[2]Puerto Limon*	10° 0.0′N	83° 2.0′W	1948–1968	17	0.187	0.057	0.65
Cuba							
[1]Cabo San Antonio	21°54.0′N	84°54.0′W	1971–1986	12	−0.097	0.216	−0.14
[1]Guantanamo Bay*	19°54.0′N	75° 9.0′W	1937-1968	26	0.190	0.048	0.63
[1]Siboney*	23°05.0′N	82°25.0′W	1966–1987	22	0.254	0.118	0.43
Dominican Republic							
[2]Barahona	18°12.0′N	71° 5.0′W	1954–1969	10	−0.229	0.106	−0.61
[2]Puerto Plata*	19°49.0′N	70°42.0′W	1949–1969	13	0.453	0.090	0.83
Guatemala							
[1]Santo Tomas de Castilla*	15°42.0′N	88°37.0′W	1964–1983	16	−0.397	0.169	−0.53
Haiti							
[2]Port au Prince	18°34.0′N	72°21.0′W	1949–1961	12	1.236	0.149	0.93
Honduras							
[2]Puerto Castilla*	16° 1.0′N	86° 2.0′W	1955–1968	13	0.307	0.105	0.66
[2]Puerto Cortes*	15°50.0′N	87°57.0′W	1948–1968	20	0.880	0.064	0.95
Jamaica							
[2]Port Royal	17°56.0′N	76°51.0′W	1954–1969	15	0.049	0.232	0.06

Mexico							
[1]Progreso*	21°18.0′N	89°40.0′W	1952–1984	27	0.492	0.037	0.94
[1]Ciudad del Carmen	18°38.0′N	91°51.0′W	1956–1966	10	−0.313	0.226	−0.44
[1]Coatzacoalcos	18° 9.0′N	94°25.0′W	1952–1961	8	0.473	0.527	0.34
[1]Alvarado	18°46.0′N	95°46.0′W	1955–1966	11	−0.303	0.282	−0.34
[1]Veracruz*	19°11.0′N	96° 7.0′W	1953–1985	31	0.153	0.045	0.53
[1]Tuxpan	21° 0.0′N	97°20.0′W	1957–1966	8	−1.101	0.375	−0.77
[1]Ciudad Madero	22°15.0′N	97°48.0′W	1962–1966	4	1.510	0.906	0.76
Panama							
Cristobal*	9°21.0′N	79°55.0′W	1909–1969	57	0.110	0.013	0.76
USA (Atlantic)							
[1]Key West I	24°34.0′N	81°48.0′W	1913–1925	13	−0.137	0.101	−0.38
[1]Key West II*	24°33.0′N	81°48.0′W	1926–1979	48	0.236	0.020	0.87
[1]Marathon Shores	24°44.0′N	81° 2.0′W	1966–1974	7	1.111	0.469	0.73
[1]Key Colony Beach	24°43.0′N	81° 1.0′W	1976–1979	2	1.111	0.469	0.73
[1]Miami Beach*	25°46.0′N	80° 8.0′W	1931–1980	43	0.243	0.020	0.89
[1]Daytona Beach*	29°14.0′N	81° 0.0′W	1925–1969	22	0.183	0.055	0.60
[1]Daytona Beach Shores	29° 9.0′N	80°58.0′W	1966–1979	9	−0.644	0.345	−0.58
[1]Mayport*	30°24.0′N	81°26.0′W	1928–1979	49	0.231	0.029	0.76
[1]Jacksonville	30°21.0′N	81°37.0′W	1953–1968	13	0.229	0.203	0.32
[1]Fernandina	30°41.0′N	81°28.0′W	1897–1986	66	0.210	0.018	0.83
[1]Fort Pulaski*	32° 2.0′N	80°54.0′W	1935–1980	43	0.270	0.031	0.81
[1]Charleston*	32°47.0′N	79°56.0′W	1921–1979	57	0.350	0.025	0.88
USA (Gulf)							
[1]Port Isabel*	26° 4.0′N	97°13.0′W	1944–1979	32	0.315	0.070	0.64
[1]Port Mansfield	26°33.0′N	97°26.0′W	1963–1986	20	0.284	0.172	0.36
[1]Padre Island*	26° 4.0′N	97° 9.0′W	1958–1978	15	0.395	0.169	0.54
[1]Port Aransas	27°49.0′N	97° 4.0′W	1958–1969	4	1.044	0.231	0.95
[1]Rockport*	28° 1.0′N	97° 3.0′W	1948–1986	20	0.413	0.107	0.67

[1]Freeport*	28°57.0′N	95°19.0′W	1954–1986	28	1.409	0.121	0.92
[1]Galveston I*	29°19.0′N	94°48.0′W	1908–1978	67	0.627	0.028	0.94
[1]Galveston II	29°17.0′N	94°47.0′W	1957–1973	14	1.156	0.225	0.83
[1]Sabine Pass*	29°42.0′N	93°51.0′W	1958–1978	15	1.161	0.358	0.67
[1]Eugene Island*	29°22.0′N	91°23.0′W	1939–1974	31	0.968	0.075	0.92
[1]Bayou Rigaud*	29°16.0′N	89°58.0′W	1947–1978	28	0.939	0.111	0.86
[1]Dauphin Island*	30°15.0′N	88° 4.0′W	1966–1979	10	0.425	0.286	0.46
[1]Pensacola*	30°24.0′N	87°13.0′W	1923–1980	54	0.241	0.026	0.78
[1]Apalachicola	29°43.0′N	84°59.0′W	1967–1978	7	0.687	0.541	0.49
[1]St. Marks	30° 4.0′N	84°11.0′W	1965–1975	7	1.438	0.545	0.76
[1]Cedar Keys I	29° 8.0′N	83° 2.0′W	1914–1925	10	0.277	0.141	0.57
[1]Cedar Keys II*	29° 8.0′N	83° 2.0′W	1938–1979	40	0.144	0.039	0.51
[1]St. Petersburg*	29°46.0′N	82°37.0′W	1947–1977	30	0.201	0.050	0.61
[1]Fort Myers	26°39.0′N	81°52.0′W	1965–1974	9	0.517	0.169	0.76
[1]Naples*	26° 8.0′N	81°48.0′W	1965–1980	15	0.200	0.168	0.31
USA (Puerto Rico)							
[2]Magueyes Island*	17°58.0′N	67° 3.0′W	1955–1986	28	0.197	0.042	0.68
[2]San Juan I*	18°27.0′N	66° 5.0′W	1962–1974	11	0.183	0.170	0.34
[2]San Juan II	18°28.0′N	66° 7.0′W	1978–1979	2	0.183	0.170	0.34
Venezuela							
[3]Puerto de Hierro	10°37.0′N	62° 5.0′W	1956–1963	8	−0.013	0.606	−0.01
[3]Guiria	10°34.0′N	62°17.0′W	1963–1967	5	−0.230	0.569	−0.23
[2]Carupano	10°40.0′N	63°15.0′W	1967–1975	9	0.292	0.236	0.42
[2]Cumana*	10°28.0′N	64°12.0′W	1954–1975	20	0.090	0.081	0.25
[2]La Guaira*	10°28.0′N	66°56.0′W	1953–1975	22	0.168	0.061	0.52
[2]Amuay*	11°45.0′N	70°13.0′W	1954–1975	19	0.105	0.110	0.22
[2]Maracaibo	10°41.0′N	71°35.0′W	1964–1975	10	0.564	0.130	0.84

Mean ± standard deviation (all stations): 62; 0.436 ± 0.632 cm/yr
(stations ≥10 years): 47; 0.338 ± 0.405 cm/yr

[1]North American Plate; [2]Caribbean Plate; [3]South American Plate

* Stations included in Fig. 9.8

petroleum/water extraction and plate tectonics, on local sea level.

Table 9.2 summarizes the PSMSL data for the Intra-Americas Sea as of 1990 (we include stations up to the latitude of Bermuda in order to have one deep-sea datum). Only data of sufficiently high quality to be placed in the PSMSL Revised Local Reference (RLR) file are used. Of the 62 stations listed, 47 have records longer than 10 years. Many of the station files have missing data or data not of RLR quality, hence the number (n) of annual means used to compute the linear trend (cm/yr) is not always equal to the time difference for each record's start and end. In addition to the linear trend (α), the standard error (SE) estimate ($\pm$cm/yr) of the linear trend, and the linear correlation coefficient (r) also are listed. Finally, as a superscript before each station entry, the tectonic plate is identified.

Linear trends in Table 9.2 are second iterations of a least-squares fit; if sea level for any year was more than ± 2 standard deviations (σ) from the first fit, it was eliminated and the fit recalculated. For the ensemble of stations, the mean linear trend is $\bar{\alpha} = 0.44 \pm 0.63$ cm/yr ($\pm 1\sigma$); for those stations with ≥ 10 years of records, the trend is $\bar{\alpha} = 0.34 \pm 0.41$ cm/yr. Ten stations have negative trends; two (Key West I and Fernandina [1898–1924]) have records that break in the mid-1920s. The remaining records with negative trends appear to be close to the boundaries of tectonic plates and/or have rather short records. For 27 stations, $r \geq 0.7$, which indicates that the long-term linear trend accounts for $\geq 50\%$ of the variance in the record at almost half of all stations.

Many of the shorter records listed in Table 9.2, those with less than 10 years of observations, have large linear trends. Linear trends in shorter records crucially depend on the decade in which they occur. For

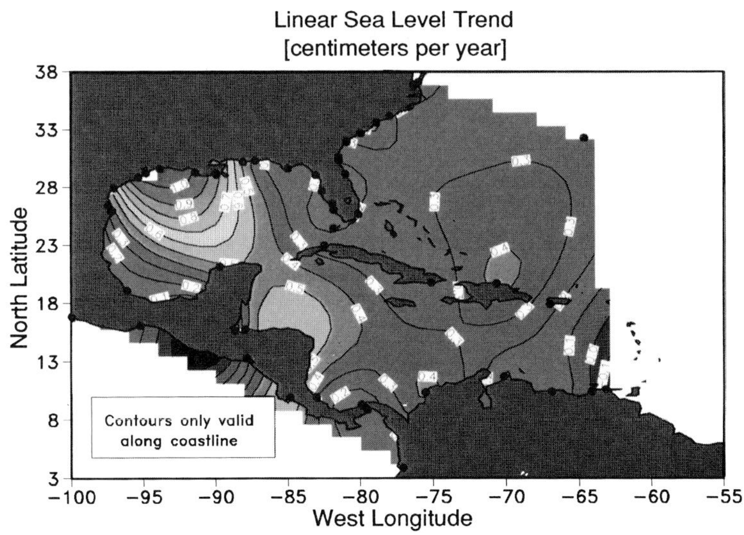

Fig. 9.8 Distribution of linear trends in annual mean sea level for stations (dots) with more than 10 years of record during the last 60 years. Of the 51 stations used, eight are from the Pacific Ocean and are not included in Table 9.2. See colour plates between pages 210 and 211.

example (Fig. 9.7), if the record was obtained from the mid-1960s to the mid-1970s (*cf.* Apalachicola, Fort Myers, Marathon Shores, and Daytona Beach Shores on the Florida coast) the rise in relative sea level is influenced by a period of rapid change compared with the entire record. While these shorter periods of rapid change are of interest in themselves, they can easily distort an estimate of the longer-term trend (*cf.* Tanner, 1992).

Fig. 9.8 attempts to put some semblance of order to the variability in linear trends for those stations with ≥10 years of records during the last 60 years. Contours in Fig. 9.8 are only meant to provide a sense of the variability. In water deeper than the continental shelf break (~100 m depth), sea-level rise is probably similar to Bermuda. Contours in deep-water regions of the Gulf of Mexico and the Caribbean Sea are artifacts of the contouring routine, and serve only to emphasize the spatial nature of the variability along the coast.

Rising relative sea level dominates the signal in the northwestern Gulf of Mexico, complimented by lowering relative sea level in the southwestern Gulf. Sea level around Florida has a relatively uniform rise of 0.2–0.3 cm/yr, but there is no information in the PSMSL record through 1987 for the Bahamas; Bermuda sea level is rising nearly 0.3 cm/yr. Jamaica seems remarkably steady compared with Haiti, where the rise at Port au Prince is about 1 cm/yr. Except for Maracaibo and Cartagena (both ~0.5 cm/yr), the southern Caribbean Sea averages ~0.2 cm/yr. Additional sea-level information for the north coast of South America is given in Chapter 6.

There seems to be little organization of sea-level change by tectonic plates, except that at or near plate boundaries the change is often negative. Only two stations (in eastern Venezuela) are on the South American Plate; both show falling relative sea level, but are not included because their PSMSL records are too short. Eight stations in the Pacific Ocean are included in Fig. 9.8 for contouring purposes, and these are located on either the Cocos Plate or Nazca Plate.

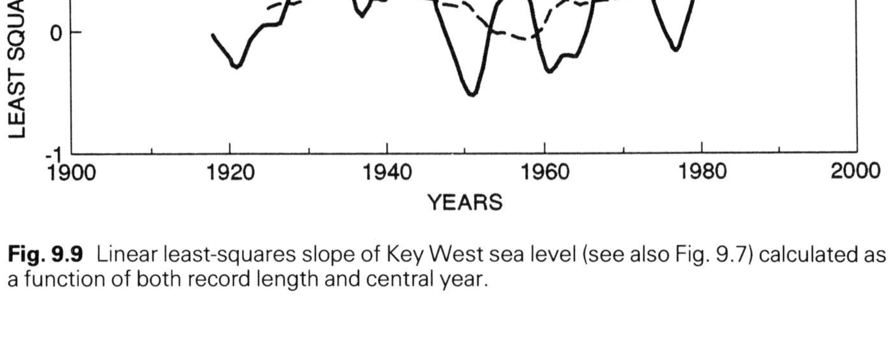

Fig. 9.9 Linear least-squares slope of Key West sea level (see also Fig. 9.7) calculated as a function of both record length and central year.

Because of the great station-to-station variability, a single estimate of the relative sea-level rise by 2025 (UNEP, 1987) for the Intra-Americas Sea is not possible. Some of the variability may be due to uncertainties in the RLR, for example due to poor levelling of the tide staff to the bench-marks. Some of the highest relative sea-level rise is in regions of significant petroleum/groundwater extraction (although Turner [1991] argues that subsidence in the northwestern Gulf of Mexico is largely non-anthropogenic), and most of the negative trends could be associated with tectonics. The average, from all records since 1930 longer than 25 years, shows a trend of $\bar{\alpha} = 0.36 \pm 0.25$ cm/yr (Fig. 1.3), but the large standard deviation (± 0.25 cm/yr) makes $\bar{\alpha}$ not statistically different from the 'global' value of 0.2 cm/yr (Warrick and Oerlemans, 1990). Clearly, many more long records are needed.

7 Key West as a Long-Term Record

The PSMSL data from Key West I and II, supplemented by other records from 1980–1986 provided by NOAA, is a 74-year record with only 1945, 1946 and 1953 missing. The plot in Fig. 9.7 uses a linear interpolation to compile a complete record for the calculations to be discussed below. Linear regression calculations, excluding the missing years, give: $n = 71$, $\alpha = 0.22 \pm 0.01$ cm/yr, and $r = 0.89$; after eliminating years $\geq 2\,\sigma$ from the trend: $n = 66$, $\alpha = 0.22 \pm 0.01$ cm/yr, and $r = 0.93$. In either calculation, $\geq 80\%$ of the variance is explained in the interannual trend.

In evaluating Fig. 9.8 and conjecturing on future sea-level rise, it is important to evaluate the stability of the linear trends. Again using Key West as an example, the linear least-squares slope has been computed for record lengths of 10 and 25 years as a function of time. Fig. 9.9 shows how the slope varies depending on when the subsample was taken. As can be seen for a 10-year record, $-0.5 \leq \alpha(\text{cm/yr}) \leq 1.0$; for a 25-year record, $-0.1 \leq \alpha(\text{cm/yr}) \leq 0.5$. Not only is the record length critical in calculating sea-level change, but the time period as well (*q.v.* Table 9.2). Evaluation of sea-level rise particularly for accelerated changes induced by climate factors (*cf.* Donn and Shaw, 1963), must consider both the length of the record, and the time frame of the observations.

The linear trend at Key West from the contemporary data can be checked using the few years in the 1846–1852 and 1857 historical record. We recalculated the linear parameters and found for 1846–1990 $\alpha = 0.19 \pm 0.01$ cm/yr and $r = 0.93$ (plot not shown). Although only five extra 19th century years are available, their inclusion in the calculation strongly supports the notion that sea level at Key West has been rising rather uniformly for almost 150 years, and that the downtrend during 1913–1925 (*q.v.* Table 9.2) may only be part of a short-time event that is not significant on the century time-scale. However, at Fernandina Beach, Florida, the continuous record from 1898–1930 is also characterized by a negative sea-level trend (Maul and Hanson, 1990); the cause is unknown.

Relative sea level is the combination of the oceanic signal plus that due to change in the land from tectonic or local effects. Near Miami, Florida, NOAA has been maintaining a Very Long Baseline Interferometry (VLBI) station (Carter *et al.*, 1986) for the purpose of determining crustal movements. Preliminary results (W.E. Carter [NOAA], pers. comm.; Eden, 1990) based on five years (1983–1987) of VLBI observations, suggest that Florida is tectonically stable, and therefore, the signal in sea level at Key West (Fig. 9.7) is probably representative of the oceanic portion of relative sea-level rise in the Intra-Americas Sea. The VLBI series is too short at this point to confirm the post-glacial rebound (pgr) modelled by Peltier (1986) of 0.8 mm/yr, but in time it is hoped that the relative sea-level signal can be separated into its geodetic and oceanographic components. For example, Maul and Hanson (1990) show for the period 1950–1987, that >80% of the long-term trend at Miami (2.7 mm/yr) can be accounted for by change in 0–1000 m dynamic height anomaly east of the Bahamas (1.4 mm/yr) plus Peltier's 0.8 mm/yr pgr.

8 Key West and Other Climate Variables

An investigation into the relationship between Key West sea level and other variables of climatic interest is summarized in Table 9.3. The upper half of Table 9.3 shows the linear correlation coefficients (r) between Key West air temperature (Air temp.), sea level (Slev.), coral growth (Coral), barometric pressure (Baro.) and precipitation (Precip.); the lower half shows the same linear correlation coefficients with linear trends least-squares removed. Except for $r = -0.67$ between sea level and coral growth at nearby Biscayne National Park (H. Hudson [USGS], pers. comm.), the correlations are small.

Detrending the data in general reduces the correlations, as can be seen in the lower half of Table 9.3. Spectral analysis (not shown) confirms

Table 9.3 Key West linear correlations of annual means 1913–1986 (n=74).

	Air temp. (°C)	*Slev. (cm)*	*Coral (mm)*	*Baro. (mb)*	*Precip. (mm)*
Air Temperature	1.00				
Sea Level	0.47	1.00			
Coral growth	−0.19	−0.67	1.00		
Barometer	0.27	0.18	−0.09	1.00	
Precipitation	−0.28	−0.01	−0.10	−0.48	1.00
Mean	25.32	708.40	8.22	1015.87	948.00
Standard deviation	0.42	5.43	0.80	0.49	212.80
Detrended Key West linear correlations					
Air temperature	1.00				
Sea level	0.31	1.00			
Coral growth	0.21	0.13	1.00		
Barometer	0.20	−0.09	0.19	1.00	
Precipitation	−0.31	−0.05	−0.14	−0.50	1.00
Mean	0.00	0.11	0.00	0.00	0.00
Standard deviation	0.39	2.53	0.49	0.47	212.80

that the coherence squared above the 95% confidence level between detrended Key West sea level and the other detrended variables occurs only at a few minor frequencies. Similarly, cross-covariance analysis (also not shown) reveals no organized lags or leads between sea level and the other variables although precipitation and surface air pressure have a significant spectral peak at six to seven years. Statistically, these calculations suggest little cause and effect amongst these variables with respect to sea level.

Since the El Niño – Southern Oscillation (ENSO) signal seems to be the major interannual climate event in the tropics, correlations were also calculated using the intensity scale 0–4 as devised by Quinn *et al.* (1987). Linear correlations between El Niño intensity and the variables shown in Table 9.3 were all $|r| \leq 0.2$, and spectrally there was no coherence squared above the 95% confidence level. However, while ENSO events affect Pacific Ocean sea level at all intensity levels, it was decided to investigate only Quinn scale 3 and 4 events to determine if Key West responded.

A superposed epoch analysis of Key West sea level was performed by sorting sea level relative to 'key years', where key years were defined as years of El Niño intensity scale 3 and 4 that were preceded by a year without an El Niño. Key West sea level was sorted for all years within ±4 years of the key years. The result is shown in Fig. 9.10, where it can be seen that the year prior to a scale 4 El Niño (five occurrences between 1913 and 1986), average Key West sea level is more than 0.7 cm below normal, and during the year of the event, is 1.0 cm above normal. Using 1000 random sortings of the Key West sea-level data set, a direct probability estimate of 0.85 is calculated that reduced sea level at year -1 followed by elevated sea level at year 0, is not due to chance occurrence of sampling a random population. A similar analysis for scales 3 and 4 El Niños (13 occurrences; not shown) indicates no such effect. However, when scales 3 and 4 El Niños that lasted more than one year are analysed (seven events) a pattern similar to that shown in Fig. 9.10 also emerges.

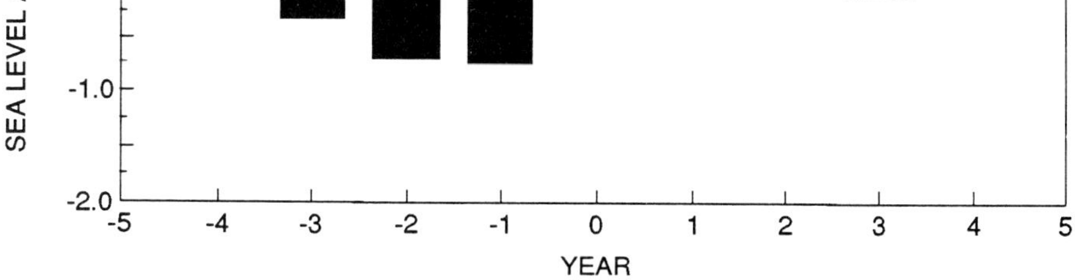

Fig. 9.10 Superposed epoch analysis of Key West sea level using scale 4 ENSO events as the key date.

9 Accelerations in Key West Sea Level

We now turn our attention to the very important question as to whether or not there is evidence for acceleration in the rise of sea level at Key West. This question is related to the issue of global warming in that if Earth is experiencing an enhanced greenhouse effect, then temperature and sea level should be increasing in a non-linear fashion. This non-linearity is referred to as acceleration in that the best-fit curve is of the form of a second-order polynomial rather than the linear trends summarized in Table 9.2.

We follow a modification of the work of Woodworth (1990) who asked a similar question of European tide-gauge records. For this question at Key West, we have joined and extended the record to cover the period 1913–1990 inclusively, or n = 78 years (not shown). First we calculate the linear regression coefficient (α_1) and compare its standard error with the standard error (ϵ) of the second-order coefficient (α_2). For Key West (Fig. 9.7; Table 9.2) we obtain $\alpha_1 = 0.217 \pm 0.0127$ cm/yr, and $\alpha_2 = -0.000548 \pm 0.000633$ cm/yr, and note that (1) the standard error for α_2 is larger than α_2 itself, and (2) that within $\pm 1\ \epsilon$, α_2 is not different from zero. This test therefore suggests that the second-order estimate is best described by the null hypothesis of no acceleration.

We carry out the question of acceleration further, in that we next calculate the correlation coefficients (r) for the first-order and second-order polynomials. Again for the 78 years of annual Key West sea level means, we obtain r^2 values of 0.792 and 0.794 for the first- and second-order equations, respectively. Although r^2 is slightly larger for the second-order fit, the 99% confidence interval of r for the first- and second-order polynomials is $0.937 \geq r_1 \geq 0.809$ and $0.938 \geq r_2 \geq 0.811$ respectively, and we again conclude that there is no statistically significant evidence for acceleration in Key West sea level.

10 Summary of Sea-Level Analysis

Based on PSMSL records for stations with records ≥25 years ($\bar{\alpha} = 0.36 \pm 0.25$ cm/yr), and ignoring a possible change in the rate of global sea-level rise, sea level throughout the Intra-Americas Sea on average will rise between 5 and 25 cm in the next 40 years (Table 9.2), but spatial variability will be large (Fig. 9.8). Using southeast Florida as an indicator that is free from local tectonic or subsidence effects, an areal average sea-level rise of ~10 cm by 2025 is expected. Linear extrapolation suggest that the Texas–Louisiana coast of the United States, Port au Prince, Haiti, and Puerto Cortes, Honduras, may experience sea-level rises more than triple that of southeast Florida if factors causing recent rises continue. Several sites, notably in proximity to tectonic plate boundaries, have negative trends and could have lowered sea level of ~20 cm by 2025 (*q.v.* Aparicio, Chapter 6). Many records, however, are short, and linear extrapolation must be applied cautiously (Fig. 9.9).

Sea level at Key West (Fig. 9.7) fell during the early part of this century and on average is rising slower than the Intra-Americas Sea average. A detailed examination of Key West shows little correlation with certain other

Fig. 1.1 Reproduction of early 18th century sea chart published by Gerard van Keulen. Translated from the French, the title reads: 'New Sea Chart of all THE COASTS OF AMERICA Showing all the Islands Bays and Rivers, as well as all the Rocks and Deeps, entirely composed from many Accounts of Very Experienced navigators by Jean Sikkena Master of Mathematics'. © 1991 Friends of the University of Miami Library; used by permission.

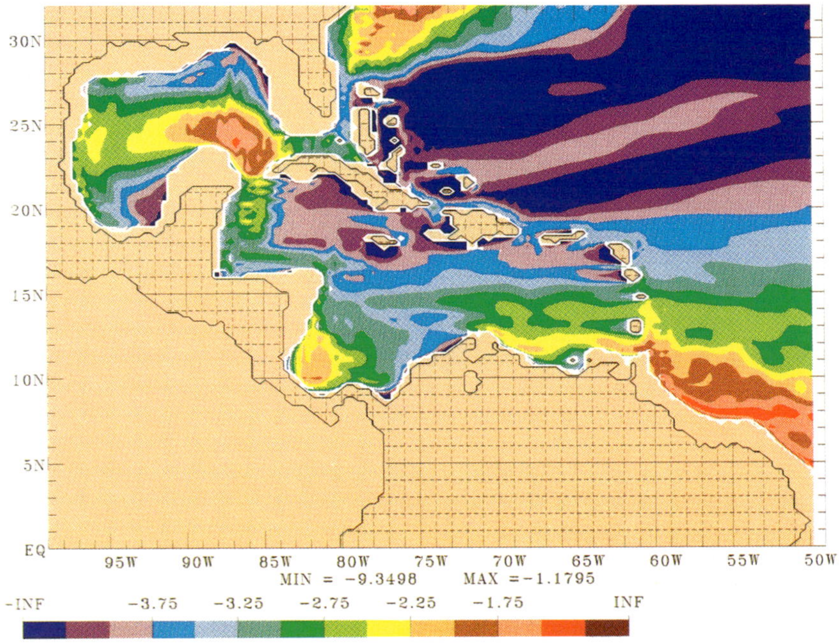

Fig. 4.1a Contours for the log of eddy kinetic energy from a 1.5-layer reduced-gravity model driven by Hellerman and Rosenstein (1983) monthly mean winds to statistical equilibrium in a closed basin. Maximum eddy kinetic energy is 0.07 m^2s^{-2}.

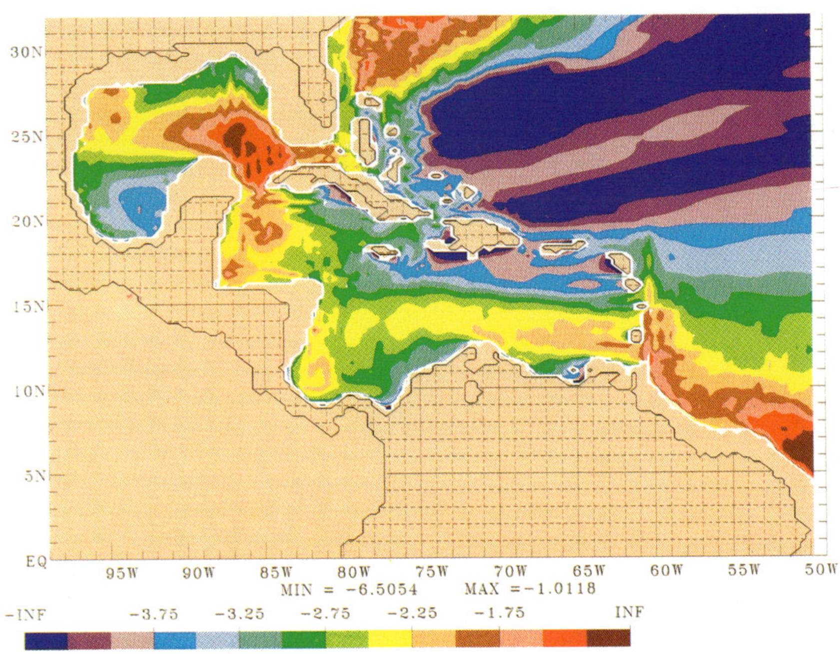

Fig. 4.1b Same as Fig. 4.1a except for an open basin with a 15 Sv South Atlantic surface inflow and a high-latitude surface outflow. Maximum kinetic energy is 0.18 m^2s^{-2}.

Fig. 5.9 Colourized visible image of Hurricane Gilbert observed from the Geostationary Operational Environmental Satellite (GOES-7) on 13 September 1988. Until 1992 with the advent of Hurricane Andrew, Hurricane Gilbert was the most intense Atlantic tropical storm on record, causing 318 deaths (202 in Mexico alone) and US$ 5,000,000,000 in estimated damages. At NOAA aircraft flight altitude (10,000 ft or 3000 m) maximum wind gusts of 173 knots (89 m/s) were recorded, and the central pressure was 888 mb.

Fig.7.6 Mud bank on the coastline of French Guiana. The gently sloping mud surface extends as far as the eye can see, its landward margin terminating against a narrow sand beach backed by wetland. The mud bank dampens wave energy across the shelf. In other locations the landward edge of the mud bank borders directly against the wetland system (*cf.* Fig. 7.5). (Photograph by Malcolm Hendry).

Fig. 7.8 The beach-wetland system at Negril, western Jamaica (see Fig. 7.7). In a setting typical of many islands this wetland is confined in a coastal basin, its landward limit defined here by a fault scarp. This boundary restricts potential for further wetland expansion under a rising sea-level scenario. Dense seagrass beds offshore are the primary source area for carbonate sediments of which the beach is composed. This proximate, general source type is quite distinct from the point sources along shorelines dominated by river discharge where longshore transport is required to nourish the coast. (Photograph by Malcolm Hendry).

Fig. 7.9 Carbonate platform islands, the Bahamas. Dissected by tidal channels, the islands are typically at elevations of 1 m or less above sea level. Bank areas produce abundant carbonate sediment but supra-tidal areas are extremely susceptible to ocean processes and storms, and shorelines in exposed areas are mobile. Developed and undeveloped islands in platform settings are susceptible to flooding and submergence as well as shoreline instability in a rising sea-level scenario, and are key locations for further impact-assessment work. (Photograph by Malcolm Hendry).

Fig. 7.10 Pocket beaches on the southeast coast of Barbados. These beach systems are characteristic of many Antillean islands: physically separated processes influencing the beach are largely confined to each bay. Beach stability in such cases is controlled to a great extent by onshore-offshore transport in a conservative sedimentary regime. An old sea-cliff defines the landward beach boundary. The long-term integrity of these beaches is under threat in rising sea-level conditions due to several factors including the upland area against which they are anchored: this may in time function as a vertical seawall, inducing wave reflection and sediment re-suspension, thus inhibiting deposition. (Photograph by Malcolm Hendry).

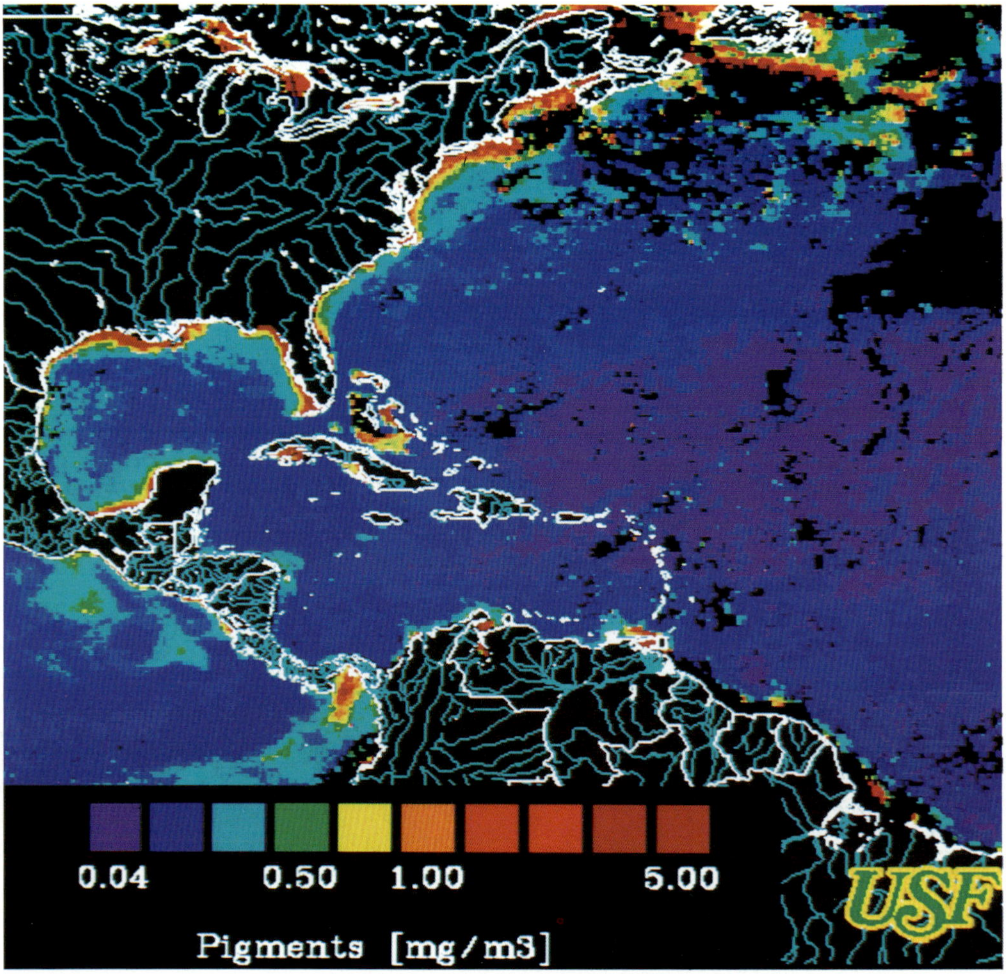

Fig. 8.11 February 1980. Monthly mean pigment concentration field obtained from CZCS satellite imagery showing the western North Atlantic Ocean in February 1980. Concentrations were colour coded, with warm colours (yellow, orange, and red) representing high phytoplankton pigment concentrations or other coloured materials in river plumes, and cooler shades (purple, blue, and green) representing low concentrations. Black represents no data, due either to cloud cover or lack of programmed coverage. Since the CZCS was an experimental instrument, rather than an operational one, data coverage was limited to specific requests and short periods of intensive data collection. This created many data gaps. Land was masked brown. Rivers that affect the region have been outlined using the World Database II files.

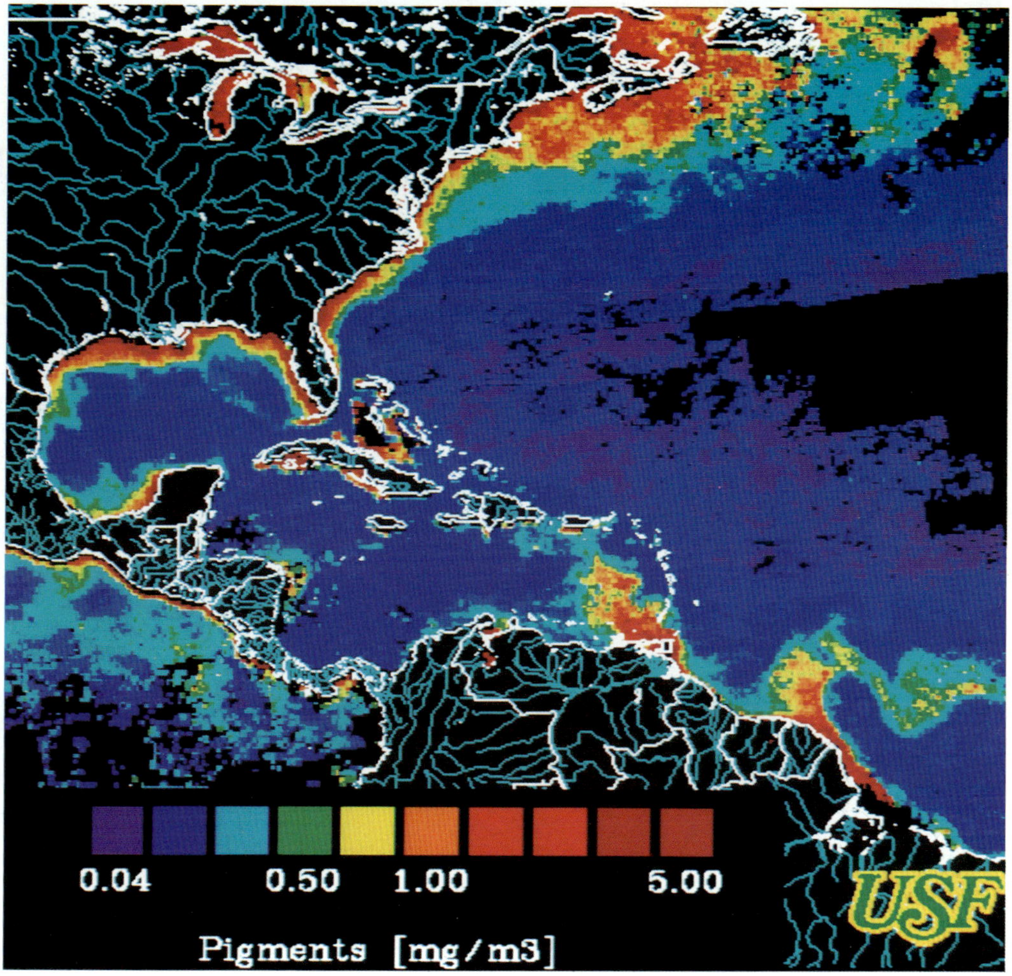

Fig. 8.12 October 1979. Monthly mean pigment concentration field obtained from CZCS satellite imagery showing the western North Atlantic Ocean in October 1980. Concentrations were coded as in Fig. 8.11.

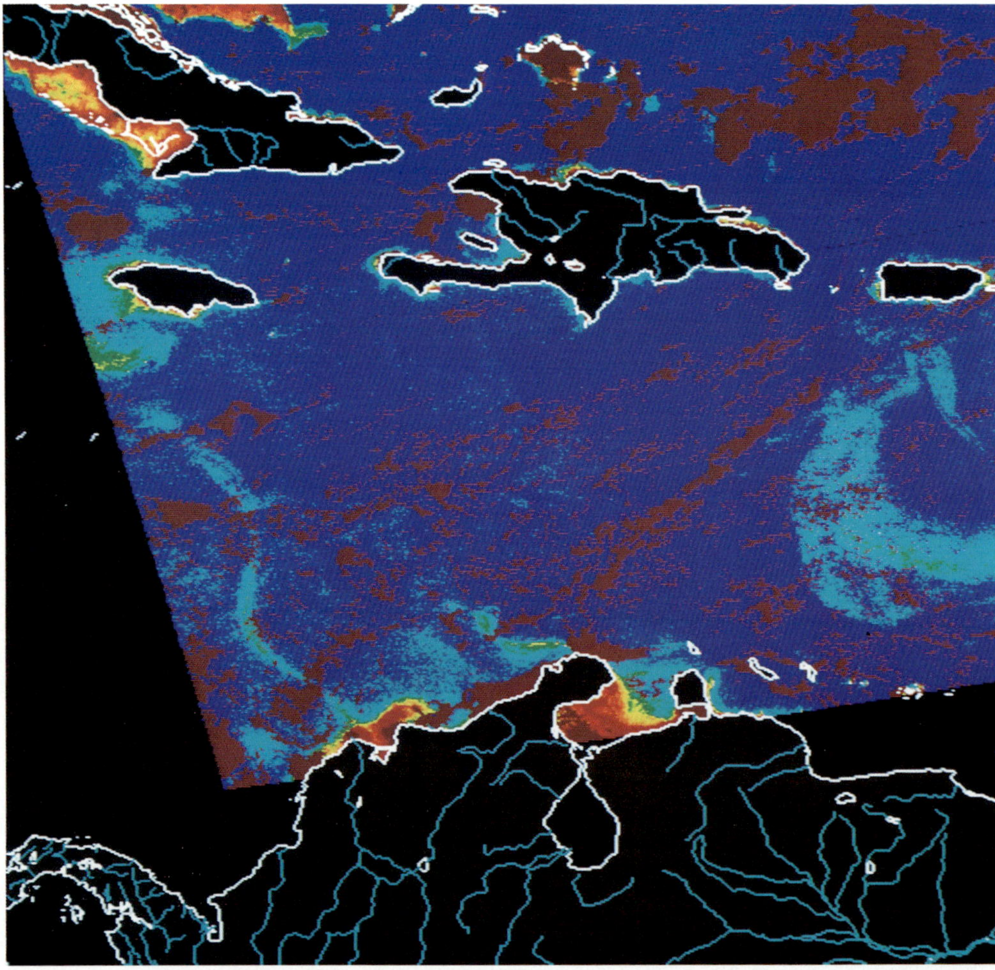

Fig. 8.13 11 October 1982. CZCS image of the western Caribbean Sea showing an extreme case of the dispersal of the Magdalena River where the plume extends across the western Caribbean Sea to the northwest and almost reaches Jamaica. Typically, the plume flows along the coast of Colombia to the west. Clouds are masked grey and areas with no data, masked black. Rivers that affect the area are outlined.

a)

b)

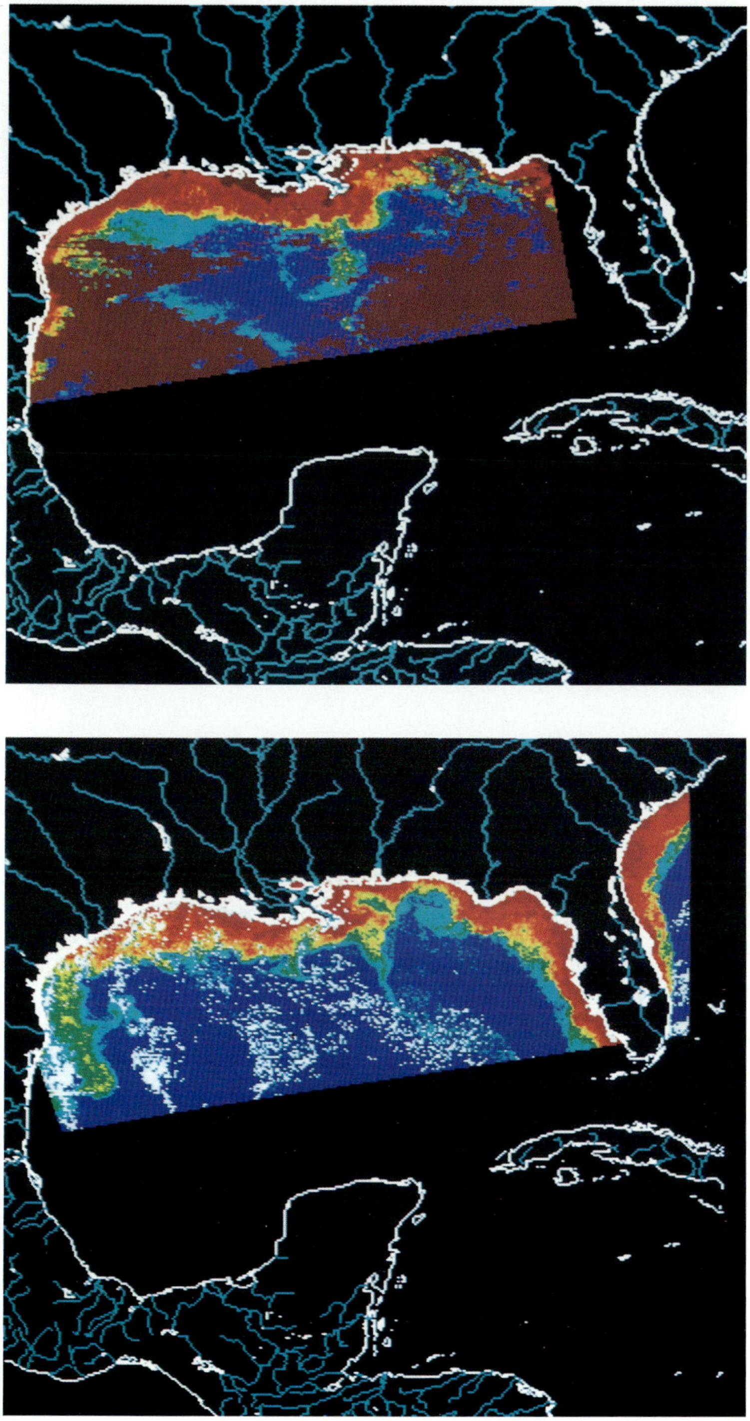

Fig. 8.14 CZCS images of the Gulf of Mexico showing the Mississippi River plume. Clouds are masked grey and areas with no data, masked black. Grey shades near the mouth of the Mississippi indicate high sediment concentrations. (a) High discharge in 1979 (24 March); (b) Low discharge in 1979 (8 October). Rivers that affect the area are outlined.

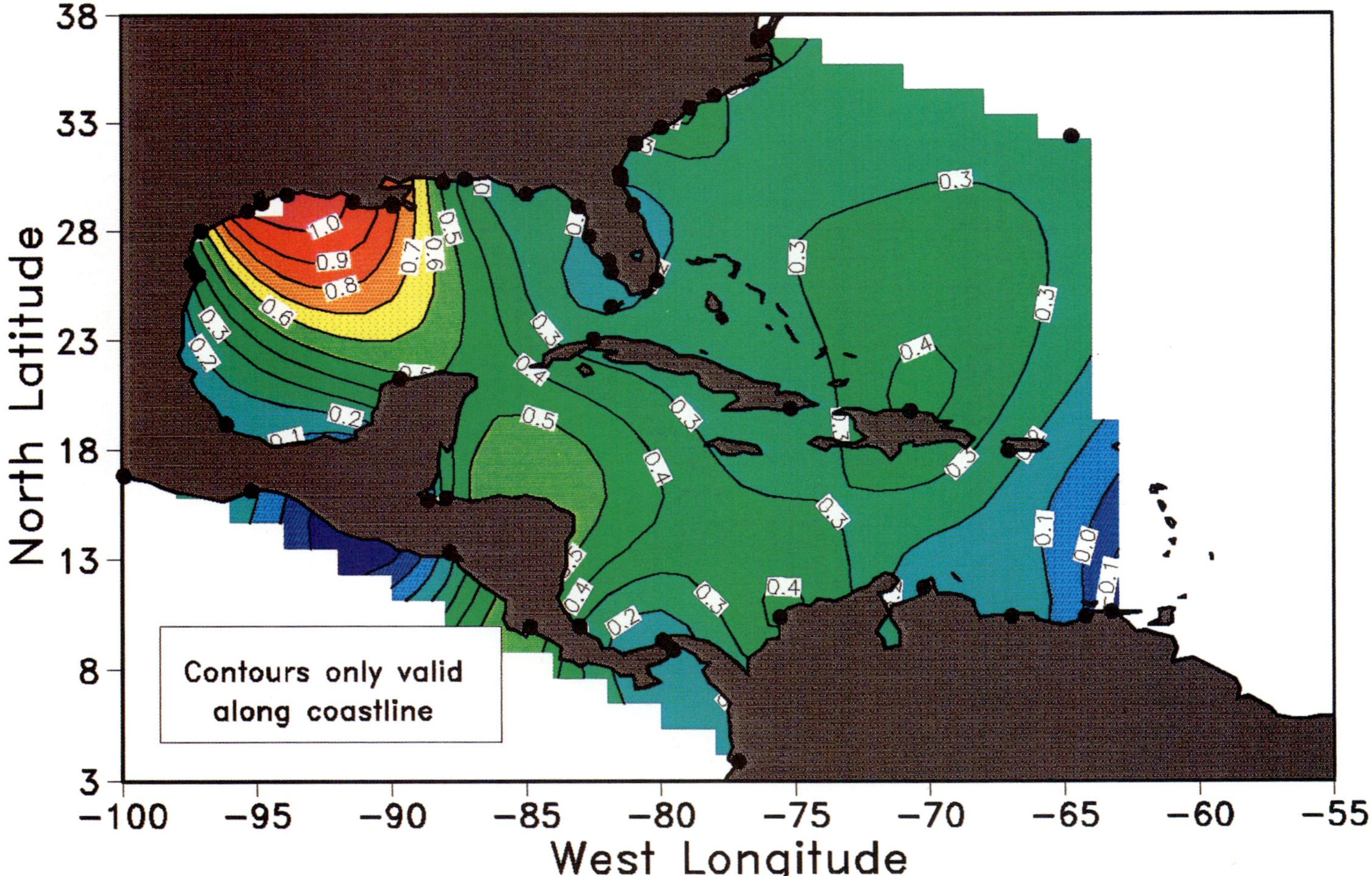

Fig. 9.8 Distribution of linear trends in annual mean sea level for stations (dots) with more than 10 years of record during the last 60 years. Of the 51 stations used, eight are from the Pacific Ocean and are not included in Table 9.2.

Fig. 12.1 The structural complexity and development of mangrove forests are greatest in coastal areas that receive fresh-water runoff from inland catchments. The input of nutrients in the runoff as well as the reduction in salinity result in high rates of primary productivity.

Fig. 12.3 *Rhizophora mangle* is frequently dominant along shorelines that are inundated on all high tides. In areas of moderate salinity, this species is also capable of high rates of root production which accumulates as peat.

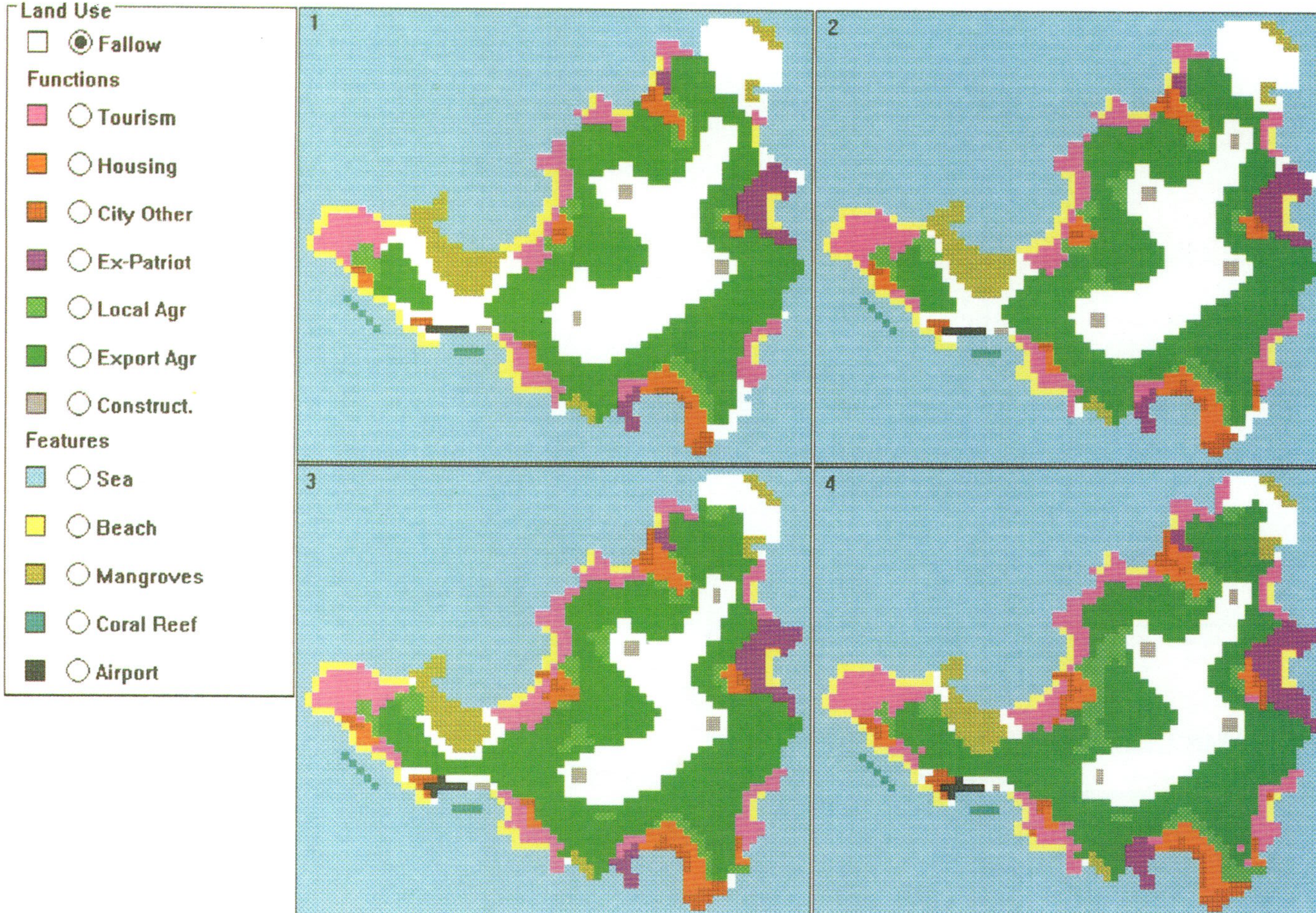

Fig. 16.4 Four stages in the evolution of a hypothetical island.

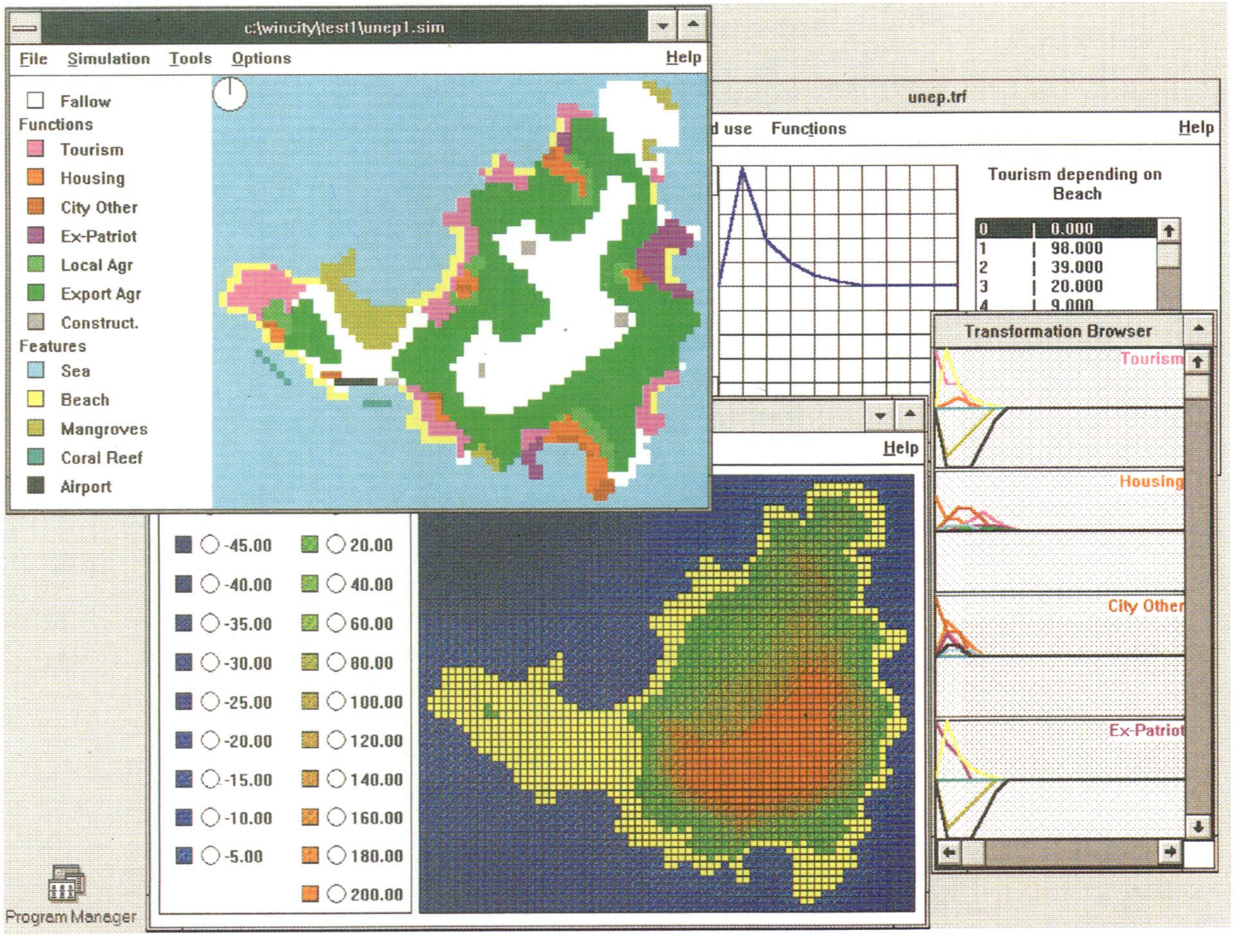

Fig. 16.5 The graphical user interface of the prototype decision-support environment.

climate indicators such as air temperature (Table 9.3). Key West sea level is known to respond to fluctuations in Gulf Stream volume transport (Maul *et al.*, 1985), and the low correlations shown in Table 9.3 could reflect other variables not considered here such as water temperature, salinity, or wind stress. On the other hand, there is a statistically significant relationship between strong El Niño events in that Key West sea level is lower the year prior to, and higher during, the year of such events (Fig. 9.10). Whether such patterns are common throughout the region is unknown, and should be the subject of further research.

We also applied two statistical tests on the Key West sea-level record to determine if there was any acceleration in the rate of sea-level rise. Both tests were conclusive in that the rate of rise over the 78-year period 1913–1990 can best be described by a linear trend. This is not to say that accelerated sea-level rise is not occurring, but that two statistical tests failed to discover any evidence for non-linearity.

11 Recommendations

The most obvious drawback in a study such as this has been the lack of reliable long-term data from which to make meaningful conclusions. We argue that an integrated sea-level/weather monitoring network receive high priority, and that the World Meteorological Organization and the Intergovernmental Oceanographic Commission *jointly* commit adequate funds for such a coastal network. In addition to the usual necessary instrumentation to measure air and water temperature, air pressure, precipitation, insolation, sea level, and salinity, the co-located instrument package should be geodetically monitored with permanent GPS receivers. We also argue that no matter how automatic the data collection and telemetering sea-level/weather system becomes, the WMO/IOC must plan to train human observers who are dedicated to keeping records of station history and operating behaviour of the instruments. Finally, we argue that a vigorous UNEP/IOC/WMO research activity that uses the records be given high priority with emphasis on strengthening existing institutions and cooperative efforts rather than creating new bureaucracies.

12 Acknowledgements

We are grateful to Mr Dennis Henize, Meteorologist in Charge of the Key West Weather Service Forecast Office for providing Fig. 9.2 and to Mr Douglas Martin, Oceanographer with the National Ocean Service's Tides and Water Level Branch for Fig. 9.3. In addition, Mr Martin discovered the 19th century Key West sea-level data that has allowed us to significantly improve our knowledge of the long-term record for the region. Finally, we wish to offer kudos to the men and women at the National Climatic Center (USA) and the Permanent Service for Mean Sea Level (UK) whose often unheralded dedication and professionalism allows one to have records such as we have used for analysis.

13 References

Carter, W.E., Robinson, D.S., Pyle, T.E. and Diamonte, J., 1986: The application of geodetic radio interferometric surveying to the monitoring of sea level. *Geophys.*

J. R. Astr. Soc., **87**, 3–13.

Cartwright, D.E., Barnett, T.P., Garrett, C.J.R., Carter, W.E., Peltier, W.R., Pyle, T. and Thompson, K.R., 1985: Changes in relative mean sea level. *EOS*, **66(45)**, 754–756.

Charney, J., 1979: *Carbon Dioxide and Climate: A Scientific Assessment*. Report of the *Ad Hoc* Study Group on Carbon Dioxide and Climate, National Research Council, National Academy Press, Washington, DC.

Donn, W.L. and Shaw, D.M., 1963: Sea level and climate of the past century. *Science*, **142**, 1166–1167.

Eden, H.F. (ed.), 1990: *Towards an Integrated System for Measuring Long Term Changes in Global Sea Level*. Joint Oceanographic Institutes, Washington DC, 178 pp.

Ellsaesser, H.W., MacCracken, M.C., Walton, J.J. and Grotch, S.L., 1986: Global climatic trends as revealed by the recorded data. *Rev. Geophys.*, **24(4)**, 745–792.

Hanson, K. and Maul, G.A., 1991: Florida precipitation and the Pacific El Niño (1895–1989). *Florida Scientist*, **314**, 160–168.

Hoffman, J.S., Keyes, D. and Titus, J.G., 1983: *Projecting Future Sea Level Rise*. EPA 230-09-007. US Environmental Protection Agency, Washington, DC 20460, 121 pp.

IPCC (Intergovernmental Panel on Climate Change), 1990: *Climate Change, the IPCC Scientific Assessment*. Cambridge Univ. Press, Cambridge, 365 pp.

Jones, P.D., Wigley, T.M.L. and Wright, P.B., 1986: Global temperature variations between 1861 and 1984. *Nature*, **322**, 430–434.

Maul, G.A., Chew, F., Bushnell, M. and Mayer, D.A., 1985: Sea level variation as an indicator of Florida Current volume transport: comparisons with direct measurements. *Science*, **227**, 304–307.

Maul, G.A. and Hanson, K., 1990: A century of southeastern United States climate change observations: temperature, precipitation, and sea level. In: *Global Change: A Southern Perspective*. Southeast Regional Climate Center, Columbia, SC, pp. 139–155.

NOAA (National Oceanic and Atmospheric Administration), 1987: Data publications, Normals, Means and Extremes, and Local Climatological Data Monthly Summaries, for Key West, Florida. National Climatic Data Center, Asheville, NC 28801.

Oort, A.H., Pan, Y.H., Reynolds, R.W. and Ropelewski, D.F., 1987: Historical trends in the surface temperature over the oceans based on the COADS. *Climate Dynamics*, **2**, 29–38.

Peltier, W.R., 1986: Deglaciation induced vertical motion of the North American Continent and transient lower mantle rheology. *J. Geophys. Res.*, **91(B9)**, 9099–9123.

Pugh, D.T., Spencer, N.E. and Woodworth, P.L., 1987: Data Holdings of the Permanent Service for Mean Sea Level. Bidston Observatory, Birkenhead, Merseyside L43 7RA, UK, 156 pp.

Quinn, W.H., Neal, V.T. and Antunez de Mayolo, S.E., 1987: El Niño occurrences over the past four and a half centuries. *J. Geophys. Res.*, **92(C13)**, 14449–14461.

Smagorinsky, J., 1982: *Carbon Dioxide and Climate: A Second Assessment*. National Academy of Sciences Press, Washington, DC.

Spangler, W.M.L. and Jenne, R.L., 1985: *World Monthly Surface Station Climatology (and associated data sets)*. National Center for Atmospheric Research, PO Box 3000, Boulder, CO 80307.

Tanner, W.F., 1992: 3000 Years of sea level change. *Bull. Amer. Meteorol. Soc.*, **73(3)**, 297–303.

Titus, J.G., 1986: Greenhouse effect, sea level rise, and coastal management. *Coastal Zone Manag. J.*, **14(3)**, 147–171.

Turner, R.E., 1991: Tide-gauge records, water level rise, and subsidence in the northern Gulf of Mexico. *Estuaries*, **14(2)**, 139–147.

UNEP (United Nations Environment Programme), 1987: Minutes of the First Meeting on Implications of Climatic Changes in the Wider Caribbean Region. Circular Letter CAR 04/1987. UNEP Caribbean Regional Coordinating Unit, 14-20 Port Royal St, Kingston, Jamaica.

UNEP (United Nations Environment Programme), 1987: Circular Letter CAR 06/1987, 11 September 1987. The Implications of Climatic Changes in the Wider Caribbean Region. 14-20 Port Royal St, Kingston, Jamaica.

Warrick, R. and Oerlemans, J., 1990: Sea level rise. In: *Climate Change, the IPCC Scientific Assessment*. Cambridge Univ. Press, Cambridge, New York, Port Chester, Melbourne, Sydney, 257–281.

Woodruff, S.D., Slutz, R.J., Nenne, R.L. and Steurer, P.M., 1987: A comprehensive ocean–atmosphere data set. *Bull. Amer. Meteorol. Soc.*, **68(10)**, 1239–1250.

Woodworth, P.L., 1990: A search for accelerations in records of European mean sea level. *Int'l. J. Climatol.*, **20**, 129–143.

Part 4
Ecological effects

10

Marine Habitats: Selected Environmental and Ecological Charts

Frank J. Gable[1]

Abstract

Marine habitats of the region are summarized in a series of 21 charts covering the following: regional definition; geography; bottom topography; ocean-surface currents; sea-surface temperature; population density; hydrocarbon resources; rivers; bays; up-downwelling areas; reefs; seagrasses; wetlands; phytoplankton productivity; phytoplankton distribution; moluscs; crustaceans; demersal fish; shrimp catch; turtles; and sea mammals. Satellite observations are the most effective means of modernizing and updating a new generation of geographical information systems.

1 Introduction

This chapter presents examples of region-wide, graphical displays of physical and biological resources pertinent to the implications of projected climate change on coastal and ocean zones of the region (Figs. 10.1–10.3). These illustrations are presented as tools to facilitate understanding of specific environmental characteristics in the region. The graphical information depicted affords new opportunities for region-wide multiple-resource planning in helping to attenuate the impacts of the projected climate changes. The illustrations are intended to improve the decision-making process and provide a basis for present and future strategic assessments. In order to illustrate fully the details of these charts Figs. 10.2–10.4 and 10.6–10.21 are on pages 224–261 following the references.

This approach is not only useful but needed to study the implications of climate change in the region because of the breadth of the area under study. The Caribbean Sea, together with the Gulf of Mexico, the world's largest gulf, covers an area of 3.48 million km^2. Dispersed over the region – the coastal and open waters of the Caribbean Sea proper, the Gulf of Mexico, and adjacent waters of the Atlantic Ocean there are 33 countries (UNIDO, 1987). Small-scale thematic maps and charts provide an excellent tool for analysing and understanding the many spatial and temporal links existing among coastal and oceanic habitats (Ellis, 1978), and the resources within the expansive region.

[1]Woods Hole Oceanographic Institution, Woods Hole, MA 02543 USA

These illustrations are not designed to show the distribution of all available local data concerning a specific physical or biological characteristic or process. For example, the population distributions (Fig. 10.4) do not display the plethora of detailed information available as specific study sites attainable from the myriad of Caribbean and Gulf governments. Rather, these charts display the regional variations at a single scale.

Fortunately, a considerable amount of baseline environmental data has been prepared for the region (IUCN, 1979; Ogden and Gladfelter, 1986), and expressly for the Gulf of Mexico. The selected synopsis of the available charts presented here deal with the coastal and ocean zones of the region. Further illustrations concerning living marine resources, oceanic parameters, and other data are also available (IUCN, 1979).

Each chart provides a topical overview of a specific characteristic of the region. As a group, the charts can be used for comparative analyses through overlays or other methods. The regional categories illustrated here may be used to identify local areas that need further investigation, in order to define the potential implications of natural and/or human-induced climatic changes (Gable and Aubrey, 1990; Gable *et al.*, 1990).

2 Examples From Figures Presented

Surface currents: Circulation patterns in the Caribbean/Gulf/Bahamas region are governed by fresh-water runoff, topography, sea-surface temperature, wind stress and primarily by the North Equatorial Current and other currents (Fig. 10.5). Tropical storms generally follow the path of these currents of the Gulf Stream System after spawning in the central Atlantic (Gray, 1990).

Sea-surface temperature: Summer sea-surface temperature is illustrated in Fig. 10.6 (see also Aparicio, Chapter 6). The fact that most tropical storms develop during the summer, with September being the month with the most storms, suggests that sea-surface temperature plays a critical role in the development and intensity of these storms. Generally, hurricane formation requires sea-surface water temperatures of 26°C, or higher (Emanuel, 1988). Since the sea-surface temperatures of the region are at 27-28°C, increased global warming might result in an extension and intensification of the tropical storm hurricane season (Emanuel, 1987; Gray, Chapter 5).

Hydrocarbons: Hydrocarbon resources are illustrated in Fig. 10.7. The extraction of fluid hydrocarbon resources can lead to subsidence, exacerbating relative sea-level rise. This anthropogenic forcing is best demonstrated in the Gulf of Mexico region in Louisiana and east Texas (Emery and Aubrey, 1991; Hanson and Maul, Chapter 9).

River systems: Rivers are major sources of fresh-water flows into the region (Fig. 10.8). Stream flow varies depending on the size of the drainage basin/watershed, amount of precipitation within the drainage basin, runoff rates, infiltration, and level of water consumption for agriculture, industry and municipalities. Modified land-use patterns in coastal river basins have produced local changes in relative sea-level by altering the

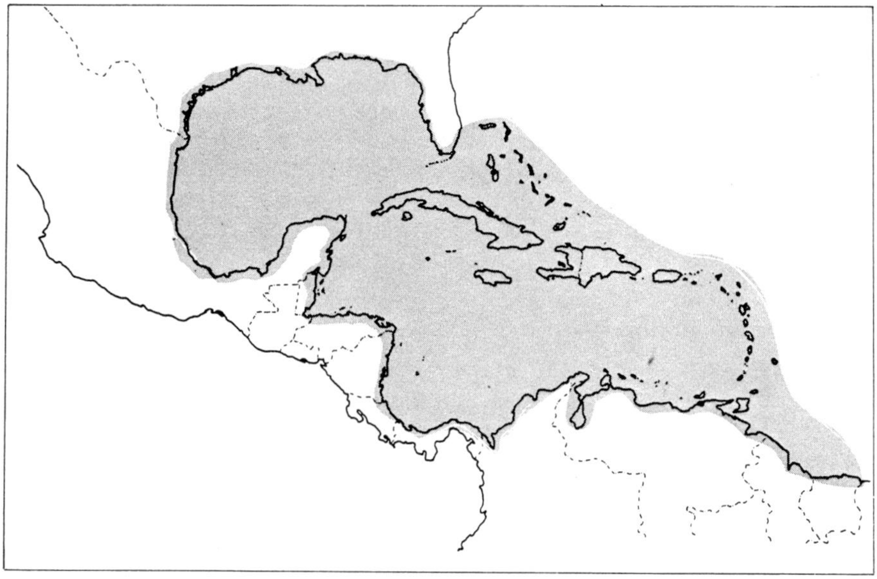

Fig. 10.1 Shaded area defined as the Wider Caribbean Region by UNEP and the Caribbean and Adjacent Regions by the IOC.

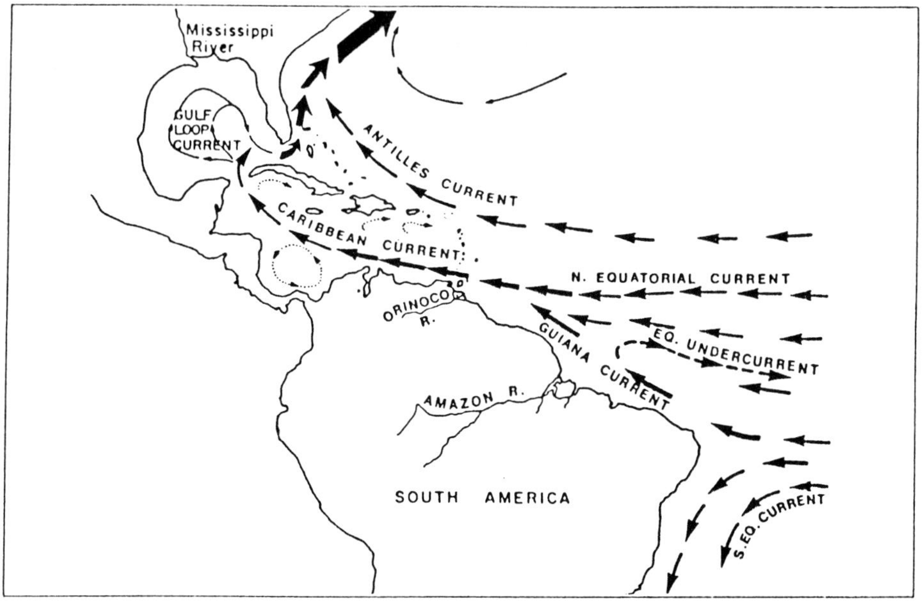

Fig. 10.5 Physical oceanography: surface currents. Source: Ogden and Gladfelter, 1986.

distribution and quantity of fresh-water flow in several areas. It has been estimated that three times more fresh-water and 15 times more sediment are introduced per unit length of coastline on the continental shelf of Nicaragua than on the east coast of the United States (Roberts and Murray, 1983). Although the Amazon River is geographically outside the bounds of the region its sediment and fresh-water influence is felt through the Guiana Current (Milliman *et al.*, 1982; Muller-Karger, Chapter 8), which, in turn, influences changes in relative sea-level rise in parts of the Caribbean Sea (Aubrey *et al.*, 1988).

Bays and estuaries: Bays, estuaries and lagoons (Fig. 10.9) provide spawning, nursery, or feeding grounds for most of the commercially and recreationally important fish and shellfish in the region (*cf*. Snedaker, Chapter 12).

Reefs: Coral and other carbonate reefs have a slow rate of growth. Because of their susceptibility to both anthropogenic and natural perturbations, they are sensitive to relative sea-level changes (Hopley and Pirazzoli, 1985). Coral reefs in particular act as buffers against shoreline degradation during periods of storm and winter wave action and as a source of sand for the beaches in the area, which are an important source of revenue from tourism (Gable, 1987). The breadth of reef systems throughout the region is illustrated in Fig. 10.10 (see also Hendry, Chapter 7; Milliman, Chapter 13).

Seagrasses: Seagrasses (Fig. 10.11) are another feature that support living marine resources and acts as both habitat and food for these resources. Anthophytes (seagrasses) are flowering marine plants that live submerged in seawater and are adversely influenced by fluctuations in salinity, turbidity and water temperature, which are projected perturbations from human-induced climate changes. In addition, sea turtle nesting beaches (Fig. 10.20) are highly correlated with nearshore seagrass beds where these reptiles feed.

Wetlands: Coastal wetlands, distributed throughout the region (Fig. 10.12), are among the most productive natural systems in the world. Human alteration of wetlands and other biomes has contributed to a decrease in the quantity and quality of these environments. In general, wetlands provide significant habitat and food for economically and recreationally valuable fishes, shellfishes and other aquatic life. Wetlands play an important role in the primary productivity of estuaries and coastal waters, while mangroves in particular buffer the shore from waves and storms (Snedaker, Chapter 12).

Plankton productivity: Generalized phytoplankton (minute floating aquatic plants) productivity is displayed in Fig. 10.13. In comparison with areas of upwelling and downwelling (Fig. 10.14), low oceanic phytoplankton is associated with areas of downwelling and higher oceanic phytoplankton is related to areas of upwelling. As can be seen in Fig. 10.13 (*cf*. Aparicio, Chapter 6), nearshore waters are generally richer in the nutrients essential for phytoplankton production. Campeche Bank, Tortuga Bank, and parts of the western Florida continental shelf offer the highest rates of species diversity (Fig. 10.15); these are areas of upwelling or adjacent to upwelling areas.

Benthic invertebrates: The coincidence of the distribution of the benthic invertebrate, queen conch (*Strombus gigas*) (Fig. 10.16) compared with reefs (Fig. 10.10) and seagrass beds (Fig. 10.11) exemplifies the usefulness of these charts for coupling and comparing different topical information. The concentration of another benthic invertebrate, the spiny lobster (*Panulirus argus*), a decopod crustacean of the family Palinuridae, is depicted in Fig. 10.17. Its adult life, like those of crabs, clams and oysters, is primarily spent in estuaries and inland waters, and this association can be seen in a comparison with Fig. 10.9.

Fishes: Demersal (bottom dwelling) fishes (Fig. 10.18) depend on the continental shelves and banks (Fig. 10.3). These areas are significantly influenced by hurricanes, wave setup, fresh-water runoff (*q.v.* Muller-Karger, Chapter 8) and nutrient upwelling (*q.v.* Aparicio, Chapter 6).

Shrimp catch: The estuarine-dependent shrimp (*Paneaus*) is a major commercially exploited species taken for the most part in the Gulf of Mexico (Fig. 10.19). This species can be detrimentally influenced by changes in salinity, which may be affected by projected climate change, (see Wigley and Santer, Chapter 2; Gallegos *et al.*, Chapter 3)

Reptiles and mammals: Graphical information on sea turtles and their important nesting beaches and feeding grounds are illustrated in Fig. 10.20. The possibility of increased beach erosion from relative sea-level rise increases the problem for many threatened and endangered species of turtles because these turtles need dry sand areas to reproduce. Lastly, Fig. 10.21 shows the areas favourable to marine mammals. For example, the West Indian Manatee (*Trichechus manatus*) is found in coastal, estuarine and fresh-water areas with vegetation, particularly in Belize and the southern Gulf of Mexico. This association can be seen when a comparison of Figs. 10.21 and 10.9 is made. This again demonstrates the value of these charts as tools for spatial relationships.

3 Concluding Remarks

The figures provided here are representative of the type of data needed to assess potential regional impacts of projected global climate change. In some cases updating or providing information not yet regionally available can be carried out through remote sensing techniques. Satellite and airborne sensors have proven valuable in the definition of coastlines and beach widths (Clark *et al.*, 1986). Shoreline characteristics have been demonstrated in Fig. 10.12 showing coastal wetlands, but more information about sand beaches and barrier islands is needed. Beaches, marshes and mangroves are associated with dissimilar shoreline accretion and erosion rates, and the nature of the shoreline affects the dissipation of severe storm effects and relative sea-level change. Both effects are anticipated to increase because of projected climatic changes.

The existence of these charts enables the widespread use of spatial region-wide geographic information for a host of resource-management and area-analysis uses, and most importantly for the UNEP/IOC Task Team, to measure the implications of projected climatic changes on the region. New charts are in the process of being digitally reproduced in a computerized Geographical Information System (GIS), which will enhance their overall usefulness. These are being prepared with support from the United Nations Environment Programme, in conjunction with interested researchers and organizations. This will allow for ease of update and revision of this GIS graphical information, as scientific research unveils expanded knowledge.

4 Acknowledgements

Support for this report was provided by a grant from the Andrew W. Mellon Foundation to the Coastal Research Center of the Woods Hole Oceanographic Institution (WHOI). Comments on earlier versions of the manuscript provided by G. Giese, J. Gentile, G. Maul and J. Milliman. WHOI Contribution No. 7942.

5 References

Aubrey, D.G., Emery, K.O. and Uchupi, E.1988: Changing coastal levels of South America and the Caribbean Region from tide-gauge records. *Tectonophysics*, **154**, 269–284.

Clark, R.K., Fay, T.H. and Terrie, G.E. 1986: Coastline Extraction and Beach Width Definition from Multispectral Imagery. *NORDA Report 149*, Ocean Science Directorate, Bay St. Louis, MS 39529, USA.

Ellis, M.Y., 1978: Coastal Mapping Handbook. US Government Printing Office, Washington, DC, 200 pp.

Emanuel, K.A. 1987: The dependence of hurricane intensity on climate, *Nature*, **326(6221)**, 483–485.

Emanuel, K.A., 1988: Toward a general theory of hurricanes. *American Scientist*, **76(4)**, 370–379.

Emery, K.O. and Aubrey, D.G. 1991: *Sea Levels, Land Levels and Tide Gauges*. Springer-Verlag, New York, 237 pp.

Gable, F., 1987: Changing climate and Carribbean coastlines. *Oceanus*, **30(4)**, 53–56.

Gable, F.J. and Aubrey, D.G. 1990: Potential impacts of contemporary changing climate on Caribbean coastlines. *Ocean and Shoreline Management*, **13(1)**, 35–67.

Gable, F.J., Gentile, J.H. and Aubrey, D.G. 1990: Global climatic issues in the coastal wider Caribbean. *Environmental Conservation*, **17(1)**, 51–60.

Gray, W.M., 1990: Strong association between West African rainfall and US landfall of intense hurricanes. *Science*, **249**, 1251–1256.

Hopley, D. and Pirazzoli, P. 1985: Late Quaternary and present sea-level changes: magnitude, causes, future applications. In: *Proceedings of the Fifth International Coral Reef Congress, Vol. 3, 77–78.*

IUCN (International Union for Conservation of Nature and Natural Resources), 1979: Planning a marine conservation strategy for the Caribbean Region. *Data Atlas*, 51 pp.

Milliman, J.D., Butenko, J., Barbot J.P. and Hedberg, J. 1982: Depositional patterns of modern Orinoco/Amazon muds on the northern Venezuelan shelf. *J. Mar.*

Res., **40(3)**, 643–657.
Ogden, J.C. and Gladfelter, E. (eds), 1986: Caribbean coastal marine productivity. *UNESCO Reports in Marine Science*, No. 41, 59 pp.
Roberts, J. and Murray, S., 1983: Controls on reef development and terrigenous-carbonate interface on a shallow shelf, Nicaragua (Central America). *Coral Reefs*, **2**, 71–80.
UNIDO (United Nations Industrial Development Organization), 1987: Industrial Development Review Series. *The Caribbean Region*, 291 pp.

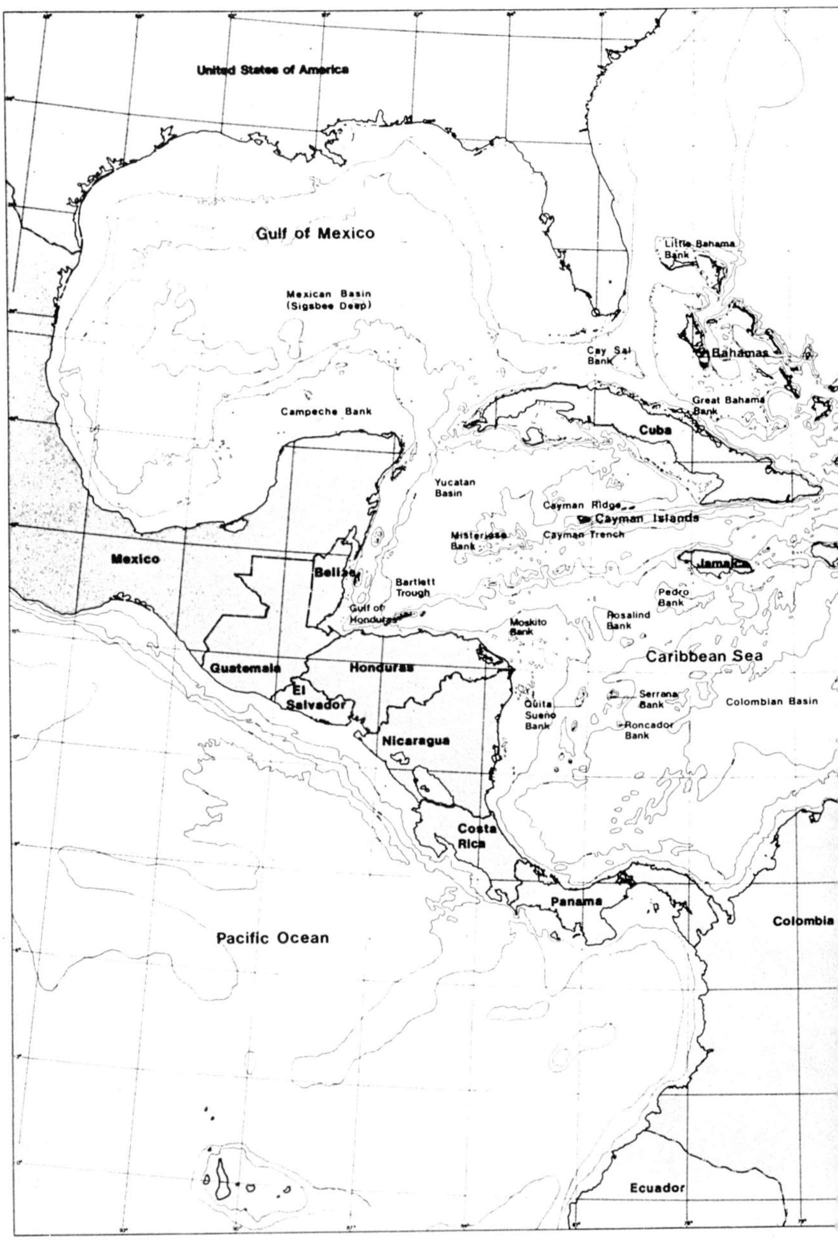

Fig. 10.2 Geography of the region. (Source: IUCN, 1979).

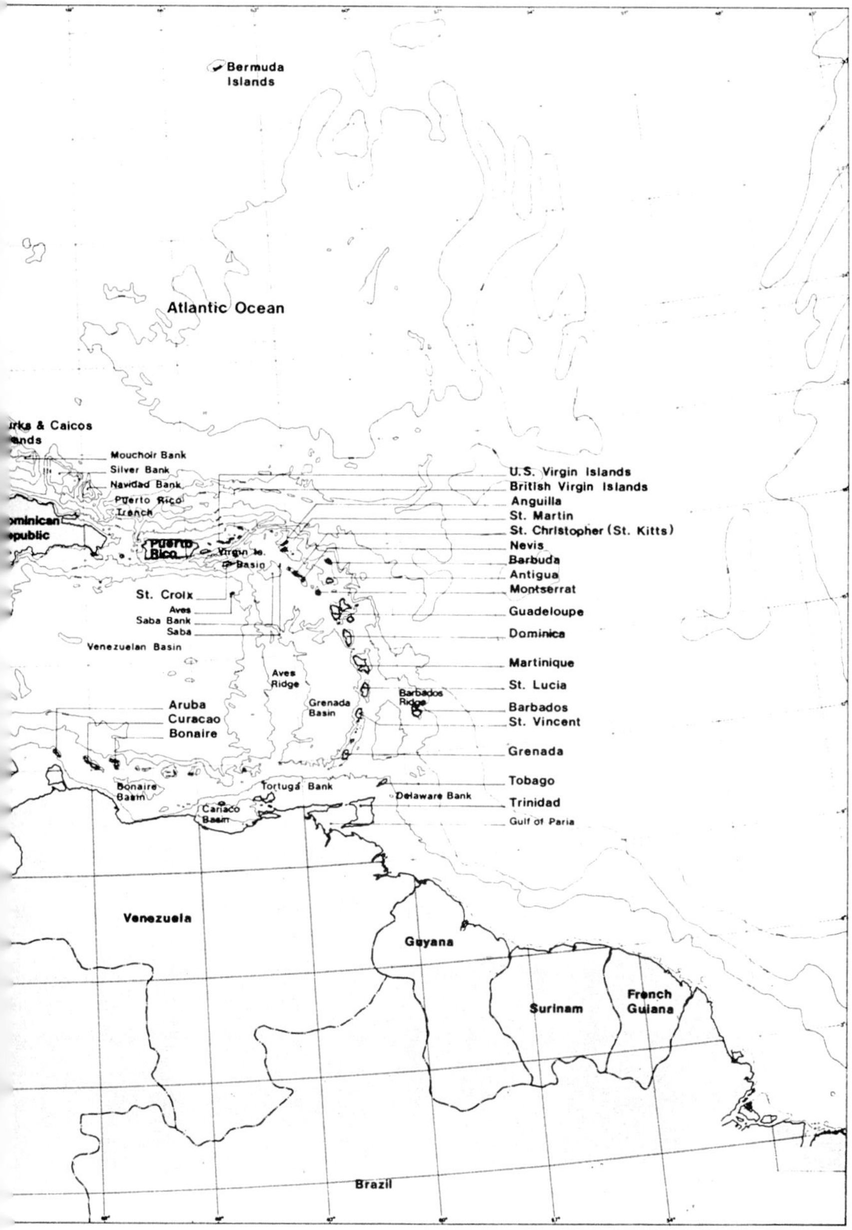
Bermuda Islands
Atlantic Ocean
Mouchoir Bank
Silver Bank
Navidad Bank
Puerto Rico Trench
Puerto Rico
Virgin Is. Basin
St. Croix
Aves
Saba Bank
Saba
Venezuelan Basin
Aves Ridge
Grenada Basin
Barbados Ridge
Aruba
Curacao
Bonaire
Bonaire Basin
Tortuga Bank
Cariaco Basin
Delaware Bank
U.S. Virgin Islands
British Virgin Islands
Anguilla
St. Martin
St. Christopher (St. Kitts)
Nevis
Barbuda
Antigua
Montserrat
Guadeloupe
Dominica
Martinique
St. Lucia
Barbados
St. Vincent
Grenada
Tobago
Trinidad
Gulf of Paria
Venezuela
Guyana
Surinam
French Guiana
Brazil

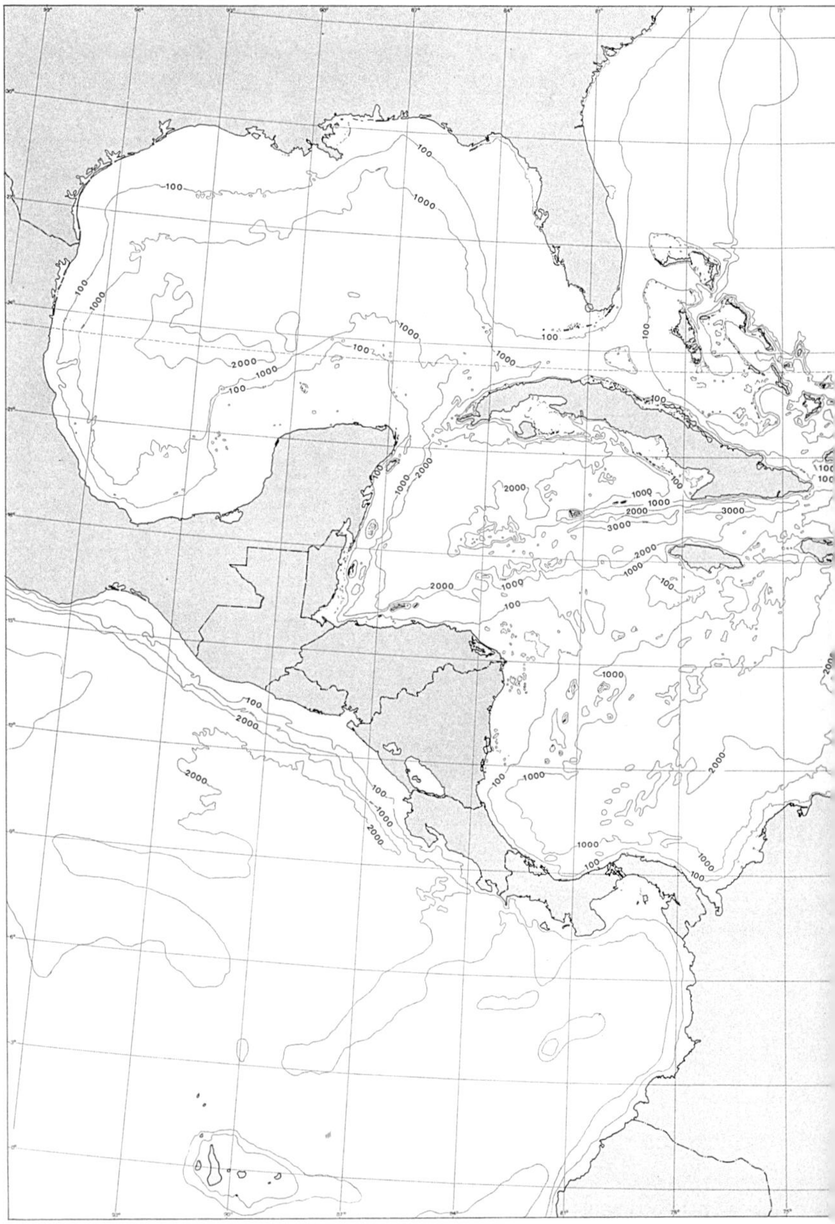

Fig. 10.3 Bottom topography in fathoms (1 fathom = 6 ft = 1.8288 m). Source: IUCN, 1979.

3000
3000
3000
3000
3000
4000
4000
3000
2000
1000
1000
1000
2000
3000
2000
2000
2000
1000
1000
1000
2000
100
1000
2000
2000
1000
100

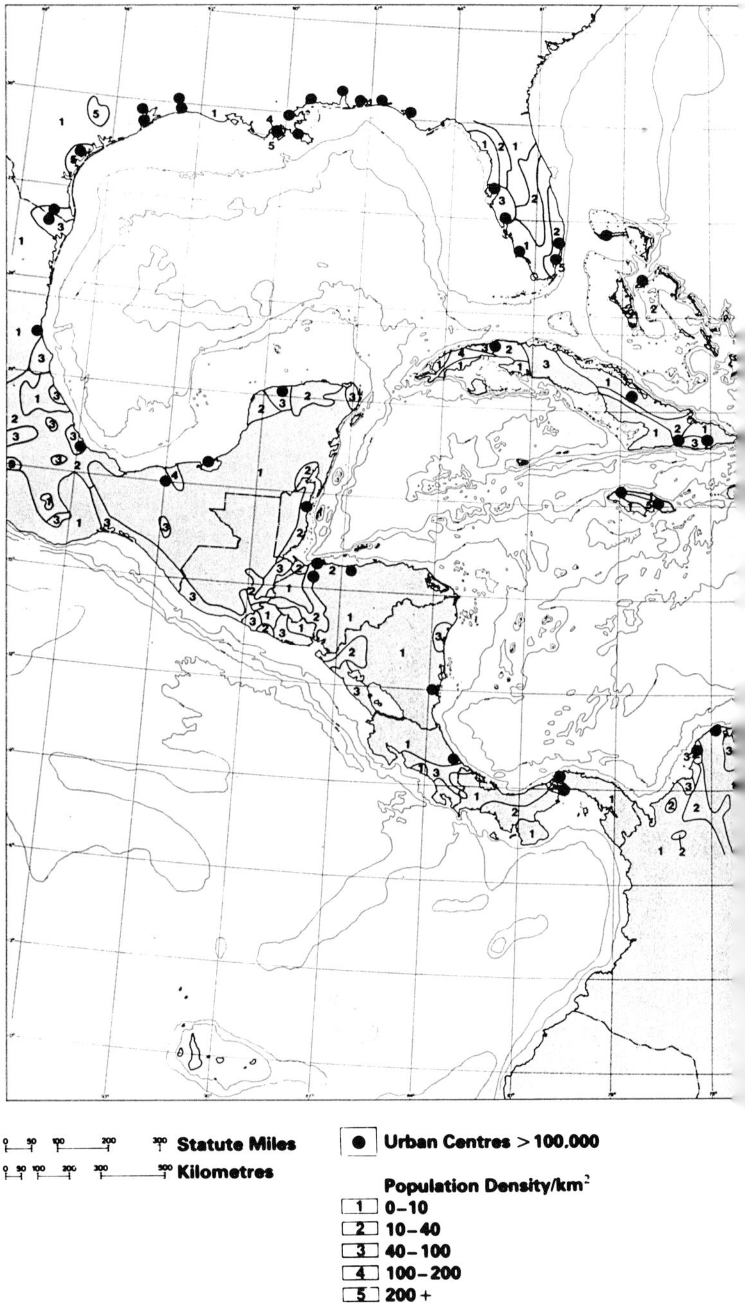

Fig. 10.4 Generalized population density (*cf.* Fig. 15.2). Source: IUCN, 1979.

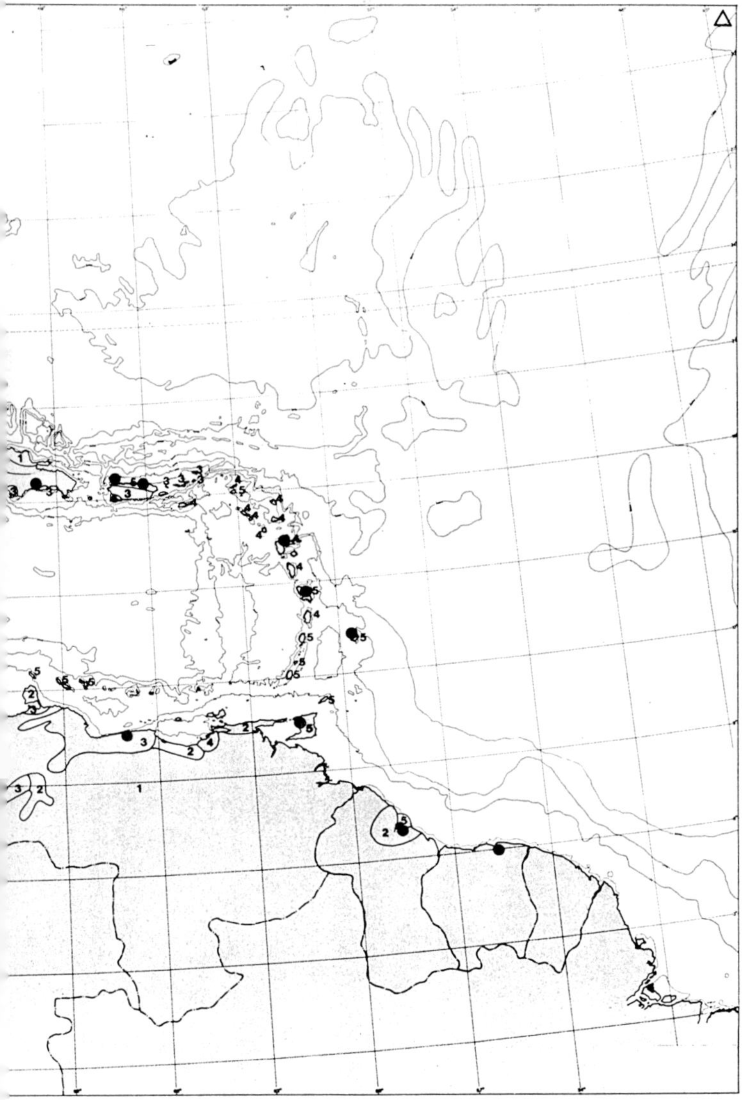

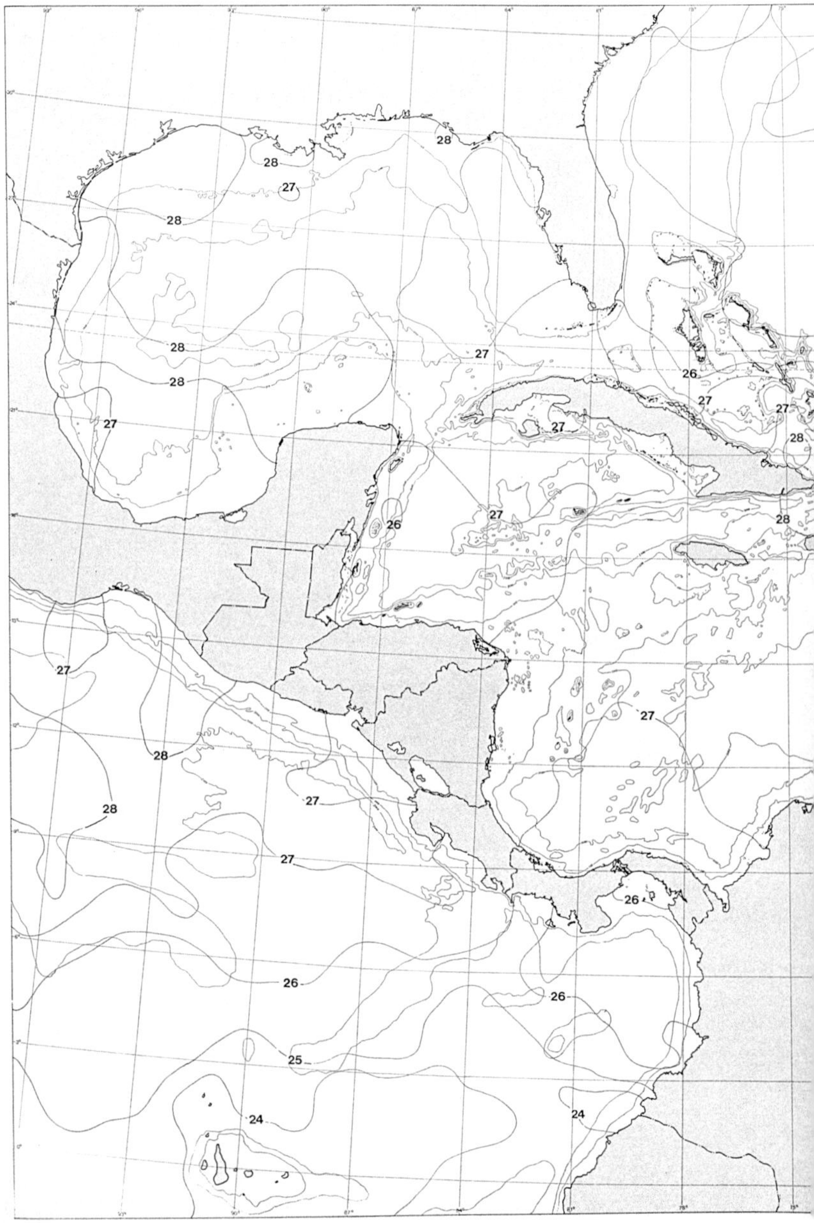

Fig. 10.6 Summer sea-surface temperatures in degrees celsius. Source: IUCN, 1979.

23
24
25
26
26
27
27
27
26
27
27

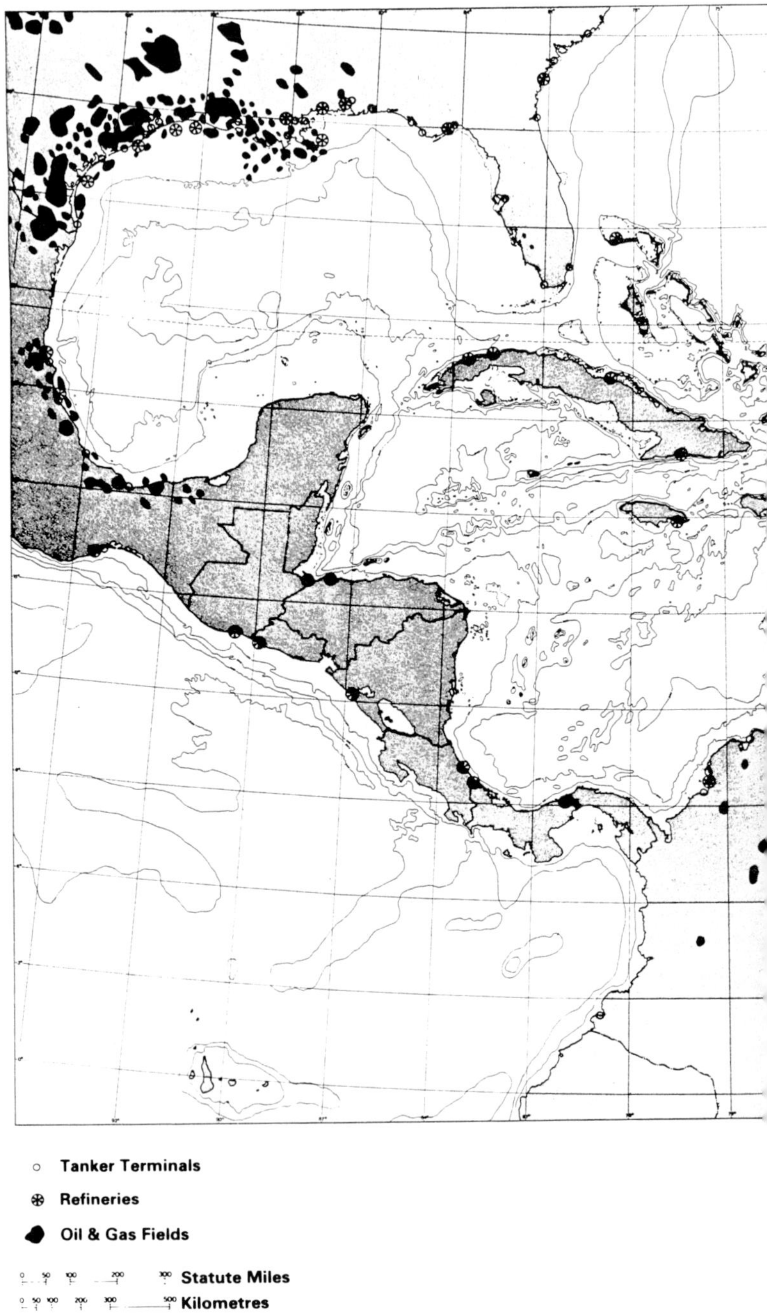

Fig. 10.7 Hydrocarbon resources. Source: IUCN, 1979.

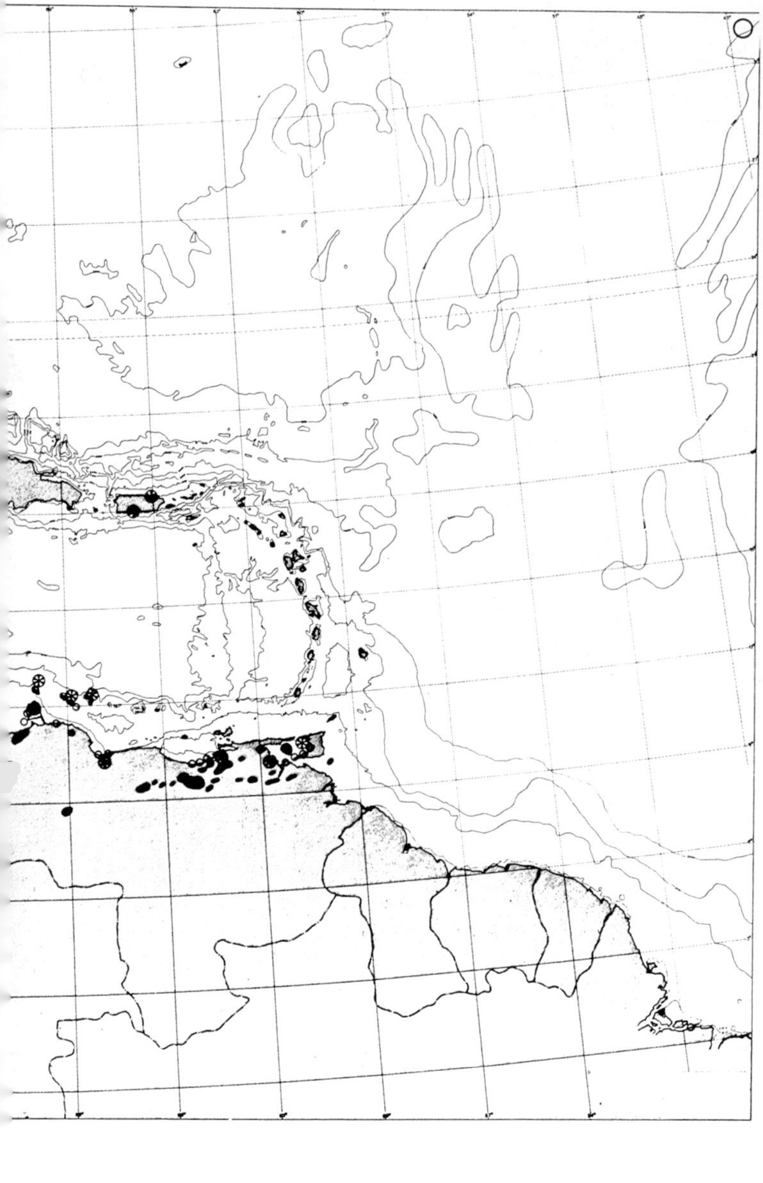

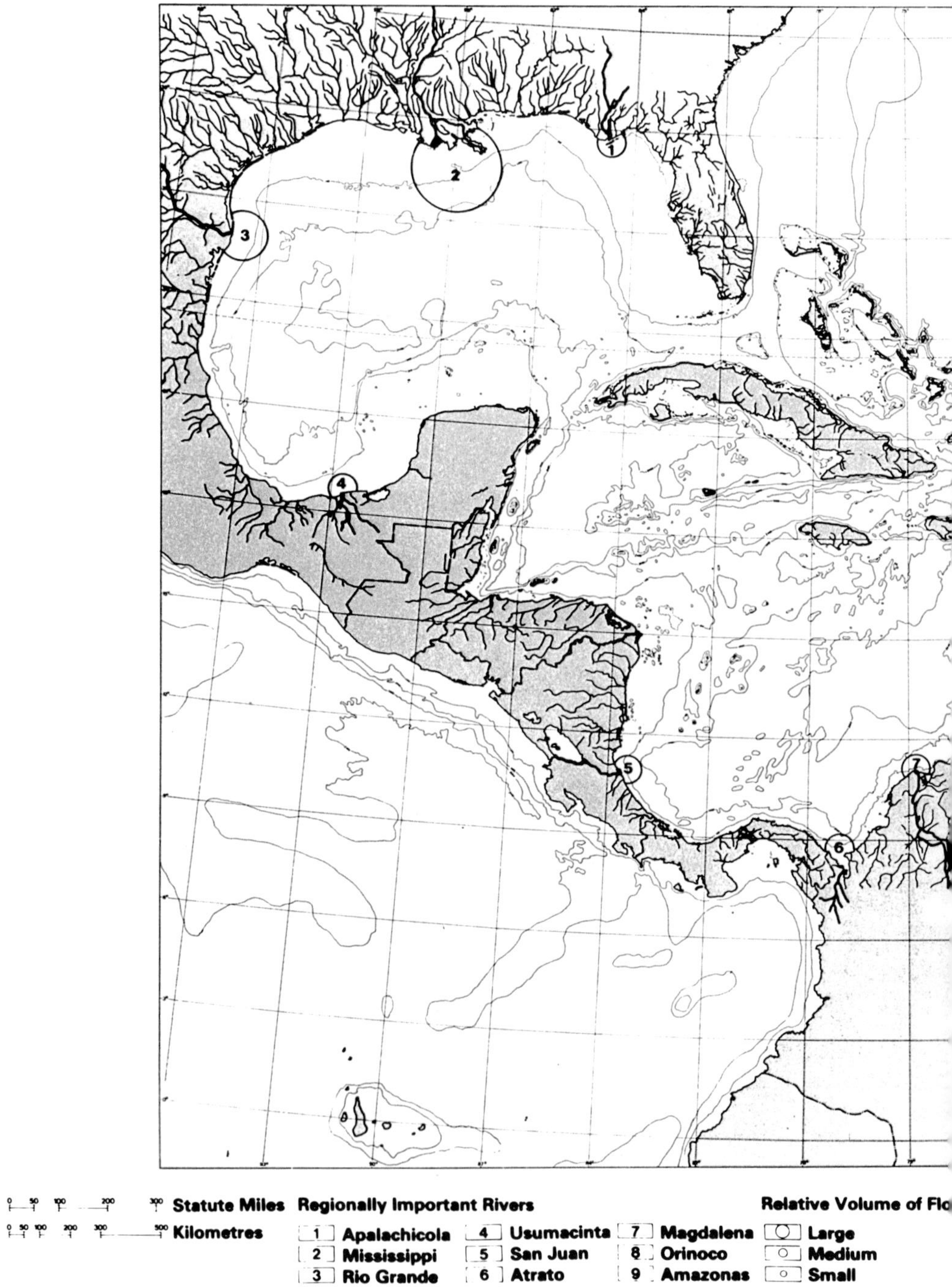

Fig. 10.8 River systems. Source: IUCN, 1979.

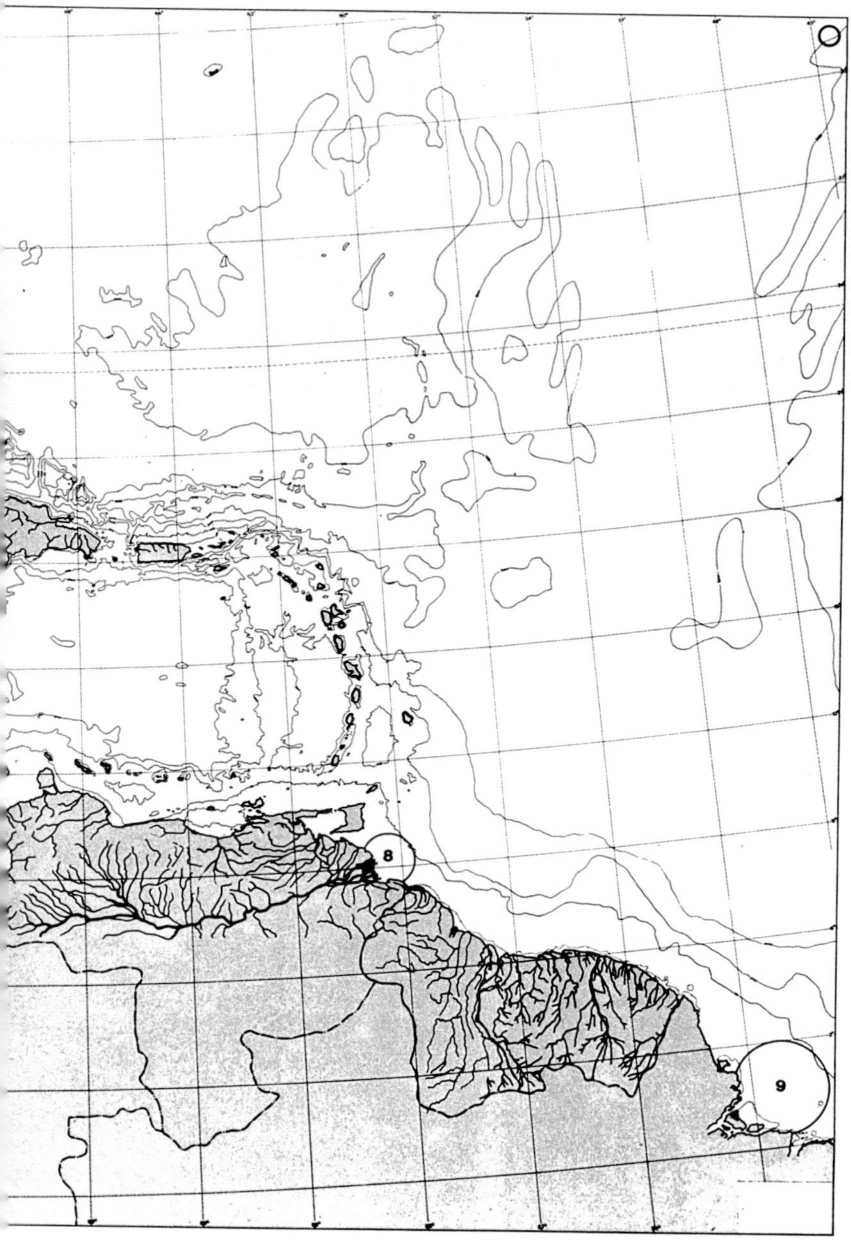
8
9

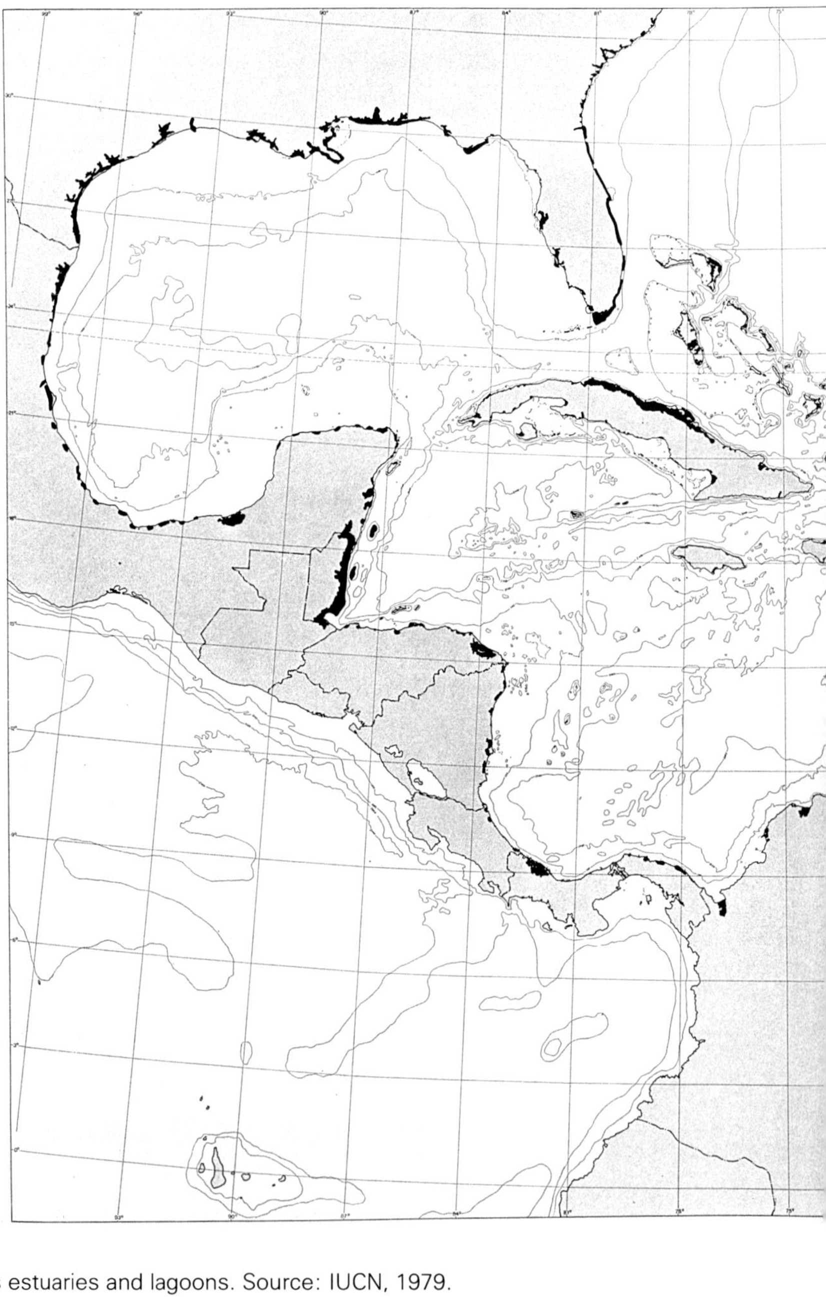

Fig. 10.9 Bays estuaries and lagoons. Source: IUCN, 1979.

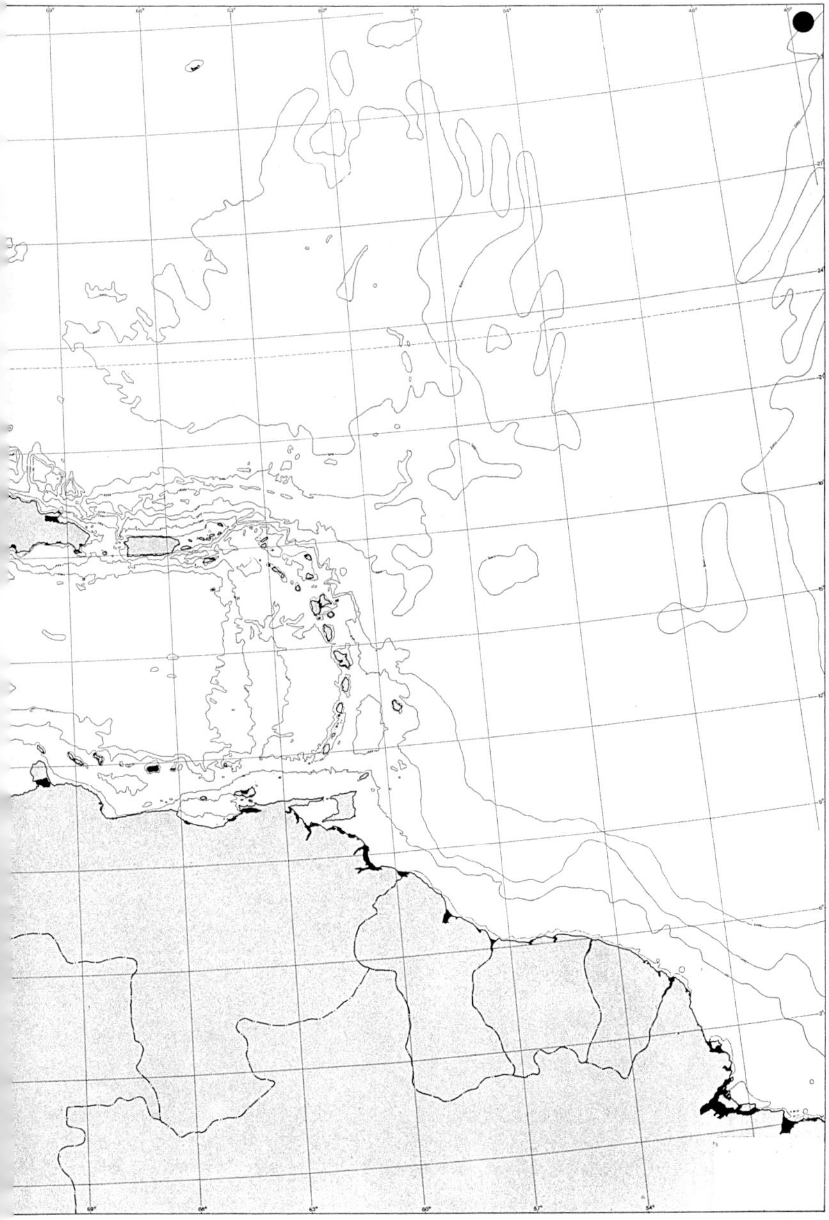

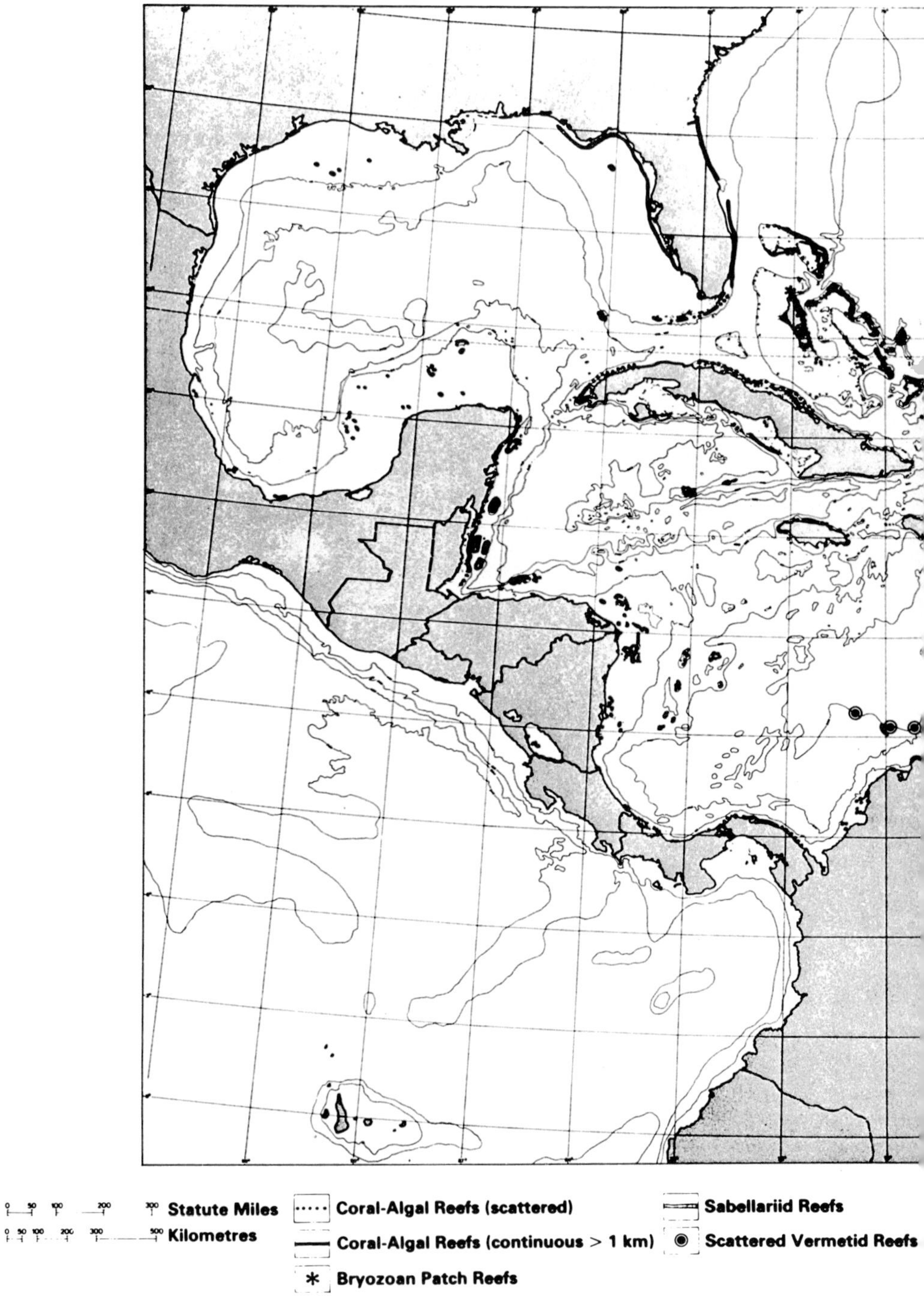

Fig. 10.10 Reef systems. Source: IUCN, 1979.

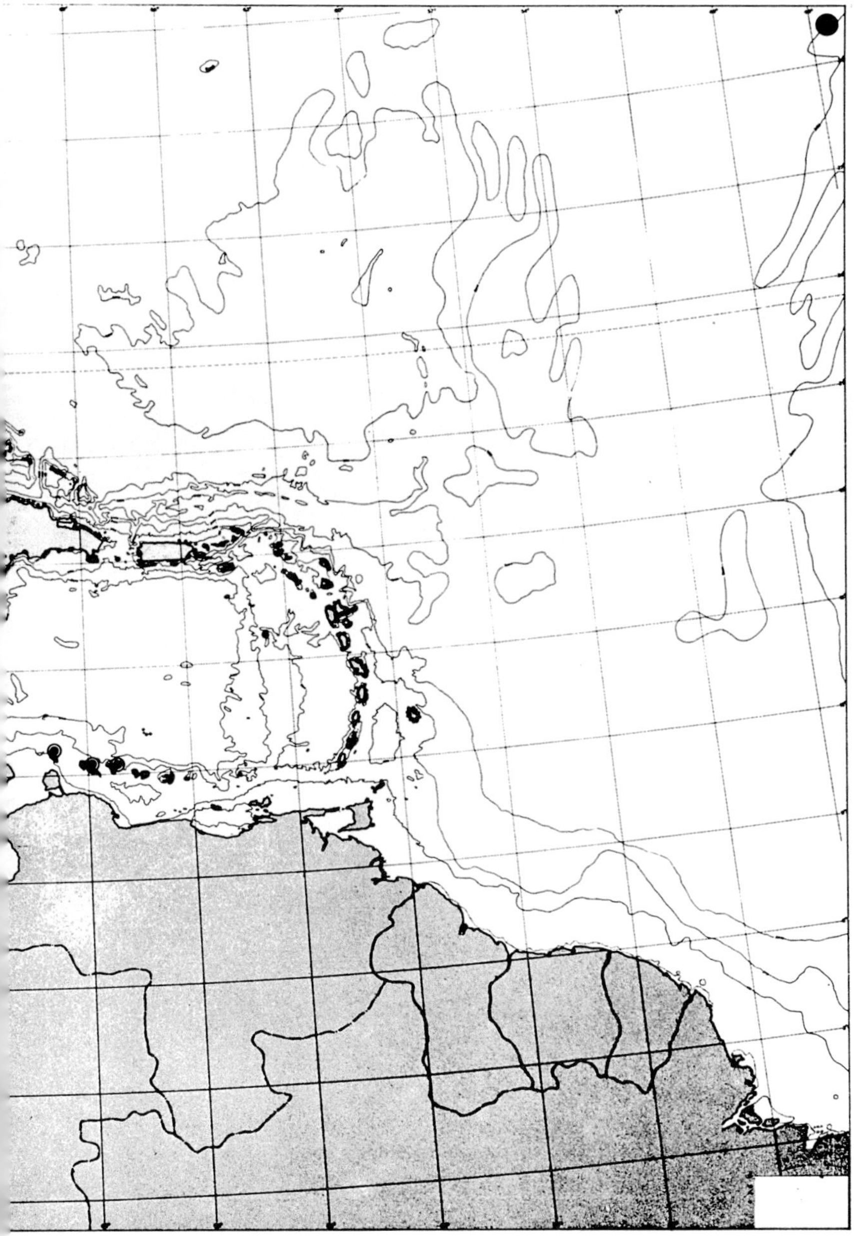

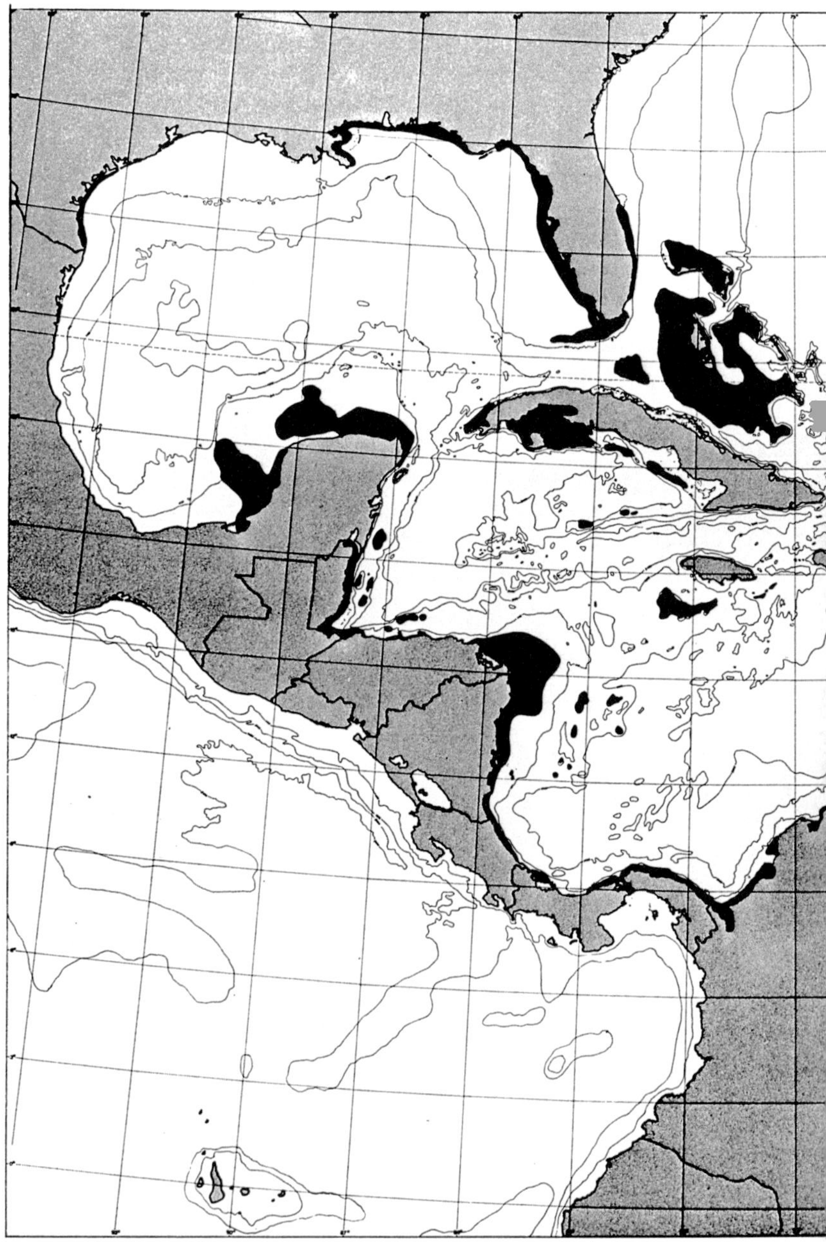

Fig. 10.11 Sea grasses. Source: IUCN, 1979.

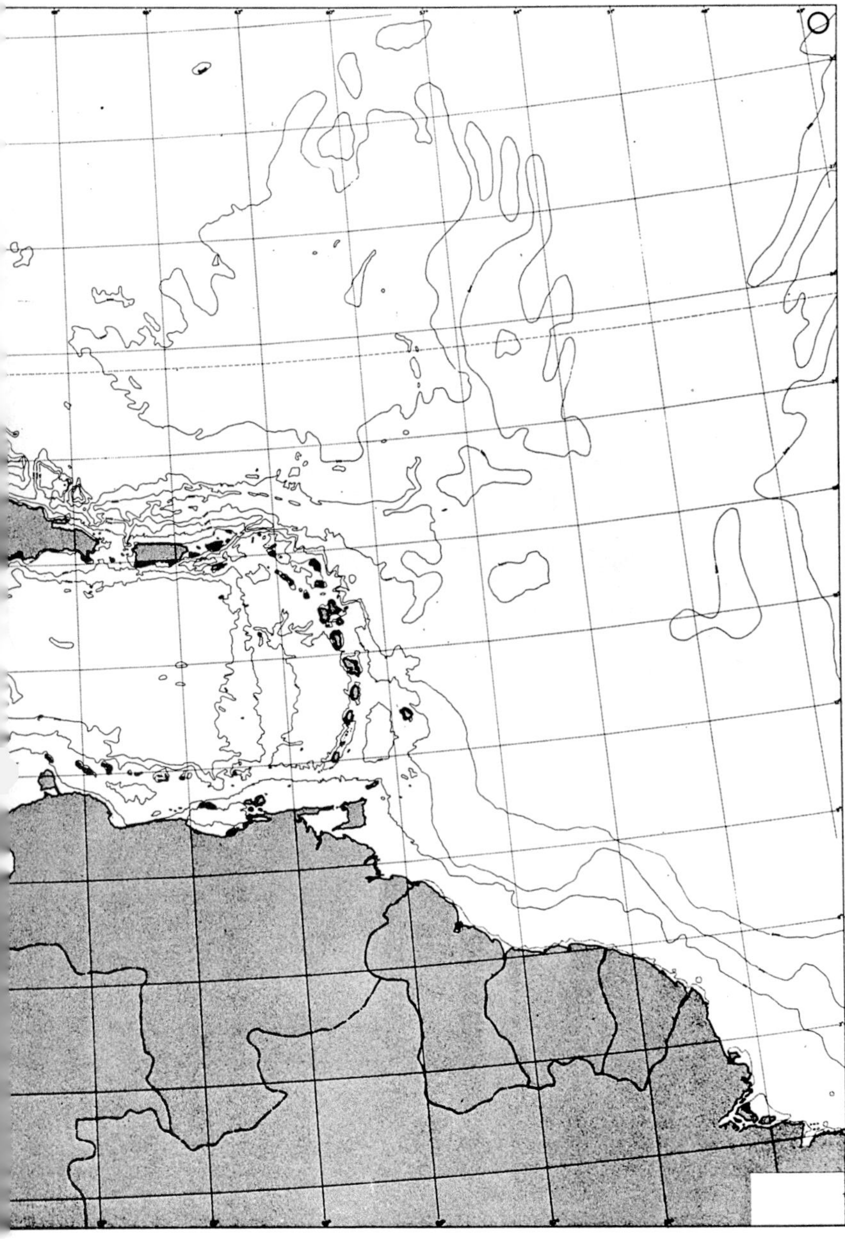

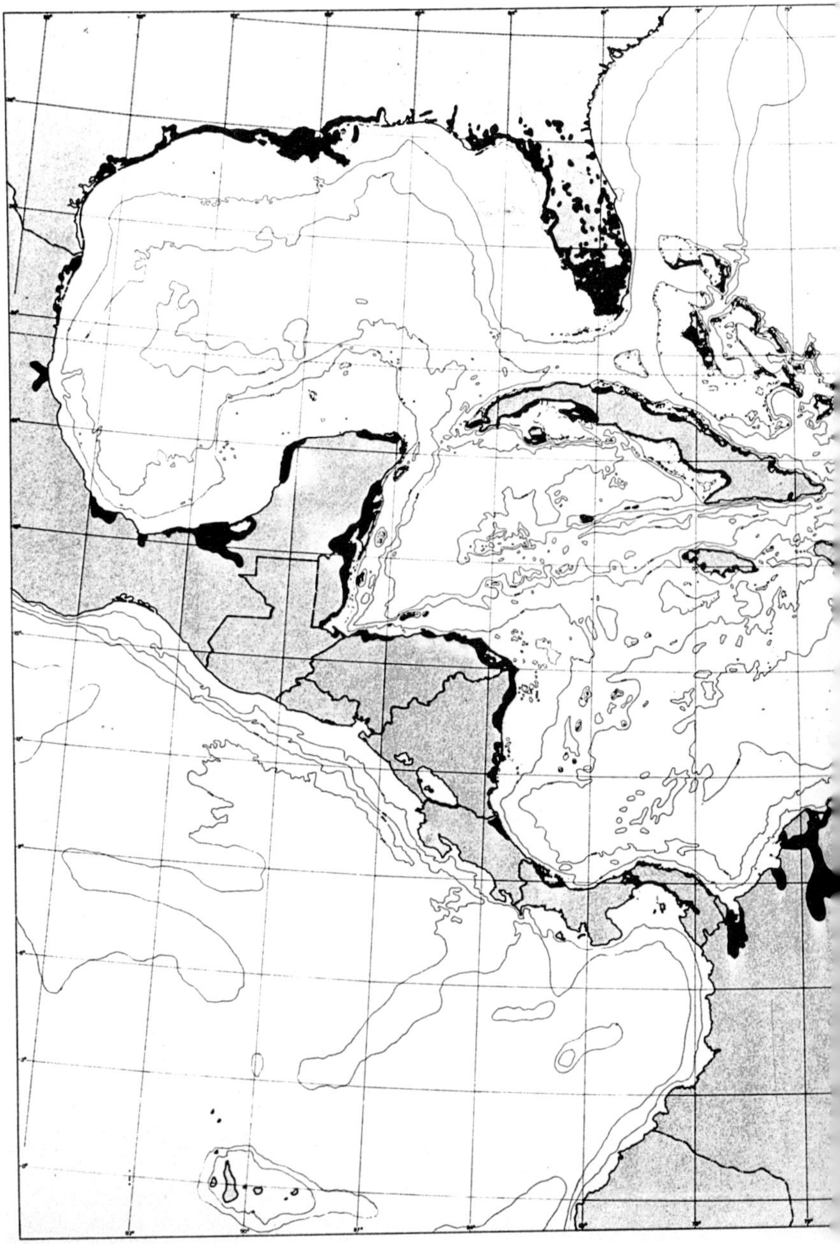

Fig. 10.12 Wetlands. Source: IUCN, 1979.

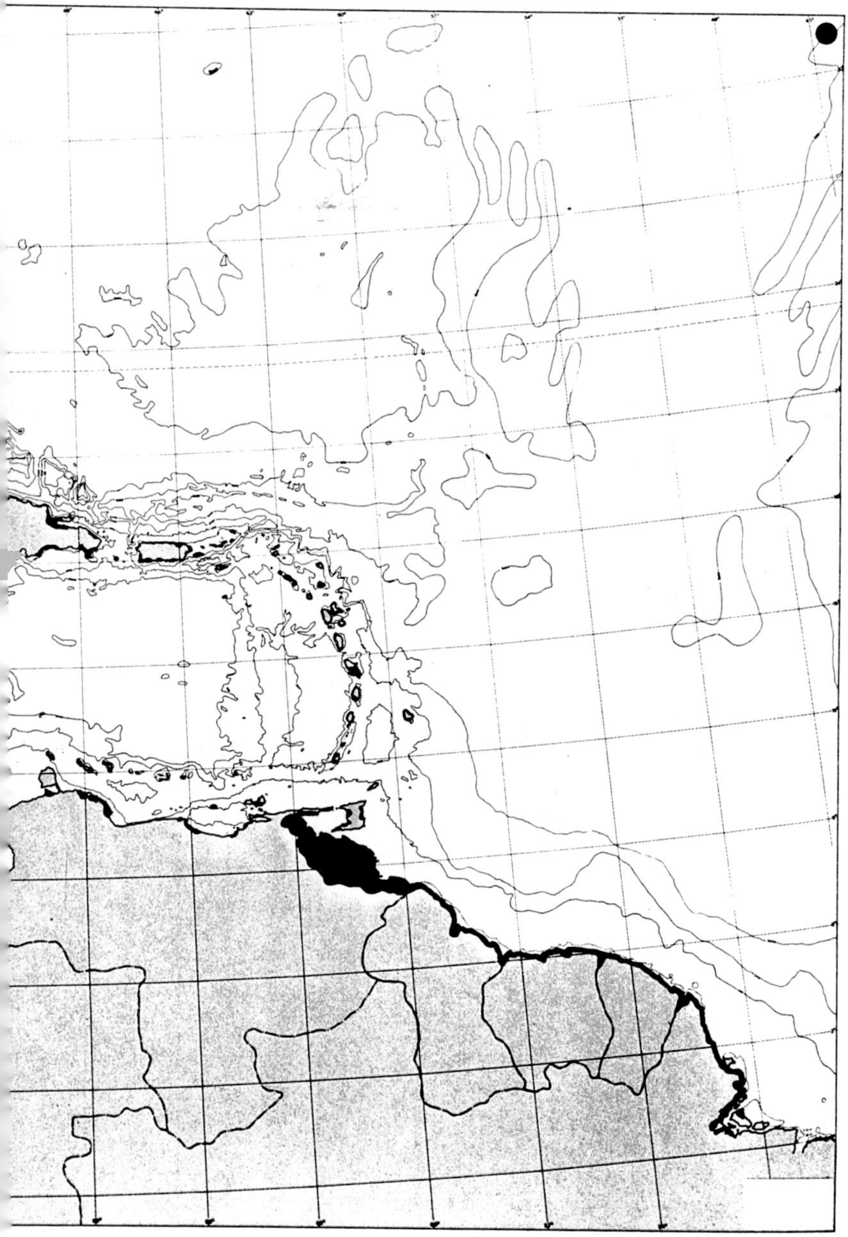

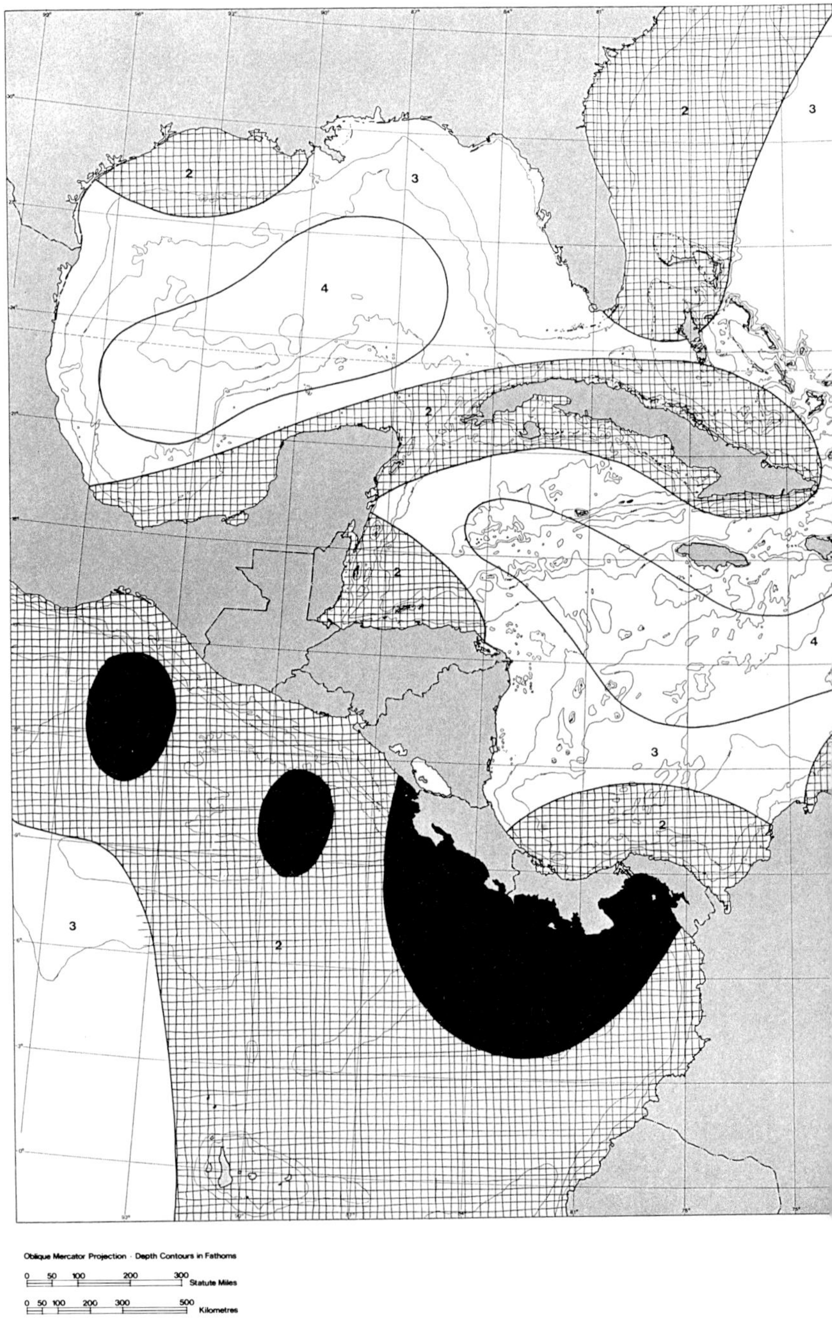

Fig. 10.13 Generalized phytoplankton productivity. Source: IUCN, 1979.

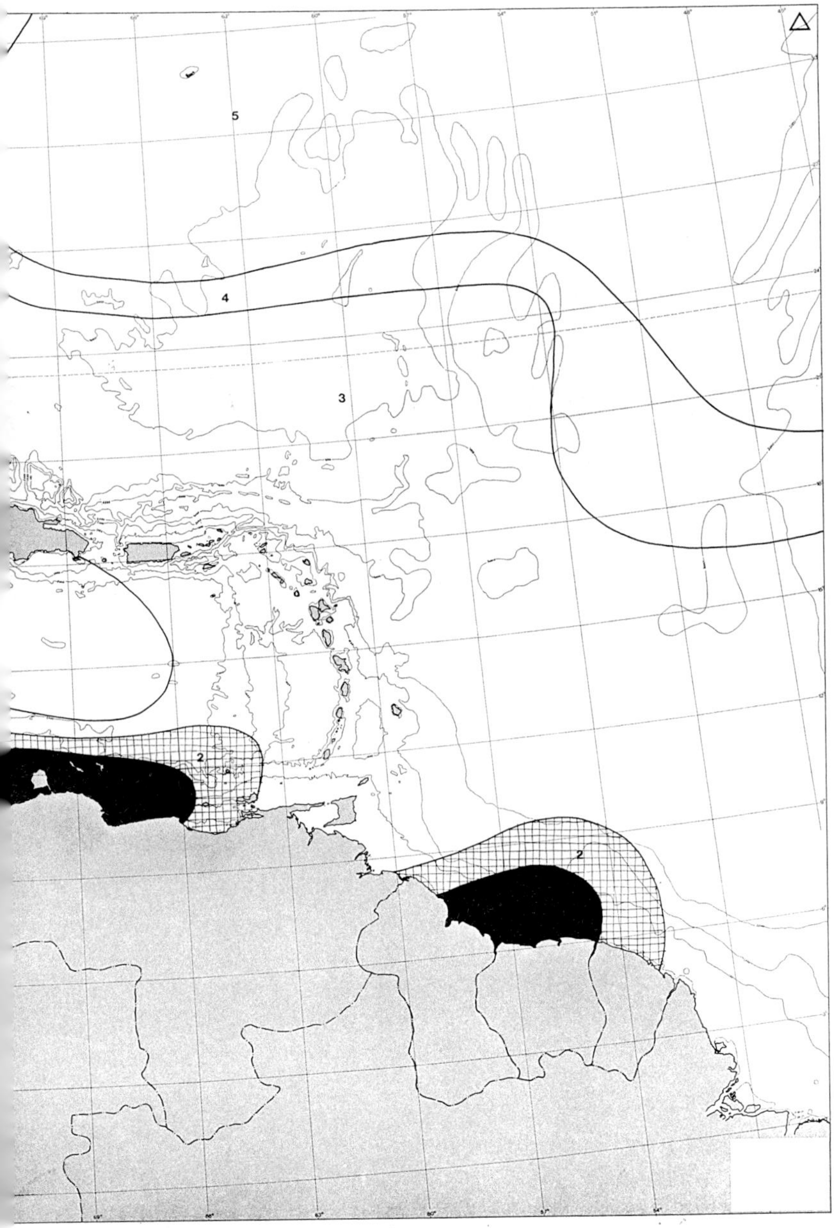

Oceanic Productivity mgC/m²/day
Productivité océanique mgC/m²/jour

■ >500
2 250–500
3 150–250
4 100–150
5 <100

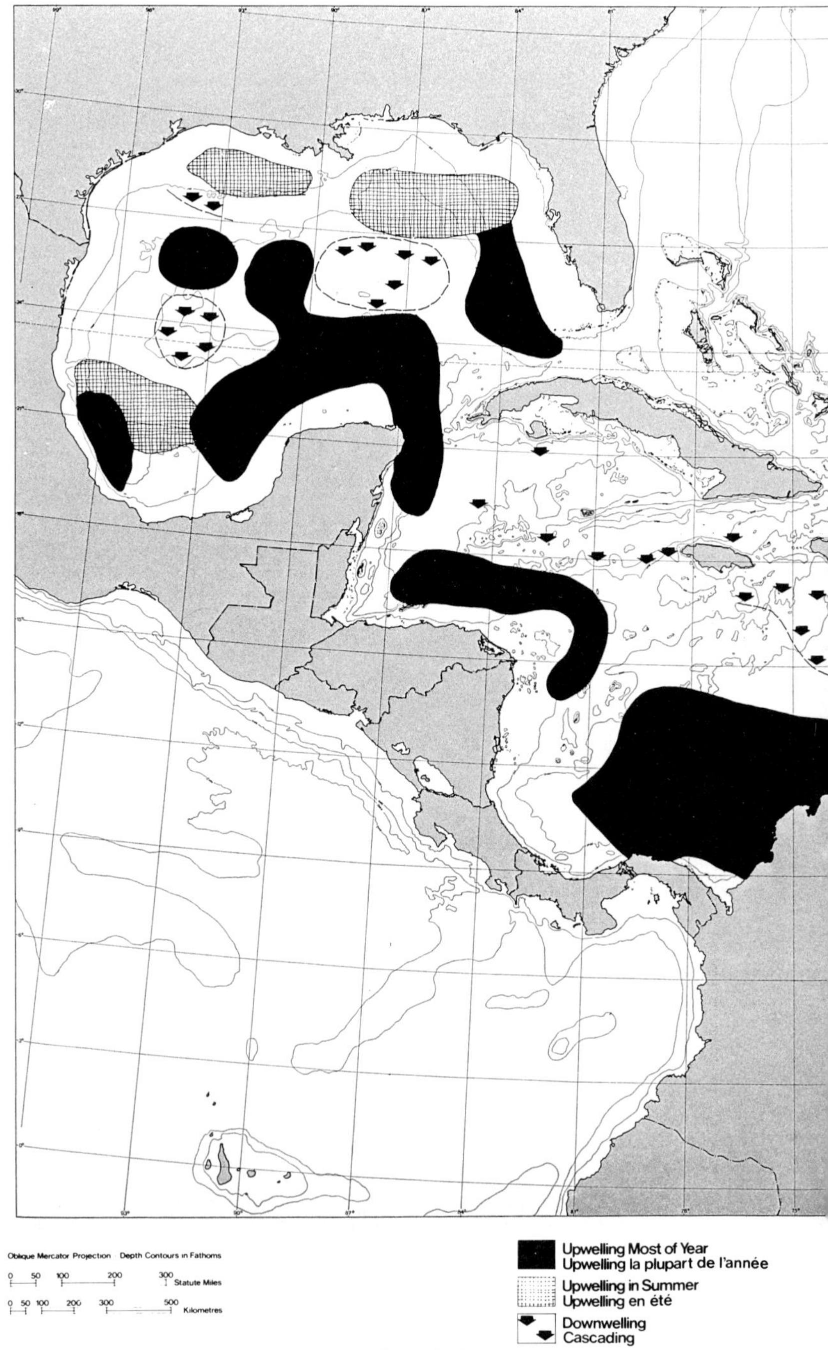

Fig. 10.14 Upwelling and downwelling. Source: IUCN, 1979.

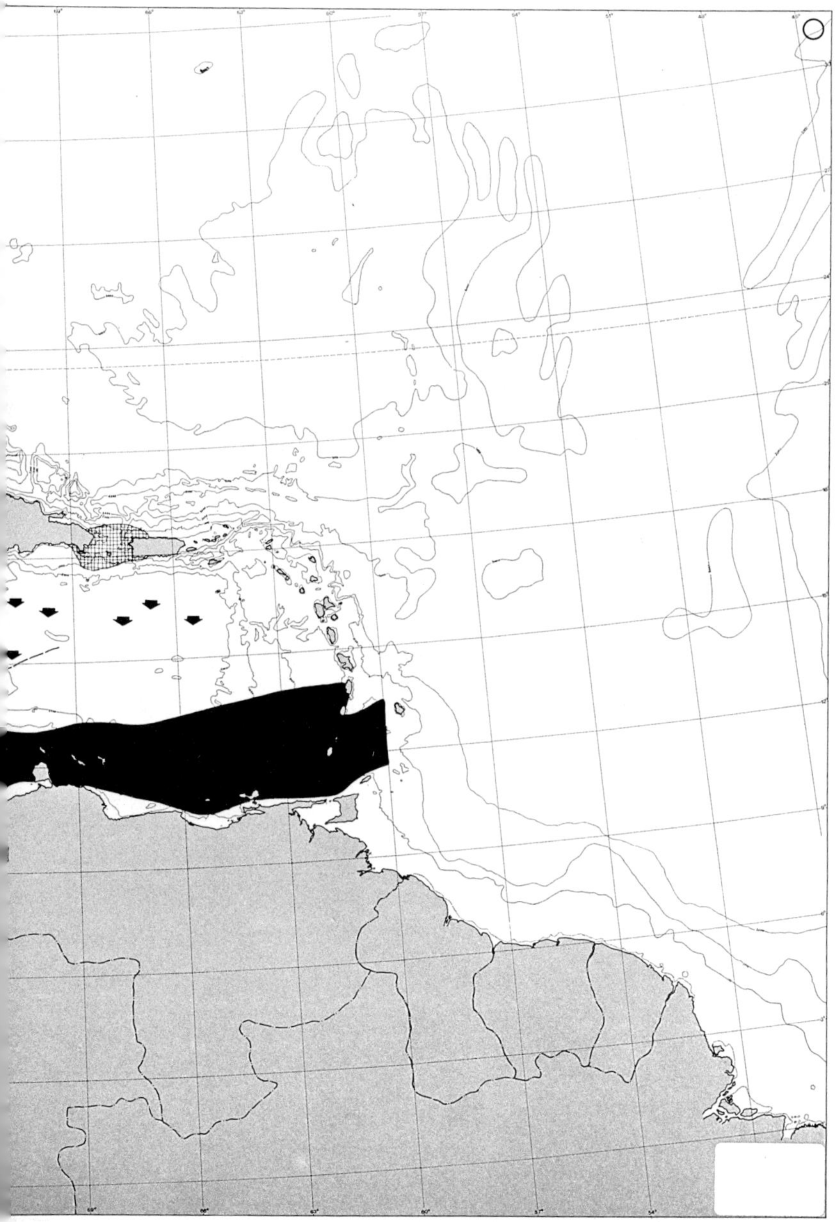

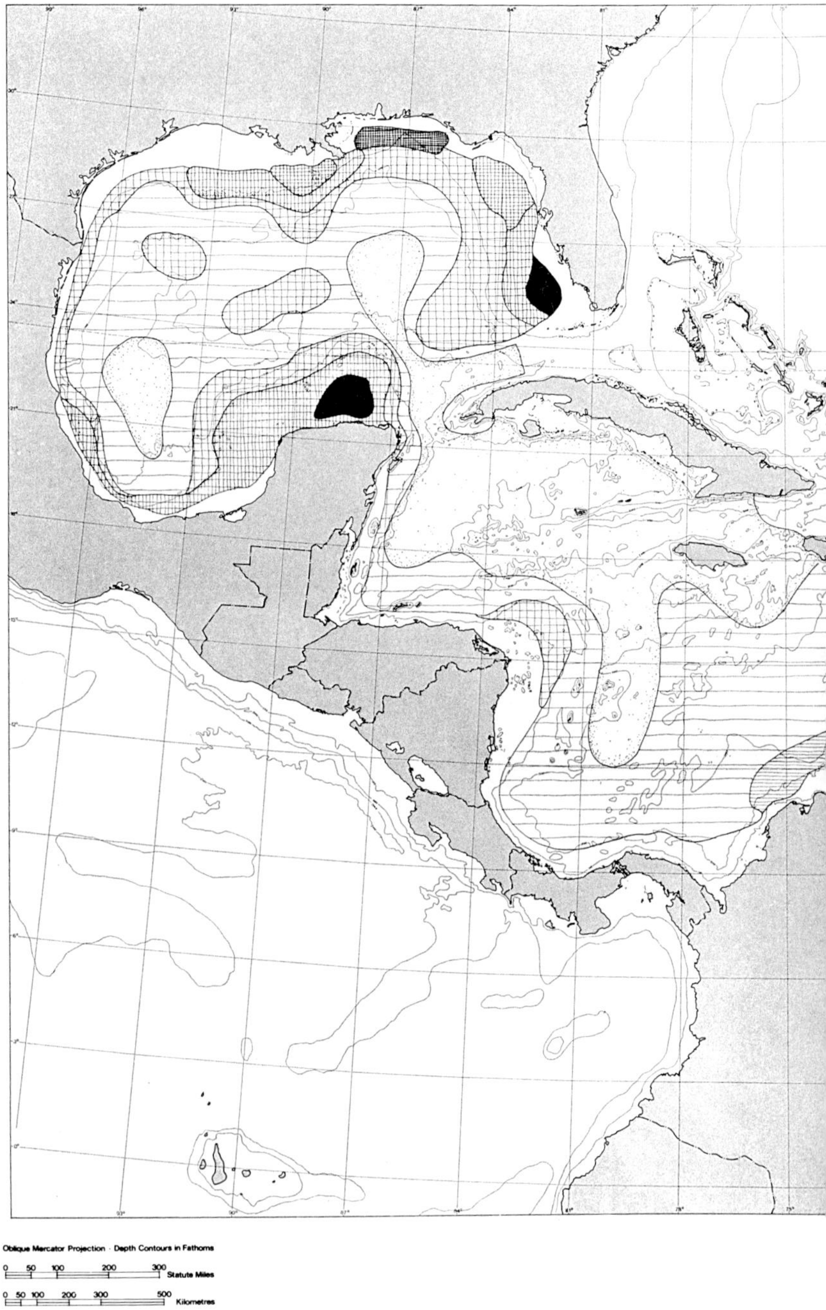

Fig. 10.15 Generalized phytoplankton distribution. Source: IUCN, 1979.

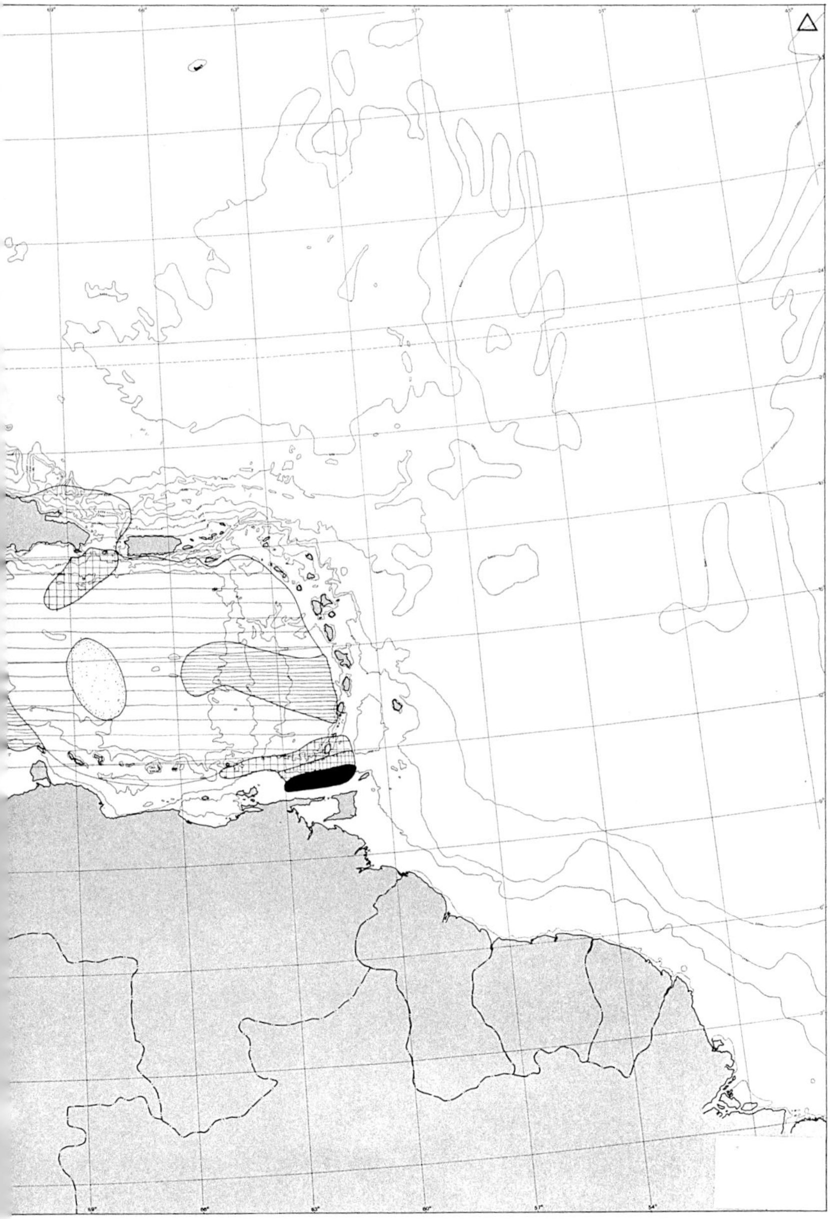

Mean Plankton Distribution in the Upper 100 m. Layer (mg./m.3 dry weight)
Répartition moyenne de plancton dans la couch supérieure de 100m. (mg./m.3 poids sec)

30-100
50-150
100-200
100-300
200-600
200-1000
100-3000
300-1000

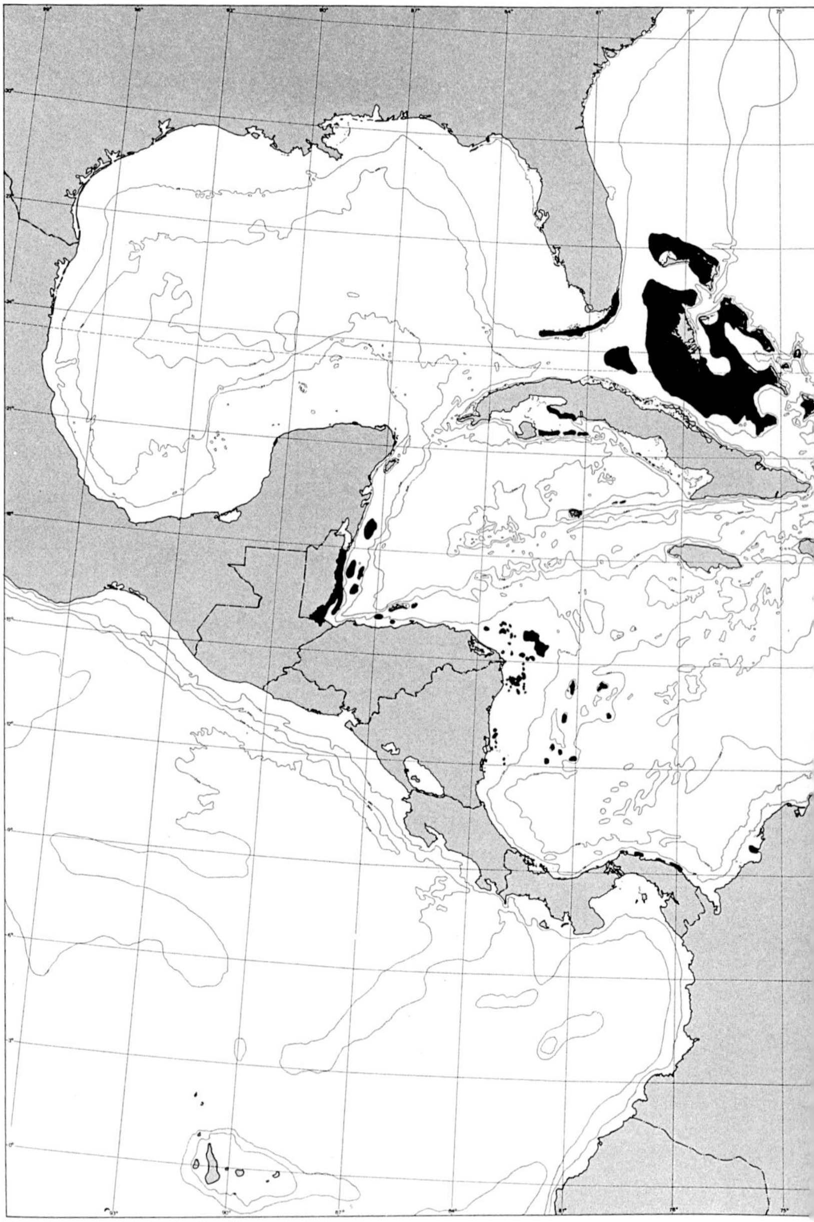

Fig. 10.16 Moluscs: conch. Source: IUCN, 1979.

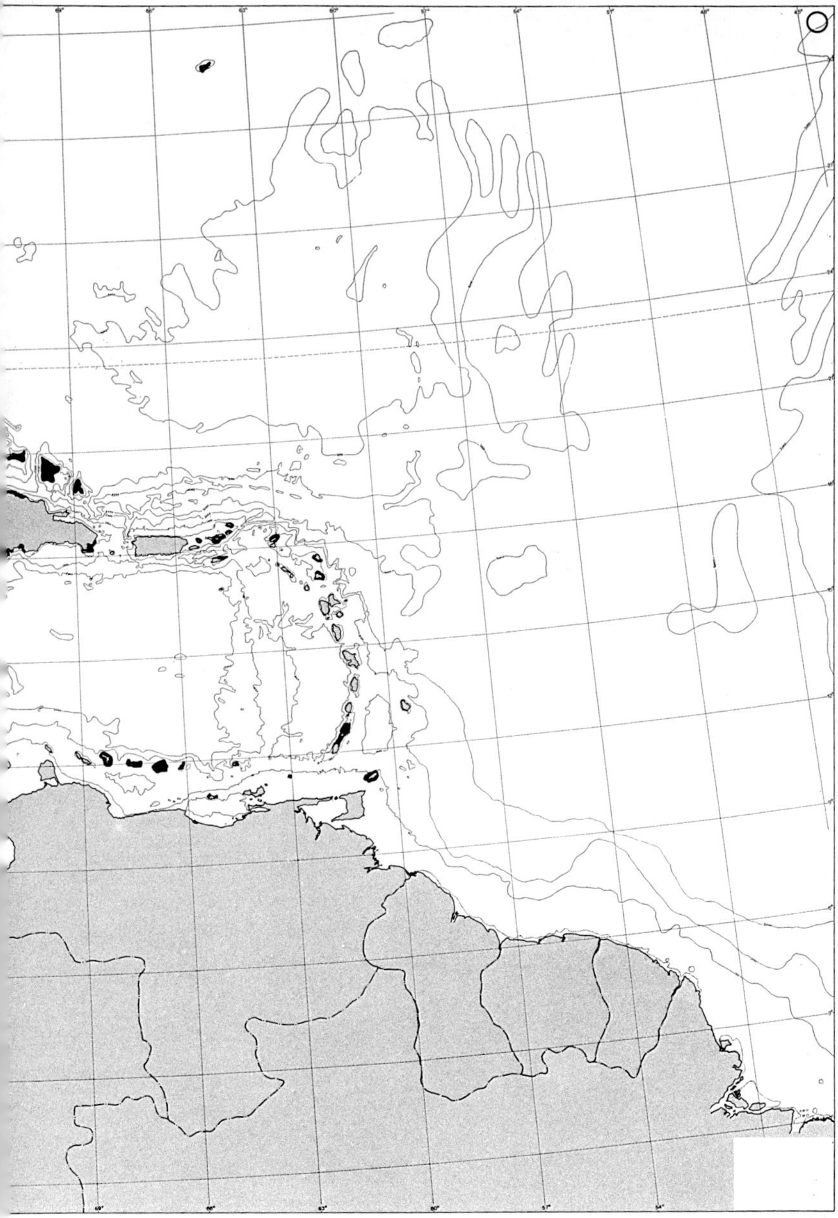
69°
66°
63°
60°
57°
54°

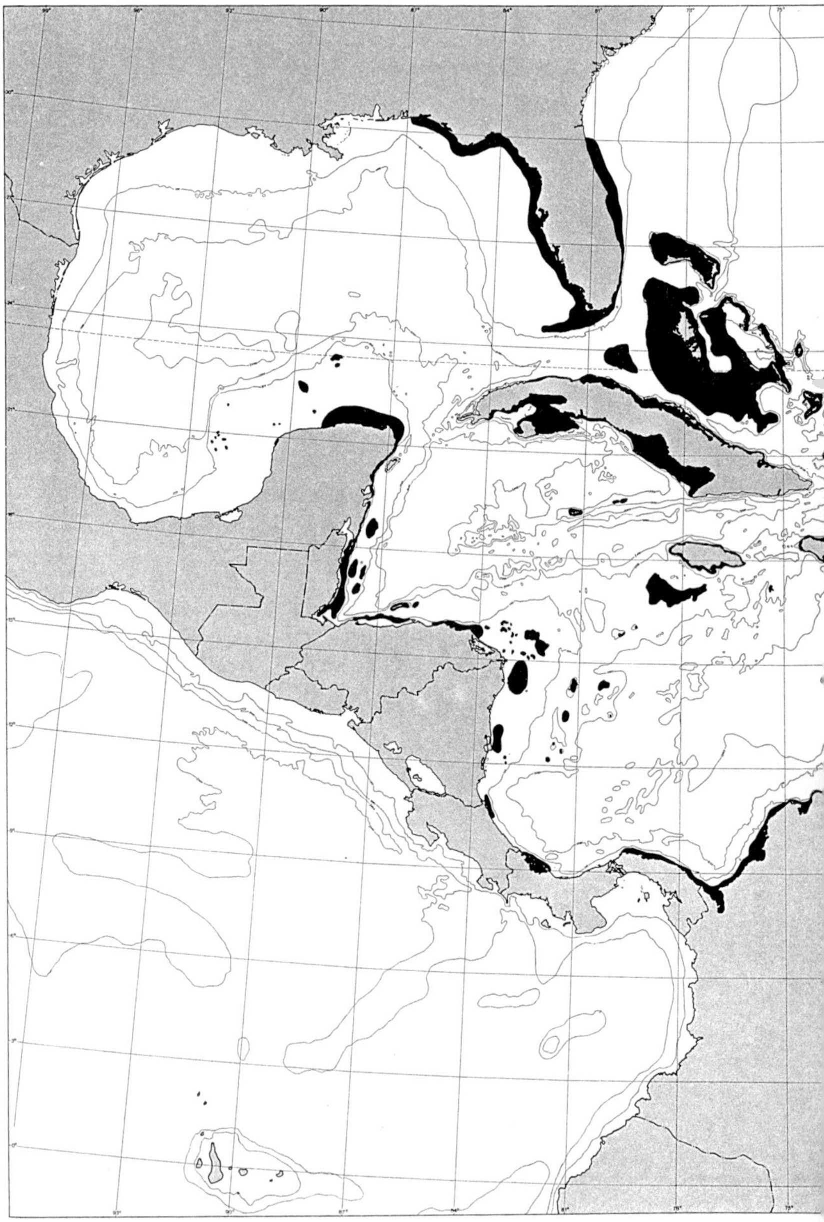

Fig. 10.17 Crustaceans: lobster. Source: IUCN, 1979.

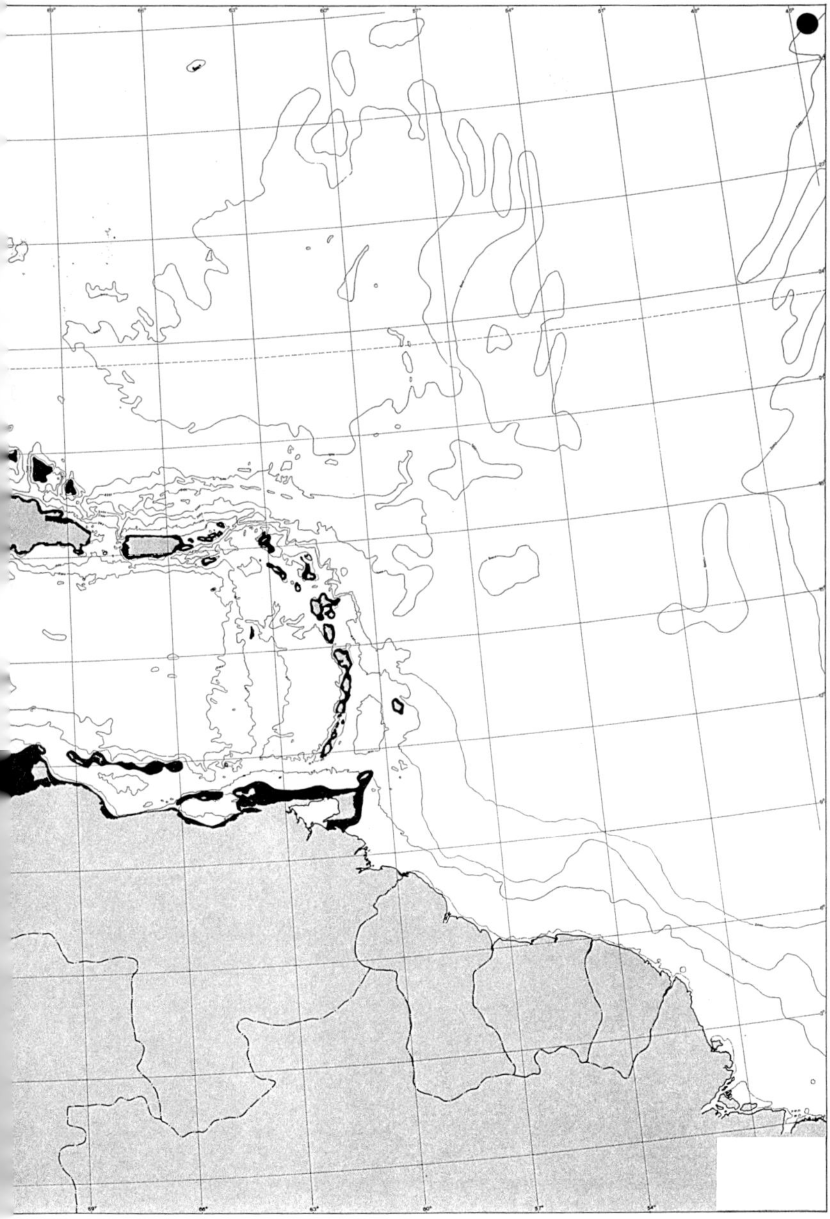

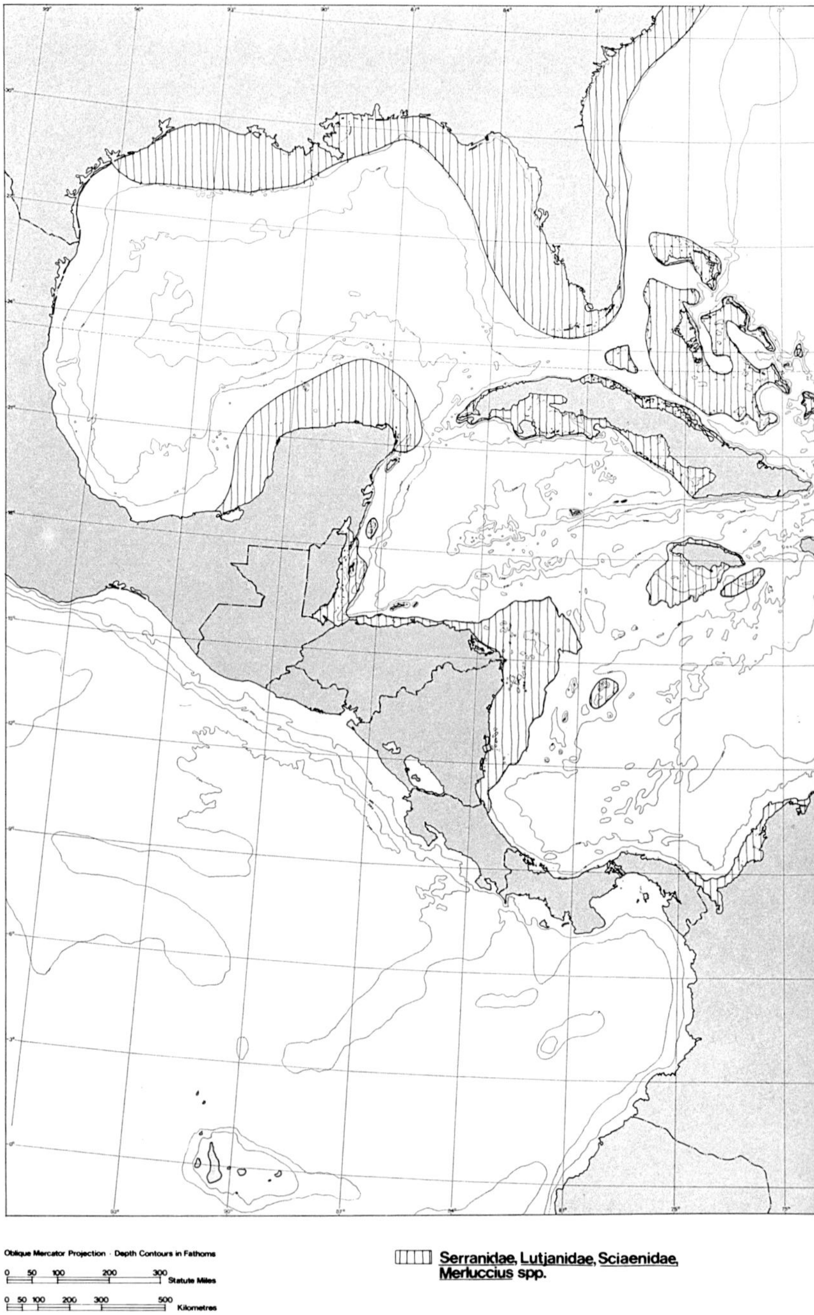

Fig. 10.18 Fishes: demersal (bottom dwelling). Source: IUCN, 1979.

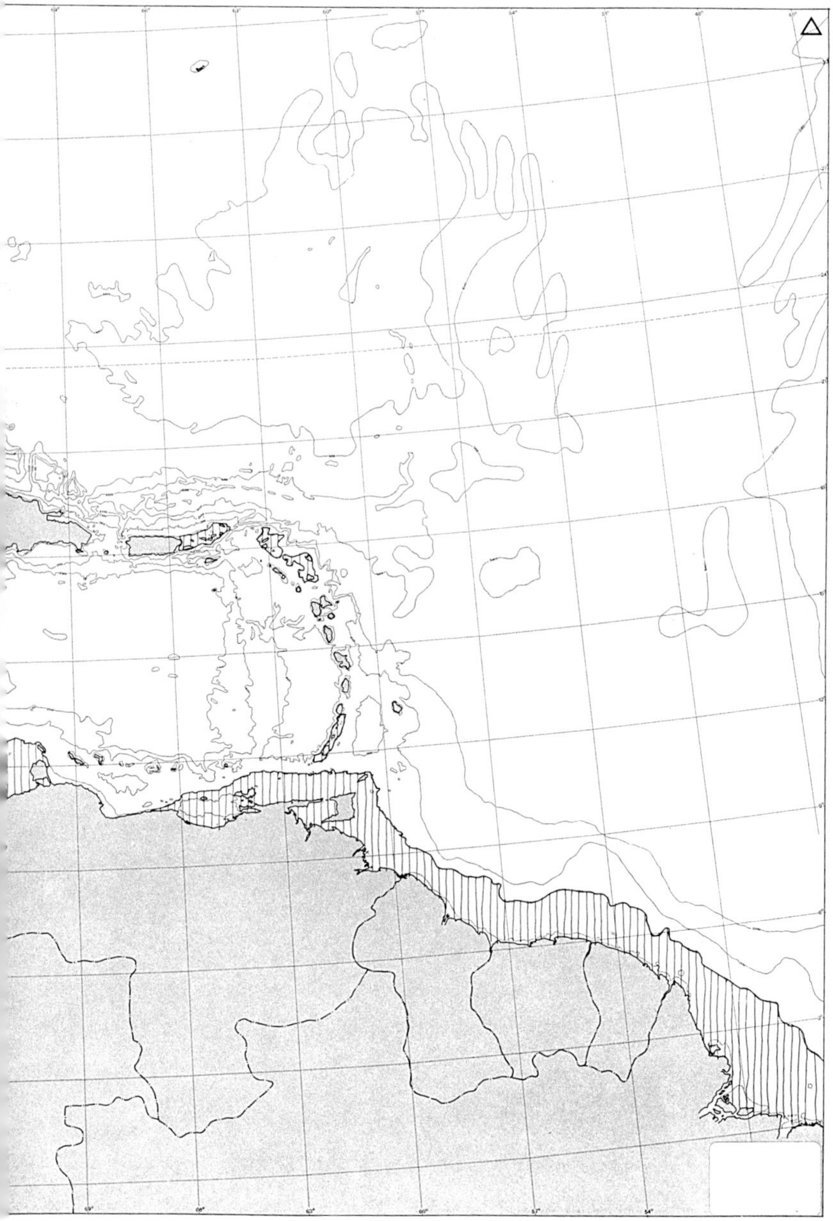

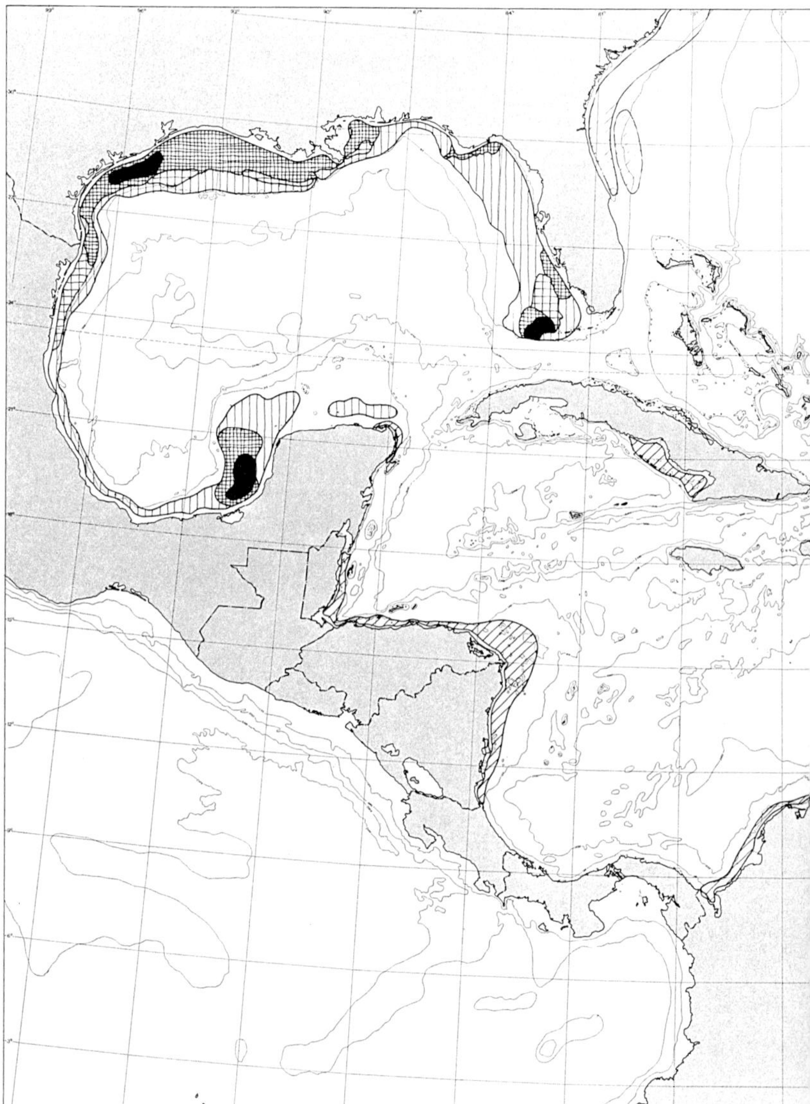

Fig. 10.19 Fisheries: shrimp catch. Source: IUCN, 1979.

63°
60°
57°
54°

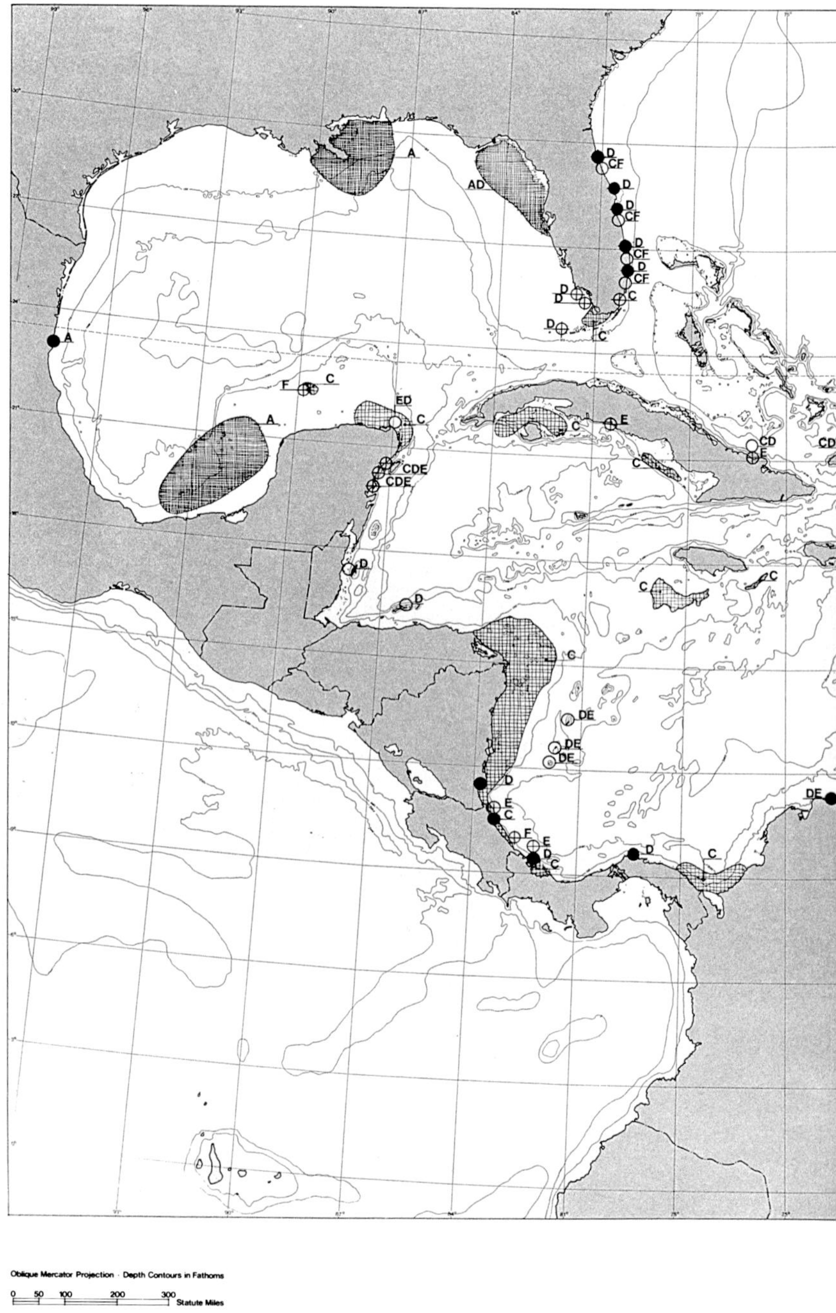

Fig. 10.20 Reptiles: sea turtles. Source: IUCN, 1979.

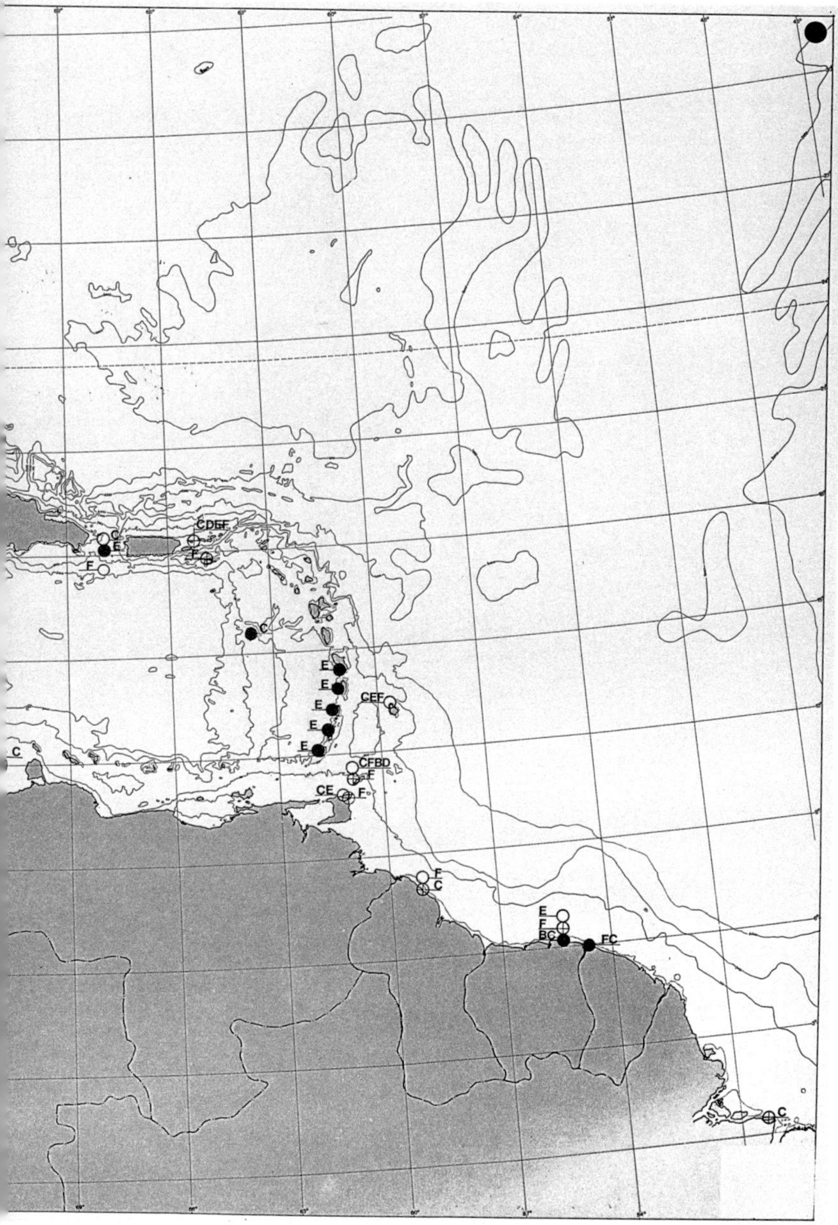

Nesting Beaches / Plages de ponte			
● Very Important / Très importantes	A Lepidochelys kempi	C. mydas: Feeding Grounds / Zones de nourriture	
⊕ Important / Importantes	B Lepidochelys olivacea	L. kempi: Feeding Grounds / Zones de nourriture	
○ Less Important / Moins importantes	C Chelonia mydas		
	D Caretta caretta		
	E Eretmochelys imbricata		
	F Dermochelys coriacea		

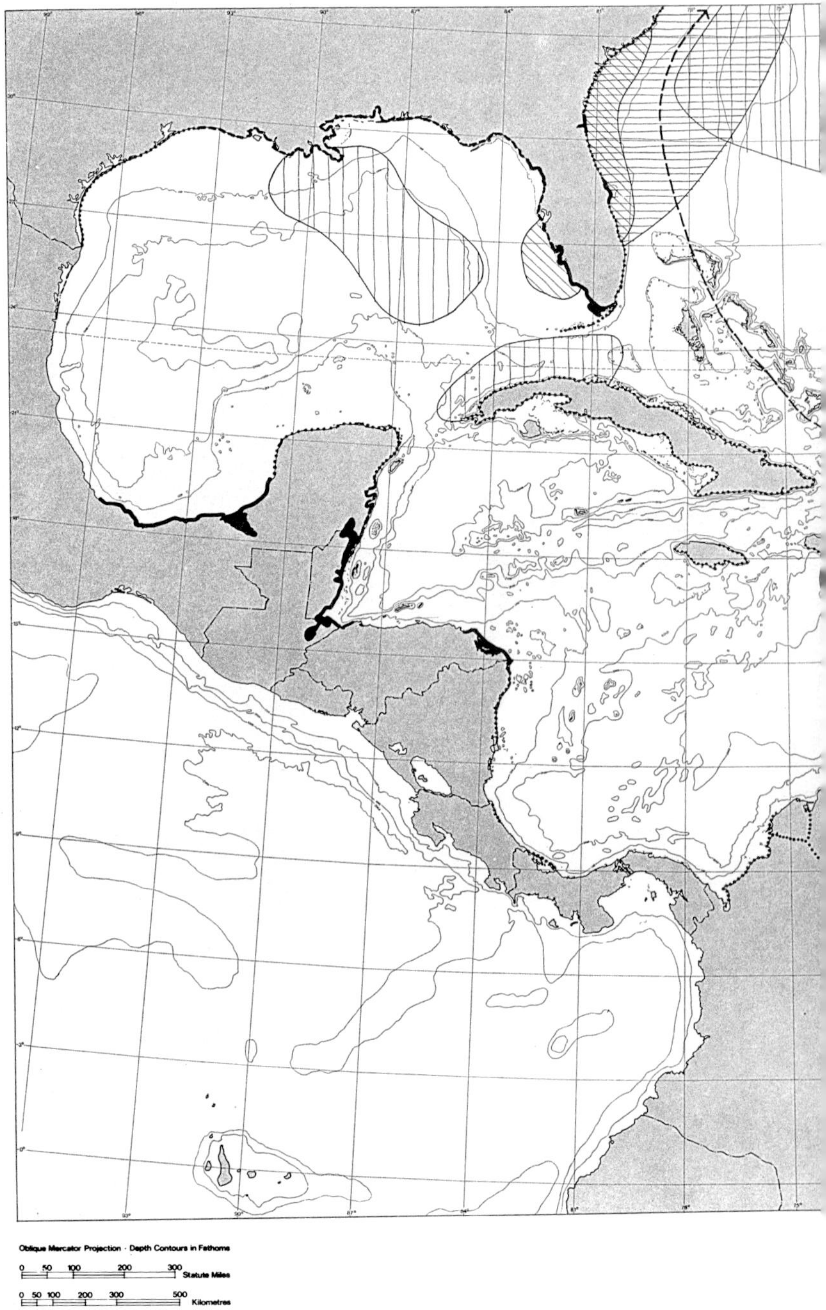

Fig. 10.21 Mammals. Source: IUCN, 1979.

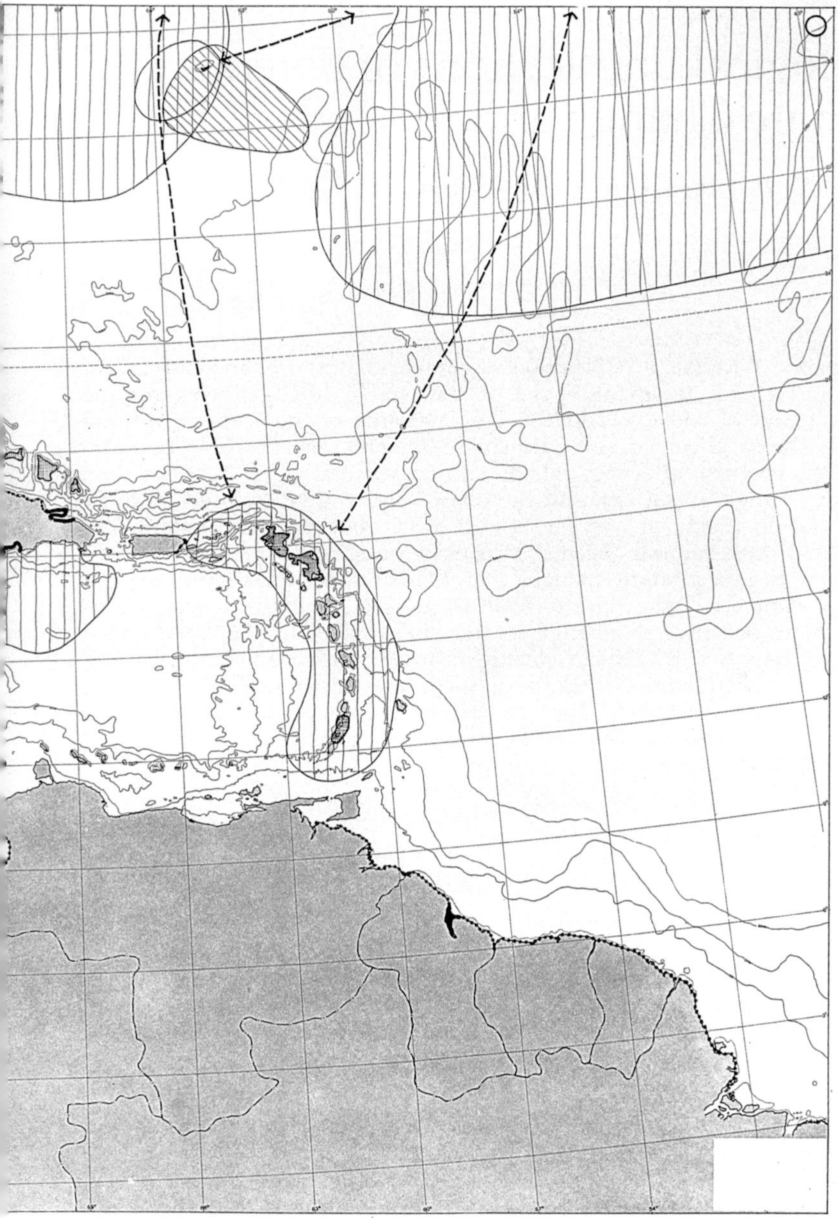

Eubalaena glacialis
Reported Sightings/Sites ayant fait l'objet de rapports

Balaenoptera physalus
Population Centres/Centres de population

Physeter catodon
Densest Population/Population la plus nombreuse

Trichechus manatus
Many/Plusieurs
Some/Un certain nombre
Few/Peu

Megaptera novaeangliae
Calving Grounds/Lieux de reproduction
Migration Routes/Routes de migration

11

Ecological Implications of Potential Climate Change and Sea-Level Rise

V. P. Vicente[1], N.C. Singh[2] and A.V. Botello[3]

Abstract

Implications of the WMO/ICSU/UNEP global climate-change scenario of a 20 cm sea-level rise and a 1.5°C temperature rise on the ecology of lands bordering the Caribbean Sea, Gulf of Mexico, and the Bahamas-area of the western tropical North Atlantic Ocean is considered. Human settlements will probably be more impacted by sea-level rise affecting potable and agricultural water supplies than by temperature change. Coral reefs on the other hand will be stressed more by temperature, not so much an average rise of 1.5°C, but the concomitant 'hot snaps' associated with natural thermal variance. Mangroves are judged to be more susceptible to precipitation changes than temperature or sea-level rise, but future regional rainfall is even more difficult to predict. Similarly, seagrass beds are probably more sensitive to change in the quality of light due to changed water turbidity than to 1.5°C/20 cm climate changes. Response strategies and management recommendations concerning these and other ecological aspects of climate change are hampered throughout the region by lack of adequate information; the most difficult task will be to separate anthropogenic effects not directly associated with climate.

1 Introduction

The region is suitably described as a community of contrasts. It is an area of wide cultural, political and ethnic diversity, rich natural endowments and scarce financial resources. It has some of the world's clearest waters and finest beaches, yet it is faced with increasing environmental stress to its marine and coastal resources. In this chapter the potential effects of climate change on human settlements, agricultural resources, and coastal systems is discussed, and recommendations are made for mitigation of future change.

1.1 Human Settlements

The region is characterized by the uneven spatial distribution of its population. Data on population density per country frequently do not show this clearly because they do not reflect the often very uneven distribution of population within a given country. This has to be taken into account when considering the apparently low density of the region as a whole.

[1] US. Department of Commerce, NOAA, NMFS, Hatorey, Puerto Rico 00918
[2] CARICOM/UNEP, Castries, St. Lucia
[3] Universidad Nacional Autonoma de Mexico, Mexico, DF, Mexico

The islands of the Antillean arc have reached such a high population density that any significant further increase in their population would endanger their carrying capacity. (All these islands have more than 100 inhabitants/km^2, with the exception of Cuba and the Dominican Republic, which have less; Barbados has over 550/km^2.) As most of these islands (with the exception of Cuba and the Dominican Republic) have few flatlands, the high population density results in a very intensive land use of hills and mountain slopes causing serious environmental degradation of their forests and of their marine coastal systems.

In contrast to islands, the continental countries of the region have a population density of slightly more than 25 inhabitants/km^2. Most of these countries therefore have relatively large territorial reserves and natural resources to accommodate an expansion of their populations. El Salvador is an exception, as its limited territory, coupled with rugged topography and a markedly high population density (169.5/km^2), make it very similar to the highly populated islands of the Caribbean Sea.

Historically, the populations of tropical Latin America settled in the valleys, plateau, and watersheds of the highlands, i.e., in the less humid conditions that favoured human habitation. In Central America these lands are either near the Pacific coast or are hinterlands far away from the coasts and separated by topographic barriers.

In recent times, the population of the coastal areas of the region has experienced a marked increase, showing a continuous trend towards the occupation of these areas. As the population pressure will probably increase in the future, if adequate administrative measures are not taken, it could result in unacceptable environmental and social consequences.

Venezuela and Colombia are the only countries with trends of population increase in the hinterlands. On the Caribbean Sea coasts of these two countries, population increases are limited to a few specific areas, influenced principally by the expansion of existing cities. Due to the very low population densities existing at present in the interior plains of these countries and expected growth rates of lower magnitude, the future stress probably will be different from that of the rest of the Central American subregion.

Large metropolitan areas make enormous demands on water resources, and sometimes sources in the nearby surrounding areas cannot suffice. Wastage due to bad maintenance aggravates the problem. Increased infrastructure using more distant sources not only increases costs but can also prejudice activities such as agriculture.

1.2 Agricultural Resources

The diverse soil resources of the region include 517,525,000 hectares, of which 9.7% are classified as arable and permanently cultivated, 22.7% pasture, 50.3% forest, and 17.3% miscellaneous. There are four main problems relating to utilization of these resources: erosion, salinization, waterlogging and chemical degradation.

The most serious problem affecting the soils of the region is erosion which is related to specific soil characteristics, type of vegetation cover, intensity of rainfall, winds, topography, and agricultural practices. The most vulnerable areas are the Greater Antilles and parts of Venezuela,

Colombia, Guyana and Trinidad and Tobago. Also, desertification due to soil erosion is high in some parts of Mexico, including the Yucatan Peninsula. It is estimated that Panama has 1 million hectares of eroded soil, Venezuela 10 times more.

Agriculture in the region is typified by three main systems of production: large-scale estate or plantation agriculture (e.g., sugar cane); small-scale sedentary agriculture (e.g., truck products); and migratory or shifting agriculture. The first of these, large-scale plantation agriculture, is the dominant system in the region. Large private estates or plantations, generally occupying most of the best land, often have been devoted to single-crop monoculture or cattle raising, traditional sources of the region's exports. The main export crops include sugar, coffee, cocoa, cotton, bananas, rice, and to a lesser extent, citrus fruits, coconuts (copra) and tobacco. With the exception of sugar cane, which is usually exported in the form of raw sugar, most of the crops are exported as unprocessed primary products.

In general, the region's agricultural production is inadequate to feed the population, in part, because so large a portion of the arable land is permanently used to produce export crops or used for purposes other than agriculture. The region is increasingly dependent on imported edible oils, cereals and dairy products. For example, imports of cereal increased from 6% of total food imports (1955–1960) to 46% (1965–1970) and then to 60% (1971–1975).

1.2.1 Soil utilization

Significant changes have been observed in soil-resource utilization. For example, since 1969, arable and permanently cultivated land increased by 4 million hectares (8.6% increase), while land for urban and industrial development, road construction and as waste land increased by 7.5 million hectares (9.1%). Some of the agricultural problems relate to availability of arable land, which is under high pressure from urban and industrial developmental activities. The per capita arable land of the Lesser Antilles is 0.13 hectares, one-third of the average for the whole region.

The large export-oriented systems, in addition to causing environmental degradation of the soil through monocultural practices, generally lead to increasing marginalization of a large section of the farming community, thus contributing to the flow of landless population into urban areas with all the environmental consequences of rapid, unplanned growth of such new areas. Agricultural practices, especially in the continental subregions, have constantly modified the agricultural frontiers by removing the protective forest cover, possibly causing unwanted environmental changes in the microclimate, availability of water resources, soil erosion, etc. In some cases, particularly in the smaller islands, there is insufficient suitable land, even where the land tenure system is satisfactory. Also, large-scale farming often is characterized by under-utilization of the resource.

Nowadays, a number of factors have contributed to the decline of agricultural production: droughts, floods and hurricanes; crop diseases, industrial disputes; labour shortages; reduced acreage; shortages of chemicals and machinery; and political instability and poor management decisions. Development agricultural plans should follow a similar approach as the one used by Vicente-Chandler (1987) for Puerto Rico.

Drought is the major climatic constraint in soil productivity. Poor soils and/or serious erosion problems limit agriculture in some areas as in Curacao and Aruba in the Netherland Antilles; Antigua, Cariacou and Nevis in the Lesser Antilles; Haiti, Jamaica, and Puerto Rico in the Greater Antilles; Colombia, parts of Venezuela, the Guyanas and Trinidad and Tobago.

1.2.2 Forest lands

The total area under forest was estimated for 1975 as 221 million hectares. Since 1966, however, 10 million hectares have been lost and, taking into account present forest-management practices, the forest area is expected to shrink to 194 to 175 million hectares by 1980 and 2000, respectively. Many areas originally covered by forests could not be reforested, since centuries of man's activities have changed the basic characteristics of soils and the topography. Barbados, once totally forested, no longer has forests; Colombia and Mexico are losing substantial forest lands. Development of commercial forests frequently leads to serious environmental damage.

The most serious ecological consequences of deforestation are erosion and the disturbance of the hydrological equilibrium. Erosion leads to destruction of the soil characteristics and fertility and, in hilly or mountainous areas, encourages landslides. Disturbance of hydrological equilibrium affects the surface-water supply of the river basins, leading to extremely exaggerated differences in river flow between seasons, reduction of underground aquifer recharge, sedimentation of rivers, estuaries, swamps and coastal areas, as well as to increased incidence of flash flooding. Also, because of changed surface-air moisture equilibria and the reduction in evapotranspiration, changes in microclimates occur, and in severe cases of deforestation major large-scale climatic changes can occur, leading to serious drought or desertification.

Deforestation has occurred throughout the region and therefore, forests are disappearing at an alarming rate. UNEP estimates that Colombia is loosing 800,000 hectares per year (ha/yr), Mexico 400,000 ha/yr, and Venezuela 250,000 ha/yr. Every year nearly 1.8 million hectares of forests are destroyed while only 34,000 hectares are replanted. In Central America, the process has been accelerated through the conversion of forest to pasture for cattle raising, mainly for export to the United States.

Other factors that have caused large-scale deforestation in the region include bad silviculture practices, 'slash and burn' agriculture practiced by landless subsistence farmers, and the felling of trees for firewood and for timber. In the insular Caribbean Sea and in particular in the Lesser Antilles, where soils suitable for food crops are very scarce, most of the best land is permanently under export-oriented plantation crops such as sugar cane, bananas, tobacco, coffee and cocoa.

The environmental effects of deforestation in the humid tropics are quite different from those in the temperate regions of the world. The humid tropics are, in general, subject to far higher annual rainfall, and this precipitation is also much more intense for longer periods. For example, Hurricane Flora reportedly caused extensive damage in deforested areas of Cuba, yet relatively insignificant losses were reported in natural forest areas. A similar situation occurred in Honduras when Hurricane Fifi struck that country.

Another significant problem associated with deforestation relates to the fact that, in the tropics in general, and the humid tropics in particular, the nutrient cycle is very rapid. Most nutrients are found in the first few centimetres of soil and in the vegetation itself. Consequently, total elimination of the forest biomass means that the majority of the nutrients are lost from the ecosystem and a poor soil is left. This can create serious obstacles to reforestation efforts if the two activities are not undertaken at the same time.

One of the prime causes of deforestation in much of the region is the migratory agricultural practice of clearing land using the 'slash and burn' technology. Much deforestation is carried out to extract mineral resources, to shift rapidly increasing urban populations, and to increase agricultural land required to feed the growing populations.

1.3 Coastal Systems

Some of the most productive and biologically complex ecosystems in the world are found within the coastal zones of the region. These include coral reefs, seagrass beds, mangrove forests, and coastal lagoons. The economic importance of these systems stems primarily from their linkage to other resources, especially fisheries and coastal tourism.

Surface waters of the Caribbean Sea are warm throughout the year, resulting in a relatively stable horizontal layering between surface and deeper waters. This stratification is indicative of a circulation system that prevents the upwelling of nutrient-rich deep water except along the northern coast of South America. Lack of nutrients limits phytoplankton production in tropical climates; the Caribbean Sea is generally clear because biological productivity is low (Muller-Karger, Chapter 8).

Nutrient limitations are overcome in nearshore marine habitats of the region through:

1) fixation of nitrogen by shallow-water, blue-green algae found on coral reefs, mud flats, seagrass beds and other environments;
2) for example, the remineralization of the organic material characteristic of recycling of certain nutrients within and between systems;
3) input of nutrients and organic material derived from terrestrial fresh-water run-off into coastal lagoons and mud bottoms.

These mechanisms result in physical, chemical and biological linkages between coastal habitats. Changes or disruptions to pathways of interaction can have some effect upon all of these habitats (Ogden and Gladfelter, 1983). The separation of nearshore resources in the following discussion is for organizational simplicity. Development and management of these resources must be based upon a perspective of the entire coastal zone rather than on individual components.

1.3.1 Coral reefs

There are approximately 60 species of corals in the region including about six dominant species that are primary reef builders. Individual corals grow at rates ranging from 1–25 cm/yr. Coral reefs, consisting of the consolidated skeletons of corals and other $CaCO_3$-depositing organisms, accumulate rapidly in geological time. Modern coral reefs in the region represent about 10,000 years of coral growth (Goodwin *et al.*, 1986).

Coral reefs require clear, clean water and relatively high wave energy to grow and flourish. Thus, they are best developed on the windward side of coasts and are absent where sedimentation from terrestrial runoff and rivers is heavy, such as most of the northern coast of Puerto Rico (Bak, 1983).

Coral reefs are one of the most important coastal resources of the region (*cf.* Milliman, Chapter 13). They are the basis of many coastal fisheries, providing food, shelter, and nursery areas for commercially valuable fishes and crustacean species. Reefs form breakwaters which protect harbours and bays and limit coastal erosion. Coral skeletons are a major source of sand and gravel resources in some islands. The beaches and visual attractions provided by coral reefs are the focus of much of the tourism in the region (Salm and Clark, 1984).

1.3.2 Mangroves

Mangrove forests are a coastal feature unique in lowlands of tropical and subtropical regions. Prop roots of some mangrove trees provide a surface for the attachment of marine organisms, reduce tidal and wave energy, and stabilize the soil, thus promoting deposition of nutrient-rich mud and silt. Mangroves provide habitat and shelter for a variety of animals such as small fishes, crabs, and birds. By breaking stormwaves and dampening tidal currents, mangroves help to maintain the coastline against erosion, and may actually extend coastal lands by trapping and binding sediments. As a result, sediment loads into coastal waters are reduced, and normally there is little, if any, resuspension of sediment through shoreline erosion. Decomposed leaves of the mangroves form the base of a food web that extends to large fishes and birds (Cintron and Schaeffer-Novelli, 1983). See Snedaker (Chapter 12) for a detailed discussion of mangroves.

1.3.3 Coastal lagoons, salinas, deltas and estuaries

Coastal areas of the region near major watersheds often contain huge lagoons of fresh or brackish water that provide important sources of organic material and nutrients as well as feeding, nesting and nursery areas for various birds and fishes. Large coastal lagoons are most prominent along the mainland and are often the breeding grounds for nearshore shrimp fisheries. Coastal lagoons buffer controlling terrestrial runoff, providing natural settling basins for suspended sediments.

Salinas, shallow ponds and lakes with limited or only tidal contact with the sea, are characteristic of many dry islands of the Caribbean Sea. Traditionally, salinas have been used as salt evaporators and have been targeted for mariculture or marina construction in more recent development schemes. These ponds also function as sediment traps and are important to the protection of nearby coral reefs from excessive sediment loading.

Mud bottoms are extremely productive, supporting commercially important shrimp and groundfish fisheries. Wide bands of relatively flat mud bottom are associated with the coasts of North, Central and South America. Mud bottoms result from seasonal runoff of fresh water carrying terrestrial sediment and are enriched in organic material by outflow from coastal lagoons.

Deltas and estuarine areas are marginal marine coastal systems that are impinged, either permanently or temporarily, by fresh-water intrusions. These systems, although usually turbid, are colonized by seagrass vegetation (especially sciophilic species), mangroves or other critical benthic habitats. Deltas are usually subsiding zones. Photosynthetic active radiation (PAR) may become limited at very shallow depths. Nonetheless, habitats within or in the vicinity of deltas and estuaries are essential to a wide diversity of euryhaline species, many of which (e.g., the oyster *Crassostrea rhizophorae* in the southern Caribbean Sea, and *C. virginica* in the northern Caribbean Sea) are of commercial importance.

1.3.4 Seagrass beds

Large meadows of seagrasses of three major species occur in close association with coral reefs in the region. Seagrasses are true flowering plants with male and female flowers and seeds borne in fruits, yet the most common form of reproduction is asexual via a rhizome growing horizontally through the sediments penetrated by the roots.

The seagrasses covering the bottom of coastal bays trap fine sediments and stabilize the bottom with interwoven rhizomes beneath the sediment surface. Sediment retention and stabilization are important for adjacent coral reefs because they prevent abrasion or burial of these reefs during storm conditions (Zieman, 1983).

Seagrass blades provide surfaces upon which many organisms attach, including a variety of algae which may produce calcareous sediment or provide food for grazing organisms (Vicente *et al.*, 1980). Seagrasses serve as nurseries for the juveniles of commercially important species (Ross, 1982) including fishes (i.e., snappers, grunts) and invertebrates (i.e., lobsters, conchs).

1.4 Coastal

Coastal resources of the entire region are under increasing impact from human activity such as:

1) Tourism.
2) Waste disposal.
3) Marine and coastal pollution.
4) Poor land-use practices.

1.4.1 Fisheries resources

Fish protein forms a significant part of the protein intake of the people in the region, and fisheries figure prominently in the national economies. This is especially true of smaller islands, which lack facilities for livestock production. Although much of the fish requirement continues to be imported, fisheries are developing and expanding.

The total estimated potential of fisheries resources for the region's continental platform ranges between 3 and 4.5 million tons per year. The theoretically sustainable exploitable potential is between 1.3 and 2.6 million tons per year. Data on the amount of these resources actually extracted are incomplete. Data on fishing by countries from outside the region are missing altogether. Incomplete data from countries of the region indicate that their present fishing practices are unlikely to result in overexploitation of their resources over the continental platform, except possibly in localized areas.

The most significant fishery activities of the region are found at Campeche Bank in the Gulf of Mexico, at Mosquito Bank in the Caribbean Sea off the coasts of Honduras and Nicaragua, in the Gulf of Paria between Venezuela and Trinidad and Tobago, and the coastal waters in the Guyana–Surinam area. Coastal and inland fishing in the region is mainly artisanal in nature.

Because of a pronounced lack of upwelling and the existence of a stable thermocline, nutrient-rich waters do not rise to the surface except along the Caribbean Sea coast of Venezuela (Aparicio, Chapter 6). This results in a generally low-level of zooplankton in the food-chain and in significantly smaller populations of exploitable fish. As a consequence, coastal mangroves, estuaries and coral-reef communities play a proportionately large role in providing nutrients and breeding grounds for many species.

Few or no statistics relating to inland fisheries or aquaculture in the region are available. Inland fishing is generally carried out in a small, unorganized, private capacity. Aquaculture is underdeveloped in the region. However, in the last few years substantial interest on mariculture has been generated in some countries. Limitations of capture fisheries, efforts to reduce imports, and the need for economic diversification have contributed to public- and private-sector interest in the culture of high-valued marine species for export and less expensive species for local consumption. The physical characteristics and resources of the region indicate considerable potential for coastal mariculture. Many countries are endowed with long coastlines dotted with protected bays, coves, harbours and lagoons; relatively fertile brackish water, estuaries, mangrove swamps and other wetlands; large areas of coastal land with marginal agricultural potential; and a tropical climate favouring year-round growth.

The presence of edible or commercially valuable species of crustaceans, molluscs, fishes, seaweeds, etc. also favours the development of coastal mariculture systems. Of special concern are marine shrimp species (*Panaeus schmitti* and *P. duorarum*), spiny lobster (*Panulirus argus*), Caribbean king crab (*Mithrax spinosissimus*), the queen conch (*Strombus gigas*), mangrove oyster (*Crassostrea rhizophorae*) and the American oyster (*C. virginica*). See Gable (Chapter 10) for the distribution of the most important species.

1.4.2 Sandy beaches

Sandy beaches are shoreline areas exposed to some degree of wave action. In contrast to rocky shorelines, the dominant substrate type is sand which may be autogenic or allochthonous in character. Sandy beaches are not biotically barren grounds. On the other hand several infaunal and epifaunal organisms inhabit the unstable bottom. They are capable of movement, both vertical (e.g., *Donax denticulatus* and *Emerica* spp) as well as horizontal movement (*Luidia* spp, *Astropecten* spp and *Mellita* spp). Their mobile character makes these species capable of adjusting to the continuously shifting substrates.

Not only do sand beaches harbour some peculiar endemic components, but also, are a major attraction of the touristic resources of the region in general (Hendry, Chapter 7).

1.5 Climate

Although the entire region lies within the tropics (2 to 30° north of the equator) there are substantial macro- and microclimatic variations due to topography and orientation with respect to the prevailing northeasterly winds. Even comparatively small islands such as Jamaica show marked differences from one area to another; thus, on the northeast coast and over the Blue Mountain range, annual precipitation averages 7600 mm, whereas it is only 760 mm on the southwest coasts, giving an island-wide reported annual average of 2000 mm. Such large variations for even a small island illustrate the lack of meaning attributable to national average rainfall figures (UNEP/CEPAL, 1980). Nevertheless, the following broad climatic zones, with reference to rainfall and temperature, can be identified within the region (*cf*. Gray, Chapter 5):

1) Humid tropics are found along most of the coastlines of Central America, eastern Venezuela, the Guyanas and the majority of the islands, as well as in the tropical rain forests below an altitude of 100 m. Here average annual rainfall is heavy (in excess of 2000 mm), although there are distinct wet and dry seasons, with the majority of precipitation occurring during a six- to seven-month period.
2) Sub-humid tropics with much lower rainfall are found in several inland areas of Central America, along the northern coast of Colombia and Venezuela and in some of the natural savannah areas of those two countries, as well as Barbados, the Netherlands Antilles, Antigua, southern Haiti and southern Jamaica. In these areas rainfall averages between 100 and 1500 mm annually.
3) Semi-arid and arid areas occur in a few locations in the region, such as northern Venezuela and northeastern Colombia. However, by far the largest area occurs in Mexico. In these regions annual rainfall averages less than 700 mm, but precipitation is highly unpredictable and can vary by as much as 40% from one year to the next.

In general, it can be said the the region does not suffer from serious water deficiency, as measured by annual rainfall minus evapotranspiration.

2 Climate-Change Impacts

Impacts on the resources and ecosystems described above would result in mainly from the rate of change of:

1) CO_2 and other greenhouse gases.
2) Climatic elements such as temperature; net precipitation; storm frequency, origin, and intensity.
3) Sea level change.

The assumed scenarios for the period leading up to the year 2025 are given in Table 11.1.

2.1 Human Settlements

These are unlikely to suffer from changes in the climatic elements except increased length of hurricane season, which can be partially planned for. Sea-level rise however, coupled with land subsistence (both natural and anthropogenic) could lead to flooding of nearshore property in island and low-lying coastal states. Damage or overwhelming of sea-defense works and saline intrusion into drinking and irrigation water could become a serious problem for low-lying coastal states such as Belize and the Guyanas.

2.2 Agricultural Resources

Saline intrusion could have severe deleterious effects on large-scale coastal agricultural operations such as rice production on the Guyana coast. An ongoing programme of varietal testing might be necessary.

2.2.1 Soils

Soil-erosion problems could increase but these should be manageable by improved conservation practices. Saline intrusion might render low-lying coastal areas (in Belize, Guyana, etc.) unfit for traditional crops or varieties.

2.2.1 Forests

Forest adaptation is slow compared to other agricultural operations and hence may suffer or benefit more from rapid climatic changes over historical norms. It is unlikely that such changes would be significant for inland and hilly forests. In any case active programmes of reforestation can only help towards a sustainable solution of the overall problem.

2.3 Coastal Systems

2.3.1 Introduction

The classical case of the *Lithophaga*-bored columns in the Temple of Jupiter Serapis near Naples (Allison and Palmer, 1980a and b) and the more recent

Table 11.1 Assumed scenarios for estimating climate-change impact by the year 2025.

Parameter	*Predicted rate of change*	*Historical average*
1. CO_2 build-up in atmosphere	50% over next 30 years (global)	8% over last 25 years (global)
2. Temperature	1.5°C over next 100 years	1.5 ± 0.5°C over last 10,000 years
3. Net precipitation	Uncertain, but likely increase in humidity	
4. Storms	Uncertain, except that hurricane season could be lengthened due to increased SST	
5. Sea-level rise	20 cm by 2025 (i.e., approx. 0.5 cm/yr)	20 to 100 cm per 100 years (0.2 to 1 cm/yr)

example of littoral faunal uplift of the fouling populations growing on a concrete bridge piling at LeJeune Road in Miami (Wanless, 1982), are two popular examples of sea-level changes (although by different causal agents) during Holocene times. The numerous coral fossils and terraces now sitting on land 400 m or more above present sea level (Horsfield, 1975), further reflect that sea-level changes have occurred on a much larger scale in the past, with significant (and perhaps at times catastrophic) consequences to the shallow benthic systems associated with these ancient shorelines. We have evidence that changes may be occurring at present (Hanson and Maul, Chapter 9), but the specific sea-level trend is still not clear because of the endemic geomorphological complexity of the region. Some of the most important factors that influence sea level are discussed below. Geoid changes are discussed in detail by Morner (1976) and will not be included in this discussion.

Predictable periodic factors that have measurable effects upon mean sea level are the tides, the 14-month cycle of the Chandler motion and the 18.6-year nodal cycle of the moon (Lisitzin, 1983). Furthermore, the complex hydrographical and meteorological character of the region can cause significant measurable effects on mean sea level. However, most of these factors have been studied to some extent and are usually considered or corrected for when determining present eustatic sea-level trends. Diastrophic movements also can cause changes in the relationship between a specific land mass and the sea. These movements, whether epeirogenic or orogenic in character influence sea level, particularly in geologically active areas like most of the boundaries of the Caribbean Plate.

An ecological evaluation of the impact of a sea-level rise on shallow benthic marine systems (e.g., coral reefs, seagrass beds, mud flats, deltas) cannot be accurate unless one considers the geological scenario of each locality. For example, if all other sea-level-influencing factors remain constant and the general environment is unchanged, a particular area that is presently becoming uplifted may partially (e.g., Barbados) or totally (e.g., the Huon Peninsula, New Guinea, is rising at a rate of 3 mm/yr) balance the sea-level rise effect on the benthic assemblages. On the other hand, the effect of the expected rise may be accentuated on benthic systems lying on top of subsiding zones (e.g., river deltas in general).

Horsfield (1975) gives a general overview of Quarternary vertical movements in the Greater Antilles by interpreting the variable elevation of raised Quarternary marine terraces. Quarternary tilting was also inferred from depth variations over shallow submarine banks. Using his estimates of tilt direction and assuming that this trend will continue unchanged for the next century, and assuming that all other physical factors remain the same, then the following localities in the Antilles will become uplifted to some extent (actual rates are not given) and therefore could at least in part locally reduce the increase in depth expected by the predicted sea-level rise: east coasts of Cayman Islands, north coast of Jamaica, southeast coast of Cuba, north coast of Bahia, and the southwest coast of Haiti. By the same token, the coasts on the opposite end of the tilt axis are more likely to become deeper or drowned.

Unfortunately, information on the diastrophic trends of islands is controversial, and qualitative. One exception may be made for the extensive geological studies done on the island of Barbados, perhaps, because of its unique composition. Unlike all of the other Lesser Antilles which are volcanic, Barbados is formed of folded sediments with a cap of reef limestone. The island is being uplifted at a rate of 0.3 mm/yr (Stoddart, 1976). Geological complexity of the region results from the entrapped position of the Caribbean Plate. Stresses along the northern plate boundary have caused uplift in many of the islands and subsidence in many others. For example, upraised limestone strata on a fault block create the spectacular cliffs of Mona Island on the west coast of Puerto Rico. Along its western boundary, the Cocos Plate is being thrust beneath the Caribbean Plate. The eastern boundary of the Caribbean Plate is a subduction zone in which the north and south American Plates are being driven under (see Fig. 7.4). Although there are some general agreements among geologists concerning the position and general eastward drift of the Caribbean Plate, there is major disagreement when it comes to modelling it in any detail. This was a major point discussed during the 7th Annual Symposium on Caribbean Geology held in Puerto Rico during 24–28 February 1988.

Florida, the Bahamas, the Bermudas, Cuba (except the southeastern tip), and the Yucatan Peninsula are not part of the Caribbean (Tectonic) Plate, and probably will remain more stable geologically in the next century than Hispaniola, Puerto Rico, the Lesser Antilles and the Caribbean Sea coast of Central and South America where strike-slip faults or subduction zones are found. Contrary to the block faulting of late Quarternary movements in the island, western Jamaica has had a history of relative tectonic stability since the Holocene (Digerfelt and Hendry, 1987). Shlemon and Capacete (1976) show that no major epeirogenic uplift has occurred along the north-central coast of Puerto Rico over the last 10,000 years. Other authors, however, suggest that up to 4 m of epeirogenic uplift over the last 4000 years has indeed occurred (= 1 mm/yr) in this region. The north coast of the Dominican Republic is being uplifted (A. Foster, pers. comm.).

Before discussing the detailed ecological implications of a real rise in sea level on shallow tropical benthic communities, areas where subsidence is occurring at significant rates should receive particular attention since they could be affected the most. Natural subsidence by tectonic tilting or just plain sinking of the coast by deposition and compaction of sediment occur in various places (e.g., Gulf of Mexico). Regional geosynclinal downwarping has been widely recognized as a subsidence process affecting deltaic regions as well as subsidence through compaction (Morgan, 1967). For example, the birdfoot delta of the Mississippi has been operative for some 500 to 600 years. The subsidence rates there are as much as 5.3 cm/yr. Furthermore following the abandonment of a deltaic distributary system, sedimentation ceases but subsidence continues. There are anthropogenic sources of subsidence which perhaps should be considered in particular areas. Some of these sources are pumping of groundwater for agriculture or industry and extraction of crude oil and natural gas. For example, a subsidence of 3.4 m in an oil field in Venezuela occurred between 1926–1954; this represents a subsidence rate of 12 cm/yr.

2.3.2 Coral reefs

The evaluation of one factor on coral-reef assemblages must be done within the context of additional variable disturbances that at present are structuring (if an ecological deterministic approach is taken) or have structured (the historical approach) the community. For example, we have evidence that both physical and biological disturbances in various scales of time and intensity have measurable effects on coral-reef populations and therefore on community structure. Some of the biological factors are: bioerosion, overgrazing, overgrowth, territorial behaviour, massive mortalities, predation and disease. Some physical factors important in determining community structure are: hurricanes, rainfall, tides, UV radiation, sediments, turbidity, erosion, sun hours, temperature, upwelling, hydrographic conditions and others. Reefs at present may be considered to be more prone to disruption in view of the recent mass bleaching event (Williams *et al.*, 1987; *cf.* Atwood *et al.*, 1992).

When evaluating what might be the result of a sea-level rise of 20 cm by the year 2025 on a reef, several factors must be taken into consideration simultaneously: the type of reef (e.g., fringing, patch, atoll barrier); the geomorphology of the coast; the ecological state of the reef and the zones and depth of the reef in question. At first one might think that the reef flat zone, if not able to grow in pace with sea-level rise, will be drowned and therefore this zone should be of primary concern. In fact, Cubit (1985) showed landward extensions and retreats of a zone of the red alga *Laurencia papillosa* that corresponded to rises and falls of yearly mean sea levels. The local temporary extinctions appeared to be the result of herbivores which expanded their grazing range when sea level was high. Also, the role of predation in limiting the distribution of intertidal invertebrates is well established in the ecological literature. On the other hand, if water level increases over the reef flat, perhaps seagrasses and algae better adapted to permanent submerged conditions, and sublittoral invertebrates will increase in abundance as a temporary result of the reduction of mortality caused by subaerial exposure (Cubit, 1985). However, due to the existing general turbid conditions generated in many reefs as the inevitable consequence of upland deforestation and poor coastal management decisions and/or as a consequence of discharges of organic pollutants and elimination of seagrass beds by thermoelectrical power plants, one should be particularly concerned with deeper zones that might be deprived of the proper photic conditions necessary to maintain the light-dependent coral and sponge populations. The potential reduction of PAR to sub-minimal levels that may significantly alter photosynthetic populations (e.g., sponges, corals and algae) could result in part, by the increase in height of the water column standing over the reef zone in question which will increase scattering and absorption of incident light. Furthermore, incursion into nearby mud flats during high tides may resuspend unstabilized fine sediments into the water column which may favour sciophilic or maybe even cryptic or fouling species considered of less ecological value than corals. Corals, both live and dead, are being bioeroded at a faster rate than Indo-Pacific scleractinians. Therefore, any change in conditions (e.g., increase in organic particle suspension) that would enhance bioeroding populations (e.g., *Lithophaga*, clionids, sprastrellids, sipunculids) or fouling populations that can transform nutrients into biomass at a fast rate, could competitively displace many corals and possibly cause more local coral-population extinctions that in turn could decrease or cease coral-reef accretion.

Accurate predictions on the effect of a sea-level rise can perhaps be made in reefs that have been physically and biologically monitored for many years. Such reefs do exist in Panama and Puerto Rico (see Vicente, 1987).

Although an increase of 1.5°C in surface-water temperature would change the normal winter temperature maxima to 27.5°C and the summer average maxima to 30.5°C, these increased temperatures in general are not high enough to impair physiological processes in marine organisms. However, potential physiological kinks could occur in photosynthetic organisms, especially at temperatures around 30°C. Furthermore, although the highest temperature endurable by West Indian and Hawaiian corals is 36°C, it is generally accepted that corals grow best between 25-29°C. Indirect long-term effects of an increase of 1.5°C are at present difficult to evaluate.

The coral reefs that would be more susceptible to becoming thermally stressed would be those occurring in the vicinity of thermal effluents from power plants and those growing in protected shallow lagoons (e.g., some patch reefs). In these particular situations temperatures of 40°C might be reached and would cause a mass mortality event, particularly in calm cloudless summer days during low tide. In the tropics, marine organisms are living closer to their maximum thermal tolerance than are temperate organisms. This is particularly true of southern Caribbean-reef biotic components which are exposed to a higher water temperature regime during the year than reef organisms in marginal tropical grounds (e.g., Florida) where even freeze-kills or cold-water stunting of marine faunal and floral elements have been documented.

It is also important to note that coral reefs, during the process of the biological calcification, removes carbon dioxide, the principal greenhouse gas, from the air–ocean system contributing to a more permanent solution.

2.3.3 *Mangroves*

Of the four common species of mangroves found in the region (*Rhizophora mangle*, *Avicenia germinans* (= *A. nitida*), *Laguncularia racemosa* and *Conocarpus erectus*), perhaps *A. germinans* should receive particular attention. For example, while the red mangrove (*R. mangle*) can send off prop roots of various dimensions in response to local hydrographic conditions, the black mangrove (*A. germinans*) has aerial roots which project only vertically for short distances above tidal muds. If the roots become permanently submerged by a relatively sudden rise in sea level, then, massive local extinctions of black mangrove with their associated floral (cyanophytes) and faunal (e.g., fiddler crabs) assemblages may occur throughout the region. The geographical extension and ecological implications of this potential demise of black mangrove forests, therefore, should receive particular attention. Furthermore, seedlings and seeds of *A. germinans* can literally drown, and will not develop or germinate under prolonged flooding conditions.

Contrary to the more (yet changing) stable coral-reef and seagrass-bed communities, the faunal and floral biocoenocis living on mangrove roots are quite ephemeral. The temporary existence of this type of community is due to periodic sloughing-off of the root itself, carrying down with it all the epibenthic biota into an anoxic-reduced mud environment. Organisms that have an early age of reproduction which helps produce high, instantaneous populations (e.g., selected species such as barnacles, oysters, some sponges and many ascidians) are usually found in this type of system. The effect of a sea-level rise on red mangrove-root community structure would probably be insignificant (everything else being held constant) because the prop roots of *R. mangle* usually stand much higher than mean sea level (sometimes more than 6 m) so that the diapores (which are adapted to colonize quickly) are not limited by lack of spatial resources. The overall effect would be a faunal-floral uplift which in a considerably short time will return to pre-existing conditions.

The littoral-supralittoral elements, furthermore, are adapted to tolerate high temperature and temporary subaerial exposure so that an increase of 1.5°C would probably cause unmeasurable changes in the population of these taxa. However, the sexual reproductive cycle of many mangrove-root organisms may undergo significant changes when the cycle is dictated primarily by temperature changes. This may cause uncoupling of sexual reproductive strategies with some other important requisite, and may become important to the biota whether of commercial importance (e.g., mangrove oysters) or not.

Although a landward extension of mangrove forests is likely to occur at first sight, this could only occur if the proper edaphic conditions exist in the new shoreline.

2.3.4 Coastal lagoons, salinas and estuaries

Marine benthic communities lying on or close to estuarine or deltaic grounds, whether these are seagrass beds, mangroves, or infaunal benthic assemblages, are, by definition, on marine marginal conditions which are hampered by periodic freshwater and terrestrial intrusive elements. These areas are generally turbid with measurable changes in salinity, and therefore are subjected to photic and osmotic stresses. Furthermore, these systems, particularly deltas, are usually sinking at measurable rates as mentioned before. As the delta fronts continue to be built forward, foreset beds are deposited over existing bottomsets, and in time, topset beds of the floodplain are extended over the foreset beds, causing subsidence. Deposition, however, has generally kept pace with subsidence but in some ancient delta deposits, marine limestone and shales are found interbedded with fresh-water sediments and soils. Therefore, benthic systems associated to deltas, particularly seagrass beds, which are already sinking under light-limited conditions as the result of the inherent nature of deltas and some estuaries, would be particularly affected or destroyed by the expected sea-level rise.

The lagoons could suffer from increased saline intrusion but depending on the rate should be able to support its usual nurseries. Salinas on the other hand could be flooded-over continuously and loose their economic value.

2.3.5 Seagrass beds

Seagrass beds are of unquestionable value not only to tropical, but also to temperate coastal systems. In order to be able to predict what will be the impact of the predicted sea-level rise on seagrass beds, baseline information about the particular seagrass bed in question is necessary. Furthermore, we need first to know the type of seagrass species in the meadow, since different species have different tolerances to physical factors such as depth and light. For example, shoal grass (*Halodule wrightii*) is a colonizing species which tolerates extreme physical stress, but sea vines (e.g., *Halophila decipiens*) is a stenotopic species if compared to *Halodule*. If the landward extension of the shore resulting from the increase in sea level includes unconsolidated aerobic substrates, then, one would expect an extension of *Halodule* beds into the new available spatial resource, assuming that enough propagules (seeds or turions) are present in the area. However, other factors, such as the wave energy of the locality, the light regime, the herbivore populations, the slope of the shoreline and in general, the history of the site, may make possible the establishment of a different seagrass species or may prevent the establishment of any seagrass species.

Although there are only six common species of seagrasses in the region (*Thalassia testudinum, Syringodium filiforme, Halophila decipiens, H. engelmani, H. baillonis, Halodule wrightii*) the seagrass species within a bed and the relative proportion of seagrass species in a given locality may be highly variable. For example, within Jobos Bay, on the southeast coast of Puerto Rico, seagrass beds within just 100 m from each other have different species and different proportions of macrofloral species. The same was found to be true of the epifaunal and infaunal invertebrates and fish associated with these seagrass beds as demonstrated by similarity indexes and polar ordination techniques. Therefore, to make predictions on what would be the impact of a sea-level rise from a seagrass-bed community structure point of view, is at present difficult. Defining a list of the specific parameters that are to be evaluated may be fruitful.

The effect of a sea-level rise on the physiology (growth rates, primary productivity, photosynthesis and respiration) of the marine phanerogams will also be complex unless we know the annual fluctuations of these parameters and how the light conditions of a particular bed, in association with other parameters, might be affected. If the sea-level rise causes a decrease in light penetration, the maximum depth of turtle grass beds could be decreased. However, this may be compensated for by an expansion of the seagrass bed into the new shore, assuming that there are adequate conditions on the new shore. Predictions related to the vertical extension of seagrass beds as a consequence of sea-level rise, however, may be complicated by the influence of certain herbivores (e.g., *Diadema antillarum*) which can also, like light, limit the distribution of turtle grass beds (Vicente and Rivera, 1982). Better predictions of the effect of a sea-level rise on a seagrass bed can be made using seagrass beds that have been monitored for a considerable length of time.

There have been interesting cases in which apparently, mangroves have not been able to keep pace with mean sea-level rise during Holocene times, and seagrass beds have established over drowned mangroves in Florida. Preliminary observations in Guayanilla Bay (south coast of Puerto Rico) may suggest a similar situation.

Only seagrass beds that occur in natural thermal-stress conditions (e.g., a shallow lagoon in the summer) or in localities receiving thermal effluents from power plants (e.g., *Thalassia* beds in Puerto Rico and Florida), could become negatively affected by a rise of 1.5°C. As discussed before, a temperature rise of this magnitude would not impair physiological processes in marine monocotyledons. Changes in biomass and productivity might be expected to change, however, since these processes are, to some degree, temperature dependent. Temperatures of 35°C or more can, for example, prevent root development in *Thalassia testudinum*.

2.4 Fisheries

In view of the widespread use of coastal tropical systems (e.g., seagrass beds, mangrove-root environments, estuaries) as breeding, mating, feeding and habitat grounds by: (1) commercially exploited species (e.g., decapod crustaceans, molluscs, shrimps); (2) species of ecological importance; and (3) endangered species (West Indian manatee, green turtle) makes these highly productive systems one of the most important resources in the region's waters. Furthermore, shallow coastal estuaries, lagoons, and embayments are utilized even by deep-sea fishes (e.g., Trichiuridae) to complete their biological cycle. The impact of a sea-level rise, and an increment in temperature on these systems will be reflected on both the artisanal and industrial fisheries of the region.

2.5 Sandy Beach Communities:

Sandy beach communities are adapted to withstand adverse environmental conditions. They can probably adapt to an increase in sea level (assuming that the integrity of the system is not lost) better than other benthic systems. Contrary to seagrasses, corals or sponges which are clonal organisms, the dominant structural components of sandy beach communities are aclonal, and furthermore are capable of limited but efficient vertical and horizontal migration to adjust to changes in depth resulting from changes in the beach profile. Now, whether the *Ocypode, Cirolana, Donax-Emerita* and *Mellita* zones (Gonzalez-Liboy, 1971) would persist after a sea-level rise will largely depend on the thickness of the sand layer, since a sandy beach can easily and quickly become transformed to a rocky coast by coastal-erosive processes. For example, Bruun's Rule states that a 1 cm rise in sea level will generally result in a 1 m shoreline retreat. Most developed beaches are narrow and therefore, with a rise of sea level of 20 cm most beaches will be inundated (i.e., no beach). Sea level together with sand extraction (e.g., Isabela, north coast of Puerto Rico), will pose a serious hazard to beaches.

3 Recommendations

1) Prepare a regional map with a classification scheme that shows regions that are more prone to become affected by the climatic changes expected as well as those less likely to become affected.

2) Prepare maps showing:
   a) those ecologically critical habitats (e.g., seagrass beds, coral reefs, mangroves, lagoons) within the region that are most likely to become affected with the sea level and temperature rise expected;
   b) supralittoral zones within each locality that will contain the following information: i) edaphic factors; ii) slope; iii) wave energy; iv) shoreline extension.
3) Analyse historical photographic records, or other historical information that may provide a clue to understanding the effect of recent sea-level rises at specific sites.
4) Prepare a detailed report on how species that are endangered and/or protected internationally, which are associated to the systems mentioned above, may become affected in any way by the proposed climatic change, particularly those that utilize the seagrass beds and/or coral reefs as feeding grounds.
5) Establish the potential impacts on environmental health including vector proliferation, waste disposal, water-quality management and toxic chemicals (including natural toxins) and their human health effects.
6) Establish the indirect impacts that the expected climatic changes will have on shallow coastal marine systems by their effect on: sewage and toxic-waste disposal and on other similar activities.
7) Establish a monitoring programme for determining saline intrusion rates into agricultural resources and drinking-water resources.
8) Prepare a report that specifically deals with the potential effect of the expected climatic changes on invertebrate populations (oysters, conchs, shrimps, crabs, lobsters) and fish populations of commercial value within the region.
9) Implement a plan of action to reforest unconsolidated shores that have a high risk to the expected climatic changes. This action should resist the erosional impact of sea-level rise.

4 References

Allison, I.S. and Palmer, D.F. (eds), 1980a: Work of streams. In: *Geology, The Science of a Changing Earth*. McGraw-Hill, Inc., pp. 247–293.

Allison, I.S. and Palmer, D.F. (eds), 1980b: Rock deformations and rock structures. In: *Geology, The Science of a Changing Earth*. McGraw-Hill, Inc., pp. 385–410.

Atwood, D.K., Hendee, J.C. and Mendez, A., 1992: An assessment of global warming stress on Caribbean coral reef ecosystems. *Bull. Mar. Sci.*, **5**(11), 115–130.

Bak, R. 1983: Aspects of community organization in Caribbean stony corals (Schleractinia). In: Ogden, J. and Gladfelter, E. (eds), *Coral Reefs, Seagrass Beds, and Mangroves: Their Interaction in the Coastal Zones of the Caribbean*. UNESCO Reports in Marine Science No. 23, Montevideo, Uruguay, pp. 51–68.

Barnett, T.P. 1983: Recent changes in sea level and their possible causes. *Climate Change*, **5**, 15–38.

Cintron, G. and Schaeffer-Novelli, Y., 1983: Mangrove forests: ecology and response to natural and man-induced stressors. In: Ogden, J. and Gladfelter, E. (eds), *Coral Reefs, Seagrass Beds, and Mangroves: Their Interaction in the Coastal Zones of the Caribbean*. UNESCO Reports in Marine Science No. 23, Montevideo, Uruguay, pp. 98–113.

Cubit, J.D. 1985: Possible effects of recent changes in sea level in the biota of the Caribbean reef flat and predicted effects of rising sea levels. Proceedings of the Fifth International Coral Reef Symposium, Tahiti, 1–8.
Digerfelt, G. and Hendry, M.D., 1987: An 8000 year Holocene sea-level record from Jamaica: Implications for interpretation of the Caribbean reef and coastal history. *Coral Reefs*, **5**, 165–169.
Gable, F. 1988: Changing Climate. *Oceanus*, **30(4)**, 53–56.
Gonzalez-Liboy, J.A., 1971. The zonation and distribution of marine beach macroinvertebrates at Playa Mani, Mayaguez, Puerto Rico. MS Thesis, Department of Marine Sciences, University of Puerto Rico, Mayaguez.
Goodwin, M., Moore, E., Nunn, T., Simpson, K. and Towle, E., 1984: An Assessment of the Mariculture Potential of Indigenous Eastern Caribbean Brine Shrimp. Report to US Agency for International Development, Grant No. DPE-5542-G SS-3054-00. Island Resources Foundation, St. Thomas, USVI, 115 pp.
Gornitz, V., Lebedeff, S. and Hansen, J. 1982: Global sea-level trend in the past century. *Science*, **215**, 1611–1614.
Hoffman, J.S., 1984: Estimates of future sea-level rise. In: Barth, M.C. and Titus, J.G. (eds), *Greenhouse effect and sea-level rise*. Van Nostrand Reinhold Co., Inc., New York, pp. 79–104.
Horsfield, W.T., 1975: Quartenary vertical movements in the Greater Antilles. *Geol. Soc. Amer. Bull.*, **86**, 933–938.
Kerr, R.A., 1988: Is the greenhouse here? *Science*, **239**, 559–561.
Lisitzin, E., 1963: Mean sea level. *Oceanog. Mar. Biol. Ann. Rev.*, **1**, 27–45.
Moore, B. and Bolin, B., 1987: The oceans, carbon dioxide and global climatic change. *Oceanus*, **29(4)**, 9–15.
Morgan, J.P., 1967: Depositional processes in deltaic environment. *Bull. Amer. Assoc. Pet. Geol.*, **50(3)**, 475.
Morner, N.A., 1976: Eustasy and geoid changes. *J. Geol.*, **84**, 123–151.
Ogden, J. and Gladfelter, E. (eds), 1983: *Coral Reefs, Seagrass Beds, and Mangroves: Their Interaction in the Coastal Zones of the Caribbean*. UNESCO Reports in Marine Science No. 23, Montevideo, Uruguay, 186 pp.
Pirazzoli, A. 1985: Sea level change. *Nature and Resources*, **XXI(4)**, 2–9.
Revelle, R.R. 1983: Probable future changes in sea level resulting from increased atmospheric carbon dioxide. In: *Changing Climate*. National Academy Press, Washington, DC, pp. 443–448.
Rind, D., 1984: Global climate in the 21st century. *Ambio*, **13(3)**, 148–151.
Ross, F., 1982: The Distribution and Abundance of Larval and Juvenile Reef Fishes in Jamaica. University of the West Indies Zoology Department, Research Report No. 6, Jamaica.
Russell, R.J., 1967: Origin of estuaries. *Science*, **83**, 93–99.
Salm R. and Clark, J., 1984: Marine and Coastal Protected Areas: A Guide for Planners and Managers. IUCN, Gland, Switzerland, 302 pp.
Scatterday, J.W., 1977: Low water emergence of Caribbean reefs and effect of exposure on coral diversity – observations off Bonaire, Netherland Antilles. *Studies in Geology*, **4**, 155–169.
Shlemon, R.J. and Capacete, J.L., 1976: Application of Holocene geological data for siting coastal nuclear power plants: An example from Puerto Rico. *Bull. Int. Assoc. Eng. Geo.*, **14**, 107–111.
Stienstra, P., 1983: Quaternary sea-level fluctuations on the Netherland Antilles – possible correlations between a newly composed sea-level curve and local features. *Mar. Geol.*, **52**, 27–37.
Stoddart, D.R., 1976: Continuity and crisis in the reef community. *Micronesica*, **12(1)**, 1–9.
Titus, J.G. and Barth, M.C., 1984: An overview of the causes and effects of sea-level rise. In: Barth, M.C. and Titus, J.G. (eds), *Greenhouse effect and sea-level rise*. Van

Nostrand Reinhold Co., Inc. New York, pp. 1–56.
UNEP/CEPAL, 1980: Overview on natural resources for food and agriculture in the Wider Caribbean Region, FAO Report. UNEP/CEPAL WG.48/INF.11.
Vicente, V.P., Aguilu J.A. and Rivera, J.A., 1980: *Thalassia* as a food source: importance and potential in the marine and terrestrial environment. *J. Agriculture (UPR)*, **64(1)**, 107–120.
Vicente, V.P. and Rivera, J.A., 1982: The depth limits of the seagrass *Thalassia testudinum* (Konig) in Jobos and Guayanilla Bays, Puerto Rico. *Caribbean J. Sci.*, **17(1–4)**, 79–89.
Vicente-Chandler, J., 1984. A new agricultural plan for Puerto Rico for the year 2000. *Technical Report*, Agricultural Experimental Station, Univ. Puerto Rico, 309 pp.
Wanless, H.R., 1982: Sea level is rising – so what? *J. Sed. Petrol.*, **52**, 1051–1054.
Williams, E.H., Goenaga, C. and Vicente, V.P., 1987: Mass bleachings of Atlantic reef corals. *Science*, **238**, 877–878.
Zieman, J., 1983: Food webs in tropical seagrass systems. In: Ogden, J. and Gladfelter, E. (eds), *Coral Reefs, Seagrass Beds, and Mangroves: Their Interaction in the Coastal Zones of the Caribbean*. UNESCO Reports in Marine Science No. 23, Montevideo, Uruguay, 186 pp.

12

Impact on Mangroves

Samuel C. Snedaker[1]

Abstract

Within the region, there are approximately 3,230,000 hectares of coastal shoreline dominated by mangrove vegetation which represents some 15% of the world inventory of mangroves. Unlike some parts of Asia, the mangroves of the region are not utilized in a sustainable manner although there are a variety of local uses, such as for timber, fuel and charcoal. In less populated areas, mangrove vegetation persists in a relatively undisturbed state. In populated areas, however, the habitat is used for the disposal of wastes, cleared for development projects, or exploited for other purposes, such as shrimp mariculture, all of which are incompatible with the sustainability of nearshore fisheries and environmental quality. In the context of global change, mangroves are more likely to be affected by changes in regional precipitation rather than by rising temperature and sea level. Specifically, mangrove areas that receive substantial precipitation and fresh-water runoff are likely to persist, whereas mangrove areas exposed to full-strength seawater may be overstepped and lost. Because of the importance of intertidal mangroves in shoreline protection, fisheries support and water quality, efforts should be taken by the appropriate authorities and organizations to curb abuses and protect the resource for both ecological and economic purposes.

1 Introduction

Although salt-tolerant mangroves form the dominant shoreline vegetation within the region, the majority of the region-specific data and information on their distribution and current status is sketchy and anecdotal. Any of the data and information obtained for this report are derived from the experience of the author in the region, from a variety of grey literature sources, and from communications with a number of regional correspondents. The resulting report focuses on the characteristics and distribution of mangroves within the region, their current status, and their ability, or not, to cope with and survive global change. The latter includes comments on certain man-related activities that take place in mangrove habitats.

2 The Mangroves of the Region

The ecological grouping of the halophytic spermatophytes known as mangroves occurs throughout the region, and includes *Avicennia germinans* L. Stearn, *Conocarpus erectus* L., *Laguncularia racemosa* L., *Gaertn.* f., *Pelliciera rhizophorae* Triana and Planchon, and *Rhizophora mangle* L. These species

[1]University of Miami/RSMAS, Miami, FL 33149, USA

Fig. 12.1 The structural complexity and development of mangrove forests are greatest in coastal areas that receive fresh-water runoff from inland catchments. The input of nutrients in the runoff as well as the reduction in salinity result in high rates of primary productivity. See colour plates between pages 210 and 211.

are ubiquitous throughout the Gulf of Mexico and Caribbean Sea, except for *P. rhizophorae* which appears to be restricted to the coasts of Colombia (Calderon, 1983, 1984) and Costa Rica (Jimenez, 1984). Other Western Hemisphere mangrove species (*A. bicolor Standl., A. shaueriana Stapf., R. harrisonii* Leechman, and *R. racemosa* G.F.W. Meyer), however, have not been verifiably reported as occurring in the region.

Whereas mangroves are adapted to saline anaerobic sediments and are commonly found along protected intertidal shorelines within the tropical latitudes (Chapman, 1976), their maximum development occurs in areas of high precipitation or fresh-water runoff from inland catchments (Pool, *et al.*, 1975; see Fig. 12.1). This phenomenon is attributed to the influence of fresh water in maintaining low-salinity regimes, and the delivery of the products of bedrock and soil weathering (Lugo and Snedaker, 1974). For this general reason, some of the largest areas of mangroves forests are associated with large river drainages (e.g., the Orinoco in Venezuela and the Grijalva-Usumacinta in Mexico), high-rainfall environments (e.g., the Boca del Toros region of Panama and parts of Cuba) and areas that receive substantial sheetflow runoff (e.g., the Atlantic coast of Nicaragua and south Florida).

Elsewhere, for example, on many of the islands, mangrove-forest development is limited by aridity, hypersalinity, and the absence of significant areas of protected intertidal shorelines. In these restrictive settings, mangrove vegetation typically consists of a relatively narrow fringe of short- to moderate-sized trees dominated almost exclusively by *R. mangle* in frost-free regions. In more temperate latitudes, (e.g., around the northern coastline of the Gulf of Mexico) winter frosts and freezes allow the existence of only *A. germinans* which has the capacity to regenerate following severe freeze damage (*cf.* Lugo and Patterson-Zucca, 1977).

3 Distribution of Mangrove Vegetation

There are a paucity of data concerning the area of the region dominated by mangrove vegetation. However, based on a variety of published and unpublished reports, and personal communications with knowledgeable individuals within the region, a partial land-area inventory has been assembled (Table 12.1); note that the coverage is incomplete, and that the data are heavily caveated. The tabular total of 3,230,000 hectares represents some 15% of a conservatively estimated total world area of mangroves of 22 million hectares (Fig. 12.2).

Table 12.1 Mangrove forest area*.

Geographic region	*Mangrove forest area (hectares)*	*Notes*
Caribbean		
Bahama Islands	233,200	(1)
Andros Island	155,500	(1)
Grand Bahama Island	51,800	(1)
Inagua Island	26,000	(1)

Barbados	12	(2)
Graeme Hall Swamp, Christ Church	8	(2)
Chancery Lane Swamp, Christ Church	<1	(2)
Cayman Islands	11,655	(3)
Grand Cayman Island	10,878	(3)
Cayman Brac Island	100	(3)
Little Cayman Island	677	(3)
Cuba	626,000	(4)
North coast	131,000	(4)
North coast islands and archipelagos	114,000	(4)
South coast	318,000	(4)
South coast islands	38,000	(4)
South coast archipelagos and Isla de Piños	25,000	(4)
Dominican Republic	23,500	(5)
Rio Yuma	6500	(6)
Bahia de San Lorenzo	2100	(6)
Lake Enriquillo	1600	(6)
remaining area	13,300	(6)
Grand Terre	4320	(7)
Guadeloupe	5700	(8)
Haiti	18,000	(9)
Jamaica	20,200	(10)
Black River	7300	(10)
Negru (Negril)	2000	(10)
Martinique	2,200	(11)
Fort de France Parish	200	(11)
Lamentin Parish	500	(11)
Ducos Parish	300	(11)
Riviere Salee	400	(11)
South Martinique (small parcels)	800	(11)
Monserrat	7	(12)
St. Anthony Parish	6	(12)
St. Georges Parish	1	(12)
St. Peter Parish	<1	(12)
Netherlands Antilles	1500	(13)
Aruba	100	(13)
Bonaire	1000	(13)
Curacao	300	(13)
St. Martin	100	(13)
Puerto Rico	6497	(14)
North central coast	475	(14)
Northeast coast	2021	(14)
East coast	1285	(14)
South central coast	937	(14)
Southwest coast	988	(14)
West coast	207	(14)
Northwest coast	48	(14)
Metropolitan San Juan	274	(14)
Culebra Island	26	(14)

Vieques Island	227	(14)
Mona Island	1	(14)
St. Kitts	20	(15)
Trinidad-Tobago	9000	(16)
Northwest	6000	(16)
Caroni Swamp	3500	(16)
Northeast	1000	(16)
Southwest	1500	(17)
Southeast	400	(16)
Tobago	100	(16)
Virgin Islands	310	(18)
Central America		
Belize	75,000	(19)
Costa Rica	35,000	(20)
Caribbean coast	400	(12)
Guatemala	16,000	(22)
Caribbean coast	8500	(23)
Honduras	145,000	(24)
Pacific coast, Bahia de Fonseca	28,000	(25)
El Salvador border to Rio Nacaome	7500	(25)
Rio Nacaome to San Lorenzo outlet	8000	(25)
Rio San Lorenzo to Rio Choluteca	6000	(25)
Rio Choluteca to Nicaragua border	6500	(25)
Caribbean coast	117,000	(26)
Nicaragua	60,000	(27)
Caribbean coast	25,000	(28)
Panama	297,532	(29)
Caribbean coast		
Bocas del Toro Province	64,010	(30)
Pacific coast		
Cocle	25,125	(30)
Chiriqui	66,645	(30)
Darien	28,225	(30)
Herrera	8450	(30)
Los Santos	8800	(30)
Panama	122,925	(30)
North America		
United States	280,594	(31)
Alabama	25	(32)
California	150	(33)
Florida	274,857	(34)
East coast	47,370	(35)
Biscayne Bay	7877	(36)
West coast	15,917	(37)
Florida Bay	14,938	(37)
Whitewater Bay	30,760	(37)
Charlotte Harbor	9504	(37)
Tampa/Hillsborough Bay	7091	(38)
Louisiana	2956	(39)
Mississippi	250	(40)

Texas	2506	(41)
Aransas County	12	(41)
Calhoun County	1500	(41)
Cameron County	400	(41)
Kenedy County	8	(41)
Kleberg County	8	(41)
Nueces County	570	(41)
Willacy County	8	(41)
Mexico	1,420,200	(42)
Caribbean coast	700,000	(43)
Laguna de Mecoacan (Tabasco)	4000	(44)
Laguna de Terminos (Campeche)	250,000	(45)
South America		
Colombia	501,300	(46)
Caribbean coast	50,000	(46)
	73,975	(47)
Islas del Rosario	297	(48)
Canal del Dique Delta	12,000	(49)
Bahia de Cartagena	30	(50)
Cienaga Grande - Magdalena Delta	46,000	(51)
French Guiana	55,000	(52)
Guyana	80,000	(53)
County of Berbice	30,000	(53)
County of Demerara	10,000	(53)
County of Essequibo	40,000	(53)
Venezuela	673,569	(54)
Western region	15,468	(54)
Central-western region	15,616	(54)
Central region	6608	(54)
Central-eastern region	138,377	(54)
Orinoco Delta	495,200	(54)
Margarita island	2300	(54)
Surinam	115,000	(55)

*See Appendix, pp 295–299, for notes and references.

4 Major Regional Problems

To a large extent, the types of problems affecting the mangrove resources are not unique from a global perspective. However, because both island land areas and mangrove habitats are limited, present and potential impacts tend to be accentuated. Two major groupings of problems affecting mangroves are identified and discussed below in sections 4.1 and 4.2.

4.1 Economic Exploitation and Direct Conversion to Other Uses

4.1.1 Non-sustainable logging

In part due to the rising global concern over the continuing loss of tropical rain forests (*cf.* Repetto, 1990), high-volume mangrove forests are increasingly being viewed as alternate wood sources, particularly for wood chips used in the pulp and paper industry, and as a source of cellulose. Up

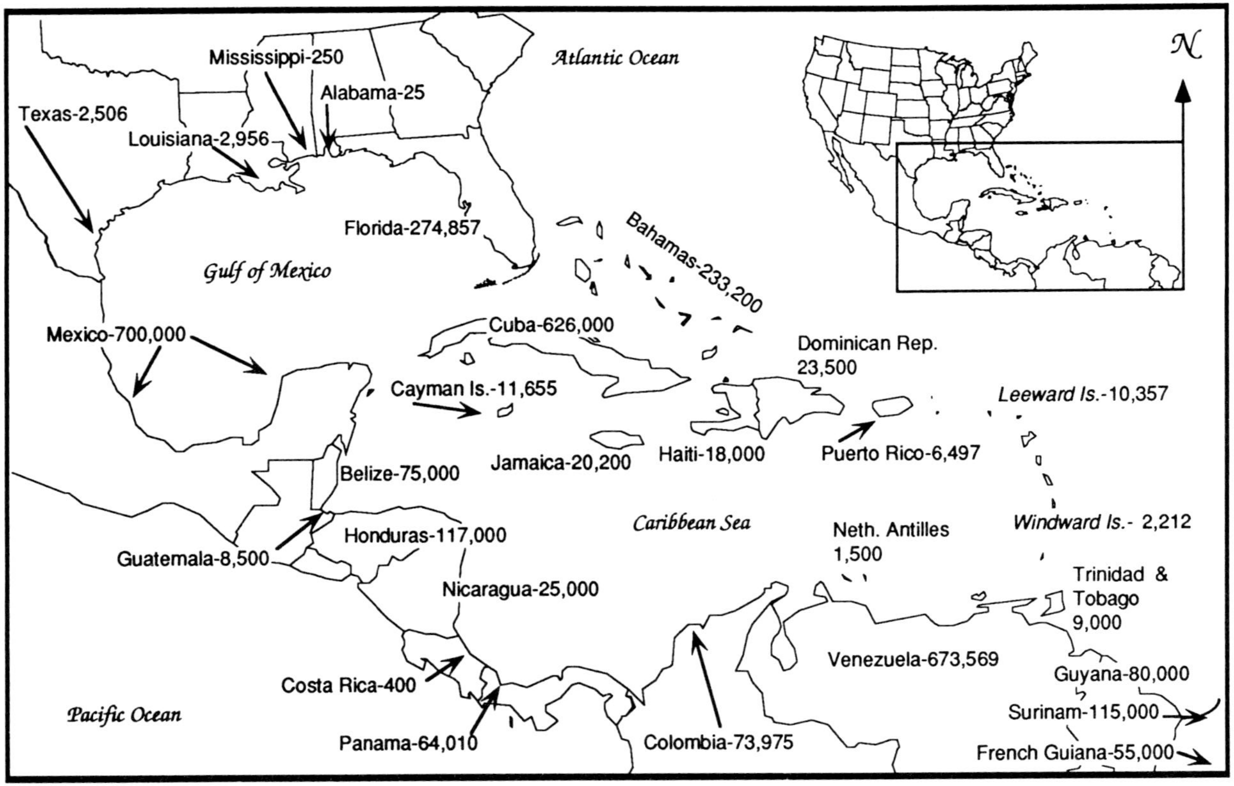

Fig.12.2 Estimates of mangrove forest area (hectares). For the Leeward Island group, estimates are available for Guadeloupe (5700), Grand Terre (4320), Virgin Islands (310), St. Kitts (20) and Montserrat (7). For the Windward Islands, estimates are available for Martinique (2200) and Barbados (12). For Mexico, and the Central and South American countries with mangrove forest on both oceanic coasts, estimates are reported for the Caribbean coast only. Refer to text and Table 12.1 for details on origin or derivation of estimates.

to now there has been no major clear-felling of mangrove forests except in two areas: (1) In Venezuela, supposedly selective extraction of tall *R. mangle* occurs in the Orinoco Delta for use as utility poles elsewhere in Venezuela (Hamilton and Snedaker, 1984). It is reported that the decision to exploit this resource reportedly was based on the quick economic gain from what had been considered a worthless forest (Saenger *et al.*, 1983). (2) In Colombia, a large *R. mangle* forest in the Canal del Dique Delta was clear-felled by a pulp and paper company. Later, after the concession was completed, the original mangrove species did not recolonize due to the invasion of *Acrostichum* (a salt-tolerant fern). However, in adjacent areas, some regrowth of *L. racemosa* has been observed (Araujo and Polania, 1985).

With the exception of some possible high-volume areas in Cuba, Mexico and Nicaragua, the only other identified potential site for potential clear-felling is in the Boca del Toros province of Panama. That area has received attention because, in addition to the mangrove species, there is also a large contiguous forest dominated by cativo (*Prioria copaifera*) and orey (*Campnosperma panamensis*).

At the present, the harvesting of mangrove wood within the region is pursued mainly for small-scale domestic and industrial use for charcoal, firewood, lumber, poles and posts. Woodcutting is a widespread practice in the mangrove forests along the Caribbean Sea coast of Colombia. In some areas (e.g., Cienaga de La Caimanera) the extraction rate is lower than the natural regrowth of trees, and thus may be considered to be a sustained-yield practice. In that situation *R. mangle* poles and sawtimber are harvested and transported to local markets in Cartagena (Laverde *et al.*, 1987). However, in most other areas of Colombia (e.g., Bahia de Cartagena, Golfo de Morrosquillo, Canal del Dique, Bahia de Barbacoas), the extraction rates exceed the annual regrowth, and the forests are being rapidly depleted and degraded.

4.1.2 Mariculture

The continuing high world demand for seafood and the apparent success of the shrimp mariculture industry in Ecuador have encouraged both governments and private entrepreneurs to pursue coastal mariculture in mangrove areas. In spite of the potential for hard-currency earnings and employment opportunities, shrimp mariculture in mangrove areas is a high-risk venture, and the incidence of failure is remarkably high due mainly to inadequate investment financing. As a result, a low-risk approach is frequently taken based on exploiting low yields in very large pond areas, as opposed to intensively managing high-yield shrimp crops in small pond areas (Snedaker *et al.*, 1986). Notwithstanding the problems and frequent failures, a bilateral US lending agency, for example, is encouraging the conversion of mangrove areas in Honduras to shrimp grow-out ponds by citing the Ecuadorean 'miracle' (Enrique Lahmann, pers. comm., IUCN/CATIE, Turrialba, Costa Rica).

4.1.3 Coastal development

A common threat to mangroves in some more-populated parts of the region is the conversion of mangrove and related coastal habitats to developments serving primarily the tourist industry, but also for upscale residential sales. In addition to the direct loss of mangroves, coastal development represents a non-point source of pollution that affects water quality and other coastal habitats, such as coral reefs, which depend on superior water quality. The principal sources of pollution result from solid-waste dumps, marinas, golf courses and the improper handling and treatment of domestic waste-water treatment (Snedaker, 1990). In spite of laws protecting mangroves, coastal development continues to be a problem around the Gulf coast of the US and in the Virgin Islands and Puerto Rico as specific examples.

In the State of Quintana Roo, Mexico, coastal development proponents call foreign tourism the 'smokeless industry', and have undertaken large coastal development projects which, like the development at Puerta Aventuras, have not been economically successful. In an attempt to control questionable coastal development practices in Quintana Roo, the Fideicomiso Caleta de Xel-ha y del Caribe has mapped all coastal habitats and identified specific areas to be protected versus those that could be developed minimally or intensively (Jorge Lopez Portillo, pers. comm., Instituto de Ecologia, Mexico). Although the government has accepted the plan in principal, strong opposing political opposition and economic pressure suggest that this first attempt at a coastal management plan for Quintana Roo may not be successful (Jorge Lopez Portillo, *op. cit.*).

4.2 Pollution

In 1989, the Caribbean Islands Directorate of the US Man and Biosphere Program sponsored a workshop on land-based sources of pollution and published a consensus statement on the major point and non-point sources. Because mangroves represent the interface between land and ocean, they frequently are exposed to more kinds of pollutants from both land and water than even most fresh-water wetland habitats. In addition to the ocean disposal of sewage and stormwater runoff, which the workshop identified as the most widespread source, other significant point sources included: on-shore refineries and petrochemical plants; sugar factories and rum distilleries; breweries, soft-drink plants and canneries; abattoirs and meat canneries; tanneries; metal and electroplating plants; textile dying industries; edible-oil production plants; cooling and scale-removal activities at powerplants; and banana washing and packing activities. The same workshop concluded that the major non-point sources of marine pollution included: agriculture and forestry; construction works; urban runoff; atmospheric fall-out; groundwater seepage; oil and other chemical spills and disposal; solid-waste disposal and its leachates; subsurface disposal of sewage and other wastes; and mining operations.

4.2.1 Sewage and stormwater runoff

With few exceptions, suitable sewage collection and treatment systems are lacking throughout the region, including parts of the United States and in such popular tourist destinations as Cancun, Mexico. In Cancun, for example, sewage is collected and partially treated, but then discharged into the adjacent mangrove-bordered lagoon (S. Campos, pers. comm., Grupo Ecologista del Mayab a.c., Cancun). Together with uncontrolled stormwater runoff in the same lagoon system, it has now become one of the most polluted mangrove lagoons in the western Caribbean Sea.

Mangrove productivity can be stimulated significantly by exposure to nutrient-rich sewage and the low salinity of the wastewater. As a case in point, one of the more luxuriant stands of mangroves in semi-arid Curaçao is located in a small bay that receives partially treated domestic-waste effluent. Whereas the trees can take up and concentrate potentially toxic materials commonly found in sanitary sewage and domestic liquid wastes, heavy metals, for example, can accumulate in leaves at levels several times above those found in the water-sediment environment (Mathis, 1973; Lindberg and Harriss, 1974; Snedaker and Brown, 1981). Because leaf detritus forms a basis for many nearshore marine foodwebs, this enrichment mechanism represents a possible direct input of pollutants into seafood consumed by man.

Largely because mangroves still carry the perception of a 'wasteland', the use of mangrove habitats as sites for unregulated dumping and solid-waste landfills is a common practice throughout the region, particularly where other options are limited. This problem also occurs in south Florida where it is otherwise illegal to even horticulturally prune mangrove branches. In addition to the permanent loss of habitat, leachates from solid-waste landfill accumulations tend to be highly enriched in toxic ammonia, and may contain a variety of other toxic materials that appear in surface runoff or that enter the groundwater, ultimately affecting the local marine life (Geohegan *et al.*, 1984). Also where there are limited alternatives, or where proper waste-disposal enforcement is lacking, landfills including those in mangrove areas are frequently used for the improper disposal of toxic industrial chemicals and infectious medical wastes. As discussed above, the ability of mangroves to take up and accumulate certain types of pollutants in leaf tissues suggests that the recycling of pollutants into nearshore fisheries may be significant problem in areas of waste dumps.

According to Towle (1982), island systems tend to have lower tolerances to excessive and recurring waste discharges. However, Saenger *et al.* (1983) note that solid-waste accumulations have increased steadily in recent years around coastal urban areas, tourist resort destinations and industrial sites. In Puerto Rico, open dump sites have been sited in mangrove areas where periodic floods and winds spread over and disperse the wastes. Solid-waste dumps also have subsided into the soft underlying substrate, and have aggravated flooding in nearby areas by acting as dikes. Spontaneous combustion and subterranean fires also occur at dump sites, particularly in dry environments (Saenger *et al.*, 1983). In urban areas, mangrove dump-site fires have sparked public controversy at the Munisport toxic dump site in north Miami in early 1990.

Elsewhere, for example, solid wastes in Venezuela are dumped at the edge of the Rio Limon mangroves, north of Maracaibo, and in many other areas along the coast (Taylor, 1988). Solid-waste disposal in uncovered dumps is also recognized as the source of environmental impacts around urbanizing areas in the Dominican Republic. In spite of the Dominican's emphasis on the development of a tourism industry, no effective mechanisms exist to control waste disposal and the ensuing environmental problems (Hartshorn, 1981). In this regard, Port of Spain, Trinidad, may be a classic example. There, a solid-waste dump and sewage-holding ponds have been constructed at the edge of the mangrove area forming the Caroni National Park (Taylor, 1988) which is visited by large numbers of local and international tourists.

A principal concern over the use of low-lying coastal areas for waste dumps is that different aspects of global change can lead to a variety of problems. For example, with increased aridity the probability of spontaneous combustion increases. Conversely, with increasing sea level and/or increased precipitation, the dispersal of solids and leaching would be accelerated. Although there are environmentally suitable techniques available for the 'management' of waste dumps (e.g., subsurface impermeable liners, surface capping, venting of volatile gases, collection and treatment of leachates, etc.), their use is extremely rare.

5 Mangroves and Sea-Level Rise

The current conventional wisdom consists of at least two primary scenarios as to how coastal mangrove wetlands might respond to sea-level rise. These are based mainly on sea-level-rise-induced changing salinity gradients with no major consideration given to other climatic-change phenomena such as warming, shifts in regional precipitation patterns and alterations in the frequency or intensity of cyclonic weather disturbances (e.g., tropical storms, hurricanes, etc.). Temperature alone can be expected to have a minimal effect except for northern range extensions around the Gulf of Mexico and Atlantic coast of the US. Precipitation, however, in terms of its influence on the intertidal seawater/fresh-water balance, is likely to have much greater importance than a simple rise in the open water elevation. Accordingly, a third scenario is proposed that incorporates precipitation and catchment area fresh-water runoff.

5.1 Mangroves Responses to Relative Slow Sea-Level Rise

The most parsimonious of the two scenarios is that mangroves would retreat progressively inland along a sea-level-rise-induced changing salinity gradient, assuming the absence of any inland barrier. Evidence for salinity-induced retreat consists mainly of reports of mangroves that have recently colonized along gradients of increasing salinity. For example, Egler (1952) discussed the historical encroachment of mangroves into fresh-water wetlands and hammocks in southeast Florida following the construction of a railroad that blocked fresh-water runoff. As the affected area became increasingly saline, the glycophytic vegetation gave way to halophytes dominated by the mangrove species. The same general phenomenon has been described elsewhere in southeast Florida, and attributed to a lowering of the water table through drainage along with related actions that promoted inland salinization (*cf.* Reark, 1975). As a conceptual model, this first scenario might apply to relatively slow rates of sea-level rise affecting low-elevation coastal plains that grade into fresh-water habitats.

With accelerated sea-level rise, the second major scenario, two other co-related factors would assume increasing importance, specifically the ability of mangrove propagules to take root and become established in intertidal areas subjected to a higher mean sea level, and the rate of sedimentation relative to the rate of sea-level rise. With respect to propagules, the research of Rabinowitz (1978a,b,c) and Jimenez (1988) in the region documents the differential survivability and establishment of propagules under varying flood levels. In general, the large propagule species (e.g., *Rhizophora* spp.; Fig. 12.3) can become established in significantly deeper water than can the smaller propagule species (e.g., *Avicennia* spp. and *Laguncularia* spp.). Based simply on propagule size, three variants of the scenario can be defined: (1) the larger propagule species become increasingly dominant in habitats occupied by the smaller propagule species; (2) the smaller propagule species retreat to, and become dominant in, any newly formed saline, shallow intertidal areas; and (3) the large propagule mangrove vegetation at the shoreline eventually disappears as a result of the failure of propagule establishment and forest regeneration. This scenario, however, is highly dependent on the rate of sedimentation. A relatively high rate of sedimentation would lead to a lower apparent rate of sea-level rise, whereas reduced sedimentation would result in a higher apparent rate of rise; based on propagule size, the species responses would vary accordingly.

Based principally on basic mangrove ecology together with supporting evidence from the Holocene stratigraphic record, it would appear that these two scenarios and their variants would largely account for mangrove responses to rising sea level. However, based on the argument below, these scenarios probably only apply to non-peat-forming mangrove forest environments. In those mangrove environments characterized by the production and accumulation of subsurface peat, a much different scenario involving precipitation and surface-water runoff must be invoked. Thus, as argued in the third scenario, any speculative projections of mangrove responses to SLR must necessarily incorporate rainfall and fresh-water runoff as principal factors.

5.1.1 Organic peat responses to changing salinity regimes

Along the Florida east coast, many mangrove forest areas are impounded by dikes and flooded with estuarine water to prevent mosquitos ovipositing on otherwise moist sediments. As long as the flooding level is relatively low (e.g., several centimetres), metabolic gas exchange can proceed normally without deleteriously affecting the mangroves (Lahmann, 1988). In effect, this periodic flooding every year mimics an apparent rise in the mean sea level over a relatively short period of time. The major problem with this practice (Jim David, St. Lucie Mosquito Control District, pers. comm.) is that mangrove productivity and net peat accumulation are accelerated to the point that ground elevations rise above practical flooding levels, thus thwarting effective mosquito control. Also, the responsible peat-building red mangrove (*R. mangle* L.) eventually dominates to the exclusion of the other woody and herbaceous halophytes. Similar vegetation changes have been reported for impounded mangroves in Puerto Rico (Vazquez, 1983).

This well-documented phenomenon in east Florida implies that red-mangrove habitats can easily keep pace with sea-level rise and not be totally overstepped and abandoned. Superficially, at least, it also explains the existing presence of extensive areas of *Rhizophora* peat, for example, around parts of the shoreline of south Florida which are exposed to abundant fresh-water runoff. However, if these observations are indeed

Fig. 12.3 *Rhizophora mangle* is frequently dominant along shorelines that are inundated on all high tides. In areas of moderate salinity, this species is also capable of high rates of root production which accumulates as peat. See colour plates between pages 210 and 211.

correct, then why are large strata or blocks of mangrove peat not recorded in the Holocene stratigraphic record of nearshore sediments on continental shelves? Some authors interpret the discrepancy as evidence that extensive areas of coastal mangrove forests were not present early in the Holocene, and only appeared over extensive areas as the rate of sea-level rise diminished later in the Holocene (*cf.* Ellison, 1990). To explain this discrepancy, it is necessary to examine the processes that result in the balance between mangrove peat production/accumulation and peat decomposition/remineralization, and therefore changes in total peat mass.

5.1.2 *Peat production and accumulation*

Although mangroves are ecologically restricted to saline intertidal environments, mangrove productivity in general increases proportionally with the availability of fresh water principally in the form of terrigenous runoff (Pool *et al.*, 1977). The main causal factors are the reduced salinity stress on the vegetation and availability of mineral nutrients in the surficial runoff. This author has likewise observed in many parts of the Caribbean Sea and elsewhere that the production and accumulation of *Rhizophora* peat also appears to be greatest in areas that receive upland fresh-water drainage during most of the year. To a limited extent, this observation is supported by the work of Lahmann (1988) who linked the low salinity of flooded mangrove impoundments with high mangrove production and the greatest abundance of mangrove peat.

5.1.3 *Peat decomposition and loss*

Rhizophora peat results primarily from root mortality, and the preservation of the organic remains under strongly reducing, or anoxic, conditions. Although anaerobic decomposition and remineralization are continually taking place (as well as aerobic decomposition of surface organic debris), the longer-term rates of loss tend to balance the continual subsurface production of peat-forming roots. Thus, as long as the balance is maintained, the volume and mass of the peat body remain relatively constant. In terms of sea-level rise, as long as there is sufficient fresh-water runoff to maintain the optimum salinity, there should be a proportional net accumulation of peat, and accordingly the mangrove zone would not retreat, be overstepped or abandoned. Conversely, however, if fresh-water runoff ceased or diminished to the point that the mangrove habitat was continually exposed to full or close to full-strength seawater, then organic production would decline. At the same time, the increased availability of sulphate (SO_4, present in seawater at $\sim$2.7 g/kg) to suffuse subsurface peat would necessarily lead to increased anaerobic decomposition by sulphur-reducing microorganisms, and thus a loss of peat mass. Theoretically, the sulphate in 1 litre of seawater is capable of causing the anaerobic decomposition and breakdown of approximately 1.9 g of organic matter.

In this regard, the apparent sea-level rise observed in temperate coastal marshes of the US has been attributed to the biological decomposition of sediment organic matter (Courtney Hackney, pers. comm., University of North Carolina, Wilmington). Although the causal factor (a fungal process) is not necessarily related to the seawater/fresh-water balance, and therefore the presence of SO_4, the rapid rate of subsidence illustrates just how quickly intertidal habitats can be degraded and lost by the loss of sediment organic matter. In addition to the habitat loss, the induced anaerobic conversion of large areas of coastal organic substrates would contribute to the total atmospheric loading of greenhouse gases, notably carbon dioxide (CO_2) and methane (CH_4). (Note, however, that methanogenesis is inhibited in the presence of sulphides such as H_2S which is produced during the sulphate-reduction process.) Although there are no reports, either published or anecdotal, of mangroves being lost to this type of subsidence, the specific spatial pattern of 'browning' and mortality of mangrove areas in southern Florida (Snedaker, unpublished observations) is similar to that described in the temperate marshes.

5.2 Research Initiatives

Because solid-waste dumps are so prevalent throughout the region, and because most are sited within or contiguous with intertidal mangrove habitats, multidisciplinary research should be initiated to determine their contribution to nearshore pollution, both directly and via recycling through mangrove detritus. Furthermore, because the physical characteristics of the underlying substrate have a substantial influence on the quantity and quality of leachates, the research should be undertaken on a comparative basis. For example, one site might be identified in a carbonate/karst environment such as in south Florida, the Bahamas or the Yucatan Peninsula, with a comparative site located in alluvial coastal plains as are found on mountainous islands such as Puerto Rico. The actual selection of study sites would necessarily depend on the local availability of qualified and interested scientists. Local access to qualified analytical laboratories, however, is not required because with proper sample handling and preservation, samples can be air-shipped and analysed at distant locations.

The research protocol should be based on a surrounding network of cased and screened monitoring wells (6-inch [15 cm] bentonite-sealed boreholes) that are suitable for measuring the rate and direction of groundwater flow, and the collection of liquid and gaseous samples for laboratory analyses. Local commercial well drillers can be employed to prepare the boreholes; the well monitoring and sampling equipment are stock items availability from a number of research equipment suppliers. The subsequent analyses should minimally include the heavy metals, organophosphates, chlorinated hydrocarbons and polycyclic aromatic hydrocarbons.

5.2.1 Sea-level rise

Based on the question relating to the spatial stability and survival of mangroves in response to sea-level rise, a combined laboratory-field research project is proposed. The laboratory protocol should have the objective of determining the rate of anaerobic/anoxic decomposition and breakdown of organic peat exposed to fresh water and seawater with varying concentrations of SO_4. The field protocol should have a dual objective. The first would be to determine whether or not the browning and mortality of non-shoreline mangroves is associated with topographic subsidence as a result of the loss of sediment organic mass by sulphate reducers or some other related process (*cf.* Padgett, *et al.*, 1986; Hackney, 1987). The second part would be to construct a field experiment in which a series of impounded mangrove forests were subjected to surficial flooding of waters of varying salinities and sulphate concentrations. Alternatively, existing and operational mosquito-control impoundments could be monitored following flooding to test the hypothesis that whereas salinity would remain relatively stable, there would be a rapid drawdown of SO_4 and a corresponding increase in H_2S.

6 Conclusions

The dominant vegetation of the intertidal zones consists of six species of salt-tolerant mangroves that collectively occupy some 3,230,000 ha. The largest and most productive expanses occur in regions of high rainfall and fresh-water runoff as compared to arid and semi-arid regions where mangroves are largely absent. Although mangroves have a number of documented ecological roles (e.g., coastal protection, fisheries maintenance, water quality, etc.) and a variety of potential economic uses (e.g., timber, fuel, charcoal, etc.), these values are poorly recognized throughout most of the region. As a result, coastal mangrove habitats are: clearcut for their quick-cash timber value, converted for coastal development and shrimp mariculture ponds, used as *ad hoc* sites for solid and liquid waste disposal, and subjected to uncontrolled exploitation for fuel wood and charcoal. Of these, the most insidious abuse is the disposal of polluting wastes, most of which do not significantly harm mangrove trees but which render the water habitat unsuitable for most forms of marine life. As population pressures increase, and land area and natural resources become more limiting, the present trend of mangrove abuse and degradation is certain to accelerate.

With respect to global change, mangrove forests are more likely to be affected by regional changes in precipitation patterns than by the more conventionally perceived consequences of temperature increase and sea-level rise. Because mangroves require substantial quantities of fresh water (via rainfall and surface runoff) to attain their maximum growth potential, a decrease in precipitation would necessarily reduce their productive potential as well as lead to increased exposure to full-strength seawater. In this event, peat substrates would subside as the result of anaerobic decomposition by sulphate-reducing microorganisms, and mangroves would eventually be eliminated at the affected sites.

To curb the current abuse and degradation of mangroves, and to plan for coastal protection in the event of accelerated sea-level rise, three explicit recommendations are made: (1) The appropriate authorities should prohibit intertidal mangrove areas from being utilized as convenient places for the disposal of wastes, particularly waste materials containing potentially toxic or harmful pollutants. (2) Under the leadership of international organizations, such as the Inter-American Development Bank, the Organization of American States, and bilateral donor agencies, research and development efforts should be organized to identify ecologically sustainable economic uses and protocols for mangroves and associated plant and animal life. (3) Because of the importance of fresh water in sustaining high rates of mangrove primary productivity and in the production and accumulation of organic peat, water management authorities should block surface-water drainage canals and divert surplus fresh water discharges into mangrove areas based on a delivery schedule that promotes mangrove productivity.

7 Acknowledgments

The published materials, personal communications, and unpublished records of the author, which formed the basis for this paper, were collated by Rafael Araujo and synthesized by Jane Snedaker who also collaborated with the author in assembling the final manuscript. Additional assistance was obtained from a large number of individuals who are acknowledged in the text.

8 References

Anonymous, 1978: Integrated development project of the eastern region of Panama-Darien. OAS, Washington, DC, 308 pp.

Araujo, R. and Polania, J., 1985: Manglares: estructura, fitosociologia y geomorfologia en el delta del Canal del Dique (Departamentos de Bolivar y Sucre, Colombia). BS Thesis, Univ. Jorge Tadeo Lozano, Bogota, 126 pp.

Beard, J.S., 1946: The natural vegetation of Trinidad. Ox. For. Mem. No. 20.

Birnhak, B.I. and Crowder, J.P., 1974: Evaluation of the extent of vegetative habitat alteration in South Florida, 1943-1970. Appendix to a South Florida ecological study. US Department of the Interior, Bureau of Sport Fish and Wildlife, Atlanta, Georgia.

Bohorquez, C. and Prada, M., 1986: Siembra y transplante de *Rhizophora mangle* L. 1773 y tipologia del manglar del Parque Nacional Natural Corales del Rosario. BS Thesis, Univ. Jorge Tadeo Lozano, Bogota, 123 pp.

Botero, L. (ed.), 1988: Estudio Ecologico de la Cienaga Grande de Santa Marta. Informe final. Inst. de Inv. Marinas de Punta de Betin 'Jose Benito Vives D'Andreis' (INVEMAR), Programa de Lagunas Costers, 47 pp.

Browder, J., Littlejohn C. and Young, D., 1976: South Florida: Seeking a Balance of Man and Nature. Center for Wetlands, Univ. Florida, Gainesville.

Calderon, E. 1983: Hallazgo de *Pelliciera rhizophorae* Triana y Planchon (Theaceae) en la costa del Atlantico, con observaciones taxonomicas y biogeograficas preliminares. *Bol. Museo del Mar*, (UJTL, Bogota), **11**, 100–111.

Calderon, E. 1984: Occurrence of the mangrove *Pelliciera rhizophorae* on the Caribbean coast of Colombia with biogeographical notes. *Bull. Mar. Sci.*, **35(1)**, 105–110.

Carrera, C.J. and Lugo, A.E., 1978: Los sistemas de mangles de Puerto Rico. Programa de la Zona Costanera, Depto. de Recursos Naturales, Puerto Rico.

Chabreck, R.H., 1972: Vegetation, water and soil characteristics of the Louisiana Coastal Region. Bull. No. 664, Louisiana State Univ., Baton Rouge.

Chapman, V.J., 1976: *Mangrove Vegetation*. Strauss and Cramer, Germany, 447 pp.

Cintron, G. and Schaeffer-Novelli, Y., 1983: *Introduccion a la ecologia del manglar*. UNESCO/Rostlac, Montevideo, 109 pp.

Cooper, G.C. and Bacon, P.R. (eds), 1981: *The Natural Resources of Trinidad and Tobago*. Edward Arnold Publishers, London, 223 pp.

De las Salas, G. and von Hildebrand, P., 1980: Esbozo de un plan de investigaciones para el ecosistema de manglar Cienaga Grande – Canal del Dique, Colombia. In: Vegas-Velez M. (ed.)., *Mem. Sem. sobre el estudio cientificao e impacta humano en el ecosistema de manglares*. UNESCO, Montevideo, pp. 387–393.

Donaldson, P.O., Macdonald, K.M., Rankin, J.W. and McKenzie, T.A., 1963: Forests of the Republic of Panama. Resource Development Industrial Development and Product Potential. In: (PIO/T 525-3-29-BG-3-20032. AID Contract No. AID la-32.) US AID and Greenacres Inc., pp. 1–59.

Egler, F.E., 1952: Southeast saline Everglades vegetation, Florida, and its management. *Vegetatio*, **3**, 213–265.

Ellison, J., 1990: Will mangroves survive? *IUCN Bulletin*, **20(10-12)**, 21–22. Greenacres, Inc., Consulting Foresters, Seattle, Washington, USA.

FAO (Food and Agriculture Organization of the United Nations), 1978a: Plan de desarrollo forestal: Estado actual del subsector. FAO/PCT/6/PAN/01/I, Panama, Informe Tecnico No. 1, 107 pp.

FAO (Food and Agriculture Organization of the United Nations), 1978b: Plan de desarrollo forestal: Estudio de las perspectivas del desarrollo forestal en Panama. FAO/PCT6/PAN/01/I, Panama, Informe Tecnico No. 2, 95 pp.

FAO (Food and Agriculture Organization of the United Nations), 1973: Inventario y Fomento de los Recursos Forestales, Republica Dominicana. Inventario Forestal FO: SF/Dom 8. Informe Tecnico No 3.

FAO (Food and Agriculture Organization of the United Nations), 1981: Tropical Forest Resources Assessment Project (in the framework of the Global Environment Monitoring System – GEMS). Forest Resources of Tropical Asia. UN32/6. 1301-78-04. Tech. Report 3. FAO and UNEP, Rome.

FAO/IVL, 1978: Estudio de la contaminacion con mercurio en la Bahia de Cartegena (Colombia). Informe preliminar, Cartagena, 57 pp.

Geoghegan, T., Jackson, I., Putney A. and Renard, Y., 1984: Environmental guidelines for development in the Lesser Antilles. Caribbean Conservation Association. Caribbean Environment Technical Report No. 3, St. Croix, USVI, 44 pp.

Hackney, C.T., 1987: Factors affecting accumulation or loss of macroorganic matter in salt marsh sediments. *Ecology*, **68(4)**, 1109–1113.

Hamilton, L.S. and Snedaker, S.C. (eds), 1984: Handbook for Mangrove Area Management. EPI(EW)C, IUCN, UNESCO, 123 pp.

Hartshorn, G., Antonioni, G., DuBois, R., Harcharik, D., Heckadon, S., Newton, H., Quesada, C., Shores J. and Staples, G., 1981: The Dominican Republic: Country Environmental Profile, A Field Study. JRB Associates; McLean, Virginia, 118 pp.

Hoi-Chaw, L. and Meow-Chan, F., 1984: Fate and Effects of Oil in the Mangrove Environment. Universiti Sains Malaysia, Pulau Pinang, 252 pp.

INDERENA/IGAC/CIAF, 1984: Los bosques de Colombia. Memoria Explicative, 206 pp.

IUCN (International Union for the Conservation of Nature), 1981: Conservar el patrimonio natural de America Latina y el Caribe. Proceedings of the 18th working session of IUCN's Commission on National Parks and Protected Areas. Gland, Switzerland, 328 pp.

Jimenez, J.A., 1984: A hypothesis to explain the reduced distribution of the mangrove *Pelliciera rhizophorae Tr*. and *Pl. Biotropica*, **16(4)**, 304–308.

Jimenez, J.A., 1988: Dynamics and dispersion patterns of two mangrove populations on the Pacific coast of Costa Rica. Ph.D. Dissertation, Univ. Miami, Coral Gables, 176 pp.

Lahmann, E.J., 1988: Effects of different hydrological regimes on the productivity of *Rhizophora mangle* L: A case study of mosquito control impoundments at Hutchinson Island, Saint Lucie County, Florida. PhD Dissertation, Univ. Miami, Coral Gables, 149 pp.

Laverde, J., Araujo, R., Vargas G. and Patino, E., 1987: Plan de monitoreo del Golfo de Morrosquillo, Primera parte, Aspecto Biologico. Asociacion 'Cravo Norte'

[OXY/ECOPETROL/SHELL], Bogota, 140 pp.

Lewis, R.R. III, Gilmore, R.G. Jr., Crewz, D.W. and Odum, W.E., 1986: Mangrove habitat and fishery resources of Florida. In: Seaman, W. Jr., (ed). *Florida Aquatic Habitat and Fishery Resources*. American Fisheries Society (Florida Chapter), Kissimmee, pp. 281–336.

Lindall, W.N. and Salomon, C.H., 1977: Alteration and destruction of estuaries affecting fishery resources of the Gulf of Mexico. *Mar. Fish. Rev.*, **39**, Pap. 1262, 7 pp.

Lindberg, E. and Harriss, R.C., 1974: Mercury enrichment in estuarine plant detritus. *Mar. Poll. Bull.*, **5(6)**, 93–94.

Lugo, A.E. and Patterson-Zucca, C., 1977: The impact of low temperature stress on mangrove structure and growth. *Tropical Ecology*, **18(2)**, 149–161.

Lugo, A.E. and Snedaker, S.C., 1974: The ecology of mangroves. *Ann. Rev. Ecol. Syst.*, **5**, 39–64.

Manrique, R. and Asociados, 1984: Esudios del efecto ambiental del oleoducto Covenas-Cartagena. ECOPETROL. Cartagena, Colombia.

Manrique, R. and Asociados, 1985: Estudio biologico muelle exportador de crudo Golfo de Morrosquillo. ECOPETROL, Cartagena, Colombia, 51 pp.

Mathis, J.M., 1973: Red mangrove decomposition; a pathway for heavy metal enrichment in Everglades estuaries. MS Thesis, Florida State Univ., Tallahassee.

McNulty, J.K., Lindall, W.N. Jr. and Anthony, E.A., 1970: Gulf of Mexico estuarine inventory project. In: Report of the Bureau of Commercial Fisheries Biological Laboratory, St. Petersburg Beach, Florida, Fiscal Year 1969. Circular 342. US Fish and Wildlife Service, Washington, DC, pp. 13–18.

McNulty, J.K., Lindall W.N. Jr. and Sykes, J.E. 1972: Cooperative Gulf of Mexico estuarine inventory and study, Florida. Phase I - area descriptions. US Dept. Commerce, NOAA Tech. Rept. NMFS Circ. 368, 127 pp.

Odum, W., McIvor, E.C.C. and Smith, T.J. III, 1982: The Ecology of the Mangroves of South Florida: A Community Profile. US Fish and Wildlife Service, Office of Biological Service, Washington DC, 144 pp.

Olexa, M.T., 1976: The distribution, etiology and importance of red mangrove diseases in Florida. PhD Thesis, Univ. Florida, Gainesville, 80 pp.

Padgett, D.E., Hackney, C.T. and de la Cruz, A.A., 1986: Growth of filamentous fungi into balsa wood panels boring in North Carolina salt marsh sediments. *Trans. Br. Mycol. Soc.*, **87(1)**, 155–162.

Pool, D.J., Snedaker, S.C. and Lugo, A.E., 1977: Structure of mangrove forests in Florida, Puerto Rico, Mexico and Costa Rica. *Biotropica*, **9(3)**, 195–212.

Rabinowitz, D., 1978a: Dispersal properties of mangrove propagules. *Biotropical*, **10(1)**, 47–57.

Rabinowitz, D., 1978b: Mortality and initial propagule size in mangrove seedlings in Panama. *J. Ecol.*, **66**, 45–51.

Rabinowitz, D., 1978c: Early growth of mangrove seedlings in Panama, and an hypothesis concerning the relationship of dispersal and zonation. *J. Biogergr.*, **5**, 113–133.

Reark, J.B., 1975: A history of the colonization of mangroves on a tract of land on Biscayne Bay, Florida. In: Walsh, G., Snedaker, S. and Teas, H. (eds), Proc. Intl. Symposium on Biology and Management of Mangroves, Vol. 2, Inst. Food & Agric. Sci., Univ. Florida, Gainesville, pp. 776–804.

Repetto, R., 1990: Deforestation in the tropics. *Scientific American*, **262(4)**, 36–42.

Renard, Y., 1975: La mangrove. Cent. Dep. Doc. Pedagogique, Pointe a Pitre.

Saenger, P., Hegerl, E.J. and Davie, J.D.S. (eds), 1983: Global Status of Mangrove Ecosystems. Commission on Ecology Pap. No. 3, International Union for the Conservation of Nature (IUCN), 88 pp.

Snedaker, S.C., 1982b: Mangrove species zonation: Why? In: Sen, D.N. and Rajpurohit, K.S. (eds), *Tasks for Vegetation Science, Vol. 2. Contributions to the*

Ecology of Halophytes. Dr W. Junk Publ., The Hague, pp. 111–125.
Snedaker, S.C., 1986: Traditional uses of South American mangrove resources and socio-economic effect of ecosystem changes. In: Kunstadter, P., Bird, E.C.F. and Sabhasri, S. (eds), *Proc. Workshop on Man in the Mangroves*. United Nations Univ., Tokyo, Japan, pp. 104–112.
Snedaker, S.C., 1990: Water quality problems and issues in the Florida Keys. Report prepared for The Nature Conservancy and the John D. and Catherine T. MacArthur Foundation. Univ. Miami, Coral Gables, 35 pp.
Snedaker, S.C. and Brown, M.S., 1981: Water quality and mangrove ecosystem dynamics. EPA 600-14-81-022, USEPA Environmental Research Laboratory, Gulf Breeze, Florida.
Spinner, G.P., 1969: The wildlife wetlands and shellfish areas of the Atlantic coastal zone. Serial Atlas of the Marine Environment Folio 18, Amer. Geogr. Soc., New York.
Stehle, H., 1945: Forest types of the Caribbean Islands. Part I. *Carib. Forest.*, **6**, Suppl. 1.
Taylor J.G., 1988: Manglares. La importancia economica de los manglares en la politica, planeamiento y manejo de los recursos naturales costeros. FUDENA/AVIPLA. Ed. Ex Libris, Caracas, Venezuela, 37 pp.
Teas, H.J., 1974: Mangroves of Biscayne Bay. Metropolitan Dade County Commission. Univ. Miami, Coral Gables.
Towle, E.L., 1976: Virgin Islands. Coastal Zone Management Marine Environment. Virgin Islands Planning Office, 191 pp.
Towle, E.L., 1982: Island Resource Management: The Virgin Islands Experience. Island Resources Foundation, St. Thomas, USVI, 70 pp.
Towle, E.L. and Teytaud, A.R., 1982: An annotated bibliography of source materials focusing on the US Virgin Islands' experience and its eastern Caribbean context. The Island Resources Foundation, St. Thomas, USVI, 40 pp.
Vazquez, M.A., 1983: The effects of impounding on a mangrove forest. MS Thesis, Univ. Florida, Gainesville, 117 pp.
Wicker, K.M., Johnston, J.B., Young, M.W. and Rogers, R.M., 1980: Mississippi Deltaic Plain Region Ecological Characterization: A Habitat Mapping Study. A User's Guide to the Habitat Maps. FWS/PBS-79/07. Office of Biological Services, Fish and Wildlife Services, US Dept. of the Interior, Washington, DC.

9 Appendix

Table 12.1 – notes and references

The following notes for Table 12.1 include the sources of the mangrove area data, and the units in which the data were reported originally. For comparative and summary purposes, all area data were converted to hectares (ha). In cases where there are several estimates for the same geographic area, the lowest estimate is reported in the absence of justification for a larger estimate. For this reason, the sum of the individual entries for a country may not equal the total area derived from an independent source(s). Revised data records for South America supercede earlier estimates in Snedaker (1986).

Caribbean

Bahamas

1) Data (in square miles) provided by Roderick Attrill (Bahamas National Trust, Nassau). Although estimates for the larger mangrove areas are reported, the total for the Bahama Island group (including Turks and

Caicos Islands) may be underestimated by as much as 50%.
Barbados
2) Data (in acres) provided by Deborah Riven-Ramsey (University of the West Indies, Bridgetown).
Cayman Islands
3) Data (in hectares) provided by the late Marco E. C. Giglioli (Mosquito Research and Control Unit, Grand Cayman). The estimates for Cayman Brac and Little Cayman are approximations.
Cuba
4) Estimates of mangrove area were extracted using an electronic planimeter by Melvin S. Brown (Law Environmental, Inc., Kennesaw, Georgia) from a private copy of the '*Atlas Nacional de Cuba*', dated 1970. Areas include salinas which may have led to overestimates for the various areas. FAO/PNUMA (1981) estimated the mangrove coverage to be 400,000 ha, based on a 1961 survey; however, that 1961 estimate includes only the larger tracts of mangroves.

Dominican Republic
5) Data (in hectares) provided by Gilberto Cintron (Department of Natural Resources, San Juan, Puerto Rico), and are based on the manual planimetry of a 1:250,000 map (unidentified).
6) Data (in hectares) obtained from FAO/UNDP (1973) and Gilberto Cintron (pers. comm.).
Grand Terre
7) Data (in hectares) obtained from Stehle (1945) reported in Chapman (1976).
Guadeloupe
8) Data (in hectares) provided by Jean Luc Toffart (Ecole Practique des Hautes Etudes, Paris, France). Renard (1975) estimated the mangrove forest area to be 9000 ha. The lower estimate is believed to be reasonable.
Haiti
9) Estimate of area (in hectares) obtained from FAO/PNUMA (1981).
Jamaica
10) Data (in acres) provided by Barry Wade (General Manager, Petroleum Corporation of Jamaica), and are based on a range of 50,000 to 75,000 acres for all wetlands (mangrove and marsh); values are therefore questionable. FAO/PNUMA (1981) reported a mangrove forest area for Jamaica of 7000 ha, but the basis of the estimate, or its accuracy, is not known.
Martinique
11) Data (in hectares) provided by Joseph Poupon (Office National des Forets, Fort de France).
Montserrat
12) Data (in acres) provided by Jay Blankenship (Department of Agriculture, Plymouth).
Netherlands Antilles
13) Data (in hectares) provided by Ingvar Kristensen (Caribbean Marine Biological Institute, Curacao). There are no mangroves on Saba and St. Eustatius.
Puerto Rico
14) Data (in hectares) obtained from Carrera and Lugo (1978).

St. Kitts

15) Data (in acres) provided by A. I. George (Ministry of Agriculture, Basseterre).

Trinidad-Tobago

16) Data (in hectares) provided by Ronald Bickram (Ministry of Agriculture, Lands and Food Production, St. James). Other estimates range from a low of 5000 ha to a high of 11,000 ha (Eugene Ramcharan and Clement Lewsey, Institute of Marine Affairs, Port of Spain, and Sheriff Faizool, Forestry Division, Port of Spain). FAO/PNUMA (1981) estimated the area of mangroves to be 4000 ha. Beard (1946) stated there were 12,670 acres (5130 ha) of mangroves on Crown Lands and in forest reserves in 1938.

17) Data (in hectares) obtained from the Management and Development Plan for the Caroni Swamp National Park and were provided by Eugene Ramcharan (Institute of Marine Affairs, Port of Spain).

Virgin Islands

18) Data (in square kilometres) obtained from the Island Resources Foundation, St. Thomas.

Central America

Belize

19) Data (in acres) provided by Janet Gibson (Fisheries Unit Laboratory, Belize City) and are in agreement with the estimates of FAO/PNUMA (1981). Other estimates range from a low of 100,000 ha (Klaus Ruetzler, Smithsonian Institution, Washington) to a high of 591,360 acres (Oscar Rosado, Ministry of Natural Resources, Belmopan).

Costa Rica

20) Estimate provided by Luis Fernando Gonzales Lopez (Direccion General Forestal, San Jose). Other estimates included a value of 45,000 ha (Ludwig Naegle, Morovia) and a value of 50,000 ha extracted from a letter to V. J. Chapman (dated 15 January 1975) from Ing. Oscar Pacheco Jimenez, Jefe, Departamento Secretaria Technica, Ministerio de Agricultura Y Ganaderia, San Jose. FAO/PNUMA (1981) gives an estimate of 39,000 ha.

21) Estimate provided by Enrique J. Lahmann (IUCN/CATIE, Turrialba).

Guatemala

22) Estimate (in hectares) provided by Enrique J. Lahmann (IUCN/CATIE, Turrialba). FAO/PNUMA (1981) gives an estimate of 50,000 ha, but is considered to be too high.

23) Estimate provided by Enrique J. Lahmann (IUCN/CATIE, Turrialba).

Honduras

24) Data (in hectares) obtained from FAO/PNUMA (1981).

25) Data (in hectares) cited from Prats-Llaurado (1958) by Gilberto Cintron (pers. comm.).

26) Estimate provided by Enrique J. Lahmann (IUCN/CATIE, Turrialba).

Nicaragua

27) Data (in hectares) obtained from FAO/PNUMA (1981).

28) Estimate provided by Enrique J. Lahmann (IUCN/CATIE, Turrialba).

Panama

29) Data (in hectares) obtained from the Ministerio de Desarrollo

Agropecuario, Direccion Nacional de Recursos Naturales Renovables (RENARE). FAO/PNUMA (1981) estimates the total area at 486,000 ha whereas PNUD/FAO (1972) gives a total mangrove area for Panama of 409,210 ha. One possible reason for the wide variation in estimates may depend on whether or not contiguous lowland forests, such as dominated by cativo (*Prioria copaifera*) and orey (*Campnosperma panamensis*) are included in the estimates. The range of estimates for Panama are: 33,700 ha (Anonymous, 1978), 104,000 ha (Donaldson, 1963), 199,000 ha (FAO, 1978a), 297,532 ha (RENARE, see below), 505,600 ha (FAO, 1978b)

30) The estimate for the Caribbean coast is based on the provincial distribution (in hectares) obtained from personnel (Thomas A. Vasquez U. and Cristina Garibaldi de Jaen) in the Ministerio de Desarrollo Agropecuario, Direccion Nacional de Recursos Naturales Renovables. They are probably representative of the relative distribution among the provinces.

Panama	*hectares*	*acres*
Caribbean coast		
Bocas del Toro	*64,010*	*15,800*
Pacific coast		
Cocle	*25,125*	*62,060*
Chiriqui	*66,645*	*164,610*
Darien	*28,225*	*69,715*
Herrara	*8450*	*20,870*
Los Santos	*8800*	*21,735*
Panama	*122,925*	*303,625*
Total	*297,532*	*734,900*

North America

United States

31) Summary total for the United States (Atlantic coast and Gulf of Mexico) obtained by summing state totals (in hectares).

32) Estimate (in hectares) made by the author.

33) Estimate is based on earlier information provided by V. J. Chapman (pers. comm.).

34) Data (in acres) obtained from Browder, Littlejohn and Young (1976), and were modified based on personal knowledge and a variety of unpublished and anecdotal information. Other estimates include 479,180 acres (Eric Heald, Heald and Associates, Miami), 500,000 acres (Lewis *et al.* 1986), and 503,700 acres (Joseph D. Carroll, US Fish and Wildlife Service, Vero Beach).

35) Data (in acres) obtained from Spinner (1969).

36) Data (in acres) obtained from Teas (1974). Metropolitan Dade County (1979), citing other sources, states that in 1972 there were 10,500 acres in the Biscayne Bay area north of Turkey Point.

37) Data (in acres) obtained from McNulty, Lindall and Anthony (1970) and McNulty, Lindall and Sykes (1972).

38) Roy R. Lewis (pers. comm.) provided a lower estimate (5630 ha) which is presumed to apply only to Tampa Bay, rather than both bays.

39) Data (in acres) from Chabreck (1972) and Wicker *et al.* (1980).

40) Estimate (in hectares) made by the author.

41) Data (in acres) provided by C. Lee Sherrod (University of Texas, Austin). Estimates were given as suggested ranges, and only the lowest value is used.

Mexico

42) Data (in square kilometres) provided by Jorge Lopez Portillo (Institute of Ecology, Mexico, DF) and Gilberto Cintron (pers. comm.). One other estimate (1,570,000 ha) was provided by Alejandro Yanez-Arancibia (Universidad Nacional Autonoma de Mexico, Mexico DF).

43) There is no reliable information concerning the portion of the total area along the eastern coastline of Mexico. For estimation purposes, one-half of total inventory is assigned to the Caribbean coast.

44) Data (in square kilometres) provided by Jorge Lopez Portillo.

45) Data (in square kilometres) provided by John W. Day, Jr (Louisiana State University, Baton Rouge).

South America

Colombia

46) Data (in hectares) provided by Jorge Hernan Torres Romero (Universidad Nacional, Bogota), and are in general agreement with the majority of other estimates for Colombia, including a report by Omar-Guauque V. (ed.) and the *Mapa General de Bosques* (1967). These sources give an estimate of 50,000 ha for the Caribbean coast. FAO/PNUMA (1981) reported a lower value of 450,000 ha of which 287,000 are located on the Pacific coast and 163,000 ha on the Caribbean coast.

47) Data (in hectares) obtained from INDERENA/IGAC/CIAF (1984).

48) Data (in hectares) obtained from INDERENA/IGAC/CIAF (1984).

49) Data (in hectares) obtained from de las Salas and Hildebrand (1980).

50) Data (in hectares) obtained from FAO/IVL (1978).

51) Data (in hectares) obtained from Botero (1988).

French Guiana

52) Data (in hectares) obtained from FAO/PNUMA (1981).

Guyana

53) Data (in hectares) provided by (C.A. Persaud, Guyana Forestry Commission, Georgetown, and Reuben Charles, Fisheries Division, Ministry of Agriculture, Georgetown). FAO/PNUMA (1981) has given an estimate of 150,000 ha, but the more conservative estimate is used in this report.

Venezuela

54) Data (in hectares) provided by Federico Pannier (Universidad Central, Caracas). FAO/PNUMA (1981) gives a total area for Venezuela of 260,000 ha, but it is believed to reflect only the larger areas of potentially commercial forest.

Surinam

55) Data (in hectares) obtained from FAO/PNUMA (1981).

13

Coral Reefs and their Response to Global Climate Change

John D. Milliman[1]

Abstract

Coral reefs represent perhaps the most important environmental (and therefore economically critical) ecosystem in the region. Climate change can affect the reefs mostly through warmer water temperatures, which may increase the corals' susceptibility to disease, such as bleaching. Accelerated sea-level rise, at least over the next several decades, probably will not retard coral growth; in fact it may locally stimulate accelerated growth.

The most immediate hazard facing reefs continues to be environmental degradation due to human activities both in the water (e.g., from boats, careless divers, sewage outfalls, dredging) and on land (e.g., increased erosion). Efforts should be made to monitor reef health more extensively and carefully. Establishment of more marine stations is needed, as well as continued financial and scientific support for established stations.

1 Introduction

Coral reefs constitute an environmental and economic keystone in the region (*q.v.* Fig. 1.1; *cf.* Fig. 10.10). Not only do they constitute a key marine biotope, but they also provide the habitat for a wide-spread artisanal fisheries, protection against wave erosion, and serve as a source of sand for natural beach nourishment, etc. Moreover, coral reefs are a major tourist attraction in the region. Changing global climate and rising sea level, therefore, could have particularly severe environmental and economic impact on such tropical areas if these changes adversely affect coral reefs (Vicente *et al.*, Chapter 11).

In this paper we evaluate the general ecology and recent geological history of Gulf/Caribbean coral reefs, as well as the present-day economic and social aspects related to reefs. The impacts of man's activities are reviewed, and possible future impacts are discussed. Some of the present-day coral-reef degradation may be blamed on climate change related to the greenhouse effect; but other degradation is related to local human activity, such as land use, harbour pollution, dredging, etc. Minimizing these local effects to some extent may help

[1] Woods Hole Oceanographic Institution, Woods Hole, MA 02543, USA

to reduce the deleterious effects that climate and sea-level change could have upon reefs.

2 Coral-Reef Ecology

Coral reefs are defined as shallow-water, hard-substrate bioherms that stand above the adjacent seafloor and in which a major constituent is hermatypic corals, that is corals containing symbiotic zooxanthellae. The exact purpose of these minute dinoflagellates is not completely understood, but it is clearly related to their utilization of the waste products from the host coral polyps (particularly nitrates; Bythell, 1990) and their subsequent production of oxygen and various enzymes that the coral can utilize. Nitrate rather than ammonium appears to be the main source of dissolved inorganic nitrogen for the zooxanthellae (Bythell, 1990). The fact that zooxanthellae assist in at least some metabolic functions allows the host polyp to put more of its energy into skeletal mineralization.

Without zooxanthellae, coral growth rates are significantly lower, a point that was forcefully proven in a series of light-dark growth experiments by the late T.F. Goreau (e.g., Goreau, 1963). Ahermatypic corals, those corals that normally do not contain zooxanthellae, as well as hermatypic corals from which the zooxanthellae have been expelled (i.e., 'bleached' corals), have calcification rates as much as an order of magnitude lower than healthy hermatypic corals. Moreover, the health of bleached corals can be impaired to the extent that they eventually die, a problem that will be discussed in greater detail later in this chapter.

Hermatypic coral growth generally is restricted to mean water temperatures between about 22 and 30°C; where wintertime temperatures fall below 18°C, coral growth is restricted, as witnessed by the few species present in waters at Bermuda or north of Miami, Florida. It has been suggested that 30°C represents the upper limit of healthy growth; at significantly higher temperatures, bleaching may occur.

In the western Atlantic Ocean, coral reefs are mostly restricted to the Caribbean Sea, where water temperatures seldom fall below 20°C. Reef development in the tropical eastern Atlantic is restricted by cool upwelled coastal waters. Corals grow best in clear waters with salinities within the normal marine range. If salinities are significantly higher or lower, coral growth (and health) is negatively affected. Some corals (such as *Porites*) can withstand greater temperature and salinity ranges than other species (e.g., *Acropora palmata*).

Most corals are not adept at removing sediment from the polyps, the result being that corals in turbid waters tend to be unhealthy or die from excessive sedimentation. Moreover, muddy areas often lack hard substrates to which coral planulae must attach. Corals and coral reefs therefore tend to be poorly developed or absent from areas in which suspended sediment is abundant, such as the Orinoco River–Amazon shelf (*cf.* Hendry, Chapter 7; Müller-Karger, Chapter 8) along northeastern South America (Branner, 1904; Milliman, 1973) and the northern coast of Puerto Rico (Bak, 1983). In areas of dredging, the waters can become sufficiently turbid to damage the corals (e.g., Wood and Johannes, 1975).

Although growth rates vary considerably with species and environment (e.g., Lewis *et al.*, 1968; Baker and Weber, 1975), they appear to be greater at lower latitudes (e.g., Glynn, 1973; Tomascik and Logan, 1990), presumably in response to warmer wintertime temperatures. Reef growth and development are also greatest in shallow waters; species diversity increases slightly with depth to about 20–30 m, below which it decreases steadily (Huston, 1985).

3 Reef Types and Distributions

Three basic types of coral reefs are commonly recognized: fringing reefs, which usually lie close to land with very shallow intervening water depths; barrier reefs, in which the intervening lagoon is wider and deeper; and atolls, where reefs surround a lagoon with no visible indication of pre-existing basement. A fourth reef type, patch reefs, are isolated reefs a few to several tens of metres in diameter. Patch reefs are particularly prevalent in some of the larger Indo-Pacific atoll lagoons (e.g., Wiens, 1962), but they also occur on the upper continental slope of the northern Gulf of Mexico (e.g., Texas Flower Garden Reef).

Fringing reefs adjacent to high-standing Caribbean Sea islands are often restricted to the leeward sides, the turbidity in the windward waters being too great to support active reef growth (Lewis, 1960; Adams, 1968). Some of the more widely cited fringing reefs, such as those off Florida and along the windward margins of the Bahamas, probably are more akin to barrier reefs, as they are separated from land by hundreds to thousands of metres and lagoonal depths can exceed 5–10 m.

Barrier reefs are not common in the Caribbean Sea, although the barrier reef off Belize, more than 200 km long and with lagoonal depths locally exceeding 20 m (Stoddart, 1962), is considered the second largest barrier-reef system in the world, smaller only than the Great Barrier Reef off eastern Australia.

There are about ten atolls in the region, although most of these cannot be considered classic atolls in the Darwinian sense, since none have volcanic cores. Most atolls form on subsiding sedimentary basement (e.g., Hogsty Reef in the southeastern Bahamas) or on structural highs (e.g., Glovers Reef off Belize). (Bermuda lies on a volcanic basement, but island development is more the result of aeolian sedimentation than reef accretion.)

4 Modern Reef Zonations

Two basic forms of reef zonations are recognized in the region: high-energy reefs, such as those on atolls and on the exposed windward portions of islands; and low-energy reefs, such as those on the protected leeward sides of islands. Part of the following discussion has been taken from Milliman (1973).

High-energy, windward reefs consist of the reef front, reef flat and patch reefs (Fig. 13.1). Depending on proximity to land, a lagoon may occur leeward of the reef flat. The morphology, ecology and distribution of corals and other organisms in these zones are generally consistent throughout most Caribbean Sea reefs, although local differences are noted.

Earlier workers tended to regard Caribbean Sea coral reefs as pale images of their more diverse and 'healthier' Indo-Pacific counterparts (e.g., Wells, 1957; Newell, 1959). In part this reflected the fact that Intra-Americas Sea corals are represented by an order of magnitude fewer species, often lack very shallow (subaerially exposed at spring low tide) reef-flat zonations, and supposedly lack the algal ridge that characterizes the seaward edge of many Indo-Pacific reef flats. Both Wells and Newell regarded modern reefs of the region as representing only thin veneers over relict Pleistocene surfaces.

Subsequent field studies, involving the acquisition and analysis of both recently exposed inner surfaces of modern reefs as well as shallow borings on the reefs and reef fronts, however, indicate that Holocene reefs have grown at far more rapid rates and thus attained greater thicknesses than envisioned by Wells and Newell (e.g., Adey, 1975; Macintyre, 1988).

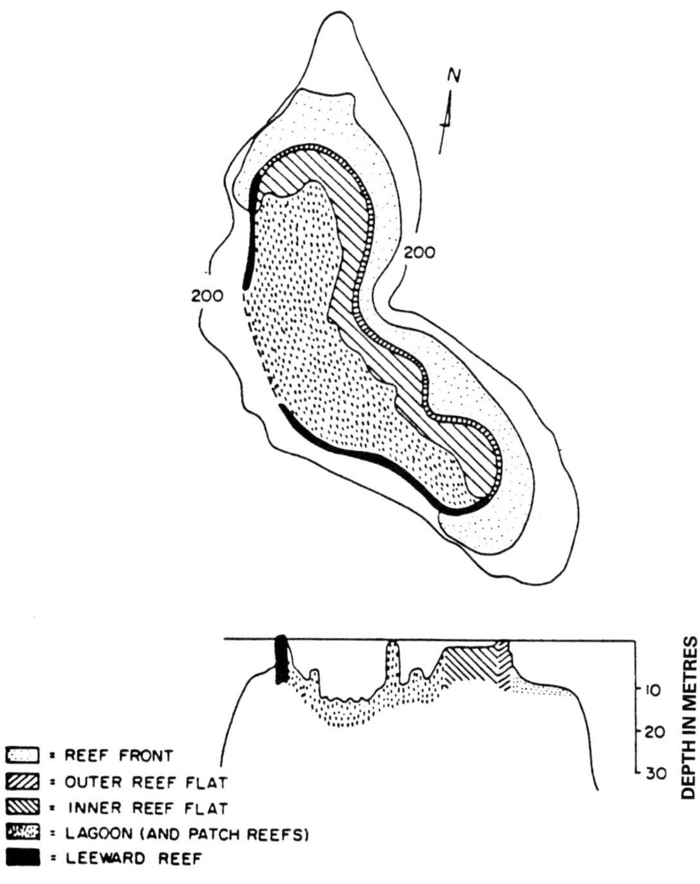

Fig. 13.1 Reef zonations at a Caribbean Sea atoll (Courtown Cays); isobath is 200 m (from Milliman, 1973).

Moreover, Milliman (1973), Glynn (1973), Adey (1975) and Adey and Burke (1976) have described healthy Caribbean Sea counterparts of the Indo-Pacific algal ridge. Thus, in many ways not only are Caribbean Sea coral reefs not 'poor cousins' to their Indo-Pacific counterparts, but they actually may be richer in terms of carbonate productivity.

The lack of hermatypic species does not seem to have hindered reef growth in the region. In part this reflects the ability of individual species to display more than a single growth form – each form more or less adapted to specific energy and light conditions. For example, both *Acropora palmata* and *Montastrea annularis*, volumetrically the two most important species in the region, display a number of growth patterns. *M. annularis*, for instance, commonly has a massive head-coral growth or columnar form in shallow waters; but it can be hemispherical and foliose in mid-depths, and it has crustose growth form in deeper waters.

4.1 Reef Flat and Reef Front

The uppermost part of some reef fronts is marked by spurs (buttresses) and grooves oriented normal to the reef. Relief can exceed 5 m, and the system can extend seaward to depths as great as 10 m (e.g., Goreau, 1959; Stoddart, 1962; Shinn, 1963).

Prolific coral growth is seen on the reef front to water depths greater than 70 m (e.g., Goreau and Wells, 1967; Land and Moore, 1977; James and Ginsburg, 1979), with crustose and platy growth forms increasing in response to greater light-utilization; decreased wave energy diminishes the need for massive growth forms. In these deeper waters, sponges and *Halimeda* often predominate. *Halimeda* is a major producer of sand-size sediment, while endolithic sponges tend to bioerode chips that can dominate the silt-size fraction (see James and Ginsburg, 1979, for a discussion).

Reef flats in the region tend to be about 1 m in depth, although in the southern Caribbean Sea reef flats can be much shallower. Corals on the windward side of the reef flat, extending down the reef front to a depth of about 5 m, are generally massive, a requirement needed to withstand continual seas and swells. *Acropora palmata* (elkhorn coral) and *Montastrea annularis* (star coral), both with robust growth forms, are the prime species, but at any particular reef one species is generally predominant. For instance, at the shallow windward reef flats at St. Croix (US Virgin Islands), *A. palmata* is the dominant coral, whereas in many western Caribbean Sea reefs, *M. annularis* dominates (e.g., Milliman, 1973). Other abundant species include *A. cervicornis* (the staghorn coral), several species of *Porites* (both finger coral and head coral), *Agaricia agaricites, Diploria* (three species of brain corals) and the encrusting coral *Siderastrea siderea*.

The hydrozoan stinging coral, *Millepora alcicornis*, is locally very common, particularly associated with crustose coralline algae in the windward algal ridge (see Milliman, 1973; Glynn, 1973). Crustose coralline algae and the codiacean green alga *Halimeda* are the dominant carbonate-producing plants, although in many areas fleshy and filamentous algae also are common. In the Florida Keys, for instance, fleshy brown, green and blue-green algae seem to be taking over many formerly healthy reef flats (see below). Locally, seagrasses (*Thalassia, Halodule* and *Syringdodium*) can also cover parts of the reef flat.

4.2 Lagoons

Leeward of the windward reef flat, water depths increase to more than 5 m, and locally depths can exceed 10–15 m. The abrupt slope between the reef flat and the lagoon, termed the 'sand cliff' by Milliman (1969), characterizes many Caribbean Sea atoll lagoons. In contrast to the Indo-Pacific, lagoon depths bear no relation to the size of the reef or the atoll. The shallow open nature of the lagoons, plus their relatively small size, means that water circulation responds directly to winds and tides, and water temperatures often reflect air temperatures.

The lagoon leeward of the Belize barrier reef is one of the few lagoons for which the Holocene development has been documented using geophysical means. High-resolution seismic profiles and drilling (Purdy, 1974; Enos, Koch and James, in James and Ginsburg, 1979) show as much as 20 m of Holocene sediment, a mixture of terrigenous (land-derived) and carbonate sediment.

Most lagoons of the region contain patch reefs, present as solitary coral mounds, low-lying reefs or high-standing pinnacles that reach sea level. Patch reefs in the Florida Keys tend to contain only a few hardy species, such as *Montastrea, Siderastrea* and *Diploria*, as well as alcyonarian and gorgonian corals, probably in response to relatively poor circulation and high turbidity. In recent years, in fact, many of these patch reefs have become unhealthy, and locally have been taken over by luxurient growth of fleshy algae (see Shinn, 1989). In southern Caribbean Sea lagoons, patch reefs are more plentiful and the species of corals often as diverse as on the reef flat and reef front. Where the lagoons are sufficiently small, deep or protected, some of the more quiet-water growth forms, such as *Acropoa cervicornis*, can predominate. In other, more open lagoons, higher-energy species and growth forms are more common. Some lagoonal reefs, such as those at Alacran Reef and Alburquerque Cays, have coalesced into cellular-like reefs (Hoskin, 1963; Milliman, 1969).

Most lagoons are floored by skeletal debris derived from both the peripheral reefs and lagoonal patch reefs. Seagrass banks, predominantly composed of *Thalassia*, trap and accrete sediments by their active root systems, thereby forming a finer-grained sedimentary environment (e.g., Ginsburg and Lowenstam, 1958). Blue-green algae can form cohesive mats on the lagoon floor, which can provide additional sediment stability (e.g., Neumann *et al.*, 1970) as well as food for lagoonal organisms.

4.3 Low-Energy (Leeward) Reefs

Reefs located on the leeward sides of islands or atolls tend to be more poorly developed than those found on the exposed windward sides. Because of the lower ambient current and wave energies, for example, high-energy corals, such as *A. palmata*, are virtually absent over much of the leeward reef tract at Barbados and St. Vincent (Lewis, 1960; Adams, 1968). Corals that can withstand warmer waters, often with more suspended matter, such as *Porites* and *Siderastrea*, are common, as are *Montastrea* and *Millepora*.

5 Coral-Reef Development During the Holocene (Last 10,000 Years)

During the last glacial epoch, which climaxed about 16–19 thousand years ago, global sea level was 80 to 120 m lower than at present. With the subsequent rapid melt of glacial ice, sea level rose quickly to about –30 m 10,000 years ago; about 5000 years ago it almost reached present-day levels. Since then, sea level has risen slowly, albeit not uniformly, throughout the region (Fig. 7.3). These sea-level curves for the region are relatively accurate, as they reflect the C-14 dating of coral species (particularly *Acropora palmata*) known to grow at near-intertidal water depths. Although coral growth in present-day reef areas began with the rise in sea level, many reefs died back, leaving the prominent submerged (dead) reefs that characterize the reef front facies throughout the region.

Macintyre (1988) has suggested that coral growth decreased as rising sea level flooded the broad shelves and thereby increased turbidity (as well as possibly seasonally cool temperatures in the northern Caribbean Sea), causing die-off of the corals. When conditions improved, about 7000 years ago, coral growth again accelerated, and this time, partly because of the diminished rate of sea-level transgression, many reefs were able to 'catch

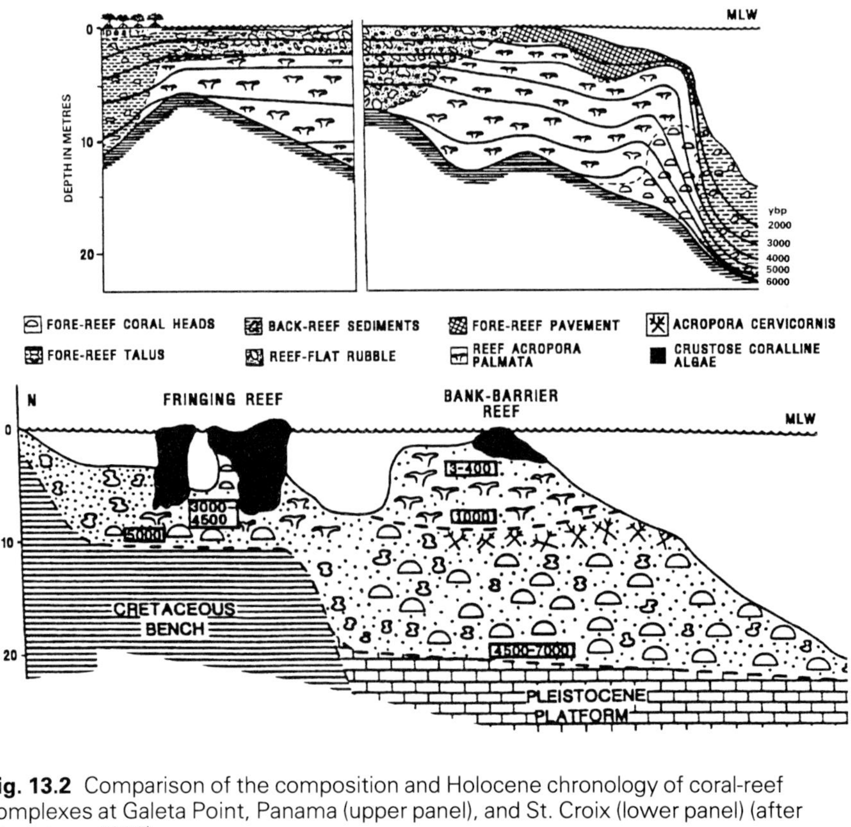

Fig. 13.2 Comparison of the composition and Holocene chronology of coral-reef complexes at Galeta Point, Panama (upper panel), and St. Croix (lower panel) (after Macintyre, 1988).

up' and 'keep up' (to use the terms coined by Neumann and Macintyre, 1985) with sea-level rise (Fig. 13.2). Accumulation rates, according to Macintyre (1988), have been higher in the western Caribbean Sea because the more quiescent environment has allowed the rapidly growing *Acropoa cervicornis* to accrete thick sequences. About 5000 years ago, an *A. cervicornis* reef at Alacran Reef (Mexico) accreted at a rate of 12 m per 1000 years, or more than 1 cm/yr; total Holocene reef thickness is 23 to 33 m (Macintyre *et al.*, 1977).

While the Alacran growth rates are the highest yet recorded in Caribbean Sea reefs, the 20–30 m of Holocene reef accumulation is not much greater than that recorded at other reefs in the region. The Holocene fringing and barrier reef on the windward side of St. Croix, for example, is 10–20 m thick, with most of its growth occurring in the last 5000 years (Adey, 1978). Similar thicknesses have been documented for Galeta Point (Panama; Macintyre and Glynn, 1976), and early Holocene growth of a now flooded reef north of Miami for a short time was greater than 10 m per 1000 years (Lighty *et al.*, 1978, 1982). Reefs that were not able to keep pace with sea-level rise, termed 'give-up' reefs by Neumann and Macintyre (1985), are now found throughout the region as relict submerged reefs on shelves and slopes.

Holocene reef growth has generally ranged between 0.1 and 1 m per 100 years, but individual coral growth rates can approach or exceed 1–2 cm/yr (see reviews by Macintyre, 1988; and Stoddart, 1990). These data, plus increased documentation of Holocene reef history in the Indo-Pacific (e.g., Hopley, 1982), indicate that, contrary to earlier opinion, Caribbean coral reefs have been as productive or more productive in terms of vertical reef accretion during the Holocene. The general lack of the region's coral species seems to have been at least partly compensated by the diversity of growth forms (e.g., Milliman, 1973).

6 Economic and Ecologic Factors

Without question coral reefs are a main factor in regional economies. Not only do they form a main (renewable) component of the beach sands upon which tourists lie and tan, but they also form a marine habitat that is particularly popular with skin divers and fishermen. Tourist arrivals increased at a 4–6% annual rate during most of the 1980s, not in small part because of the access to beaches and coral reefs (Gable, 1990). Tourism accounts for foreign income greater than US$ 6 billion annually (Heyman, 1988; Alm, *et al.*, Chapter 15); while it is difficult to calculate how much of this amount is a direct result of coral reefs, if the reefs were removed or degraded, tourism almost certainly would decrease. Protecting the reefs, therefore, means more than just environmental protection; in many countries (such as the Bahamas and Antigua) it also means economic survival.

Reefs not only serve as a main focus for tourism/diving, they also are a primary habitat for commercially important fishes. US AID (1987) lists more than 30 species that inhabit Caribbean Sea coral reefs, most important of which are snapper, grouper and lobster. Overfishing already has almost completely eliminated some stocks from many reefs (e.g., the conch, *Strombus gigas*) whereas deeper water reefs may have been under-utilized

(US AID, 1987, p. 32). The decline in many of these species is of concern not only for commercial and artisanal fisheries but also for recreational fisheries.

Unfortunately, the flux of tourism also can lead to conditions that are deleterious to the health of coral reefs. Increased onshore construction, for example, may necessitate the dredging of reef sands that not only results in turbid waters but also may cut off the source of beach-sand renewal (e.g., Gable, 1990). The use of fresh water and the direct introduction of pollutants to coastal waters through sewage outfalls (such as hotel resorts located near reefs) can degrade the quality of reef waters (e.g., Beekhuis, 1981).

Problems with coral reefs can be dramatically accentuated by other human-induced environmental problems. For example, mining of coral reefs in many Pacific and Indian Ocean islands (e.g., UNEP, 1987) has impacted the health and growth of local reefs (*cf.* Hendry, Chapter 7). At the same time, poor agricultural practices and coastal construction (e.g., Craik and Dutton, 1987) have increased erosion of hilly and mountainous areas adjacent to many reefs; the result can be turbid waters that make coral growth (or even survival) difficult (*cf.* Vicente *et al.*, Chapter 11).

Nutrient loading is a major problem for Gulf of Mexico and Caribbean Sea coral reefs, particularly in view of the fleshy algae that grow preferentially in nutrient-rich, polluted waters, thereby displacing corals. LaPointe (1989) points out that an increased discharge of sewage into reef waters in the Florida Keys may have accounted for the increased abundance of microfilamentous blue-green algae (*Phormidium*) that causes black-band disease in corals, particularly in *Montastrea*, and white-band disease in branching corals (e.g., Shinn, 1989). Shinn argues that most corals can survive all but the most severe abuse from anchors, boat groundings and dredging, as well as hurricanes (although certain species, such as *Acropora palmata*, are susceptible to careless snorklers and boat anchors; Rogers *et al.*, 1988). Rather, Shinn argues, it is increased suspended sediment in the reef waters, temperature change and disease (such as black- and white-band diseases) that have the greatest capacity to kill entire reefs.

The interaction between habitat change and its effect on the reef, however, is not always straightforward (although, sometimes in hindsight it may appear obvious). For instance, many grazing animals (both benthic invertebrates and fish) remove fleshy algae that otherwise would overgrow the corals and possibly introduce disease that would adversely affect the corals (LaPointe, 1989). The effect of the decreased number and activity of these primary grazers was seen in the artificial removal of all sea-urchins (*Diadema antillarium*) from the reef tract at St. Croix, which led to an order of magnitude increase in fleshy algae in just four months (Sammarco *et al.*, 1974). The subsequent basin-wide disease and depopulation of *Diadema* in the past few years may explain why fleshy algae have thrived on some reefs, at the expense of coral and calcareous algal growth.

7 Climate Change

Most models of global change in climate predict a general warming of the Earth's atmosphere by as much as 1–2°C over the next 100 years (*cf.*

Wigley and Santer, Chapter 2; and Gallegos *et al.*, Chapter 3). Two effects of global change are seen to be particularly important for coral reefs: coral bleaching and sea-level rise. These concerns are addressed in the following sections.

7.1 Climate Warming and Coral Bleaching

Environmental pressures may increase the frequency of both coral-related diseases and predators. Of particular concern in recent years has been coral bleaching, a condition that results in the loss of symbiotic zooxanthellae from the coral polyps, thereby weakening the coral's capacity to survive. Other types of reef organisms also appear to be affected by bleaching events. P. Hallock-Müller (1991, pers. comm.) reports bleaching in the benthic foraminifera, *Marginopora*, and some local sponge bleaching was reported during the 1987 bleaching event in Puerto Rico (Vincente, 1990).

Particularly well documented in the Caribbean Sea (e.g., Williams *et al.*, 1987; Williams and Bunkley-Williams, 1988; Goreau, 1990), coral bleaching may be caused by rapid increase or fall in temperature (e.g., Porter *et al.*, 1978; Steen and Muscatine, 1987), either the result of ENSO events (e.g., Glynn, 1984) or possible climatic change (Bunkley-Williams and Williams, 1990). Recovery of the bleached corals by gradual accumulation of zooxanthellae within reef corals can occur within six months, with little difference between it and previously unbleached coral tissue (Hayes and Bush, 1990).

Most coral species cannot tolerate temperatures greater than about 30°C, and even a rise in seawater temperature of 1–2°C over the next century could cause severe impact or even extinction of many shallow-water coral species. To some degree, this might be balanced by a poleward shift of hermatypic species as higher latitude waters become warmer. Decreased coral health/growth may prove especially troublesome if combined with continued or accelerated sea-level rise. However, based on the geological record, many other species probably would survive such global warming (Glynn, 1991).

There is a general agreement that at temperatures greater than about 30°C, corals begin bleaching. T. Goreau (Kingston) and R. Hayes (Washington) recently have concluded the temperature at a number of Caribbean Sea locations over the past decade has warmed between 0.4 and 1.0°C; corals may be a 'canary in the mine' – that is, that they may provide an initial warning of wide-spread global warming (e.g., Roberts, 1990).

Goenaga and Canals (1990) show that the bleaching events in 1987 and 1990 occurred at times when mean monthly sea-surface temperatures were significantly higher than normal mean temperatures, although they took pains to point out that temperature increase alone may not explain such bleaching events. Rather, it appears that bleaching probably is not related to a single temperature warming but rather than the cumulative effect of population, land use and resource exploitation (D'Elia, in Roberts, 1991). In the Exumas (Bahamas), bleaching in 1987 occurred more in waters 10–60 m in depth than in shallower habitats, suggesting that more than mere temperature increase causes bleaching; Lang *et al.* (1988) suggest that prolonged contact with colder upwelled water masses may account for this bleaching. Other agents, such has increased solar UV radiation (with depletion of the ozone layer; Shick and Lesser, 1991), also may be at least partly responsible for coral bleaching.

One indication of the degree of concern over coral bleaching was the international workshop held in Miami on 17–21 June 1991 to address these problems and propose future studies (D'Elia *et al.*, 1991). It was felt by the workshop attendees that a rise in sea-surface temperature represents the greatest threat from climate change, although at present no such systematic increase can be delineated. The participants endorsed the development of a global-scale coordinated programme of coral-reef monitoring as well as more interdisciplinary and collaborative research on present-day and near-present coral-reef systems.

7.2 Sea-Level Rise

Because the region contains many coral-reef systems, one of the most critical questions asked by both concerned scientists and politicians (as well, of course, as by island inhabitants) is what will be the fate of associated low-lying coral islands with future sea-level rise? A 2–3 m rise of sea level, for instance, would flood much of the Bahamas and Florida Keys (*cf.* Hanson and Maul, Chapter 9). When Gable and Aubrey (1989) rated the potential impacts of rising sea level for various Pacific nations, not surprisingly, a direct correlation was made with mean elevation above sea level. The groundwater resources on low-lying islands may be particularly susceptible to sea-level changes, with the intrusion of salt water into the water table (e.g., Oberdorfer and Buddemeier, 1988).

In their review of sea-level rise and coral-reef growth over the past 6000 years, Buddemeier and Smith (1988) and Stoddart (1990) have concluded that most reefs could not withstand the highest predicted rates of sea-level rise, and that even much lower rates could have disastrous effects on both reef flat and lagoonal reefs. As sea level rises, however, carbonate production in shallow-water reefs should increase and many present-day reef flats would be recolonized (Kinsey and Hopley, 1991). A rise in sea level may actually enhance coral-reef carbon production if it means that corals are replaced by algae; on the other hand, increasing water temperatures to or beyond the limit of survival may reduce production. Overall, therefore, it is not clear whether greenhouse-related changes will significantly change the productivity of most present-day reefs (Crossland *et al.*, 1991).

8 Coral-Reef Protection

Adequate coral-reef protection, both for the corals and the habitat, clearly requires a number of protective and corrective measures. Some of these involve mitigating the greenhouse-related rise in temperature, change in precipitation patterns and/or sea-level rise. But other concerns involve human impacts that are local in origin and have local impact. Such factors include increased turbidity, sand dredging, coral and limestone mining (although these are of greater concern in the Indo-Pacific), coastal-zone construction, and excessive recreational utilization of the reefs (e.g., propeller wounds, removal of 'souvenirs', spearfishing, etc.).

Davidson and Gjerde (1989) have recommended five actions that could alleviate these problems:

1) Nations in the Wider Caribbean Region should ratify and implement international efforts to protect the marine environment.
2) All reef systems facing possible threat should be placed under the protection of the Specially Protected Areas and Wildlife Protocol of the Cartagena Convention.
3) Nations of the Caribbean and Adjacent Regions should request that the region be classified as a Special Area under Marpol annexes I, II and V, which would restrict further dumping or discharge of hazardous substances into the coastal waters.
4) Reefs could be nominated for designation as World Heritage Sites.
5) Reef corals (and presumably other reef-associated organisms) could be included in Appendix II of the CITES, which would provide a monitoring of the commercial trade of these items.

New approaches to coral-reef management are required, ones that take into account the multiple uses of coral reefs, both for local populations as well as tourists, and ones that consider both the economic and ecologic impacts of reef presence (e.g., Kenchington, 1988). Van't Hof (1988) gives three main reasons why reef management so far has been ineffective: (1) legislative and legal enforcement has been inefficient at best; (2) the public and resource users (e.g., fishermen) do not understand sufficiently the need for effective reef management; and (3) most management efforts to date have been local and isolated, either regionally or with respect to single or few species.

9 Concluding Statements

Human impact from sheer population increase over the next 50 years may have far greater impact than any greenhouse-related changes (e.g., Buddemeier and Hopley, 1988). Pollution runoff, increased fishing, etc. can have more immediate impact than sea-level rise or temperature rise. Traditional reef use has been generally conservative, whereas channel blasting, dredging, dynamite and poison fishing, and sewage runoff have led to accelerated reef degradation (Dahl and Salvat, 1988). While it does not seem to be as critical problem as in the Pacific, where entire reefs are being destroyed (e.g., Gomez, 1983), the ban on coral trade and the restriction of dredging are strongly endorsed as two activities that will lessen the impact of climate change by helping to maintain healthy reefs that can therefore withstand increased climatic stress.

Clearly more permanent study sites, such as those at Discovery Bay, Jamaica (e.g., Gates, 1990) are needed – ones in which both coral health and various environmental parameters can be monitored – in order to delineate long-term changes in reef 'health' as well as perturbations in the reef environment. At present, however, even the established marine stations are under financial pressure. Perhaps the most widely used marine laboratory – the Fairleigh Dickinson University Laboratory on St. Croix – recently closed its doors after 20 years of extensive research and teaching; the laboratory simply had become too expensive for a relatively small and financially restricted university to maintain. Demise of such important laboratories will only make coral-reef research more difficult in the future.

10 References

Adams, R.D., 1968: The leeward reefs of St. Vincent. West Indies. *J. Geol.*, **76**, 587–595.

Adey, 1975: The algal ridges and coral reefs of St. Croix: their structure and Holocene development. *Atoll Res. Bull.*, **187**, p. 67.

Adey, W.H., 1978: Coral reef morphogenesis: A multidimensional model. *Science* **202**, 831–837.

Adey, W.H. and Burke, R., 1976: Holocene bioherms (algal ridges and bank barrier reefs) of the eastern Caribbean. *Geol. Soc. Amer. Bull.*, **87**, 95–109.

Bak, R., 1983: Aspects of community organization in Caribbean stony corals (Scleractinia). In: Ogden, J. and Gladfelder E. (eds), Coral Reefs, Seagrass Beds and Mangroves: Their Interaction in Coastal Zones of the Caribbean. *Unesco Rept. Mar. Sci.*, **23**, 51–698.

Baker, P.A. and Weber, J.N., 1975: Coral growth rate: Variation with depth. *Earth Planet. Sci. Lett.*, **27**, 57–61.

Beekhuis, J.V., 1981: Tourism in the Caribbean: Impacts on the economic, social and natural environments. *Ambio*, **10**, 325–331.

Branner, J.C., 1904: The stone reefs of Brazil, their geological and geographical relations, with a chapter on the coral reefs. *Bull. Mus. Comp. Zool., Harvard Univ.*, **44**, 285 pp.

Brown, B.E., 1990: Coral bleaching. *Coral Reefs*, **8**, 153–232.

Buddemeier, R.W. and Hopley, D., 1988: Turn-ons and turn-offs: Causes and mechanisms of the initiation and termination of coral reef growth. *Proc. 6th Int'l. Coral Reef Symp.*, Vol. 1, 253–261.

Buddemeier, R.W. and Smith, S.V., 1988: Coral reef growth in an era of rapidly rising sea level: predictions and suggestions for long-term research. *Coral Reefs*, **7**, 51–56.

Bunkey-Williams, L. and Williams, E.H. Jr., 1990: Global assault on coral reefs. *Natural History*, **4**, 47–54.

Bythell, J.C., 1990: Nutrient uptake in the reef-building coral *Acropora palmata* at natural environmental concentrations. *Marine Ecol. Prog. Series*, **68**, 65–69.

Chambers, G., 1988: Coastal change and its links to tourism development. *Caribbiana*, July-Aug., 11–14.

Craik, W. and Dutton, I., 1987: Assessing the effects of sediment discharge on the Cape Tribulation fringing coral reefs. *Coastal Manag.*, **15**, 213–228.

Crossland, C.J., Hatcher, B.G. and Smith, S.V., 1991: Role of coral reefs in global ocean production. *Coral Reefs*, **10**, 55–64.

Dahl, A.L. and Salvat, B., 1988: Are human impacts, either through traditional or contemporary uses stabilizing or destabilizing to reef community structure? *Proc. 6th Intern. Coral Reef Symp.*, Vol. 1, 63–69.

Davidson, L. and Gjerde, K., 1989: An evaluation of international protection offered to Caribbean coral reefs and associated ecosystems. Unpubl. document from Greenpeace International and Woods Hole Oceanographic Inst.

D'Elia, C.F., Buddemeier, R.W. and Smith, S.V., 1991: Workshop on coral bleaching, coral reef ecosystems and global change: Report of Proceedings, Maryland Sea Grant College, 36 pp.

Dunstan, P. and Halas, J.C., 1987: Changes in the reef-coral community of Carysfort Reef, Key Largo, Florida: 1974–1982. *Coral Reefs*, **6**, 91–106.

Gable, F.J., 1990: Caribbean coastal and marine tourism: Coping with climate change and its associated effects. In: Miller, M.L. and Auyong, J. (eds), Proceedings of the 1990 Congress on Coastal and Marine Tourism., 25–31 May

1990, Honolulu, Nat. Coastal Resour. Res. Devel. Inst., Newport, OR, Vol. 1, 248–258.

Gable, F.J. and Aubrey, D.G., 1989. Changing climate and the Pacific. *Oceanus*, **32(4)**, 71–73.

Gates, R.D., 1990: Seawater temperature and sublethal coral bleaching in Jamaica. *Coral Reefs*, **8**, 193–197.

Ginsburg, R.N. and Lowenstam, H.A., 1958: The influence of marine bottom communities on the depositional environment of sediments. *J. Geol.*, **66**, 310–318.

Glynn, P.W., 1973: Aspects of the ecology of coral reefs in the western Atlantic region. In: Jones, O.A. and Endean, R. (eds), *Biology and Geology of Coral Reefs*. Academic Press, London, Vol. 2, pp. 271–324.

Glynn, P.W., 1984: Widespread coral mortality and the 1982–83 El Niño warming event. *Environmental Conservation*, **11**, 133–146.

Glynn, P.W., 1991: Coral reef bleaching in the 1980s and possible connections with global warming. *Trends Ecol. Evol.*, **6**, 175–178.

Goldberg, W.M., 1973: The ecology of the coral-octocoral communities off the southeast Florida coast: geomorphology, species composition, and zonation. *Bull. Mar. Sci.*, **23**, 465–488.

Goenaga, C. and Canals, M., 1990: Island-wide coral bleaching in Puerto Rico: 1990. *Carib. J. Sci.*, **26**, 171–175.

Gomez, E.D., 1983: Perspectives on coral reef research and management in the Pacific. *Ocean Manag.*, **8**, 282–295.

Goreau, T.F., 1963: Calcium carbonate deposition by coralline algae and corals in relation to their roles as reef builders. *Ann. N.Y. Acad. Sci.*, **109**, 127–167.

Goreau, T.F. and Wells, J.W., 1967: The shallow-water *Scleractinia* of Jamaica. Revised list of species and their vertical distribution range. *Bull. Mar. Sci.*, **17**, 442–453.

Goreau, T.J., 1990: Coral bleaching in Jamaica. *Nature*, **343**, 417.

Heyman, A.M., 1988: Self-finance resource management in the Caribbean: A direct approach to maintaining marine biological diversity. Waterlog, Mississippi-Alabama Sea Grant Consortium.

Hopley, D., 1982: *The Geomorphology of the Great Barrier Reef*. John Wiley and Sons, New York, 453 pp.

Hoskin, C.M., 1963: Recent carbonate sedimentation on Alacran Reef, Yucatan, Mexico. *Nat. Acad. Sci.-Nat. Res. Council Publ. 1089*, 160 pp.

Huston, M.A., 1985: Patterns of species diversity on coral reefs. *Ann. Rev. Ecol. Syst.*, **16**, 149–177.

James, N.P. and Ginsburg, R.N., 1979: The seaward margin of Belize barrier and atoll reefs. *Spec. Publ. Int'l. Assoc. Sedimentologists*, **3**, 191 pp.

Kaufman, L.S., 1983: Effects of Hurricane Allen on reef fish assemblages near Discovery Bay, Jamaica. *Coral Reefs*, **2**, 43–47.

Kenchington, R.A., 1988: Managing reefs and inter-reefal environments and resources for sustained exploitive, extractive and recreational uses. *Proc. 6th Int'l. Coral Reef Symp.*, Vol. 1, 81–87.

Kinsey, D.W. and Hopley, D., 1991: The significance of coral reefs as global carbon sinks – response to greenhouse. *Paleogeog., Paleoclim., Paleoecol.*, **89**, 363–377.

Land, L.S. and Moore, C.H., 1977: Deep forereef and upper island slope, north Jamaica. *Amer. Assoc. Petrol. Geol., Studies in Geology*, **4**, 53–65.

Lang, J.C., Wicklund R.I. and Dill, R.F., 1988: Depth- and habitat-related bleaching of zooxanthellate reef organisms near Lee Stocking Island, Exuma Cays, Bahamas. *Proc. 6th Int'l. Coral Reef Symp.*, Vol. 3, 269–274.

LaPointe, B.E., 1989: Caribbean coral reefs: are they becoming algal reefs? *Sea Frontiers*, **35**, 83–91.

Lewis, J.B., 1960: The coral reefs and coral communities of Barbados, WI. *Can. J. Zool.*, **38**, 1133–1145.
Lewis, J.B., 1984: The *Acropora* inheritance: a reinterpretation of the development of fringing reefs in Barbados, West Indies. *Coral Reefs*, **3**, 123–130.
Lewis, J.B., 1989: The ecology of *Millepora*. *Coral Reefs*, **8**, 99–107.
Lewis, J.B., Axelson, F., Goodbody, I., Page C. and Chislett, G., 1968: Comparative growth rates of some reef corals in the Caribbean. Mar. Sci. Rept., 10, McGill Univ., 26 pp.
Lighty, R.G., Macintyre, I.G. and Stuckenrath, R., 1978: Submerged early Holocene barrier reef, southeast Florida shelf. *Nature*, **276**, 59–60.
Lighty, R.G., Macintyre, I.G. and Stuckenrath, R., 1982: *Acropora palmata* reef framework: A reliable indicator of sea level in the western Atlantic for the past 10,000 years. *Coral Reefs*, **1**, 125–130.
Macintyre, I.G., 1988: Modern coral reefs of western Atlantic: New geological perspective. *Bull. Amer. Assoc. Petrol. Geol.*, **72**, 1360–1369.
Macintyre, I.G., Burke, R.B. and Stucknerath, R. 1977: Thickest recorded Holocene reef section, Isla Perez core hole, Alacran Reef, Mexico. *Geology*, **5**, 749–754.
Macintyre, I.G. and Glynn, P.W., 1976: Evolution of modern Caribbean fringing reef, Galeta Point, Panama. *Bull. Amer. Assoc. Petrol. Geol.*, **60**, 1054–1072.
Milliman, J.D., 1969: Four southwestern Caribbean atolls: Courtown Cays, Albuquerque Cays, Roncador Bank and Serrana Bank. *Atoll Res. Bull.* No. 129, 41 pp.
Milliman, J.D., 1973. Caribbean coral reefs. In: Jones, O. and Endean, R. (eds), *Biology and Geology of Coral Reefs*. Academic Press, New York.
Muscatine, L., 1980: Productivity of zooxanthellae. In: Falkowski, P.G. (ed), *Primary Productivity in the Sea*. Plenum Press, pp. 381–402.
Neumann, A.C., Gebelein, C.D. and Scoffin, T.P. 1970: The composition, structure and erodability of subtidal flats, Abaco, Bahamas. *J. Sed. Pet.*, **40**, 274–297.
Neumann, A.C. and Macintyre, I.G., 1985: Reef response to sea-level rise: keep-up, catch-up or give-up. *5th Intern. Coral Reef Congress Proc.*, Vol. 3, 105–110.
Newell, N.D., 1959: The coral reefs. *Natural History*, **68**, 119–131.
Oberdorfer, J.A. and Buddemeier, R.W., 1988: Climate change: Effects on reef island resources. *Proc. 6th Intern. Coral Reef Symp.*, **3**, 523–527.
Porter, J.W., Battey, J.F. and Smith, G.J., 1982: Perturbation and change in coral reef communities. *Proc. Nat. Acad. Sci.* (USA), **79**, 1678–1681.
Purdy, E.G., 1974: Karst determined facies patterns in British Honduras: Holocene carbonate sedimentation model. *Bull. Amer. Assoc. Petrol. Geol.*, **58**, 825–855.
Roberts, H.H. and Murray, S.P., 1983: Controls on reef development and the terrigenous-carbonate interface on a shallow shelf, Nicaragua (Central America). *Coral Reefs*, **2**, 71–80.
Roberts, L., 1991: Greenhouse role in reef stress unproven. *Science*, **253**, 258–259.
Rodriquez, A., 1981: Marine and coastal environmental stress in the wider Caribbean region. *Ambio*, **10**, 291–294.
Rogers, C.S., McLain, L. and Zullo, E., 1988: Damage to coral reefs in Virgin Islands National Park and Biosphere Reserve from recreational activities. *Proc. 6th Intern. Coral Reef Symp.*, **2**, 405–410.
Shick, J.M. and Lesser, M.P., 1991: An overview of photooxidative stress and defenses against it in coral reef symbiosis. In: D'Elia, C.F., Buddemeier, R.W. and Smith, S.V., Workshop on coral bleaching, coral reef ecosystems and global change, Miami, Florida.
Shinn, E.A., 1963: Spur and groove formation on the Florida reef tract. *J. Sed. Pet.*, **30**, 471–480.
Shinn, E.A., 1976: Coral reef recovery in Florida and the Persina Gulf. *Environ. Geol.*, **1**, 241–254.

Shinn, E.A., 1989. What is really killing the corals? *Sea Frontiers*, **35**, 72–81.
Steen, G.R. and Muscatine, L., 1987: Low temperature evokes rapid exocytosis of symbiotic algae by a sea anemone. *Biol. Bull.*, **172**, 246–263.
Stoddart, D.R., 1962: Three Caribbean atolls: Turneffe Islands, Lighthouse Reef and Glovers Reef, British Honduras. *Atoll Res. Bull.* No. 87, 147 pp.
Stoddart, D.R., 1990: Coral reefs and islands and predicted sea-level rise. *Progress in Phys. Geogr.*, **14**, 521–536.
Tomascik, T. and Logan, A., 1990: A comparison of peripheral growth rates in the recent solitary coral *Scolymia cubensis* (Milne-Edwards and Haime) from Barbados and Bermuda. *Bull. Mar. Sci.*, **46**, 799–806.
US AID, 1987: *Caribbean marine resources*. 80 pp.
Van't Hof, T., 1988: Management of coral reefs and associated systems and resources in the western Atlantic: A status review. *Proc. 6th Intern. Coral Reef Symp.*, Vol. 1, 89–95.
Vincente, V.P., 1990: Response of sponges to autotrophic endosymbionts during the coral-bleaching episode in Puerto Rico. *Coral Reefs*, **8**, 199–202.
Wells, J.W., 1957: Coral reefs. In: Hedgpeth, J. (ed), *Treatise on Marine Ecology and Paleoecology*. Geol. Soc. Amer. Mem. 67, 609–631.
Wiens, H.J., 1962: *Atoll Environment and Ecology*. Yale Univ. Press, New Haven.
Williams, E.H. and Bunkey-Williams, L., 1988: Bleaching of Caribbean coral reef symbionts in 1987–1988. *Proc. 6th Intern. Coral Reef Symp.*, Vol. 3, 313–318.
Williams, E.H. Jr, Goenaga, C. and Vincente, V., 1987: Mass bleachings on Atlantic coral reefs. *Science*, **238**, 877–888.
Wood, E.J.F. and Johannes, R.E., 1975: *Tropical Marine Pollution*. Elsevier Oceanography Series, 12, 192 pp.

Part 5
Socio-economic effects

14

Human Health

Donald P. de Sylva[1]

Abstract

Climate-change-induced temperature rise is judged as having a potentially detrimental effect on human health in the region, but definitive environmental information needs to be measured. A regional biometeorological data base should be established, with easy PC-based access by the various public health agencies within the area's countries. In addition, epidemiological data are needed on the relation between nutrition and disease, particularly focused on potential long-term changes in dietary habits associated with changes in the food supply.

1 Introduction

It is well known that weather and climate affect all living things. Meteorological conditions influence our way of living, our food habits, our socio-economic relationships, and thus agents causing or transmitting parasites and disease (Tromp, 1980). This chapter, which draws heavily on the exhaustive review by White and Hertz-Picciotto (1985), will concentrate on the latter effects of climate change.

As climate becomes warmer, the boundaries of the tropics may extend into the present subtropics, and current temperate areas may become subtropical. Hence, presently extra-tropical regions will more resemble the region climatologically. As air temperatures gradually increase, human morbidity rates should increase from a variety of reasons to be discussed, while human mortality rates will increase significantly (Kalkstein *et al.*, 1986). Ways in which climate change may affect human health are shown in Fig. 14.1.

2 Weather Variables and Human Health

Acute effects of high temperature, heat stroke, and heat edema can cause death if not treated quickly (White and Hertz-Picciotto, 1985). Persons with chronic diseases, especially the elderly, are particularly susceptible to aggravation of disease from very hot weather. Temperatures in the tropical Gulf/Caribbean are ideal for survival and propagation of

[1]University of Miami/RSMAS, Miami, FL 33149, USA

causative agents for some bacterial, viral, and parasitic diseases (Kellogg and Schware, 1981; Robinson, 1985). The effects of high temperature on human health are modified by humidity, which is high throughout the low-lying maritime countries, especially along the coast. Certain levels of humidity are ideal for reproduction and survival of pathogenic bacteria, viruses, parasites, and their vectors (White and Hertz-Picciotto, 1985; Tromp, 1980).

Precipitation may increase humidity and may modify the ecological habitats of parasites, their hosts, and insect vectors. Wind, together with increased temperatures and humidity, may affect human ability to thermoregulate, and can help in spreading the causative agents of disease, insect vectors, and allergens.

2.1 Mortality and Morbidity Related to Climate and Weather

A warm climate may be more favourable for the propagation of airborne and waterborne communicable diseases (White and Hertz-Picciotto, 1985). In areas where such diseases occur, mortality rates among the young are high, thus leaving a smaller elderly population to die of diseases common to the elderly. If temperature rises much above 27°C, there may be upsurges in heart and cerebrovascular mortality (Tromp, 1980). Fortunately, the proximity of many nations to the ocean, with its high heat capacity, prevents extremely high air temperatures.

Mortality from cancer includes skin melanoma, thought to be associated with the amount of ultraviolet radiation, which is not expected to change with a greenhouse warming, unless a change in cloud cover occurs concomitantly. On the other hand, any decrease in the protective ozone

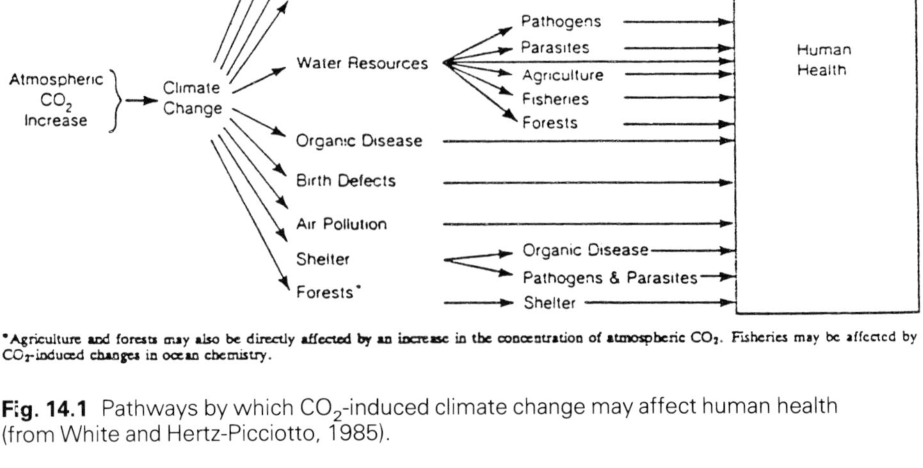

*Agriculture and forests may also be directly affected by an increase in the concentration of atmospheric CO_2. Fisheries may be affected by CO_2-induced changes in ocean chemistry.

Fig. 14.1 Pathways by which CO_2-induced climate change may affect human health (from White and Hertz-Picciotto, 1985).

layer could increase skin disease considerably. Burkitt's lymphoma is a cancer found in areas where the temperature exceeds 15°C and annual rainfall exceeds 500 mm. It occurs where malaria is highly endemic, and thus as global warming increases, new geographic regions should be increasingly subjected to malaria-carrying mosquitoes (Burkitt, 1983).

Hotter summers in tropical climates will probably increase heat-related deaths (White and Hertz-Picciotto, 1985). Humidity, precipitation and wind act synergistically with temperature to modify thermal stress. Socio-economic factors are a component in the death rate during heat waves, those in the lower status generally having the largest increase in deaths (Buechley *et al.*, 1972). These persons tend to live predominantly in the inner cities, frequently in high-density, inadequately ventilated housing. More heat-stroke victims occur in upper levels of multi-storey buildings (Clarke, 1972; Kilbourn *et al.*, 1982).

2.1.1 *Extreme weather events*

Droughts affect human health, mainly through failure of crops and forage, consequently affecting nutrition and economics. Floods also can affect nutritional status by destroying crops, while hurricanes can destroy crops, kill and injure people, and may also increase the spread of disease. Hurricanes are expected to increase in frequency (de Sylva, 1986) and intensity (Emanuel, 1987), but there is no way to predict this change at the present time (*cf.* Gray, Chapter 5).

2.1.2 *Airborne materials related to human health*

Numerous airborne fungi, viruses, bacteria and allergens causing human illness are distributed in the atmosphere, and factors such as temperature, precipitation, humidity and wind may affect their multiplication, dispersal and survival. For example, the tetanus bacterium *Clostridium tetanus* persists longer in warm, moist soil than in dry, cool climates, and consequently people in the tropics are at particular risk (Robinson, 1985). Dust at high humidity can serve as a site for bacterial survival (Lidwell, 1964). The effect of temperature on bacteria is modified by relative humidity, and high relative humidity may protect some bacteria from the lethal effects of ultraviolet radiation (White and Hertz-Picciotto, 1985).

Several viral diseases in animals have been shown to cause epidemics, quite far from the source; almost certainly transmittal was airborne. By implication, this could happen in human diseases, although indoor spread of viral diseases is more likely, especially where temperature and humidity are high (Robinson, 1985). Hepatitus B virus is more easily transmitted in the tropics. Similarly, plant pollens are one of the most common causes of human allergies and are affected by precipitation, temperature and soil moisture.

2.1.3 *Seasonal diseases caused by microorganisms*

Most airborne infectious diseases transmitted from one person to another are prevalent in cold or changeable weather. However, epidemic cerebral meningitis is associated with warm, humid air masses (Tromp, 1963). Poliomyelitis is more prevalent during the summer (Tromp, 1980), and is associated with higher relative humidity (Armstrong, 1952), especially with the rainy season. It can be assumed that with a greenhouse warming, summer diseases would increase and winter diseases would decrease (White and Hertz-Picciotto, 1985), these being associated with changes in humidity and precipitation.

Of the human-carrier diseases influenced by climate and weather, waterborne cholera is prevalent during the summer. It becomes dormant when the water is cold and increases with warming (May, 1958), its incidence also being affected by rainfall and humidity. In addition, bacillary dysentery prevails in tropical countries during the rainy season (Sangster, 1977).

Periodic outbreaks of encephalitis are associated with temperature and rainfall (Burnet, 1952; Hess *et al.*, 1963), the rainfall and flooding providing breeding ponds for mosquitoes. Temperature increases, when the viruses and mosquitoes are present, increase the probability of epidemics. Encephalitis would be expected to increase accordingly with an increase in rainfall and temperature.

2.1.4 *Parasitic diseases*

These occur more frequently in tropical and subtropical climates, sometimes in more humid areas (Robinson, 1985). Such infections frequently contribute to decreased work ability, malnutrition and greater susceptibility to other diseases (White and Hertz-Picciotto, 1985; Robinson, 1985). For example, malaria may be associated with an increase or decrease in rainfall which modify the habitats of mosquitos, which have different proclivities to aquatic habitats (Learmonth, 1977; Tromp, 1980). The rate of development of the parasite increases with temperature (Garnham, 1964).

For parasites requiring an intermediate host, schistosomiasis (bilharzia) is an extensive problem in parts of the region (Robinson, 1985). Snails, the intermediate host of this disease, thrive in warm waters up to 35°C, above which is lethal; the most favourable temperature for their life cycle is 26–28°C (Purnell, 1966). The snail population peaks during the rainy season (Anderson and May, 1979), but artificial water resources (dams, irrigation canals and water impoundments) are ideal for living and breeding of the snails. Human emigration/immigration may cause a spread of the parasite, associated with increased temperature and precipitation.

Parasites not needing intermediate hosts as vectors include hookworms, whose eggs are very sensitive to climate. The optimal temperature for egg hatching is 27–32°C, plus at least 1270 mm of annual rainfall (Gilles, 1984). Accordingly, increased temperature and rainfall may expand the geographic range of hookworms and helminths, while changed human migration patterns resulting from climate change may spread hookworm incidence (Robinson, 1985).

2.1.5 *Nutrition*

Agriculture, both plant and animal, will be affected by climate changes (Decker *et al.*, 1985). These changes will depend upon regional changes in rainfall, humidity and soil-moisture patterns and their effects upon terrestrial and aquatic plants and animals (Newman, 1975). The need for more food per person may increase with a climate warming, as more calories are needed for heavy work by humans at temperatures higher than 30°C (Consolazio *et al.*, 1961). Low nutritional levels in humans may also increase susceptibility to parasites and disease (Scrimshaw *et al.*, 1968; Mata *et al.*, 1972).

2.1.6 Water

Climate change may have serious effects on the quality and quantity of fresh water (Callaway and Currie, 1985). Fisheries may be affected by the amount of water from estuaries and the role which fresh water plays in the production of estuarine fish and shellfish (de Sylva, 1986; Snedaker and de Sylva, 1987), which are so vital to human nutrition. Forests are dependent upon rainfall for their survival (Solomon and West, 1985). Clearly, water needed for sanitation, drinking and cooking depends upon rainfall; where water is not abundant, human health is endangered. Similarly, fresh water must be clean for drinking and cooking. Climate change will involve changes in groundwater availability; as sea-level rises, salt-water intrusion will reduce the mixing of fresh water with salt water, thus increasing the local effects of estuarine pollution and the intrusion of salt water into estuaries (de Sylva, 1986). This is also associated with a decrease in the potability of fresh-water outflows due to increased levels of sodium chloride. This saltier water contaminates aquifers for agricultural and industrial purposes (Hull *et al.*, 1986), increases its danger by causing hypertension in humans (Braun and Florin, 1963), and, because of the widespread and common phenomenon of hypertension in warm countries (Robinson, 1985), makes salinization of fresh-water supplies a serious problem due to climate warming and sea-level rise. Hazardous-waste sites in coastal areas containing metal drums would be subject to damage because of the highly corrosive nature of seawater on metal (Flynn *et al.*, 1984), as sea-level rise encroaches on the coastal waste sites.

2.1.4 Changes in the marine environment

Some changes in the marine environment, which have not been covered elsewhere in this volume, may also occur. Sharks are especially attracted to warmer waters (de Sylva, 1969) and may become more numerous as coastal waters warm, while toxic red tides, which kill off marine life from the proliferation of dinoflagellates, may increase with increasing temperature and coastal eutrophication (Anderson *et al.*, 1985). Similarly, ciguatera is a tropical human disease caused by the ingestion of tropical fishes that, in turn, have eaten various organisms which have ingested dinoflagellates at several levels in the marine food chain of coral reefs (Bagnis, 1984; Lewis, 1986).

3 Recommendations

1) Information is needed on the rate of climate change in the region, as measured by long-term information on air and water temperature throughout the area and contiguous lands and waters.

2) Epidemiological records should be upgraded wherever possible, and a central collecting agency be established. This could be carried out together with the establishment of biometeorological data collection agencies (Tromp, 1980). These should include records of mortality and morbidity data on organic diseases, and information on thermoregulation and acclimation as related to organic diseases such as heart and cerebrovascular diseases. Data also are needed on birth defects and infant and foetal deaths, and on the incidence, geographic occurrence and seasonality of diseases from bacteria, viruses, parasites and allergens, especially in relation to biometeorological factors, including a careful monitoring programme of these.
3) Epidemiological data are needed on the relationship between nutrition and disease, long-term change in dietary habits, and long- or short-term changes in the kinds and quantity of seafood eaten, as related to coastal marine climate change. A regional United Nations centre could supply data from cooperating nations.

4 References

Anderson, D.M., White, A.W. and Baden, D.G. (eds), 1985: *Toxic dinoflagellates*. Elsevier, New York, 561 pp.

Anderson, R.M. and May, R.M., 1979: Prevalence of schistosome infections within molluscan populations: observed patterns and theoretical predictions. *Parasitology*, **79**, 63–95.

Armstrong, C. 1952: Poliomyelitis and the weather. *Proceedings of the National Academy of Science*, **38**, 613–618.

Bagnis, R., 1984: Ciguatera fish poisoning: an objective witness of the coral reef stress. In Salvat, B. (ed), *Human Impacts on Coral Reefs: Facts and Recommendations*. Antenne Mus., Polynésie franç., Moorea, 241–252.

Braun, P., and Florin, A.A., 1963: Drinking water and congestive heart failure sodium concentration of selected New Jersey water supplies. *J. Med. Assoc., New Jersey*, **60**, 504–509.

Buechley, R.W., Van Bruggen, J. and Trippi, L.E., 1972: Heat island = death island? *Environmental Res.*, **5**, 85–92.

Burkitt, D.P., 1983: The discovery of Burkitt's lymphoma. *Cancer*, **51**, 1777–1786.

Burnet, F.M., 1952: Murray Valley encephalitis. *Amer. J. Public Health*, **42**, 1519–1521.

Calloway, J.M. and Currie, J.W., 1985: Water resource systems and changes in climate and vegetation. In: White, M.R. (ed), *Characterization of Information Requirements for Studies of CO_2 Effects: Water Resources, Agriculture, Fisheries, Forests, and Human Health*. US Dept. Energy, DOE/ER-0236, pp. 23–68.

Clarke, J.F., 1972: Some effects of the urban structure on heat mortality. *Environmental Res.*, **5**, 93–104.

Consolazio, C.F., Shapiro, R., Mastinson, J.E. and McKinzie, P.S.L., 1961: Energy requirements of men in extreme heat. *J. Nutrition*, **73**, 126–134.

Decker, W.L., Jones, V. and Achutuni, R., 1985: The impact of CO_2-induced climate change on US agriculture. In: White, M.R. (ed), *Characterization of Information Requirements for Studies of CO_2 Effects: Water Resources, Agriculture, Fisheries, Forests, and Human Health*, US Dept. Energy, DOE/ER-0236, pp. 69–94.

de Sylva, D.P., 1969: Theoretical considerations of heated effluents on marine fishes. In: Krenkel, P.A. and Parker, F.L. (eds), *Biological aspects of thermal pollution*. Vanderbilt Univ. Press, pp. 229–293.

de Sylva, D. P., 1986: Increased storms and estuarine salinity and other ecological impacts of the greenhouse effect. In: Titus, J.G. (ed), *Effects of Changes in Stratospheric Ozone and Global Climate, 4, Sea-Level Rise*. UNEP/EPA, pp. 153–164.

Emanuel, K.A., 1987: The dependence of hurricane intensity upon climate. *Nature*, **326(2)**, 483–485.

Flynn, T.S., Walesh, S.G., Titus, J.G. and Barth, M.C., 1984: Implications of sea-level rise for hazardous waste sites in coastal floodplains. In: Barth, M.C. and Titus, J.G. (eds), *Greenhouse Effect and Sea-Level Rise*. Van Nostrand Reinhold, New York, pp. 271–294.

Garnham, P.C.C., 1964: Factors influencing the development of protozoa in their arthropodan hosts. In: Taylor, A.E.R. (ed), *Host-Parasite Relationships in Invertebrate Hosts*, second symposium, British Society for Parasitology. Blackwell Scientific Publications, Oxford, United Kingdom, pp. 33–50.

Gilles, H.M., 1984: Intestinal nematode infections. In: Strickland, G.T. (ed), *Hunter's Tropical Medicine*. W.B. Saunders, Philadelphia, PA.

Hess, A.D., Cherubin, E.C. and Lamotte, L.C., 1963: Relation of temperature to activity of western and St. Louis encephalitis viruses. *Amer. J. Tropical Medicine*, **12**, 657–667.

Hull, C.H.V., Thatcher, H.L. and Tortoriello, R.C., 1986: Salinity in the Delaware River Estuary. In: Hull, C.H.J. and Titus, J.G. (eds), *Greenhouse Effect, Sea-Level Rise, and Salinity in the Delaware Estuary*. US Environmental Protection Agency, EPA 230-05-86-010, pp. 15–39.

Kalkstein, L.S., Davis, R.E., Skindlov, J.A. and Valimont, K.M., 1986: The impact of human-induced climatic warming upon human mortality: A New York case study. In: Titus, J.G. (ed), *Effects of Changes in Stratospheric Ozone and Global Climate, 4: Sea-Level Rise*. UNEP/EPA, pp. 275–293.

Kellogg, W.W. and Schware, W., 1981: *Climate Change and Society: Consequences of Increasing Atmospheric Carbon Dioxide*. Westview Press, Boulder, CO, 178 pp.

Kilbourne, E.M., Choi, K., Jones, T.S., Thacker, S.B. and the Field Investigation Team, 1982: Risk factors for heatstroke – a case-control study. *J. Amer. Med. Assoc.*, **247**, 3332–3336.

Learmonth, A.T.A., 1977: Malaria. In: Howe, G.M. (ed), *A World Geography of Human Diseases*. Academic Press, New York, 61–108.

Lewis, N.D., 1986: Disease and development: ciguatera fish poisoning. *Soc. Sci. Med.*, **23(10)**, 983–993.

Lidwell, O.M., 1964: Microbiology of the atmosphere and airborne infection. In: Licht, S. (ed), *Medical Climatology*. E. Licht, New Haven, CN, pp. 131–158.

Markell, E.K. and Voge, M., 1981: *Medical Parasitology*. W.B. Saunders, Philadelphia, PA.

Mata, L.J., Ursutia, J.J., Albertazzi, C., Pellecer, O. and Arellano, E., 1972: Influence of recurrent infections on nutrition and growth of children in Guatemala. *Amer. J. Clinical Nutrition*, **25**, 1267–1275.

May, J., 1958: *The Ecology of Human Disease*. MD Publications, New York.

Newman, M.T., 1975: Nutritional adaptation in man. In: Damon, A. (ed), *Physiological adaptation*. Oxford University Press, London, pp. 210–259.

Purnell, R.E., 1966: Host-parasite relationships in schistosomiasis. *Annals of Tropical Medicine and Parasitology*, **60**, 90–93.

Robinson, D. (ed), 1985: *Epidemiology and the Community Control of Disease in Warm Climate Countries*, 2nd ed. Churchill Livingston, Edinburgh, 767 pp.

Sangster, G., 1977: Diarrhoeal diseases. In: Howe, G.M. (ed), *A World Geography of Human Diseases*. Academic Press, New York, pp. 145–174.

Scrimshaw, N.S., Taylor, C.E. and Gordon, J.E., 1968: *Interaction of Nutrition and Infection*. World Health Organization, Monograph Series No. 57. World Health Organization, Geneva, Switzerland.

Snedaker, S C. and de Sylva, D.P., 1987: Impact of climate change on coastal resources: implications for property values, commerce, estuarine environments, and fisheries, with special reference to south Florida. In: Meo, M. (ed), *Symposium on Climate Change in the Southern United States: Future Impacts and Present Policy Issues*. US Environmental Protection Agency, Washington, DC, pp. 187–215.

Solomon, A.M. and West, D.C., 1985: Potential responses of forests to CO_2-induced climate change. In White, M.R. (ed), *Characterization of Information Requirements for Studies of CO_2 Effects: Water Resources, Agriculture, Fisheries, Forests, and Human Health*. US Dept. of Energy, DOE/ER/00236, pp. 145–170.

Tromp, S.W., 1963: *Medical Biometeorology*. Elsevier Publishing, New York.

Tromp, S.W., 1980: *Biometeorology: the Impact of Weather and Climate on Humans and their Environment*. Heyden and Son Ltd., London.

White, M.R. and Hertz-Picciotto, I., 1985: Human health: analysis of climate-related to health. In: White, M.R. (ed), *Characterization of Information Requirements for Studies of CO_2 Effects: Water Resources, Agriculture, Fisheries, Forests, and Human Health*. US Dept. Energy DOE/ER-0236, pp. 173–206.

15

Climatic Changes and Socio-Economic Impacts

Anders Alm[1], *Erik Blommestein*[2] *and James M. Broadus*[3]

Abstract

Many of the physical effects of climatic changes in the Gulf of Mexico/Caribbean Sea region will have socio-economic impacts. The presumption is that, for most cases, the net effect will be costly, but this will depend on the facts of each case. For several reasons, prior work has been mostly on the potential impacts of sea-level change, but other climatic changes in the region may have impacts which equal or surpass those. A matrix that relates types of climate-change effects to qualitative impacts on economic sectors is presented. Potential impacts on fisheries, agriculture and forestry, tourism, settlements and structures, the immediate coastal zone, public health, and waste disposal and drainage are discussed. Descriptive and prescriptive issues of public response are examined, along with attention to techniques and research issues in efforts to estimate socio-economic impacts.

1 Introduction

Industrialization and combustion of fossil fuels has increased the atmospheric level of carbon dioxide (CO_2). A continuation of this increase is expected to modify the world climate and has given rise to speculation that varies from fears of impending disaster, to the belief that there are no problems (WMO/ICSU/UNEP, 1980; National Academy of Sciences, 1983; Morrisette, 1987; Ausubel, 1991).

The increased level of carbon dioxide changes the atmosphere's energy balance and may result in a temperature rise at the Earth's surface and lower atmosphere; an intensification of the so-called 'greenhouse' effect. The warming of the Earth's surface will have different implications. This warming is associated with several effects. Increased sea-surface temperature for instance, may change the global-scale oceanic circulation, with influence on weather systems, rainfall patterns, wind trends, storm activities, etc. Worldwide concern has risen regarding the effects of significant warming at the poles. Increased temperature could gradually melt the polar ice-caps, resulting in addition of water masses to the

[1]IOC Secretariat for IOCARIBE, Cartagena, Colombia, SA
[2]ECLAC, Port of Spain, Trinidad and Tobago
[3]Woods Hole Oceanographic Institution, Woods Hole, MA 02543, USA

oceans and accelerated sea-level rise. Indeed, even without melting glaciers and ice-caps, thermal expansion of the oceans' water volume would be expected to raise sea levels.

Sea-level change is not a new phenomenon (Cartwright *et al.*, 1985; Pirazzoli, 1985). The level of the oceans also has been changing in the more recent past with great geographical differences (Aubrey and Emery, 1983; Barnett, 1984; Emery and Aubrey, 1985; Solow, 1987). However, the human-induced climatic changes are expected to accelerate the global trend of sea-level rise that is currently believed underway. This acceleration may have serious consequences.

2 Climate-Change Impacts

Significant sea-level rise and increased sea-surface temperature are two complex effects of human-induced climatic changes, the implications of which are difficult to survey. The changes are supposed to be lesser in the tropics than at the high latitudes (*q.v.* Fig. 1.6), with great local differences arising from such factors as in coastal geomorphology. In those places where coastal areas are being uplifted by tectonic forces, for example, an acceleration in global sea-level rise could work to stabilize local shorelines. Or, in areas with very steep coastlines, rather large changes in sea level might be accommodated with little geographic alteration. Because of the complex geology and poorly understood tectonic behaviour of the region, such local differences are an added significance. In any case there is, due to local effects, considerable variation in the magnitude and direction of sea-level rise within the region (Hendry, Chapter 7; Hanson and Maul, Chapter 9), that will necessitate site-specific studies.

Much of the literature on the socio-economic impacts of climatic changes has been focused on (1) the United States and (2) on the impacts of sea-level rise (Chen *et al.*, 1983; Titus, 1986a). The former reflects the availability of resources and the existence of many highly developed areas which are already exposed to the effects of a rising sea level. The concentration on sea-level rise also can be explained by the already apparent impacts of sea-level rise on some coastal communities and ecosystems.

Another reason for preoccupation with sea-level rise is that climate models are not yet of a sufficient precision to predict the extent or even direction of some of the other anticipated changes. Such changes as altered precipitation patterns or increased storm frequency, however, may have impacts that equal or surpass those from a sea-level rise (Gray, Chapter 5). A case in point is that existing climate models cannot yet forecast precipitation changes in specific regions (Titus *et al.*, 1987). In this region, with its dependency on rain-fed agriculture, changes in precipitation may drastically alter the livelihood of large segments of the population. In contrast, it is generally agreed that the rate of sea-level rise will accelerate, although estimates are still wide ranging and are subject to revision in the light of better understanding of global processes (Brooks, 1985; Maul, Chapter 1).

Many studies advocate the incorporation of the anticipated sea level rise into public planning processes (Titus, 1986b). The high level of

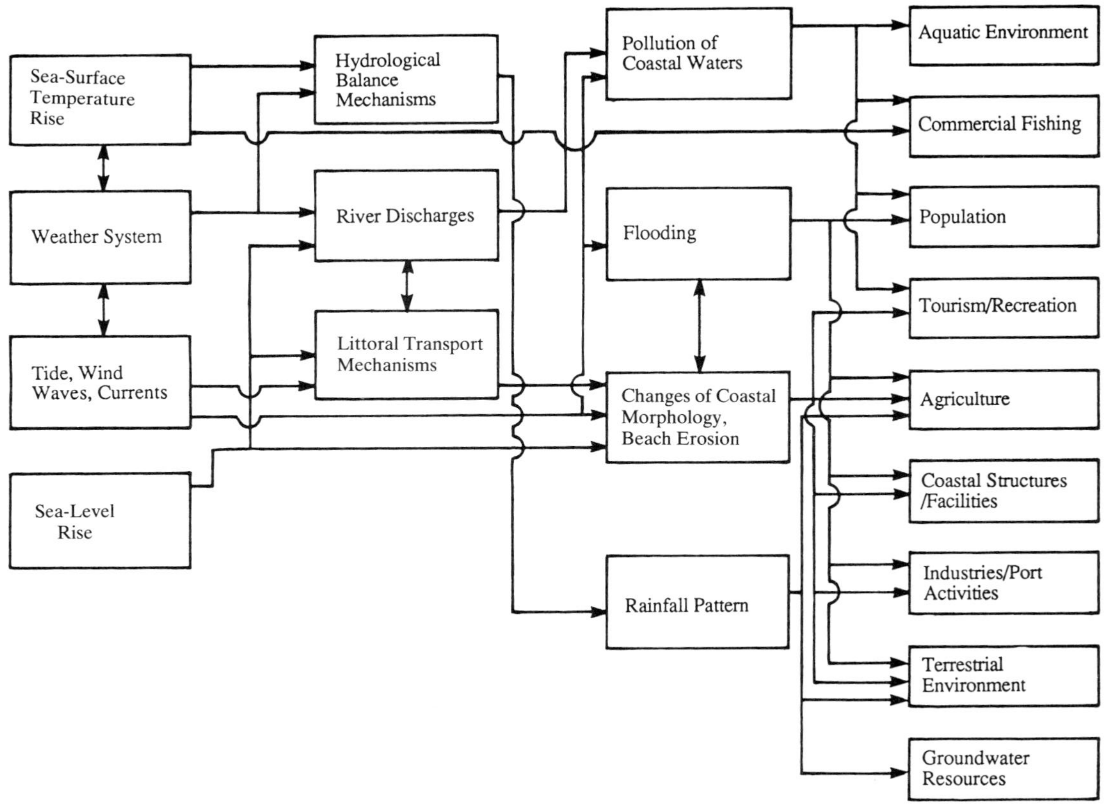

Fig. 15.1 Schema depicting implication of climatic changes on important socio-economic activities.

uncertainty about the magnitude and direction of some of the changes, however, reduces the willingness of decision-makers to initiate responses to the anticipated climatic changes. Therefore, there is a need to reduce significantly the degree of uncertainty about the likelihood, the extent and the directions of these climatic changes (*q.v.* Wigley and Santer, Chapter 2; Gallegos *et al.*, Chapter 3). Nonetheless the potential socio-economic impacts of the largest predicted changes are enormous; therefore, governments and the private sector of the region should include the evaluation of likely and possible impacts in their planning efforts and project design. This is particularly valid for the incorporation of sea-level rise, about which there is less uncertainty, and of which the impacts are already visible in much of the region.

2.1 Types of Impact

Many, if not most, of the physical effects of climatic changes will have socio-economic impacts. Such impacts may be essentially of a micro-economic or localized nature but also will have macro-economic, or economy-wide, implications. In relation to sea-level rise it will be generally true that the socio-economic impacts will have greater macro-economic implications on the smaller and the more coastal-zone-oriented economies (IPCC, 1990). In this respect the smaller islands with a high dependence on tourism or fisheries or both, seem to be particularly vulnerable.

Socio-economic impacts of climatic changes may be derived directly from the physical changes (for example, loss of property due to inundation, increased storm or hurricane damage) or indirectly through attendant changes in ecosystems (such as changing nursery grounds). The complex of underlying interactions, feedbacks and internal dependencies is still not well understood (Engelen *et al.*, Chapter 16), and many of the physical and ecological changes result also from human-induced effects other than climatic changes (sea-level rise is only one of the factors that contribute to beach erosion). Such effects will often reinforce the impacts of climatic changes, and indeed the total effect of all impacts may be greater than the sum of the individual impacts. This will make it difficult to establish clear cause-and-effect relationships, which again will not facilitate acceptance by policy-makers of advice for action.

Fig. 1.4 is a generalized presentation of the interactions between physical, ecological and socio-economic changes, while Fig. 15.1 represents some of the interactions in more detail.

The socio-economic impacts of climatic changes are typically portrayed as a sum of costs (lost or spent resources). Obviously, this can be misleading since some climatic changes will also result in benefits (say, increased agricultural productivity) in some areas. What is of interest initially is the net effect of these combined costs and benefits. The presumption is that, for most cases, the net effect will be costly. This will depend, however, on the facts of each case. In every case, it is important not to confuse *transfers* of benefits and costs within an economy with the *creation* of new benefits and costs by a climatic change. The construction sector seems a likely beneficiary of a decision to construct sea-defense works, for example, but this apparent benefit is just a transfer of resources from other uses. The cost of the defenses may be smaller than the cost of land lost to inundation, but the diversion of resources to the defense project is still a true cost to the economy that would not be incurred in the absence of climatic change.

If the above qualifications are accepted then the economic impacts could be categorized as:

1) Losses (improvements) in the existing stock of land and capital. Examples would be the loss of property caused by an increased frequency and intensity of tropical storms and hurricanes (this implies a loss in the capacity to produce future goods and services) or an improved productivity of arable land as a result of a higher concentration of CO_2 and of more rainfall, in those, as yet unknown, areas where an increase in precipitation will be one of the effects of climatic change.
2) Increase (reduction) in costs associated with maintaining the production of goods and services at the current (i.e., pre-change) level. Examples would be (a) expenditures on sea-defense works, (b) cost associated with a retreat from the seashore.

(The above categories are related to the efficiency aspects of the economy. A third category, below, would focus on the distributional or equity aspects of the economy.)

3) Which groups in a society will bear the costs of any adjustment of the economy in response to climatic changes? Or alternatively, which groups will benefit from any such changes? Eventually climatic changes may involve changes in the allocation of labour and capital among sectors. While a reallocation may be relatively insignificant for large, highly developed and diversified economies, such a process is likely to be more difficult in smaller, less diversified economies. With respect to sea-level rise, such issues will be brought into focus when governments of the region have to make decisions on the retreat or defense of low-lying areas which alternatively are being used for peasant agriculture, plantation agriculture, housing of the urban poor or that of the middle classes, industrial sites, etc. Having made a decision, the next question will then relate to who will bear the cost for relocation or for the construction of sea-defense works.

A corresponding issue to that of the distribution of national impacts is the occurrence of possible cross-boundary effects. Such effects may be relevant when impacts of climatic change have an effect in more than one country. In such cases international cooperation and possibly co-funding is likely to be cost-effective and beneficial to all parties. Examples of cross-boundary effects could be shifting nursery and fishery areas or the existence of continuous, cross-border, low-lying areas which may warrant some form of sea defense.

A related issue, but considerably more complex and likely to be subject to extensive international discussion and negotiation, pertains to the responsibility for the causes of global climatic changes and eventually to the question of the distribution of the cost adjustment to the changes in climate between the lesser developed countries and the industrial economies, both market and planned. These questions will become relevant since the lesser developed countries, as discussed below, are likely to lack the resources, technology and organization to implement the most technically efficient responses. These concerns were of major importance at UNCED (United National Conference on Environment and Development) held in Rio de Janiero in June 1992.

2.2 A Matrix of Impacts

The economic repercussions of the anticipated climatic changes may be severe, although not all sectors will be influenced to the same degree. And so the impact of some changes will be limited to one or a few sectors, while others may have an influence on many sectors. The existing literature either concentrates on a few changes and their impacts or describes a whole array of likely, possible or even remote economic impacts stemming from a large set of anticipated climate changes (Russell, 1970; Nordhaus, 1982; Heal, 1984; Kellogg, 1984; McFadden, 1984; Gibbs, 1986; Smith, 1987; Snedaker and de Sylva, 1987; Ausubel, 1991).

The assessment of the impacts of climatic changes is essentially one of close cooperation between physical and social scientists. To facilitate the translation of physical and ecological changes into socio-economic impacts, matrix methods are one of the techniques that can be usefully applied (Neuman, 1979). Matrices perform at least two functions:

1) they illustrate and help conceptualize the interrelationships between the physical, ecological and socio-economic variables;
2) for specific studies they provide a checklist in determining whether all relevant variables and implications have been included in the analysis.

Table 15.1 is a first attempt to relate the physical and ecological changes to economic impacts in a systematic manner. As a preliminary presentation it does not claim comprehensiveness in the sense that not all possible physical and ecological changes nor all economic implications have been included.

The individual cells of the matrix in Table 15.1 depict whether a particular physical or ecological change will have an anticipated impact on a certain sector (denoted by X), have no anticipated impact (denoted by –) or whether there will be an impact is unknown (denoted by ?). For example, in the first column, an increase in the concentration of CO_2 is anticipated to have a direct impact on agriculture, forestry and health. It is unknown whether there will be an impact on fisheries, while no direct impacts are anticipated for the other sectors. (The emphasis here is on direct impacts, as indirect impacts such as sea-level rise resulting from increased CO_2 are shown on a different column as having impacts on these same sectors.) The anticipation of an impact in a particular cell does not indicate whether such an impact is positive or negative, or in other words it does not signify a direction of impact.

At a later stage, the matrix can be improved by including more relevant physical and ecological changes and a broader array of relevant economic and social sectors. Further improvement also can be made by indicating the direction of the impacts and, whenever possible, by attempting to quantify the impacts ordinally.

Table 15.1 Matrix of socio-economic climate change impacts

	Agriculture	*Fisheries*	*Forestry*	*Tourism*	*Construction*	*Transportation*	*Water*	*Health*	*Settlement*
Physical changes									
Increase concentration CO_2	X	?	X	–	–	–	–	X	–
Increased concentration of other greenhouse gases	?	?	?	–	–	–	–	X	–
Increase cloud cover	X	?	X	X	–	-	X	X	–
Temperature increase	X	X	X	X	–	–	X	X	–
Precipitation	X	X	X	X	X	?	X	X	X
Wind regime	X	X	X	X	X	X	X	?	?
Increase sea temperature	–	X	–	X	–	–	?	?	–
Increased frequency of tropical storms	X	X	X	X	X	X	X	X	X
Wave/coastal current patterns	–	X	–	?	X	X	?	X	X
Sea-level rise									
Shoreline retreat/inundation	X	X	X	X	X	X	?	X	X
Beach erosion	–	–	–	X	X	–	–	–	X
Saltwater intrusion	X	X	X	–	X	–	X	X	X
Increased soil and water salinity	X	–	X	–	–	–	X	?	X
Soil moisture	X	–	X	–	–	–	X	?	X
Tides	–	X	–	–	X	X	?	?	?
Other (to be specified)	?	?	?	?	?	?	?	?	?
Ecological changes									
Coral reefs	–	X	–	X	–	–	–	–	?
Wetlands (mangroves)	–	X	X	–	–	–	–	X	?
Beaches	–	–	–	X	–	–	–	–	?
Estuaries and lagoons	–	X	–	X	X	–	–	?	X
Seagrass beds	–	X	–	–	–	–	–	–	?
Lowlands	X	–	X	–	–	–	X	X	?
Forest 1	–	–	X	–	–	–	X	X	?
Forest 2	–	–	X	–	–	–	X	X	?
Forest 3	–	–	X	–	–	–	X	X	?

3 Some Specific Impacts

3.1 Fisheries

Several major fisheries depend on upwelling of nutrient-rich waters with high biological primary production (*q.v.* Aparicio, Chapter 6). These phenomena, which are sensitive to changes in the wind regime and ocean circulation, may also affect the emigration pattern of commercially important species and the transport and distribution of eggs and larvae. Glantz (1992) edited a recent review of the issue of fisheries and climate on a global scale, which stresses the importance of this issue, one that is particularly relevant in the Intra-Americas Sea.

In estuarine and coastal lagoons the balance of fresh water and oceanic water may be altered by increased sea level. This may lead to a hypersalinity, which in combination with increased sea-surface temperature, may affect productivity. As these ecosystems serve as nursery areas for the sensitive early life stages such as fish larvae and juveniles, recruitment of commercial species may be affected. However, in some places an opposite effect of inundation of nearshore land could be to increase the total suitable reproduction area.

The impact on large-scale or artisanal fisheries of a moderate increase in sea-surface temperature and sea-level rise (1.5°C, 20 cm) is not expected to be very serious. However, displacement in traditional fishing sites may be observed to have negative effects for some local artisanal fisheries.

3.2 Immediate Coastal Zone

The most obvious effect of sea-level rise and increased sea-surface temperature in the immediate coastal zone may be the submersion and loss of low-lying land. Coastal wetlands, important sanctuaries for flora and fauna, will be lost by sea-level rise. However, every variation in sea level is accompanied not only by emergence and submersion of part of the coast but also changed littoral transport mechanisms which will initiate changes of coastal morphology. Beach erosion, which is already a serious problem in many areas in the region, may increase and various long-term shoreline stabilization and beach-defense systems might have to be constructed. The distribution pattern of river discharges, sediments and pollutants may change due to different coastal currents.

Tides, wind regime, wave and coastal current patterns may change due to increased sea-surface temperature. As sea-level rise permits tides and waves to operate from a higher base the effects of flooding will be more frequent, more serious, and threaten new areas. The economic implications due to hurricane damage may be increased, and construction of costly sea-defense systems may be necessary.

3.3 Agriculture and Forestry

Increased sea-surface temperature affects the formation, routes and frequency of storms. Sea-surface temperature is a key element controlling the evaporation process and associated heat transfer over the ocean. Changes in wind regime, levels and distribution of rainfall, and other climatic effects are of greatest importance for agriculture and forestry.

Agriculture production also may suffer from land loss in coastal areas due to permanent inundation or frequent flooding. Saline intrusion in groundwater resources may affect cultivation and living conditions in coastal areas. Changes in the hydrological balance affect river runoff (Muller-Karger, Chapter 8) and erosion processes in riverside agriculture areas.

3.4 Tourism

Tourism is a major economic activity in most of the region, and negative impacts on tourism will have severe repercussions on the prospects for economic growth and development. Changes in rainfall patterns, tropical storms and warming of the climate in temperate countries may affect the comparative advantages of tourism. However, there are still insufficient data on such implications; as an immediate policy concern, this is less pressing than beach erosion which already affects many beaches (Blommestein and Singh, 1987). Due to its nearshore character, the tourism industry will be affected by sea-level rise. Beach stabilization and other sea-defense systems may be needed to protect beach-front hotels and other tourism facilities. Important tourism beaches may have to be continuously supplied with sand and stabilized to prevent erosion. Sea-level rise is only one factor that contributes to beach erosion (Bruun, 1962; Bird, 1986) and its relative significance is not always known. Within planning efforts for recreational beaches, attempts should be made to estimate rates of erosion in the form of shoreline retreat per unit of rise through the application of Bruun's Rule or similar adjusted formula and evaluate the cost and benefits of possible beach-nourishment programmes.

3.5 Settlements and Structures

Increased beach erosion threatens buildings and other construction. Nearshore roads, bridges and other structural facilities may have to be reinforced and reconstructed. Wilcoxen (1986) concluded for San Francisco '. . . structures with long operating lives should not be built in erodible sediment adjacent to recreational beaches. Sea-level rise will make protection for such a structure expensive and preservation of the beach difficult. Sound planning for future projects must consider the erosion that will be caused by sea-level rise'.

Navigation and port facilities like wharfs and piers will be affected, and clearance under bridges will decrease. Increased maintenance dredging may be needed due to increased beach erosion, changed coastal currents and sedimentation pattern. However, as port structures normally have to be continuously reconstructed and maintained, the implications for shipping are not expected to be serious.

Saline intrusion in groundwater resources and upstream migrations of saline fronts in rivers may affect municipal fresh-water supplies to coastal communities. In low-lying cities the capacity of draining and sewage systems may be reduced. Changed coastal currents may affect the transport and dilution of urban and industrial waste waters. In nearshore poor communities these impacts may affect public health and sanitary conditions in general.

3.6 Public Health

Water is a principal agent for many diseases in tropical countries, either directly through the contamination of water or indirectly through insects which depend on water at some stage in their life cycle. In both instances it appears that water temperature is a significant factor, with higher incidences related to higher temperatures (Snedaker, Chapter 12; de Sylva, Chapter 14). The impact of the climatic changes could be both a northward shift into the area in which tropical diseases are prevalent and increased incidence in those areas already subject to some or all of these diseases. The economic impact of such a consequence on public health would be through a lower productivity of the labour force as a consequence of increased health-related absenteeism (person-days lost) and through increased expenditures on public health.

3.7 Waste Disposal and Drainage

In much of the region, solid-waste disposal sites are located in low-lying areas near the coast (such as mangroves or salt ponds) which may be at risk with increased flooding or permanent inundation during a sustained or accelerated process of sea-level rise. Sewage and drainage systems may also be affected. In many areas sewerage (raw or partially treated) is eventually discharged into the ocean. A rise in the sea level may imply a reduced efficiency of sewage and drainage systems. Sewage and drainage systems typically have a long lifespan of 30–50 years. In other words, the time is approaching when there is a risk that installed sewage and drainage systems may not satisfy design specifications well within their estimated lifespans and that retrofitting or reconstruction may become necessary. Gibbs (1986) has described cases in which near-term investment to accommodate future sea-level rise is efficient in constructing new projects.

4 Responses and Measurement

Two questions about the human responses to climatic change seem most relevant, one descriptive and one prescriptive. The descriptive question is, what will be the responses to climatic changes and how will these affect the socio-economic impact of the changes? The prescriptive question is, what should be done now and at each future point along the way? The answers for both questions will be shaped by two different views of how effective responses emerge. One view stresses that people have a strong talent for adjusting to change, and that their adjustments are most effective when crafted little by little to incremental changes spread over time (Schelling, 1983). This view often also holds that decision-makers closest to the facts of each choice are likely to devise the most appropriate incremental responses (McFadden, 1984). In a prescriptive sense, this view tends to be non-interventionist and favours letting things sort themselves out as the facts emerge. In contrast to such atomistic incrementalism, the other view of human responses stresses the value of collective actions. Whether for reasons of political ideology or the failure of free exchange to account for effects that are spread across large social groupings, this view highlights the value of governments and other collective institutions to anticipate and help engineer effective responses. Both views share the assumption that economic decision-makers act rationally, that they act in such a way as to do the best they can with the information and other resources available.

Predicting and prescribing human responses, whether incrementalist or discrete, atomistic or centralized, require estimates of what will happen under different scenarios and how much it will cost. Thus, methods of measurement and estimation are important issues here. For estimating socio-economic effects of climatic change, however detailed or sophisticated the effort, two basic approaches are virtually unavoidable. The first depends on physical scientists and is the description of expected changes and their effects, a kind of dose-response specification often framed as a set of scenarios about what might happen. The basic economic approach is a cost-benefit analysis built on the scientific scenarios. Such an analysis seeks to compare the flow of costs and benefits over time arising from the different scenarios in order to estimate the expected net impact from each. Obviously, accurate estimation of benefits and costs will depend on the expected effects of human responses. By the same token, the choice and timing of responses will be affected by estimates of costs and benefits.

The reasonable assumption of rational decision-making leads to a fairly strong expectation about human responses. It is, that responses will be selected to reduce the expected costs and maximize the expected benefits of climatic change. Of course, mistakes may be made, but, in general, human responses can be expected to lower the net cost (or raise the net benefit) of climatic change. If they did not, they would not be undertaken. Nor is it likely that responses will be confined to the areas most affected by the physical changes. If a nation loses its low-lying coastal rice production to sea-level rise, for example, this will alter the allocation of resources in the economy in ways that extend beyond the loss of rice production. The resources formerly applied to rice (including materials, mobile capital and labour) will tend to be redirected into other activities that eventually will help replace at least part of the economic benefits lost with the rice fields. How quickly and how completely this economic adaptation will occur clearly depends on the suddenness of the initial loss, on the organization of the economy and on the availability of other opportunities for productive employment of the displaced resources.

Because the timing, magnitude and physical effects of climatic changes are still highly uncertain, potential near-term responses are more in the nature of risk management. Scientific predictions of sea-level rise reveal an increased riskiness of certain coastal activities that, as it becomes more widely known, should automatically tend to curtail those activities. For example, as developers become more concerned about the risk of sea-level rise, a likely reaction is to increase the rate of return required to justify an investment. This will tend to reduce the number of projects deemed profitable and will eventually result in reduced investment. This market behaviour would be in line with and reinforce any policy of abandonment or retreat since it will gradually reduce development activities in the exposed area. At the same time, this increased risk will lend greater value to research efforts that increase information and to arrangements that permit diversification and pooling of risks.

One important class of the near-term responses presents special problems. These are responses that, if invested in now, will reduce the cost of future climatic changes that turn out to be severe. If the future climatic changes turn out to be less severe, however, then the near-term investment will have been wasted. The point is that the investment decision must be made before the severity of climatic change has been revealed. This is the so-called 'retrofit' problem, and Gibbs (1986) and Wilcoxen (1986) have analysed sea-level-rise scenarios under which certain costly near-term public works investments are economically superior to the alternative of waiting to observe future sea level and retrofitting to meet it. Because the probability of sufficiently severe change is unknown, however, the prescriptive value of these results is still subject to question.

As a near-term aid to governments of the region attempting to manage the risk of climate change, preparation of a region-wide first-order assessment of the magnitude, timing, and distribution of economic impacts would be valuable. At the same time, for sharper focus on the details involved, a small number of case-study areas should be selected for more intensive study. These more detailed studies would be most useful if they took account of possible responses and adjustments. The studies will require interactive inputs from natural scientists.

As already mentioned, the starting point for any assessment of the socio-economic impact of climatic change is a scientific prediction or scenario describing the expected change and associating it with natural, physical impacts. For example, virtually all economic analyses of future sea-level change have begun with a statement of the hypothetical rate of sea-level rise and a topographical portrayal of the resulting area of inundation in the absence of defensive measures (Schneider and Chen, 1980; Barth and Titus, 1984; Broadus *et al.*, 1986; Broadus, 1988). Where property-value data are available, they can be projected into the future and provide a reasonable estimate of the value of land lost to inundation (Barth and Titus, 1984). In poor data cases, or where institutional arrangements render market values irrelevant, other methods of estimating the lost land's productivity must be sought (Broadus *et al.*, 1986; Broadus, 1988). (Generally, more work is required to develop practical and appropriate techniques to evaluate socio-economic impacts in cases of limited financial and data resources.) The value of lost fixed capital (such as buildings, highways, bridges, etc.), and of losses associated with salt-water intrusion, increased exposure to storm damage, etc., can also be estimated from the base provided by the scientific scenarios. The sum of such costs, and any identifiable benefits, provides a starting point against which to compare estimates based on hypothetical projected responses (though, in fact, little work has been reported that extends the estimates to include alternative-response scenarios).

Aside from the technical judgements involved in choice of growth rates and interpretation of other economic data, several controversial issues must be encountered in this basic cost-benefit approach. The most noteworthy of these include:

1) *Discounting future values:* Most people or nations tend to display a preference for a given sum of income now over that same amount in the future, if for no other reason than that the income can be put to beneficial use in the meantime. This 'time preference for income' implies that values observed in the future should be reduced by some 'discount rate' for a meaningful comparison with present values. The choice of discount rate can seriously affect estimates of the present value of future costs and benefits. While there is growing agreement among economists on the proper choice of discount rate (Lind *et al.*, 1982), this remains a difficult issue and especially where intergenerational choices are involved (D'Arge *et al.*, 1982).
2) *Welfare measures:* Different forms of cost-benefit analysis use different measures of human welfare, but most assume unrealistically that the marginal value of income is the same for all people. Also, all such analyses have difficulty with economic values not recorded in market prices, and even the use of market prices is potentially misleading. For example, transactions prices for property and other assets do not account for the so-called 'consumers' surplus' received by purchasers, and so may underestimate the benefits generated by those assets.
3) *Comparative systems:* Although all economic systems attempt the same functions of organizing production, providing employment and delivering services, there are cultural, ideological, political, organizational and distributional differences among systems that can affect the performance of different estimation methodologies These change, but cooperative international efforts should provide ideal forums for progress on this topic.

Beyond the fundamental, scenario-linked cost-benefit approach to climatic impact estimation, a number of further approaches have been attempted or show promise. One class of such approaches is the 'holistic systems models,' which may be contrasted with a diverse array of more analytical economic models. The holistic systems models (Warrick and Riebsame, 1983; Riebsame, 1985) are extremely ambitious in their attempt to capture all effects of climatic change comprehensively by describing the full range of linkages and feedbacks between causal inputs and affected activities. A major drawback of such models is that their complexity can make their results opaque. That is, difficulty often arises in discerning exactly which relationships within the model are most important in generating the model results. Perhaps most damaging, however, is the predictive failure of such models in empirical applications. Eventual perfection of holistic systems models, on the other hand, would strengthen the prospects for coupling socio-economic impact models directly with physical process models (*q.v.* Engelen *et al.*, Chapter 16).

The more analytical economic models are less comprehensive in intent, but they may also be more manageable and more instructive about the precise nature of certain key relationships. Although the cost-benefit estimations based on scientific scenarios are straightforward and natural first cuts at the impact problem, several other approaches from the arsenal of modern economics should be fruitful. For example, general equilibrium analysis may help to predict and account for indirect or secondary effects occurring outside the economy's immediately affected sectors (Smith, 1987). Smith has shown, however, that excellent approximations of the general equilibrium welfare loss arising from a shock like sea-level rise can be achieved in principle with partial equilibrium valuation estimates, especially when there are only small indirect effects outside the coastal sector. Other promising approaches include: search theory, to help model and guide responses under uncertainty; portfolio analysis, to help balance risks and returns among a variety of risky activities and responses; and identification of potential winners as well as losers and of possible institutional mechanisms (such as new markets or opportunities for exchange) through which their risks can be shared and traded.

5 Conclusion

It would be a considerable error of omission in these deliberations not to focus momentarily on a socio-economic issue that may eclipse climate change by an overwhelming amount. The human population of the earth is growing at an astounding rate, and the region with which this report is concerned is no exception. Fig. 15.2 shows a comparison between the

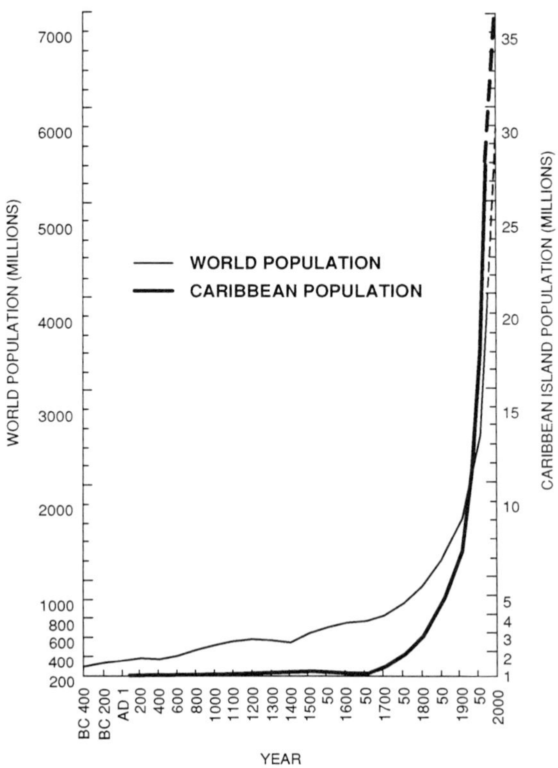

Fig. 15.2 Human population as a function of time, based on data to 1975 and projected (dashed) to 2000, showing the global curve and that of the Greater and Lesser Antilles, Bahamas and Bermuda (redrawn from McEvedy and Jones, 1978).

global human population trend for the last few thousand years, and the trend on the islands of the Antilles (Lesser and Greater), the Bahamas, and Bermuda. Although the two curves are similar, a striking difference is seen after about 1700, when the rate of population change in the 'Caribbean Islands' (McEvedy and Jones, 1978) is increasing faster than the global population change. Not shown are curves for other subregions, all of which share the characteristics presented in Fig. 15.2.

In a very thoughtful editorial entitled, *People and Global Warming: A Critical Link*, Fisher (1992) notes that 'While population has been considered in the context of environmental problems such as soil degradation and deforestation, policy-makers have neglected its relevance to global warming.' Thus, it is argued, that population growth between 1950 and 1985, 2.5 to almost 5 billion people, was responsible for about two-thirds of the global CO_2 increase and methane increase in this time period is even more closely linked with population growth. So while industrialized nations have been primarily responsible for greenhouse-gas production in the past, it is developing nations that will contribute most in the future. Fisher, reflecting on a joint statement of the Royal Society of London and the US National Academy of Sciences (Atiyah and Press, 1992), argues for a global solution in which all nations cooperate in this aspect of climate change.

Institutional responses and institutional differences will play a critical role in shaping the socio-economic impacts of climatic changes globally as well as in the Gulf/Caribbean/Bahamas/Guyanas region. The region is an area with differing institutions and levels of economic development, and the response to such climatic changes as accelerated sea-level rise will vary with political and economic factors. The more highly developed countries will have access to a larger and more expensive array of mitigation opportunities. Less developed countries may be constrained by their resource endowments to a narrower selection that may not include the technically optimal responses. Capital scarcity, in other words, may force abandonment of areas whose defense would otherwise be economically justified. Even within the poorest countries, however, there may be low-lying areas with dense concentrations of human settlement and economic activity where, for social and economic reasons, efforts may be undertaken to maintain the existing shoreline (Bird, 1987). These situations will call for difficult tradeoffs in the allocation of scarce public resources and, where foreign capital is proportionately important, in the management of international financial relationships.

6 References

Atiyah, M. and Press, F., 1992: *Population Growth, Resource Consumption, and a Sustainable World*. National Academy of Sciences, Washington, 8 pp.

Aubrey, D.G. and Emery, K.O., 1983: Eigenanalysis of recent United States sea levels. *Continental Shelf Res.*, **2**, 21–33.

Ausubel, J.H., 1991: A second look at the impacts of climate change. *Amer. Sci.*, **79**, 210–221.

Barnett, T.P., 1984: The estimation of 'global' sea-level change: A problem of uniqueness. *J. Geophy. Res.*, **89**, 7980–7988.

Barth, M. and Titus, J. (eds), 1984: *Greenhouse Effect and Sea Level Rise: A Challenge for This Generation*. Van Nostrand Reinhold Co., New York, 325 pp.

Bird, E.C.F., 1986: Potential effects of sea-level rise on coasts of Australia, Africa and Asia. In: Titus, J.G. (ed.). *Effects of Changes in Stratospheric Ozone and Global Climate, 4: Sea Level Rise,* UNEP/EPA, pp. 83–98.
Blommestein, E. and Singh, N., 1987: The Impact of the Climatic Changes on Tourism. Paper presented at the XXI Annual Meeting of the Caribbean Conservation Association, Tortola, British Virgin Islands.
Broadus, J., 1988: Possible Impacts of and Adjustments to Sea-Level Rise: The Cases of Bangladesh and Egypt. Presented at International Workshop on the Effects of Climatic Change on Sea Level, Severe Tropical Storms and their Associated Impacts, Climatic Research Unit, University of East Anglia, Norwich, UK, 1–4 September, 1987. Unpublished manuscript.
Broadus, J.M., Milliman, J.D., Edwards, S.F., Aubrey, D.G. and Gable, F., 1986: Rising sea-level and damming of rivers: possible effects in Egypt and Bangladesh. In: Titus, J.G. (ed.), *Effects of Changes in Stratospheric Ozone and Global Climate, 4: Sea Level Rise,* UNEP/EPA, pp. 165–189.
Cartwright, D.E., Barnett, T.P., Garett, D.J.R., Carter, W.E., Peltier, W.R., Pyle, T. and Thompson, K.R., 1985: Changes in relative mean sea-level. *EOS, Trans. Amer. Geophy. Union,* **66(45)**, 754–756.
Bruun, P., 1962: Sea level rise as a cause of shore erosion. Amer. Soc. Civil Engr. Proc., *J. Waterways and Harbors Div.*, **88**, 117–130.
Chen, R.S., Boulding, E. and Schneider, S. (eds), 1983: *Social Science Research and Climate Change: An Interdisciplinary Appraisal.* D. Reidel Publishing Co., Boston.
D'Arge, R., Schulze, W.D., Brookshire, D.S., 1982: Carbon dioxide and intergenerational choice. *American Economic Review Papers and Proceedings,* **72(2)**, 251–255.
Emery, K.O. and Aubrey, D.G., 1985: Glacial rebound and relative sea levels in Europe from tide-gauge records. *Tectonophysics,* **120**, 239–255.
Fisher, M., 1992. People and global warming: a critical link. *Christian Science Monitor,* 17 April, 1992, pp. 19.
Gibbs, M., 1986: Planning for sea-level rise under uncertainty: A case study of Charleston, South Carolina. In: Titus, J.G. (ed.), *Effects of Changes in Stratospheric Ozone and Global Climate, 4: Sea Level Rise,* UNEP/EPA, pp. 57–72.
Glantz, M. (ed.), 1992: *Climate Variability, Climate Change and Fisheries.* Cambridge Univ. Press, Cambridge, 458 pp.
Heal, G., 1984: Interactions between economy and climate: a framework for policy design under uncertainty. In: Smith, V.K. and Witte, A.D. (eds). *Advances in Applied Micro-Economics,* **3**, 151–168.
IPCC (Intergovernmental Panel on Climate Change), 1990: *Climate Change, the IPCC Impacts Assessment.* Australian Gov't. Publ. Serv., Canberra, 245 pp.
Kellogg, W., 1984: Modelling the prospects for climatic change: current state of the art and implications. In: Smith, V.K. and Witte, A.D. (eds), *Advances in Applied Micro-Economics,* **3**, 109–132.
Lind, R. C. and Arrow, K.V., 1982: *Discounting for Time and Risk in Energy Policy. Resources for the Future.* John Hopkins University Press, 468 pp.
McEvedy, C. and Jones, R., 1978: *Atlas of World Population History.* Penguin Press, New York, 368 pp.
McFadden, D., 1984: Welfare analysis of incomplete adjustment to climate change. In: Smith, V.K. and Witte, A.D. (eds), *Advances in Applied Micro-Economics,* **3**, 133–149.
Morrisette, P. M. (ed), 1987: *A selected annotated bibliography of climate and society research.* Natural Hazards Research and Applications Information Center, University of Colorado, Boulder, CO, 103 pp.
NAS (National Academy of Sciences), 1983: *Changing Climate: Report of the Carbon Dioxide Assessment Committee.* National Academy Press, Washington, DC.

Neuman, L. D., 1979: Methodologies for Coastal Management: Matrix Approaches. In: Valencia, M. J., (ed.), Proceedings of the Workshop on Coastal Area Development and Management in Asia and the Pacific, Manila, Philippines, 3–12 December 1979. East–West Center, Univ. Hawaii, Honolulu.

Nordhaus, W., 1982: The Global Commons I: Costs and climatic effects. How fast should we graze the global commons? *American Economics Review Papers and Proceedings*, **72(2)**, 242–246.

Pirazzoli, P.A., 1985: Sea-level change. *Nature and Resources*, **XXI(4)**, 2–9.

Riebsame, W.E., 1985: Climate hazards: lessons from recent events in the United States. *Disasters*, **9(4)**, 295–300.

Russell, C. S., 1970: Losses from natural hazards. *J. Land Econ.*, **46(4)**, 383–393.

Schelling, T. C., 1983: Climatic change: implications for welfare and policy. In: *Changing Climate: Report of the Carbon Dioxide Assessment Committee.* National Academy Press, Washington, DC, pp. 449–482.

Schneider, S.H. and Chen, R.S., 1980: Carbon dioxide warming and coastline flooding: Physical factors and climatic impact. *Annual Review of Energy*, **5**, 107–140.

Smith, V. K., 1981: CO_2, climate, and statistical inference; a note on asking the right questions. *J. Environ. Econ. Manag.*, **8**, 391–394.

Smith, V. K., 1987: General Equilibrium Welfare Measures and Economy-wide Shocks: The Case of Sea-Level Rise. Presented at Workshop on Science, Economics, and Sea-level Change, Marine Policy Center, Woods Hole Oceanographic Institution, Woods Hole, MA, 7–8 May 1987. Unpublished manuscript.

Snedaker, S.C., and D.P. de Sylva, 1987: Impact of climate change on coastal resources: implications for property values, commerce, estuarine environments, and fisheries, with special reference to south Florida. In: Meo, M., (ed.), *Symposium on Climate Change in the Southern United States: Future Impacts and Present Policy Issues.* US Environmental Protection Agency, Washington, DC, pp. 187–215.

Solow, A. R., 1987: The application of eigenanalysis to tide-gauge records of relative sea level. *Continental Shelf Res.*, **7(6)**, 629–641.

Titus, J.G. 1985: Sea-Level Rise and Wetland Loss. Coastal Zone '85, 1979–1990.

Titus, J.G., 1986a: *Effects of Changes in Stratospheric Ozone and Global Climate*, Vols. 1–4, UNEP/EPA.

Titus, J.G. 1986b: Greenhouse effect, sea-level rise and coastal zone management. *Coastal Zone Manag. J.*, **14(3)**, 147–171

Titus, J.G., Kuo, C.Y., Gibbs, M.J., LaRoche, T.B., Webb, M.K. and Waddell, J.O., 1987: Greenhouse effect, sea-level rise and coastal drainage systems. *J. Water Res. Plan. Manag.*, **113(2)**, 216–227.

Warrick, R.A. and Riebsame, W.E., 1983: Societal response to CO_2-induced climate change: opportunities for research. In: Chen, R.S., Boulding, E. and Schneider, S. (eds), *Social Science Research and Climate Change: An Interdisciplinary Appraisal*, D. Reidel Publishing Co., Boston, pp. 20–60.

Wilcoxen, P. J., 1986: Coastal erosion and sea-level rise: implications for Ocean Beach and San Francisco's Westside Transport Project. *Coastal Zone Manag. J.*, **14(3)**, 173–191.

WMO/ICSU/UNEP, 1980: Joint Meeting of Experts on Assessment of the Role of CO_2 in Climate Variations and their Impact. Villach, Austria, November 1980.

16

Exploratory Modelling of Socio-Economic Impacts of Climatic Change

Guy Engelen[1], Roger White[2] and Inge Uljee[1]

ABSTRACT

Socio-economic systems may be influenced by climatic change in ways ranging from minor or very local to drastic and nation-wide. Any such changes will be superimposed on trends already present in these evolving systems. Therefore, it is vital to anticipate dangers, as well as new opportunities, as soon as possible. To allow governments and policy-makers to play an active role in managing these socio-economic systems effectively, they should be provided with tools that will permit them to explore impacts in their full holistic, spatial and temporal contexts. Decision-support systems are designed to assist in such tasks. The most essential part of such systems is a set of tools, mostly quantitative models and methods, which at relatively low cost, allow the user to analyse and evaluate a range of possible futures resulting from different scenarios and hypotheses.

In this chapter we propose, as part of a larger decision environment, a two-level mathematical modelling framework, geared to study the effects of climatic change on the level of the individual island or mainland state. The long-range mechanisms of change are modelled in a classic, non-equilibrium spatial interaction model. This model then feeds regional growth coefficients into a low-level cellular model that deals with the short-range location and interaction mechanisms. This technique of linked models is necessary in order to capture successfully the effects resulting from climatic change on the appropriate scales. The prototype presented is a first, mostly conceptual, step towards a system for use in real-world applications.

1 INTRODUCTION

This closing chapter of the book is an indication of the increased interest by UNEP and the IOC in climatic change: with the understanding gained from the multi-disciplinary study of the physical and biological impact of climatic change, emphasis will be put on studying the impact on human societies and their responses to the challenges raised. The ultimate aim is to provide the threatened peoples, their policy analysts and governments,

---

[1]Research Institute for Knowledge Systems, 6200 AL Maastricht, The Netherlands
[2]Memorial University, St. John's, Newfoundland, Canada

with the necessary tools to anticipate the obstacles well in advance in order to minimize the effects on their well-being and to secure their future.

This chapter is a first (and mostly conceptual) step towards the design of a model-based decision-support system, capable of integrating the knowledge gathered by different specialists. It describes the development of a modelling framework, integrating physical, ecological, economic and social characteristics of nations in the region, within which the complex multiple consequences of public policies and actions can be examined as completely as possible. The development of such a system is the subject of a long term and possibly ambitious project. However, the rapid evolution of the information technology in the past few years, together with evolving scientific paradigms and derived modelling frameworks, offer new perspectives for the management of socio-economic systems, which make us confident that such projects can lead to successful outcomes.

Practically, the chapter will propose, as part of a decision-support system for public-policy exploration, a simple version of a complex dynamic model of a purely hypothetical island with characteristics typical of those in the region. This model ultimately will evolve into a more generic tool to study important effects of climatic change on individual islands and mainland nations. As such it may serve as a discussion piece to concert further research by the task team.

2 Posing the Problem: Climatic Change in the Region.

If the link between greenhouse-gases emission and climate change continues to be confirmed, then the 'consensus view' on the future of the planet could come at great cost (Davis, 1990). This cost might be tolerable for the rich industrial countries, having the technology and financial resources to cope with the immediate effects of a changing climate (Ausubel, 1991). For the developing world, however, the cost might lie beyond the bearable (Ominde and Juma, 1991). The 'sustainable growth' world view, which emphasizes adaptability and voluntarily reduced exploitation of resources and strives to match the need for economic growth with a viable, environmentally sound planet for the generations to come, is certainly more adequate in this respect. Technology will propel society toward sustainability most quickly if policy-makers can agree on the appropriate global and local guide-lines (Reddy and Goldemberg, 1990). In order to attain such agreement, a good understanding of the mechanisms linking human activity and greenhouse-gases production, and of the subsequent effect of greenhouse gases on the ecological and socio-economic systems, is an essential first step.

2.1 An Evolutionary Context

Tourism and agriculture, both major sources of income in the region, rely on natural systems. Any effects of climatic change on these sectors may have a long term, large scale and irreversible impact on the physical and socio-economic environment, thereby undermining the sustainability of the economic development. Thus, climatic changes will alter the evolutionary path of the socio-economic systems. Socio-economic systems are highly dynamic, but at the same time laden with inertia. *Dynamic*, because no such system has been found to be in equilibrium; rather it is perpetually in a transition phase between an old and a new form of organization. *Inertia*, because of the long time it takes for all effects of actions and disturbances to propagate through the system and to run out in a fully reorganized system. Hence, the dynamics driving the system today may have been set in motion, deliberately or not, a long time ago, and may continue to do so for many years to come.

2.2 A Global Context: Changing World Markets

Much along the lines of the approach of meteorologists, exemplified by Wigley and Santer (Chapter 2), effects of climatic change on the social and economic system need to be studied in their full spatial extent. Meteorologists are not able to separate the Intra-Americas Sea from the global weather and climatic system. The region fits like a single piece in a world-wide jigsaw and is only one of the many interacting points in the General Circulation Models. Likewise, the region is firmly embedded in the world's demographic, social and economic systems. If we really are interested in the effects of global warming in the region, some of our attention will need to be on the way the region interacts with the rest of the world. Indeed, the worst socio-economic problems in the region could well find their origin far outside the area, because a substantial part of its income is generated externally. Tourists are being attracted for reasons strongly related to its climate: year-round sunny weather, warm seas, coral reefs, exotic fruits, beaches; they typically come from regions lacking such properties: the northern USA, Canada, northern and western Europe. If climatic change affects more drastically higher latitude regions (Vincente *et al.*, Chapter 11), possibly with temperature increases as high as 4.5°C by 2025, the drive of the traditional tourist to leave his home could weaken considerably, thus depriving the region of substantial income. Other places could acquire advantages similar to the ones now unique to the area and could start acting as intervening opportunities, thus diverting important streams of tourists. Although slightly out of the scope of this book, we imagine that further depletion of stratospheric ozone with its attendant effects on human health, may completely alter peoples' attitude to sunny beaches, causing even further stress on the economic system.

In the course of history the region has become an important, specialized exporter of agricultural staple products (coffee, fruits, sugar cane). If climatic bands start shifting, bringing more precipitation to dry areas and higher temperatures to cold ones, this specialization might become questionable. Transport costs make products sold at distant markets very sensitive to competition because the distance between origin and destination allows for intervening opportunities to pop-up. By making reasonable assumptions about the likely competition for agricultural or other products, however, early detection of vanishing comparative advantages as well as the appearance of new openings on local or world markets might give the system sufficient time to transmute smoothly from the old production base to a new one. Hence, important questions regarding investments in the region need to be addressed to prevent capital from leaving and actions will have to be taken accordingly.

2.3 A Local Context: Areas and Activities at Risk

Countries whose economic activity is strongly concentrated in the coastal areas, on lowlands, deltas, etc., doubtlessly will be strongly affected by climatic change (Alm *et al.*, Chapter 15; and Vicente *et al.*, Chapter 11). Without adequate policy interventions, activities with low value added per unit area, such as subsistence agriculture, will suffer most from changing climate or rising sea level. Firstly, on a cost-benefit basis, it will be considered non-economical to build sea-defence structures to protect this land against intrusion. Secondly, if land used by activities with higher value added, such as commerce or industry, is endangered, these activities will preferentially relocate in the immediate neighbourhood, in order to enjoy similar location advantages irrespective of the land's value for its initial use. This has to be viewed in its historical context. If, for example, commercial activities would invade subsistence agricultural land, it has to be assumed that farmers have cultivated the land they work today because of its high productivity value within the constraints of the technology at hand. Losing it to any other usage will automatically force them to farm less adequate fields, with a reduction of productivity per unit area and labour as a consequence. To the immediate clearing and building costs resulting from such relocation will be added the cost of reduced productivity. Furthermore, social tensions are likely to be amplified as the socially weak are pushed out and often have to settle for second best. They also will face the inconvenience of travelling longer distances to services and to work. Hence, deteriorating working and living conditions will worsen an already difficult financial situation. Therefore, timely anticipation and active intervention on behalf of governments will be necessary in order to keep social peace in the system.

The above points are certainly not an exhaustive enumeration of conceivable effects of global warming on the socio-economic structure of the region. They are merely meant to widen the discussion of the possible impacts on the intensity and resilience of the socio-economic linkages between the region and the external world. Even if we are only interested in what happens in the region itself, narrowing down the study to the region as a self-contained unit, thus pretending that climatic change does not exist beyond its frontiers, in a system culturally as diverse and with an economy as export oriented, runs the risk of missing the problem at hand and excluding the driving forces of the system from the study. As explained later, successfully exploring future challenges in the area calls for modelling it as an open system.

3 Towards an Integrated Framework for Policy Evaluation

The Intra-Americas Sea has a unique variety of peoples, cultures and political systems, representing countries with different types and stages of economic development. Its physical, geomorphological and ecological diversity is remarkable as well (*cf.* Chapters 5–12). Hence, defining the effects of climatic change on the socio-economic systems should take into account both the diversity among countries and the reliance of their economies on more or less distant and intercontinental markets. We therefore propose an integrated multi-disciplinary approach on the level of the individual member states.

Decision-making in social systems is a formidable task because of the dynamic character, the multi-dimensionality, the spatial character and the complex interactions that typify them. Processes in these systems are strongly interrelated and indivisible. Consequently, situations which at first glance seem well defined and limited in scope turn out to have deeper dimensions and repercussions, ones that can not be untangled or comprehended easily, if at all. Decision-makers therefore need tools that *accumulate* intelligently the historical knowledge about the system, tools to *explore* the behaviour of the systems when stressed, and tools to assist in *selecting* the best policy alternatives given the objectives and constraints. Typically, such tools are made available in decision-support systems.

3.1 Decision-Support Systems

Spatial Information Systems have evolved in the past to cover a number of rather different computer-based systems, despite their common purpose of integrating data from different sources to provide the necessary base for decision-support in a spatial context. They include systems varying from data-centred systems such as DBMS (Database Management Systems), information-centred systems such as GIS (Geographical Information Systems) and intelligence-centred systems such as Decision Support Systems (DSS). The intelligence and knowledge content increases gradually from DBMS to DSS, and the type of decision-support changes from passive support (enabling the formulation of the decision alternatives) to active support (helping to choose and implement the 'best' decision). If we define intelligence as 'the ability to catch the essential factors from complex information and data' (Catanese, 1979), we should try to provide the planner with intelligence rather than data or information. Thus, we should focus our attention on DSS rather than on DBMS or GIS. Both DBMS and GIS perform essential sub-tasks in the decision process, but separately they fall short of dealing adequately with temporal and spatial dynamics (Marble and Amundson, 1988) or in providing the decision-maker with the required active support.

Decision Support Systems (Fig. 16.1) are principally made up of three components (Han and Kim, 1990): (1) *a user interface* enabling easy interaction between the user and the system; (2) *a data base* containing the raw and processed data of the domain and area at study; and (3) *a model base* with relevant models, methods and techniques.

3.1.1 The user-computer interface

Decision-support systems are intended for use by high-level decision-makers to solve ill-structured problems. Although often specialized in their domain, these decision- and policy-makers may be unfamiliar with information science and technology. The user interface, the vehicle of interaction between the user and the computer, takes into account differences in the cognitive styles and relative knowledge of the users. It is designed to hide the complexities of the internal computer system without hampering its flexibility and provides insight into the structure of the mathematical models, methods, variables, parameters and mechanisms, the underlying theoretical assumptions, the boundary conditions and other constraints. It allows the user to address the different components of the DSS (tools, models, data, etc.), translates the user input into appropriate computer instructions, and reports back the results of the computations. To provide maximal user-friendliness, state-of-the-art interactive graphic techniques are being applied extensively (Cleal and Heaton, 1988).

3.1.2 The data base

A thorough data base serves as an input for the models used, and is appropriate to the management or policy issues dealt with in the DSS. There is a growing trend to store spatial, social and ecological data in GIS data bases. Consequently, a good interface linking GIS and DSS is of utmost importance for policy-making and planning.

3.1.3 The model base

The convenience, richness and scope of a DSS is primarily determined by the spectrum of tools available in its model base. Typically decision models, statistical and operations research methods, as well as tools to describe, portray, compare and evaluate different policy strategies, are part of it. Even more essential in the model base are the domain-specific (simulation) models capable of grasping the complexities of the system and the problems at hand. The elements in the model base are of a formal nature, and exclude decision-making solely based on common sense or intuition. Good formal decision-support methods will assume that the decision-maker desires to make decisions on the basis of a consistent line of reasoning (Holtzman, 1989) and will suggest solutions that make intuitive sense. Elaborate model bases will contain both mathematical and rule-based techniques, often playing complementary roles in decision-making processes.

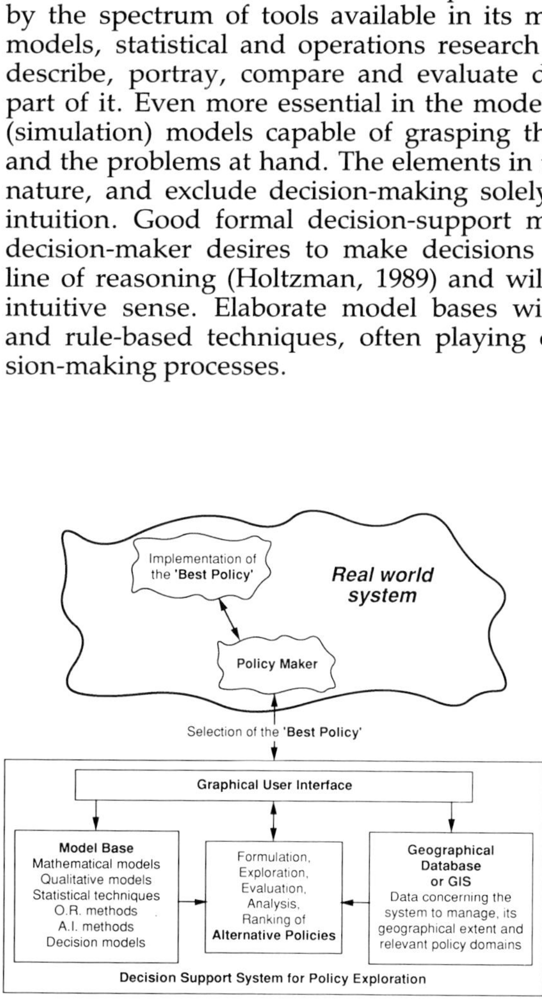

Fig. 16.1 A decision-support system for public-policy exploration consists of three essential components: a model base, a data base and a user interface, and assists the policy analyst in the evaluation of different alternatives.

Rule-based tools are often expensive to develop and to maintain, as they require a number of experts and domain specialists to build or to update them when knowledge is outdated or becomes erroneous. Mathematical models usually will be more generic in nature. Their creation will oblige the precise, detailed and explicit formulation of assumptions and relations. Hence logical contradictions will be avoided and, contrary to qualitative models, solutions will not change imperceptibly with different instances of the same problem. Mathematical models will run comparatively fast and cheaply on the computer, thus allowing the analyst to discover and fill gaps in their formulation, to test the consequences of hypotheses, and to evaluate efficiently the impacts of a number of scenarios and strategies. In the past, large models of socio-economic systems have had a chequered record, partly because of naive expectations about their ability to make detailed and precise predictions, and partly because of inapt modelling paradigms.

The highly interconnected and intertwined social systems, the human creativity and intelligence that they embody, their holistic nature, their explicit temporal and spatial dynamics, and their irreversible evolution towards greater complexity, indicate that studying partial questions is at best debatable. Rather, an integrated approach is required, as represented in the concept of holistic management, referring to the notion that 'one must seek to understand the greater whole in order to understand its parts, not vice versa' (Savory, 1988). This same view is also important in systems analysis generally and is most essential in the 'Systems Dynamics' paradigm (Forrester, 1961). It further postulates that the understanding of the relations, causalities and feedbacks in the system is more essentially an aim of the modelling exercise than the exact description of the system.

These observations are also at the roots of chaos, fractals and self-organization. Prigogine (1981) proved that thermodynamically 'open systems', far from equilibrium conditions, obey different laws of thermodynamics in the sense that they can order themselves. They are said to self-organize; they become unstable and change their macroscopic structure due to microscopic fluctuations. Hence, the great strength of the paradigm resides in the explanation of the macroscopic structure of a system as the result of its internal microscopic characteristics. When applied to social systems, which by definition are open systems, this is a clear recognition of the role of the individual in structuring the greater system of which he or she is part.

The constant exposure of a socio-economic system to fluctuations, the uncertainty as to the moment when these will hit the system and the limited knowledge about their impact, the lack of knowledge of future values and norms, and the lack of insight in the evolution of the meta-system, not treated or explained by the model, suggest that it is nearly impossible to predict fully and exactly the future of the system. This belief contrasts with the assumptions of normative and predictive modelling techniques (Simmonds, 1986) based on equilibrium conditions and deterministic relations. It leans towards instrumentalism (Casti, 1989), a view that 'models should be judged by their usefulness as algorithms for correlating observations and ordering experiences, rather than the exact description of reality'. They remain valuable tools for examining complex systems as holistic units to clarify the important elements and linkages, to point to how the system may be most critically bounded, and to examine the relative merits of various scenarios and strategies. They can at best be used as tools to explore the different possible futures of the system, in order to take the necessary actions to pilot the system past the worst outcomes and to set a more desirable course. In such a way an interactive session between the decision-maker and the modelled system can avoid catastrophes and expensive experimentation with the real system. In principle, models consisting of a set of interconnected dynamic equations with non-linear positive feedback loops have the characteristics of chaos and self-organization, and a number of schools have developed regional dynamic models based on these paradigms (for examples see among others: White, 1977; Allen and Sanglier, 1979; Wilson, 1981; Engelen and Allen, 1986; Allen and McGlade, 1987; IERC and RIKS, 1992).

4 Modelling the Impact of Climatic Change on Socio-Economic Systems in the Region

In line with the assumptions accepted in 1985 at the International Conference in Villach (*cf.* Preface), we set out to construct a model that would predict the effects of increases of 1.5°C temperature and 20 cm sea-level by 2025. As mentioned before, the model should (1) take into account larger temperature increases at higher latitudes, and (2) it should capture the main social and economic characteristics and relations of nations in the region. Further, it should look at effects that (3) happen on a time horizon of at least 35 years (1985–2025), and are (4) influenced by world-scale external factors, both climatological and economical. It should deal with (5) the macro-economic effects of climatic change that directly affect the entire economic and social system, but (6) also with the direct effects (flooding, devastation, etc.) on a fairly narrow band of land adjacent to the sea as well as (7) the 'ripple' effects inland. The geographical resolution at which we should study the effects therefore must be sufficiently fine to see all these various influences. As part of a spatial decision-support system, the model also should allow the planner to change the characteristics of spatial zones, such as change of local topography, extract building material from the land or coral reefs, develop land for a specific purpose, etc. Moreover, the model should show the human-induced effects on the ecological system, to understand better how this will change with climatic change.

From the above we conclude that to model both long- and short-range interaction in a spatial system, as well as the associated dynamics, there should be a symbiosis of techniques such that the growth due to long-range mechanisms of interaction can be allocated in a spatially detailed way on the basis of short-range interaction. That is why we propose the coupling of a dynamic spatial interaction model with a cellular automaton model, both running on the same geographical data base, ideally a GIS. We have shown (White and Engelen, 1992) that such coupled models successfully simulate urban land-use dynamics, and believe that the technique can be extended to handle the dynamics of land use and land coverage in Gulf/Caribbean settings equally well.

In this approach, the system operates on two levels. Each level takes into account the interactions and dynamics that typify it, and thus requires specific spatial information. A substantial amount of this information is obtained from or passed on to the other level in a continual interchange as the system dynamics unfold. In general this sort of two-level model operates as follows (Fig. 16.2). At each time step:

1) The basic geographical data needed by the high-level, long-range model is retrieved from the data base. This is aggregated for the regions used in the high-level model and then passed to that model.
2) The high-level model calculates the derivatives (changes) for each variable in each region modelled and passes them to the cellular model.
3) The low-level cellular model receives the derivatives for each variable in each region as external input and allocates them on the basis of short-range interaction mechanisms. To do so, it may call up additional information from the data base.
4) The results of step [3] are used to update the data base, and they thus serve as an input at the next time step (return to step [1]).

While in general these models should be fully regionalized on both levels, in our prototype model only has a single region at the high-level, meaning that we assume no long-range spatial interaction. To test the framework presented so far, we have chosen to model a theoretical island having the general characteristics of several nations. The choice of a purely theoretical case allows us to concentrate on some general mechanisms before adding specific details. The size of the island is small (about 35 by 30 km), allowing us to assume that the spatial interactions on the short-range dominate the island dynamics. For larger islands or regions, this assumption will no longer hold, and the prototype model will have to be regionalized in its more generic versions.

The state variables, the set of variables describing the fundamental properties of the island at each particular point in time of the simulated period, have a one-to-one mapping in the high-level and the low-level model. In the high-level model, they refer to activities, residential or economic, while in the cellular model they are expressed in terms of the amount of land taken in by each function. The relation between the two is simple: each person requires a fixed amount of residential space, and to be productive each type of job requires a minimum amount of space. The state variables are:

	High-level model	*Low-level model*
1)	Total population	Residential land use
2)	Jobs in the local agriculture sector	Local agriculture land use
3)	Jobs in the export agriculture sector	Export agriculture land use
4)	Jobs in the tourist sector	Tourist land use
5)	Jobs in expatriate population sector	Expatriate land use
6)	Jobs in the construction sector	Quarries and gravel pits
7)	'Other' sector (retail, wholesale and manufacturing)	'Other' land use

Additional land uses in the low-level model include: coral reefs, beach area, mangrove area, fallow land and sea. 'Sea' is the no-land state,

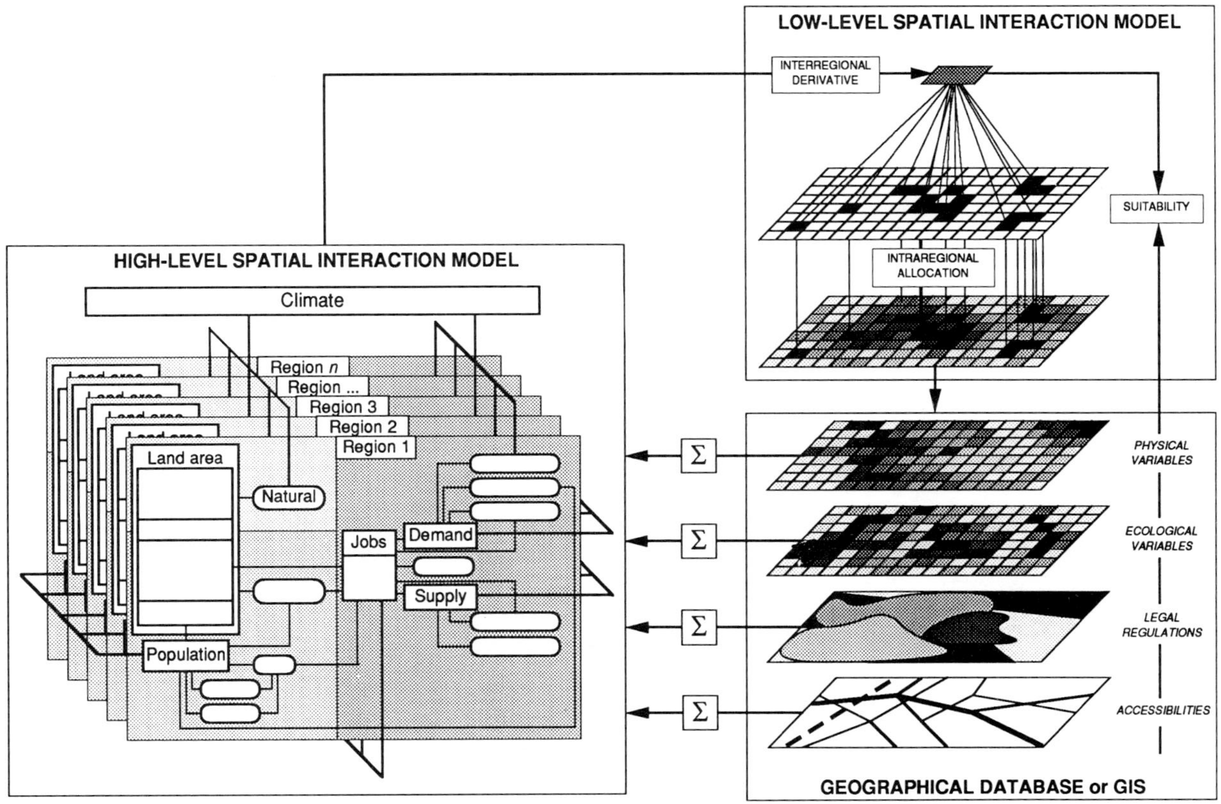

Fig. 16.2 A two-level model. On the high-level, long-range interactions are modelled by means of a dynamic spatial interaction model. The regional growth coefficients are fed into a low-level cellular model to perform the allocation on a detailed scale based on short-range interaction mechanisms. Both models will store and retrieve information from the same geographical data base.

determined by the absolute height of the sea level. The extent of 'coral reefs', 'beaches' and 'mangroves' is strongly influenced by natural factors related to climate (precipitation, temperature, storms and sea level) and by anthropogenic factors (economic activity and social welfare and behaviour of inhabitants). 'Fallow land' is the no-use state; it represents (other) natural ecosystems (e.g., tropical forest) as well as land not being cultivated at the moment.

The high-level model (Fig. 16.3) essentially consists of three coupled subsystems each represented by sets of linked variables: the natural subsystem, the social subsystem and the economic subsystem. The natural subsystem consists of a set of relations that express the evolution in time of temperature, sea level, precipitation, and storm activity. The social subsystem deals with the demographics of the island and the welfare of the population. The population of the island grows in a non-linear way; both natural growth and migrations depend on the well-being of the island population as represented by a function of the employment rate and by physical health (*cf.* de Sylva, Chapter 14). The economy of the island is represented by an input-output model, which is evaluated iteratively. At each step, extra demand (which might be negative) for products from each of the modelled economic sectors (subsistence agriculture, export agriculture, construction, tourism, services to expatriates, and 'other'

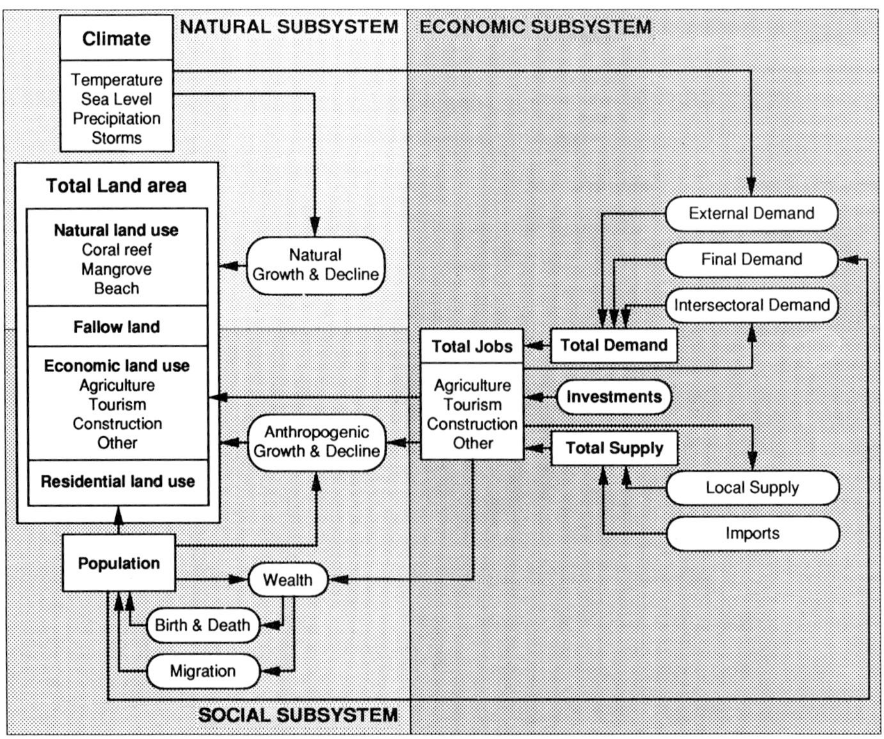

Fig. 16.3 Schematic representation of the high-level interaction model showing the loops linking the natural, the social and economic subsystem.

sector), whether exogenous (external demand), endogenous (intersectoral demand) or due to population growth or consumer behaviour (final demand), is translated into new job opportunities. New jobs require suitable land for the associated production activities, and so the demand for new jobs in each sector is translated into sectoral demands for land, using current levels of land productivity for each sector. These demands are then passed onto the low-level model.

A number of feedback loops are built into the high-level model, the most essential of which are shown in Fig. 16.3. As an example, notice the influence of 'climate' (e.g., increase in temperature in the northern USA and Canada of 3°C and sea-level rise of 10 cm by 2025; *cf.* Gallegos *et al.*, Chapter 3) on the 'external demand' (a drop in number of tourists) and its effect on the 'total demand,' (for hotel facilities and services), as opposed to the existing 'supply' (existing hotel facilities and services), the imbalance of which will influence 'investments' (e.g., in construction and maintenance of hotels, and creation of new services) and the number of 'jobs' (in the tourism and construction sector). Notice further the influence of jobs on 'wealth,' influencing 'birth and death' and'migration' rates (e.g., a drop in the number of jobs in the tourist sector leads to higher unemployment rates, which influence people's decision to emigrate which may lead to a decreasing total population that in turn decreases the amount of residential land required). Notice also the loop indicating the natural loss of available land due to the rising sea level.

These relations are written out in mathematical form. The evolution of each of the state variables is governed by a non-linear differential equation. Due to the non-linear character of the equations and the high degree of linkage among the variables, the sets of equations can not be solved numerically; rather their simultaneous solution is simulated on the computer as a time trajectory for the state of the system.

5 Cellular Automata for Short-Range Spatial Interaction

The representation of spatial interaction common in models similar to our high-level model has the great merit of dealing with the intrinsic complexity of the systems modelled in the most explicit and direct way. Unfortunately, these models have a number of drawbacks, which limit their applicability in decision-making at a detailed geographical scale: (1) computational requirements make them impractical if the number of interacting spatial actors and geographical units is increased; (2) they are subject to the so-called 'scale problem' (Huggett, 1980), thereby limiting their validity to the spatial interactions at one specific geographical scale; and (3) they are incapable of dealing with the morphological aspects of growth on the inter- and intra-regional scale (Batty, 1991).

In contrast with the conventional dynamic interaction models in which the number of interacting regions must be severely limited, in the cellular automaton approach it is natural and feasible to use a very large number of cells or regions and thus to achieve great spatial detail. A cellular automaton consists of (Langton, 1986): (1) a Euclidean space, normally two-dimensional, divided up into an array of unit squares called cells;

(2) each cell is surrounded by a neighbourhood of the same size and geometrical shape; (3) each cell is in one of n discrete possible states (land-uses in our application); (4) transition rule(s) describe the new state of a cell as a function of its own state and the states of the cells in its neighbourhood; and (5) time progresses uniformly, and at each discrete time step, all cells simultaneously change state as defined in the transition rule(s).

Cellular automata are best known as games (Couclelis, 1985). One of the simplest but most widely studied models is certainly Conway's 'Game of Life' (Gardner, 1970). Although Tobler (1979) called cellular automata geographical models *par excellence*, they have hardly been used for modelling socio-economic phenomena, but have been applied more extensively to model spatial flow processes: surface and subsurface water flows (Maidman, 1991) or lava flows (Young and Wadge, 1990).

The approach has several advantages in the study of spatial phenomena. In the context of this chapter we mention:

1) Cellular automata allow extreme spatial detail. For many applications, such detail is crucial.
2) They tend to produce complex (Wolfram, 1986; Langton, 1986) and frequently fractal patterns.
3) They show, at least at the functional level, apparent similarities with the cartographic modelling known in GIS (Tomlin and Berry, 1979), and their linkage with GIS seems feasible.
4) Expanding on the previous point, the approach permits a straightforward integration of physical and environmental qualities in economic and social modelling; in contrast, virtually all regional models currently postulate uniform background conditions. This integration of socio-economic and environmental variables is certainly of major importance in the actual planning context.
5) Building a cellular automaton model and calibrating it are done in one and the same process. They are both accomplished with the definition of the correct rules for the transition functions. This type of direct manipulation makes the cellular automaton model into an excellent decision-support and learning tool, such as referred to in terms of Modelling by Example (MbE; Angehrn, 1991) or Visual Interactive Modelling (VIM; Bell *et al.*, 1984).

Since the model is designed primarily to investigate basic principles of regional spatial form, it is kept very simple. Nevertheless, since it must give a reasonable representation of diverse, competing land uses on an island, it is more complicated and more specific than the highly generic models like 'Forest Fire' or 'Game of Life'. It is specified as follows:

The island grows and evolves, in a non-isotropic space, as cells are converted from one state (type of land use) to another. Each cell is in one of 12 states (y), each representing a land use: sea, coral reef, beach, mangrove, fallow land, subsistence agriculture, export agriculture, tourism, expatriate housing, residential housing, construction, other (retail, wholesale, industry, office, etc.). The net number of cells (N_z) required by each non-vacant state (z) at each time step is determined by the high-level model. The fate of a cell at each iteration depends on the state of the cell itself and the cells in its neighbourhood. For the island model, the neighbourhood consists of 113 cells, each of which falls within one of 19 discrete distance categories. The transition potentials for a cell are calculated as weighted sums:

$$P_z = f(s_z) * \sum_d \sum_i w_{z,y,d} * I_{d,i} + \epsilon_z \qquad (1)$$

where P_z is the transition potential to state z, $f(s_z)$ is a function expressing the suitability of cell for function z ($0 \leq f(s_z) \leq 1$), $w_{z,y,d}$ is the weighting parameter applied to cells with state y in distance zone d ($0 \leq d \leq 19$), i is the index of cells within a given distance zone d, $I_{d,i} = 1$ if the state of cell i in distance zone $d = y$, $I_{d,i} = 0$ if the state of cell i in distance zone $d \neq y$, and ϵ_z is a stochastic disturbance term.

Thus, cells within the neighbourhood are weighted differently depending on their state y and also depending on their distance d from the cell for which the neighbourhood is defined. Since different parameters can be specified for different distance zones, it is possible to build-in weighting functions that have distance-decay properties similar to those of traditional spatial-interaction equations. The function $f(s_z)$ is a measure of the physical suitability of the cell to receive the function z. Hence the cellular model retrieves from the data base (GIS) data on factors such as slope, elevation, soil conditions, legal constraints on use and access to transport, all of which affect the suitability of the cell for use by each function. To reflect the unknown factors in locational decisions, the deterministic transition potentials are subjected to a stochastic perturbation (ϵ_z). To select the N_z cells to receive the function z at each iteration, the potentials calculated for each cell for transition to a particular state are ranked. The N_z cells with the highest potentials are identified and the transformations executed.

The suitability factors for all cells actually occupied by a particular activity are monitored, and changes in average suitability for the activity are passed back to the high level which results in further changes in productivity parameters. In other words, the detailed land suitabilities and land-use patterns in the lower level of the model are reflected in changes in the specification of the high level. The micro-scale geography thus affects the global dynamics directly and continuously.

The model currently is being verified, and a sensitivity analysis is being performed to investigate the behaviour of the model under various combinations of parameter values. Thus, for example, the effects of different proportions of the 11 functions, of different growth rates, and of different levels of stochastic disturbance are being examined. Most of these results are unexceptional and will not be described here. No calibration as such has been performed since no particular island is being modelled, rather, general principles of spatial reorganization induced by climatological changes are being explored. Nevertheless, some sets of parameter values give more reasonable results than others, while some yield patterns that bear no conceivable resemblance to any actual island. It is therefore encouraging that patterns of parameter values which represent known locational preferences are also the ones which generate the most realistic looking islands. This does not guarantee, however, that the prototype model can be applied directly to a real Caribbean-region island or mainland state. To prove its applicability to any particular case, an exhaustive validation and calibration process will be required. Fig. 16.4 shows four stages in the growth of one cellular island.

6 Implementation of the Decision-Support Environment

As stated earlier, our long-term aim is to develop a spatial decision-support system allowing policy analysts to anticipate, explore and counter the risks associated with climate change for island and coastal nations in the region. As part of the model base of such a system, the prototype island model should evolve into a more generic tool. It should be clear, however, that one model does not constitute a full decision-support system. Indeed, we mentioned that additional tools must be incorporated to ease the job of the user. A number of additional models could be added into the model base. These should perform tasks complementary or supplementary to the ones done by the (enhanced) island model; for example, calculation of storm damage, soil-erosion sedimentation in deltas, which could include climatological or hydrological models and an age-cohort model for detailed demographic evolution. Typical decision models, allowing us to formulate, compare, evaluate, rank and select solutions among a set of alternatives and graphical interface tools supporting the input of data, the retrieval and presentation of output data and model results, should also be added. Already in the prototype, effort has gone into the implementation of such tools.

The system is currently being developed in MS-Windows 3 on top-range IBM-PC compatible computers, equipped with 80386 or 80486 processor and 8 Mb of RAM, and has a fully graphical interface (Fig. 16.5). This platform has the great advantage of being widely available and is sufficiently powerful to perform the calculations required within reasonable time limits. The system will allow the user to specify interactively his or her own applications. This will include decisions on the number of functions to be modelled as well as the resolution and size of the grid (currently a grid of 80 by 80 cells is being used). To start a new application, the user will import a land-use matrix from an existing application (e.g., GIS), or will enter it by means of the built-in 'land-use editor' through mouse-clicking. Similarly, data will be entered to calculate the suitabilities. The transition functions are introduced in a graphical way as well. The user disposes of a tool, mainly consisting of an X-Y graph, in which he can draw, by mouse-clicking, the distance-decay function defining the probability of transition of one type of land use into any other type, or he can select a specific distance-decay function from a library-list previously stored. The DSS environment is being enhanced with a number of dedicated tools for easy and interactive definition of policy strategies (e.g., changing land uses deliberately, influencing the suitabilities or constituent factors, etc.), and to compare and evaluate the outcomes of different scenarios.

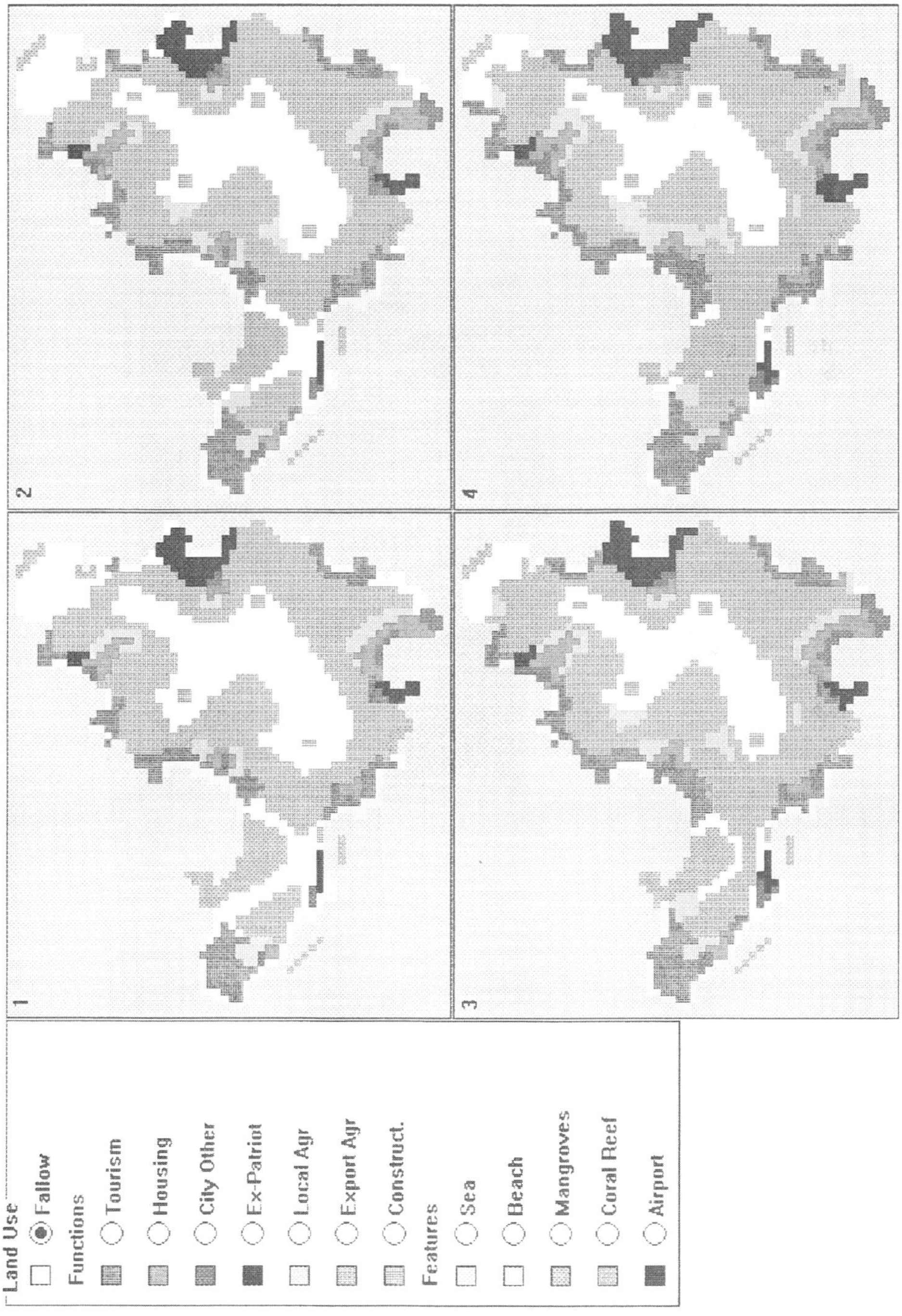

Fig. 16.4 Four stages in the evolution of a hypothetical island. See colour plates between pages 210 and 211.

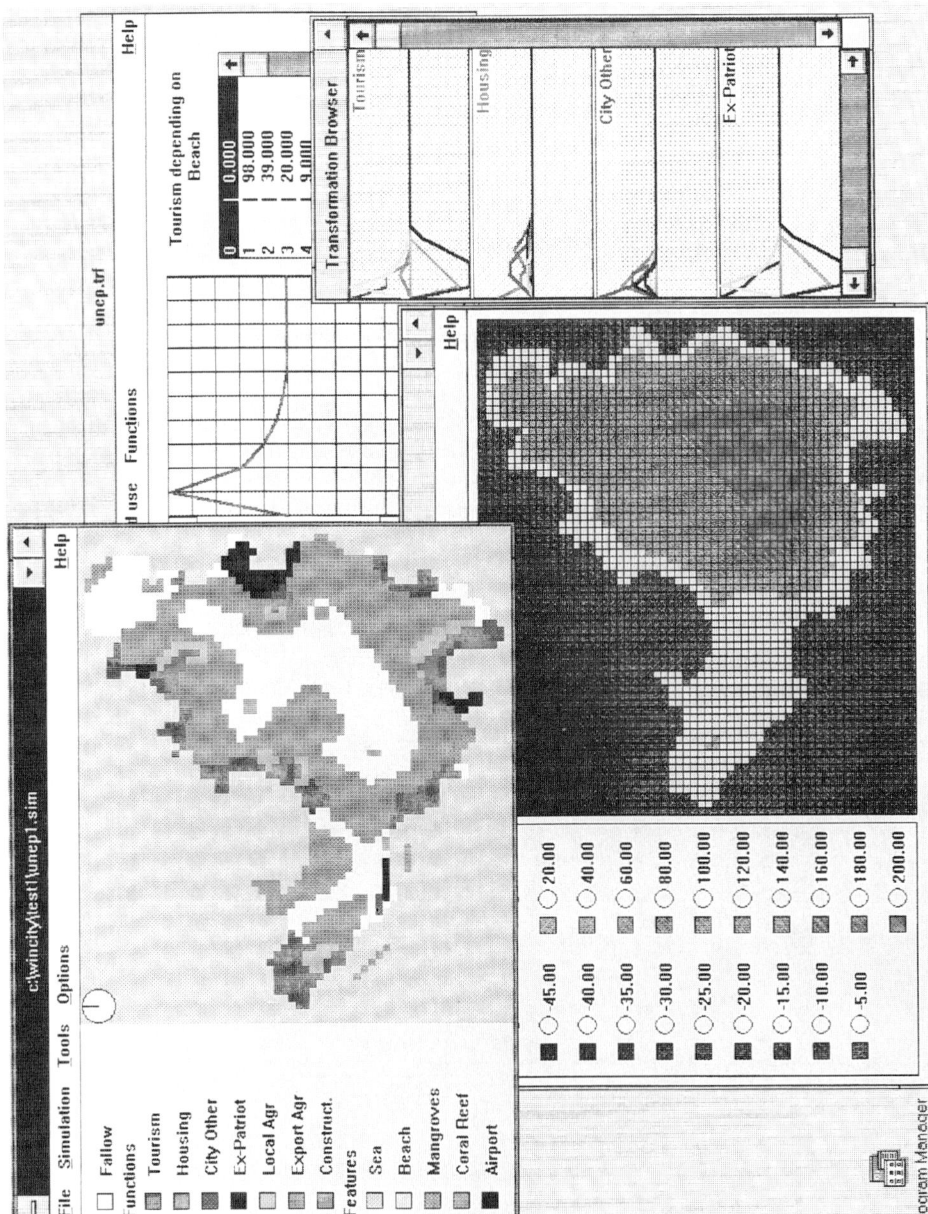

Fig. 16.5 The graphical user interface of the prototype decision-support environment. See colour plates between pages 210 and 211.

7 Concluding Remarks

A model-based decision-support framework should lead to a strategic policy tool that allows decision-makers to explore and to evaluate the possible effects of a changing climate, thereby helping them to formulate policy actions for countering any baleful influence. Among the essential requirements of such a framework it should treat climatic change as a new problem exacerbating existing or potential problems. As a world-wide phenomenon influencing social and economic behaviour, climatic change is likely to have economic effects on a macro- as well as a micro-scale in the mostly small and export oriented societies of the region. These societies, therefore, should be modelled as holistic systems, with their working and organization being subject to perturbations from abroad, but also from within the system. The time horizon of the problem at hand, the different geographical scales at which effects will be visible, as well as the importance of rather detailed physical features and ecosystems of the societies studied, lead us to propose a two-level modelling approach, in which at the high level a classical, regionalized spatial-interaction model deals with the long-range social and economic interaction mechanisms and exchanges in a cyclic process information with a low-level, short-range cellular model. The cellular model reads data such as physical and environmental characteristics from a geographical data base, ideally an existing GIS. Based on local interaction mechanisms, these serve to translate the growth coefficients received from the high-level model into detailed land-use changes. When two models are integrated in this way, their joint behaviour typically will be more realistic and complex than that of either component in isolation.

We have set out to develop a prototype model of a purely theoretical island, but have tried to incorporate in it as much as possible the typical characteristics of island nations. Initially this allows us to concentrate most of our attention on the variables, mechanisms and causalities to incorporate in the model at the most generic level. It also should allow us to test our hypotheses and to come to define better data requirements, further scientific analysis, and the desired functionality of the decision environment we set out to develop.

8 References

Allen, P.M. and McGlade, J.M., 1987: Modelling complex human systems: A fisheries example. *European J. of Operational Res.*, **30**, 147–167.

Allen, P.M. and Sanglier, M., 1979: A dynamical model of growth in a central place system. *Geographical Analysis*, **11**, 256–272.

Angehrn, A.A., 1991: Modelling by example: A link between users, models and methods in DSS. *European J. of Operational Res.*, **55**, 296–308.

Ausubel, J.H., 1991: Does climate still matter? *Nature*, **350**, 649–652.

Batty M., 1991: Generating urban forms from diffusive growth. *Environment and Planning*, **A23**, 511–544.

Bell P.C., Parker, D.C. and Kirkpatrick, P., 1984: Visual interactive problem solving – A new look at management problems. *Business Quarterly*, **49(1)**, 14–18.

Casti, J.L., 1989: *Alternate Realities*. J. Wiley & Sons, New York.

Catanese, A.J., 1979: Information for planning, International City Management Association: Washington DC.

Cleal, D.M. and Heaton, O., 1988: *Knowledge-Based Systems, Implications for Human–Computer Interface*. Ellis Horwood, Chichester.
Couclelis, H., 1985: Cellular worlds: a framework for modelling micro-macro dynamics, *Environment and Planning*, **A17**, 585–596.
Davis, G.R., 1990: Energy for planet earth. *Scientific American*, **263**, 21–27.
Engelen, G. and Allen, P.M., 1986: Modelling the spatial distribution of energy demand for the Province of Noord Holland: towards an integrated approach. *Sistemi Urbani*, **2/3**, 241–261.
Forrester, J.W., 1961: *Industrial Dynamics*. MIT Press, Cambridge.
Gardner, M., 1970: The fantastic combinations of John Conway's new solitaire game Life. *Scientific American*, **223**, 120–123.
Han, S.-Y. and Kim, T.J., 1990: Intelligent urban information systems: Review and prospects. In: Kim, T.J., Wiggins, L.L., Wright, J.R. (eds), *Expert Systems: Applications to Urban Planning*. Springer Verlag, New York, 241–264.
Holtzman, S., 1989: *Intelligent Decision Systems*. Addison-Wesley, Reading.
Huggett, R., 1980: *Systems Analysis in Geography*. Clarendon Press, Oxford.
IERC and RIKS, 1992: An integrated strategic planning and policy framework for Sénegal, European Community DG VIII, Final Report, IERC Cranfield, England.
Langton, C.G., 1986: Studying artificial life with Cellular Automata. *Physica*, **22D**, 120–149.
Maidman, D.R., 1991: GIS and hydrologic modelling. Proceedings of the First International Conference on Integrated Geographic Information Systems and Environmental Modelling, Boulder, Colorado.
Marble, D.F. and Amundson, S.E., 1988: GIS in urban and regional planning. *Environment and Planning*, **B15**, 305–324.
Ominde, S.H. and Juma, C., 1991: *A Change in the Weather, African Perspectives on Climatic Change*. ACTS Press, Nairobe, Kenya.
Prigogine, I., 1981: *From Being to Becoming*. Freeman, San Francisco.
Reddy, A.K.N. and Goldemberg, J., 1990: Energy for the developing world. *Scientific American*, **263**, 63–72.
Savory, A., 1988: *Holistic Resource Management*. Island Press, Washington.
Simmonds, D., 1986: Modelling contexts and purposes: defining alternative paths of model development. In: Hutchinson, B. and Batty, M. (eds), *Advances in Urban Systems Modelling*. North Holland, Amsterdam, pp. 75–90.
Tobler, W.R., 1979: Cellular geography. In: Gale, S. and Olsson, G. (eds), *Philosophy in Geography*. Reidel, Dordrecht, pp. 279–386.
Tomlin, C.D. and Berry, J.K., 1979: A mathematical structure for cartographic modelling in environmental analysis. Proceedings of the 39th Symposium, American Congress on Surveying and Mapping, Falls Church Virginia, 269–283.
White, R., 1977: Dynamic central place theory – Results of a simulation approach. *Geographical Analysis*, **9**, 226–243.
White, R. and Engelen, G., 1992: Fractal urban land use patterns: A Cellular Automata approach. To appear.
Wilson, A., 1981: The evolution of spatial structure. In: Bennet, R.J., *European Progress in Spatial Analysis*. Pion Books, London.
Wolfram, S., 1986: Approaches to complexity engineering. *Physica*, **22D**, 385–399.
Young, Ph. and Wadge, G., 1990: Flowfront: Simulation of a lava front. *Computers & Geosciences*, **16(8)**, 1171–1191.

17

Glossary of Scientific Terms

George A. Maul[1] *and Stephen R. Baig*[2]

Abstract

Scientists use terms that can make a treatise such as *Climatic Change in the Intra-Americas Sea* inaccessible to non-specialists, and in particular to management and policy makers not trained in the natural sciences. To overcome this difficulty, this glossary of terms covering some aspects of oceanography, geodesy, meteorology, ecology, and climatology has been included. The terms and abbreviations selected for inclusion are mostly those that have appeared in the text; the meanings are meant to be scientific rather than legal. We have freely paraphrased many of the definitions from more recent sources (Baker et al., 1966; Bates and Jackson, 1980; NOAA, 1988) without explicit acknowledgement in order to minimize text, and we have also cross-referenced some of the older texts (Sverdrup et al., 1942; Mitchell, 1948; Huschke, 1959). The reader is referred to the several sources listed below for more information, but is cautioned that terms and meanings change with time (Maul, 1988). For the future, it is urged that a multi-lingual glossary be created for the region, one that includes more information on socio-economic and public health terms.

References

Baker, B.B., Jr., W.R. Deebel, and R.D. Geisenderfer [1966]. *Glossary of Oceanographic Terms, 2nd edition*, SP-35. U.S. Naval Oceanographic Office, Washington, DC 20390, 204 pp.

Bates, R.L., and J.A. Jackson (editors), [1980]. *Glossary of Geology*. Amer. Geol. Inst., Falls Church VA, 749 pp.

Huschke, R.E. (editor), [1959]. *Glossary of Meteorology*. Amer. Meteorol. Soc., Boston, MA, 638 pp.

Maul, G.A. [1988]. Guest Editorial, Special Issue: Pacific Ocean Sea Level Variability. *Mar. Geod. J.*, *12*(3-4), 229–233.

Mitchell, H.C. [1948]. *Definitions of Terms used in Geodetic and other Surveys*. U.S. Dept. of Commerce, Coast and Geodetic Survey SP-242. U.S. Government Printing Office, Washington 25, DC, 87 pp.

NOAA (National Oceanic and Atmospheric Administration) [1988]. *Geodetic Glossary*. NOAA, National Geodetic Survey, Rockville MD 20852, 274 pp.

Sverdrup, H.U., M.W. Johnson, and R.H. Fleming [1942]. *The Oceans*. Prentice-Hall, Englewood Cliffs, NJ, 1087 pp.

[1]Atlantic Oceanographic and Meteorological Laboratory, Miami, Florida 33149, USA
[2]National Hurricane Center, Coral Gables, Florida 33146, USA

Glossary and Abbreviations

ADCP (**A**coustic **D**oppler **C**urrent **P**rofiler) A shipboard or moored ocean current measuring system that measures the Doppler shift of backscattered sound from particles in seawater, and computes the current speed and direction as a function of depth (300 meter vertical range typically).

Air-Sea Interaction The scientific study of the rates of energy transfer from the atmosphere to the ocean, and from the sea to the air; considerable feedback exists between the two fluids.

Albedo The ratio of total radiant energy reflected from a body to the total radiant energy incident upon it; Earth's albedo is typically 0.4, but cloud albedos can be 0.9 or greater.

Altimetry The art or science of determining sea surface heights (SSH) by remote sensing from above the sea surface.

Altithermal Period A dry, warm postglacial period 7500–4000 ybp (*cf*. Little Climatic Optimum).

Anthropogenic Involving the impact of humankind on nature; induced or altered by the presence and activities of humans.

Anticyclonic Fluid flow having a sense of rotation (clockwise in the northern hemisphere) opposite to Earth's, rotation, and associated with an oceanic or atmospheric high pressure cell.

Aquifer A subterranean layer of unconsolidated material containing water, which may be connected to the sea.

Atoll A torus-shaped organic reef enclosing a lagoon in which there is no pre-existing land, and being surrounded by open sea.

AVHRR (**A**dvanced **V**ery **H**igh **R**esolution **R**adiometer) Five channel scanning visible and infrared radiometer operational on the NOAA series of polar orbiting meteorological satellites.

Barrier Island (or **bar**) An emerged (or submerged) embankment of sand, shell, or gravel, that is formed in shallow water by waves and currents.

Baroclinicity The state of stratification in a fluid in which surfaces of constant pressure intersect surfaces of constant density.

Barotropy The state of a fluid in which surfaces of constant density are coincident with surfaces of constant pressure; it is the state of zero baroclinicity.

Bathymetry The science of measuring ocean depths to determine sea floor topography.

Beach A zone of unconsolidated material that is in alongshore or onshore-offshore *active* transport.

Benthic A zone of ocean bottom always under water; pertaining to organisms that live there.

Boundary Conditions In numerical modelling, boundary conditions are mathematical expressions defining the interaction of the fluid with the limits of the model regime (*e.g.*, the frictional coupling between the ocean and the sea bottom).

Brunt-Väisälä Frequency One of the most important natural oscillation frequencies of a vertical column of stratified fluid, given by

$[(g/\rho)(\partial\rho/\partial z)]^{1/2}$ where g is gravity, ρ is density, and z is the vertical coordinate.

Bruun Rule An empirical rule of beach erosion for a closed system in which there is no external source or sink of sand, given by $W \cdot \Delta Z/D$ where W is the active beach width, ΔZ is sea level rise, and D is tide range; W/D is often found to be ~100.

Canalization The restricting of river flows into channels or canals; oftentimes this prevents flooding in deltas but leads to a loss of the very sediment that builds deltas.

Carbon Dioxide (CO_2) The primary radiatively active atmospheric gas responsible for the greenhouse effect; its concentration is approximately 350 parts per million (ppm).

Carbonate ($CO_3^{=}$) A salt or ester of carbonic acid; the most common form in the Intra-Americas Sea is calcium carbonate ($CaCO_3$), *i.e.*, limestone.

Caribbean Current A branch of the North Equatorial Current (NEC) passing through the Lesser Antilles and flowing into the Gulf of Mexico through the Yucatan Channel.

Cartagena Convention An agreement within the Wider Caribbean Region to mitigate and prevent pollution of the sea.

Catchment Area Geographical region from which rainfall drains into a common river system; also called drainage area.

Cellular Automata A numerical modelling approach where a large number of small areas (cells) are used to represent details in a physical or socio-economic process, and where the cells are mathematically connected in a computer algorithm.

CFC (**c**hloro**f**luor**c**arbon) A non-radiatively active anthropogenic compound made of one or two carbon atoms, chlorine, and fluorine; CFC's interact with UV radiation to chemically destroy ozone in the stratosphere.

Chlorophyll A green organic molecule active in photosynthesis; it can be detected in ocean waters by satellite sensors such as the CZCS and (if successful) SeaWIFS.

Ciguatera A poison caused by ingestion of various tropical fish, typically larger carnivores such as barracuda.

Circulation The three-dimensional change associated with the motion of the ocean or the atmosphere, and giving rise *e.g.* to surface currents and winds respectively.

Climate The statistical collective of weather for a given region over a specified length of time, typically 30 years.

Climatology The scientific study of climate.

Cloud Feedback Radiation from Earth at visible and infrared wavelengths is absorbed and re-emitted by clouds and can interact again with Earth's surface; this interaction is termed feedback.

COADS (**C**ombined **O**cean-**A**tmosphere **D**ata **S**et) A compilation of global surface observations from merchant and naval ships covering the last 150 years or so, from the National Center for Atmospheric Research (NCAR).

Coastal Zone The area of Earth's surface from near the coast to the continental shelf break (~200 m isobath) where marine species (saltwater) dominate the ecology.

Compaction The geological process whereby a soil mass loses pore space, becomes more dense, and thus increases its bearing capacity.

Conveyor Belt A description of the three dimensional oceanic circulation involving sinking waters from the arctic North Atlantic flowing at about 3000 m depth into the Southern Ocean, rising again to the surface in equatorial zones and completing the global circuit as cross-equatorial surface waters into the Caribbean Sea and the Gulf Stream System.

Coral Bleaching A condition of corals when photosymbiotic organisms, e.g. dinoflagellates such as zooxanthellae, are forced to leave the host coral animals.

Coriolis Force An apparent force on moving particles when measured relative to a rotating coordinate system; the force is to the right in the Northern Hemisphere and is proportional to $2\Omega\sin\phi$ where Ω is Earth's angular velocity and ϕ is latitude.

Cryosphere The region of Earth dominated by ice.

Cyclone A closed circulation cell surrounding a low pressure area, either atmospheric or oceanic; the fluid flow is counter clockwise in the Northern Hemisphere, *i.e.* in the same direction as Earth's rotation.

CZCS (**C**oastal **Z**one **C**olor **S**canner) An experimental global ocean color sensing (visible wavelength) multispectral scanning radiometer launched in 1978 on the NASA NIMBUS 7 satellite (no longer operating).

Datum In surveying, it is the reference for calculation of horizontal or vertical position. Vertical control datums are usually referenced to mean sea level at a number of tide gauge sites along the coast; horizontal control datums are usually with respect to a point on the spheroid where the geoid height is assumed to be zero.

DBMS **D**ata-**B**ase **M**anagement **S**ystem.

Density In fluid mechanics, it is the mass per unit volume, symbolized by the Greek letter ρ; in population dynamics it is the number of humans per unit area.

Diapirism The geological process of piercing of uplifted rocks by core material, through tectonic stress or geostatic loading, such as in salt domes or igneous intrusions.

Diastrophism Processes through which Earth's crust is deformed.

Downwarping Subsidence of a broad region of Earth's crust.

Downwelling In oceanography, a downward movement of water, usually caused by fluid convergence; in radiation, the radiant energy incoming or received at a locale; in meteorology, the term used is "sinking".

DSS (**D**ecision **S**upport **S**ystem). Intelligence centred method of decision-making consisting of a model base, a data base, and a user interface.

Dynamic Height (or geodynamic height) The amount of work done when a water parcel of unit mass is moved vertically from one (pressure) level to another.

Dynamic Height Anomaly (ΔD) (also called anomaly of geopotential difference) The excess of the actual geopotential difference between two given isobaric surfaces, over the geopotential difference in a homogeneous water column of salinity 35 psu and *in situ* temperature 0°C; $\Delta D = \int \delta \cdot dp$ where δ is the specific volume anomaly and p is pressure.

Dynamic Number The work done to raise a unit mass from sea level to a given point, expressed in absolute units. In the metric system, the dynamic number of a point is the work required to raise a mass of 1 kilogram against the force of gravity from sea level to the level surface passing through the point, the work being measured in standard kilogram-meters at sea level in latitude 45°.

Dynamic Topography The configuration formed by the geopotential difference or dynamic height anomaly between a given isobaric surface and a reference surface.

Ecosystem A community of animals, bacteria, and plants, and their interrelation with their immediate chemical and physical environment.

Eemian Period North European climatostratographic and floral stage equivalent in time to the Riss/Würm interglacial about 125,000 ybp.

EEZ **E**xclusive **E**conomic **Z**one.

Ekman Layer Vertical zone in the ocean in which Ekman transport occurs, typically the upper 100 meters; more generally it refers to the upper or lower oceanic boundary layer.

Ekman Transport The volume flow that occurs at right angles to the wind in the Northern Hemisphere; it arises from the balance of the Coriolis force and wind stress. Alongshore coastal winds lead to upwelling or downwelling.

Ellipsoid An ellipsoid of revolution may be defined by stating the lengths of the semiaxes (a and b) of the generating ellipse, or the length of one of the semiaxes and the flattening (or ellipticity, $[(a-b)/a]$) of the ellipse.

El Niño A condition, originally applied to the surface waters of the eastern equatorial Pacific Ocean, wherein the usual cold, marine life supporting surficial waters are replaced by unusual warm waters; it was thought to occur around Christmas, hence the Spanish name for "The Child".

Energy The capacity for doing work, expressed as a result of motion (kinetic), of availability to cause motion (potential), of being associated with light (radiant), of the exchange of heat and cold (thermal), *etc.*

ENSO (**E**l **N**iño – **S**outhern **O**scillation) A term that connects the oceanic phenomenon (El Niño) with the atmospheric phenomenon (Southern Oscillation) leading to the realization of the global consequences of air-sea interaction (see TOGA).

EOS (**E**arth **O**bserving **S**ystem) A NASA proposed system for spacecraft measurements of Earth including the systems for validation and archival of data.

Epeirogeny The uplift or subsidence of broad continental or ocean-basin areas, and the physical causes of such motion.

Epibenthic Associated with the zone between the low tide line and the continental shelf break (typically the 200 m isobath).

Equilibrium Temperature The temperature Earth would achieve in due time after an increase in greenhouse gases.

ERS-1 The first earth resources satellite of the European Space Agency; launched in 1991.

Estuary A tidal bay formed from the drowning of a non-glaciated river valley juxtaposed to the ocean.

Eustatic Simultaneous world-wide change in sea level associated with growth or melting of continental and/or mountain glaciers (*cf.* Glacial).

Evapotranspiration The sum of water loss from soil through evaporation plus that from plants through transpiration through their surfaces.

FAO Food and Agricultural Organization of the United Nations.

Feedback The interaction between components of a mechanical, thermal or radiant system; can be negative or positive depending on the definition of the directional vector.

Ferrel Cell Weak atmospheric circulation cell with descending air in mid-latitudes, poleward surface flow to about 70° latitude, rising air to about 200 mb and equatorward flow to about 40° (*cf*. Hadley Cell).

Fishery The business of catching and selling fish (*e.g*. being associated with single boats: artisanal), or type of fish (*e.g*. bottom feeders: demersal); also the study of fishing in a particular environment (*e.g*. estuarine) or with an age class (*e.g*. juvenile).

Florida Current That branch of the Gulf Stream System flowing through the Straits of Florida from Dry Tortugas to Cape Canaveral.

Forecasting In geophysical fluid dynamics, the art of predicting future states of the atmosphere or ocean through computer models (numerical forecasting), through projection of observed trends (persistence forecasting), or through comparisons with prior epochs (scenario forecasting).

GIS Geographic Information System.

GCM General Circulation Model.

GCS Global Climate System.

Gelbstoff Humus-like dissolved yellow substance found in the sea; from the German for "yellow substance".

Geoid An equipotential (level) surface (*i.e*. one to which, at every point, the plumb line is perpendicular). Specifically, the figure of the earth considered as the level surface of a motionless $\alpha_{35,0,p}$ ocean, where $\alpha_{35,0,p}$ is the specific volume of a 35 psu, 0°C, ocean, at a particular time.

Geoid Height The vertical distance between the ellipsoid and the geoid; ranges approximately ± 100 m globally.

Geopotential Number The work required to raise a unit mass from the geoid to height h; *i.e*. $\int g \cdot dh$, where g is actual gravity and h is along the plumbline.

Geomorphology The study of the form of the earth and the general configuration of its surface.

Geostrophic The balance between the Coriolis force (f) and the horizontal pressure gradient force ($\alpha \cdot \partial p/\delta x$) that neglects the other terms in the equations of motion for the atmosphere and ocean, but which accounts for most of the current velocity (v) and winds (i.e. $fv=\alpha \cdot \partial p/\delta x$); *cf*. Specific Volume Anomaly.

GFDL NOAA Geophysical Fluid Dynamics Laboratory, located in Princeton, New Jersey USA.

GISS NASA Goddard Institute for Space Studies located in New York, New York USA.

Glacial Pertaining to ice that is on land, thus affecting eustatic sea level; sea ice does not contribute to changes in sea level.

GLOSS (Global Level Of the Sea Surface) The Global Sea Level Observing System of the Intergovernmental Oceanographic Commission of UNESCO.

GPS (**G**lobal **P**ositioning **S**ystem) U.S. satellite navigation system capable of ± 10 meter dynamic positioning accuracy globally, and ± 1 centimeter three-dimensional accuracy locally when used in differential mode with VLBI (*q.v.*).

Greenhouse Effect The analogy to the atmosphere of how a greenhouse works, *i.e.*, that the Earth is some 30°C warmer because certain gases (notably water vapor and CO_2) allow shortwave radiation (sunlight) to pass through to the surface, but absorb some of the longwave (infrared) radiation emitted by the ground and sea.

Groundwater Water below the water table (minimum water-well depth) as distinguished from interflow and soil moisture.

Gulf Stream System The primary oceanic current system of the western North Atlantic Ocean including the Caribbean Current, the Yucatan Current, the Gulf Loop Current, the Florida Current, and the "Gulf Stream" proper.

Hadley Cell Atmospheric vertical circulation cell of ascending air over the equator, poleward flow aloft to about 30° north and south latitude, descending flow at 30°, and equatorward flow thereafter; an explanation of the Trade Winds.

Halocarbon A molecule of carbon and any of the five chemical elements fluorine, chlorine, bromine, iodine, and astatine.

HF Radar A high frequency (25 MHz) shore-based radar system for mapping ocean surface currents over an area of ~10,000 km^2 based on the Doppler shift of backscattered HF radio waves from ocean surface waves.

Holocene The most recent geological epoch, usually taken as the last 10,000 years or so.

Hot Snap A short (several day or week-long) period of elevated temperature; the opposite of a cold snap.

Hydrograph A graphical presentation of river or stream discharge as a function of time; *cf.* Marigram.

Hydrography The science of the physical description of the oceans, seas, rivers, and lakes, with particular emphasis on bottom depths and features.

Hydrology The science of Earth's waters, especially concerning evaporation, precipitation, and the character of water in streams, lakes, and under ground.

Hydrostatic A fluid where pressure (p) is a function of density (ρ), gravity (g) and depth (z) only; *i.e.* $p = \int \rho g \cdot dz$.

Hypersalinity Extremely high salinity condition, often leading to detrimental conditions for marine organisms.

IADB **I**nter-**A**merican **D**evelopment **B**ank.

ICSU **I**nternational **C**ouncil of **S**cientific **U**nions.

IGCP **I**nternational **G**eological **C**orrelation **P**rogramme.

IGBP **I**nternational **G**eosphere-**B**iosphere **P**rogramme of the ICSU.

IMO **I**nternational **M**aritime **O**rganization of the **U**nited **N**ations.

Infrastructure The basic facilities upon which a modern human community is dependent such as roads, schools, ports, dams, *etc.*

Interglacial The periods in Earth's history when glaciers are a minimum in size and extent, and the temperatures are warm, such as at present.

Intra-Americas Sea The region of the tropical and subtropical western North Atlantic Ocean that includes the Guianas coast of South America, the Caribbean Sea, the Gulf of Mexico, the Bahamas, and Bermuda; *approximately* bounded by 0°–30°N latitude and 50°W–100°W longitude.

IOC **I**ntergovernmental **O**ceanographic **C**ommission of UNESCO.

IOCARIBE IOC Subcommission for the Caribbean and Adjacent Regions (*cf.* Intra-Americas Sea; Wider Caribbean Region).

IPCC **I**ntergovernmental **P**anel on **C**limate **C**hange of the WMO, ICSU and UNEP.

ITCZ (**I**nter-**T**ropical **C**onvergence **Z**one) The dividing line between the Northeast Trade Winds and the Southeast Trades, along portions of which the converging Trades' air rises.

Lagoon A shallow pond generally separated from the open sea (*cf.* atoll, coral reef).

LANDSAT An operational satellite system designed for observing landform, measuring (primarily) visible radiation reflected from plants, fields, cities, *etc.*

Larve *Archaic form of larva;* the early form of any animal that at birth is fundamentally unlike the adult stage.

Leveling The art of determining the vertical height difference between two points on Earth. Geodetic leveling is extremely precise and is always referenced to the vertical control datum.

Little Climatic Optimum The period in Earth's history, approximately 8,000 to 5,000 ybp, when mean air temperatures were 1–2°C warmer than today (*cf.* Altithermal Period).

Little Ice Age The period in Earth's history, approximately the years 1300 to 1800 AD, when mean surface air temperatures were 1–2°C colder than today.

Littoral The nearshore zone, typically encompassing the high tide line to below the low tide line (in some usage, out to the continental shelf break, *i.e.* ~200 m).

Loop Current That portion of the Gulf Stream System flowing into the Gulf of Mexico beyond the Yucatan Channel, turning (or looping) anticyclonically within the Gulf, and exiting through the Straits of Florida.

Manatee A marine mammal found throughout the coastal regions of the tropical and subtropical Atlantic Ocean; also known as the sea cow.

Mangrove One of several genera of tropical and subtropical trees and shrubs that have prop roots and that grow in the shallow waters of the coastal zone.

Mariculture The art of using seawater to cultivate crops of fish or plants; a subset of aquaculture.

Marigram A graphical presentation of the height of the tide as a function of time (*cf.* hydrograph).

Mean Sea Level The mean surface water level determined by averaging heights at all stages of the tide over (traditionally) a 19-year period. Mean sea level is not an equipotential surface (*i.e.* not the geoid).

Medieval Warm Epoch That period in Earth's history, approximately the years 800–1200 AD, when air temperatures were 1–2°C warmer than the Little Ice Age, and similar to today's.

Mesoscale A length dimension associated with a circulation feature, typically of the order of 100–1000 km; the scale upon which geostrophic ocean eddies and hurricanes occur.

Methane (CH_4) A colorless naturally occurring and anthropogenically generated atmospheric greenhouse gas that contributes to the radiative warming of Earth (*cf.* carbon dioxide).

Milankovitch Cycle Regularly changing insolation due to variations in Earth's orbit around the sun, theorized to cause ice ages; for the next 5,000 years Earth will undergo decreasing insolation due to orbital parameters.

Model A mathematical or heuristic description of a physical, chemical, biological, geological or socio-economic process, which can be prognostic (forecasting) or diagnostic (analytical).

Monsoon A term for seasonal winds, not restricted to Asia, but also those that dominate the climate of Mexico and the southern U.S.A.

Morphology The scientific study of form and structure, especially in biology and geology.

Mudbank Migratory shallow water ocean bottom feature primarily of fine grain clay-sized particles, but sometimes including sands; those off Guiana have amplitudes of tens of meters and move at rates of meters per day.

Nitrous Oxide (N_20) A colorless naturally occurring and anthropogenically generated atmospheric greenhouse gas that contributes to the radiative warming of Earth (*cf.* carbon dioxide).

NASA U.S. **N**ational **A**eronautics and **S**pace **A**dministration.

NCAR **N**ational **C**enter for **A**tmospheric **R**esearch located in Boulder, Colorado USA.

NCDC NOAA **N**ational **C**limatic **D**ata **C**enter located in Ashville, North Carolina USA.

NEC (**N**orth **E**quatorial **C**urrent) A westward flowing ocean current driven by the northeast Trade Winds; in the Atlantic Ocean it enters the Caribbean Sea primarily through the Windward, Mona, and Anegada Passages.

NECC (**N**orth **E**quatorial **C**ounter **C**urrent) An ocean current flowing eastward near the equator; in the Atlantic Ocean it seems to be related to the retroflection zone off the Guianas.

NOAA U.S. **N**ational **O**ceanic and **A**tmospheric **A**dministration.

Orogenic Associated with the regional mountain forming processes of folding, faulting, and thrusting.

OSU **O**regon **S**tate **U**niversity located in Corvallis, Oregon USA.

Ozone (O_3) Layer The stratum of the atmosphere between 10 and 50 km above the surface where O_3 is highly concentrated; also called the ozonosphere. O_3 is a faintly blue gaseous form of oxygen formed photochemically when ultraviolet light interacts with oxygen. Ozone reacts with CFC's which reduces the ozonosphere's UV absorbing properties.

Paleoclimate The climate of a time in the geological past.

PAR (**P**hotosynththically **A**ctive **R**adiation) Visible and UV wavelengths of light important in the process of photosynthesis.

Pegasus An acoustically-tracked oceanographic free-falling instrument for determining current velocity with depth; oceanic analog to atmospheric rawinsondes.

Phaeopigment Chlorophyll-like plant molecules that can contribute to water discoloration. They can be sensed by satellite systems such as the CZCS.

Phosphorite Non-crystalline natural calcium phosphate; same as phosphate rock.

Pixel Contraction of the words picture and element; the smallest cell in a digital image.

Population Density The number of humans per unit area.

Pressure Gradient The primary driving force per unit mass per unit volume in fluid flows, commonly expressed as $\partial p/\partial x$ where p is pressure and x is horizontal distance on an equipotential (level) surface.

Productivity In oceanography, it is the amount of carbon fixed by living organisms per unit area per unit time.

PSMSL **P**ermanent **S**ervice for **M**ean **S**ea **L**evel located in Bidston, Merseyside, U.K.

PSU (**P**ractical **S**alinity **U**nits) The modern units of salinity; very approximately the same as *per mille* (‰) or *parts per thousand*, or grams of salt per kilogram of seawater.

Quaternary The latest period of geological time, typically the last 1,000,000 years.

Radiation The energy output of an object, usually expressed with the wavelength band such as infrared (3–20μm), microwave (1–5 cm), net (all wavelengths), ultraviolet (2000–4000 Å), visible (400–700 nm), *etc*.

Reef Zonation The condition of coral reefs being formed in zones, *e.g*. the distribution of fossils more or less parallel to the reef's biostratigraphic zones.

Retroflection An ocean current zone off the Guianas where the northwestward moving cross-equatorial flow turns anticyclonically and continues out into the open Atlantic Ocean (see the book cover for a CZCS image of the retroflecting flow).

Rossby Wave (or **β-wave**) Wave motion in a rotating fluid, such as the ocean or atmosphere, when the change in the Coriolis parameter (f) with latitude (y) is considered a constant ($\beta=\partial f/\partial y$; β=constant is sometimes called the β-plane approximation).

RSL (**R**elative **S**ea **L**evel) The long-term change in ocean water level measured by a tide gauge, including the (usually unknown) vertical motion of the gauge plus the change in the water due to eustatic, steric, and/or wind-driven effects.

Salina A shallow salt marsh or pond separated from the ocean but flooded by high tide.

Salinity The grams of salt in a kilogram of seawater (*cf*. psu); specifically, the total amount of dissolved solids by weight when all the carbonate has been converted to oxide, the bromide and iodide to chloride and all organic matter is completely oxidized.

Saltwater Intrusion The inflow of saltwater into a normally freshwater aquifer; associated with sea level rise and/or groundwater extraction.

Seagrass Members of either the Hydrocharitaceae or Zosteraceae families of bottom growing grass-like spermatophytes, usually found in waters less than 10 m deep, and important in stabilizing unconsolidated bottoms.

Sea Level (or water level) The height of the surface of the sea at any time. In surveying and mapping the term "sea level" should be avoided, but if used it should be with the meaning of mean sea level. Sea level is *not* an equipotential surface (*cf.* Geoid).

SeaWIFS (**S**ea **W**ide-**F**ield **S**ensor) Follow-on ocean color multispectral scanner planned to be orbited by NASA in 1993.

Sigma-t (σt) An abbreviated value of the density (ρ) of seawater at normal atmospheric pressure, as a function of salinity (s) and temperature (t) given by the formula $\sigma t = [\rho_{s,t}-1]\times 10^3$; if $\rho_{s,t}=1.02754$, then $\sigma_t=27.54$.

SLOSH (**S**ea **L**ake **O**verland **S**urges from **H**urricanes) Storm-surge model for predicting the height of water levels associated with a tropical storm or hurricane.

SLR (**S**atellite **L**aser **R**anging) Precision point-positioning system using ground-based lasers reflected off satellites.

SOI (**S**outhern **O**scillation **I**ndex) Atmospheric measure of the strength of the Walker Cell given by the surface barometric air pressure at Tahiti minus that at Darwin (Australia); *cf.* ENSO.

Specific Volume Anomaly (δ) The excess of the actual specific volume (reciprocal of density, $\rho_{s,t,p}$) of seawater at any point in the ocean ($\alpha_{s,t,p}$) over the specific volume of seawater of salinity s=35 psu and *in situ* temperature t=0°C at the same pressure ($\alpha_{35,0,p}$); $\delta=\alpha_{s,t,p}-\alpha_{35,0,p}$.

Spheroid A mathematical figure closely approaching the geoid in form and size, and used as a reference for geodetic surveys.

SSH (**S**ea **S**urface **H**eight) The vertical difference of the height (h_s) of an altimetric satellite above the ellipsoid minus the vertical distance (h_a) from the altimeter to the sea surface, *i.e.*, SSH=h_s-h_a.

SSH Anomaly The departure of the instantaneous SSH from the mean SSH calculated over a long time interval (typically one year or longer) at a particular location (*cf.* Dynamic Height Anomaly).

SST (**S**ea **S**urface **T**emperature) The temperature of the upper meter of the water column, usually measured with a thermometer from a bucket sample.

Steric In oceanography, steric refers to the expansion or contraction of the water column due to the distribution of temperature and salinity.

Stratosphere The region of the atmosphere above the troposphere or from about 10 to 25 km, characterized by ozone in addition to the normal gases.

Sulphate A salt or ester of sulphuric acid (H_2SO_4), that occurs naturally and anthropogenically in the ocean and the atmosphere (where it is associated with acid rain).

Superposed Epoch Analysis A mathematical technique whereby all segments of a time series showing repeats of a phenomenon (*e.g.* El Niños) are organized one above the other centered on the same epoch (*e.g.* El Niño year).

Sverdrups (Sv) A commonly used measure of oceanic volume transport equal to 10^6 m^3s^{-1} or 1 gigaliter/second; Florida Current volume transport is 30 Sv, but the Amazon River is only 0.2 Sv.

Sverdrup Transport The depth-averaged mass transport of seawater that is the fundamental description of oceanic flows, arising from wind stress curl acting on a β-plane.

Tectonic The investigation of the origin and evolution of the broad structural features of Earth, particularly associated with the motion of crustal plates (plate tectonics).

Thermocline The oceanic zone below the (isothermal) wind-wave mixed layer where the temperature decreases rapidly with depth; in the tropics it is usually the zone from about 50–700 m depth.

Tidal Benchmark A bronze disk firmly cemented into a solid foundation that serves as the vertical (leveling) reference for a tide gauge or tide staff.

Tide Gauge A device for measuring water level as a function of time, and which is referenced to tidal benchmarks through periodic leveling surveys.

TOGA (**T**ropical **O**cean – **G**lobal **A**tmosphere) A term describing the relationship between the Pacific Ocean's interaction with (and forcing of) the whole atmosphere; also the name of a scientific research programme.

TOPEX/Poseidon A joint France/USA altimetric satellite for precise measurement of SSH, launched in 1992.

Topography The configuration of a surface including its relief. In oceanography the term is applied to a surface such as the sea bottom (*cf.* Bathymetry) or a surface of given characteristics within the water mass.

Trade Winds Westward flowing winds associated with the Hadley Cell and the ITCZ; the northeast Trades are the dominant wind system of the Intra-Americas Sea.

Transport In fluid flows, it is a measure of amount per unit time; used to define volume transport (liters per second), heat transport (watts), mass transport (grams per second), *etc.*

TRMM (**T**ropical **R**ainfall **M**easuring **M**ission) Planned NASA multispectral microwave radiometer to be orbited for measuring precipitation rates.

Troposphere The atmospheric shell closest to Earth's surface, extending approximately 10 km upwards; the zone where weather occurs.

Turbidity A measure of the amount of suspended matter in water or aerosols in air; a more precise term is attenuation which is the sum of scattering and absorption.

UNCED **U**nited **N**ations **C**onference on **E**nvironment and **D**evelopment, held in Rio de Janeiro in June 1992.

UNDP **U**nited **N**ations **D**evelopment **P**rogramme.

UNEP **U**nited **N**ations **E**nvironment **P**rogramme.

UNESCO **U**nited **N**ations **E**ducational, **S**cientific, and **C**ultural **O**rganization.

Upwelling The vertical motion of seawater (as distinct from rising air) often associated with Ekman Transport in the ocean.

USAID **U.S.** **A**gency for **I**nternational **D**evelopment.

USGS **U.S.** **G**eological **S**urvey.

UV (ultraviolet) Radiation invisible to the naked eye, with wavelength shorter than violet (i.e., less than 400 nm).

Varve A sedimentary deposit, bed, or lamination deposited in one season; usually distinguished by differences in color or composition.

VLBI (**V**ery **L**ong **B**aseline **I**nterferometry) A space-based geodetic measuring system to determine (to within ± 1 centimeter or better) the location of sites on Earth's surface.

Walker Cell An atmospheric circulation cell along the equator with ascending air over Asia, descending air over the central Pacific Ocean, and westward flowing surface winds (*cf.* SOI).

Wetland Areas that are covered with fresh surface water for some period of each year on a recurring seasonal basis.

Wider Caribbean Region UNEP term for the Gulf of Mexico/Caribbean Sea/Bahamas/Guianas region; *cf.* IOCARIBE, Intra-Americas sea.

Wind Stress The force per unit mass per unit area exerted by atmospheric flow on the ocean's surface; units are dynes per square centimeter.

Wind Stress Curl The torque applied to the water from the spatial variation of wind stress; *cf.* Sverdrup Transport.

WMO **W**orld **M**eteorological **O**rganization of the United Nations.

WOCE **W**orld **O**cean **C**irculation **E**xperiment.

XBT (expendable bathythermograph) A 20 cm long torpedo-shaped expendable oceanographic probe launched from ship or aircraft (AXBT); used to measure temperature versus depth.

ybp **Y**ears **B**efore the **P**resent.

Zooxanthellae An algal cell living symbiotically in the cells of certain invertebrate animals such as corals.

Index

Note page numbers in italic refer to figures and those in bold refer to tables.